國輝兄，

祝考試成功！

弟國強

22/8/72.

ELECTRONICS: BJTs, FETs, AND MICROCIRCUITS

McGRAW-HILL ELECTRICAL AND ELECTRONIC
ENGINEERING SERIES

FREDERICK EMMONS TERMAN, *Consulting Editor*
W. W. HARMAN AND J. G. TRUXAL,
Associate Consulting Editors

BROOKLYN POLYTECHNIC INSTITUTE SERIES

ANGELO · Electronic Circuits
ANGELO · Electronics: BJTs, FETs, and Microcircuits
LEVI AND PANZER · Electromechanical Power Conversion
LYNCH AND TRUXAL · Signals and Systems in Electrical Engineering
    Combining Introductory System Analysis
            Principles of Electronic Instrumentation
MISHKIN AND BRAUN · Adaptive Control Systems
SCHILLING AND BELOVE · Electronic Circuits: Discrete and Integrated
SCHWARTZ · Information Transmission, Modulation, and Noise
SHOOMAN · Probabilistic Reliability: An Engineering Approach
STRAUSS · Wave Generation and Shaping

ELECTRONICS: BJTs, FETs, AND MICROCIRCUITS

E. JAMES ANGELO, JR.
Formerly Professor of Electrical Engineering
Polytechnic Institute of Brooklyn

INTERNATIONAL STUDENT EDITION

McGraw-Hill Book Company
New York St. Louis San Francisco Düsseldorf
London Mexico Panama Sydney Toronto
Kōgakusha Company, Ltd.
Tokyo

Electronics: BJTs, FETs, and Microcircuits

INTERNATIONAL STUDENT EDITION

Exclusive rights by Kōgakusha Co., Ltd., for manufacture and export from Japan. This book cannot be re-exported from the country to which it is consigned by Kōgakusha Co., Ltd., or by McGraw-Hill Book Company or any of its subsidiaries.

I

Library of Congress Catalog Card Number 68-55260

TOSHO PRINTING CO., LTD., TOKYO, JAPAN

Preface

This book is intended for a first course in electronics for electrical engineering students in the junior year. The principal prerequisite for the course is a one-year course in electric circuits; however, for the first half of the book alone the prerequisite can in many cases be reduced to the electricity and magnetism portion of the college physics course.

With so many well-written books on electronics available, it might fairly be asked why another book on the subject is needed. The answer is that important breakthroughs in the technology of solid-state electronics come so rapidly that the useful life of any textbook is no more than about 5 years. As a matter of historical interest and to gain some perspective for the future, it is worthwhile to recall some of the breakthroughs that have occurred. The interval from 1950 to 1955 produced the alloy-junction germanium transistor and thus made the mass production of transistors possible. The interval from 1955 to 1960 saw first the development of the diffused-base transistor and then the development of the silicon planar technology for making transistors; this latter development was destined to supplant the germanium alloy-junction technology and to bring monolithic integrated circuits into the realm of feasibility. Between 1960 and 1965, junction field-effect transistors and integrated digital circuits were put into large-scale production, and finally, in the interval between 1965 and 1970, integrated analog circuits and MOS field-effect transistors came into widespread use. In order to keep up with this changing technology, textbooks must be brought up to date frequently, and that is the reason for this book. To achieve this objective, the book contains a reasonably detailed treatment of junction and MOS field-effect transistors, and it is oriented toward integrated circuits as much as possible throughout.

However, although the technology of solid-state devices has been changing rapidly, the fundamentals of electronic circuit analysis and design have not changed very much. For example, Kirchhoff's laws, Thévenin's theorem, gain-and-phase frequency characteristics, and the concepts of small-signal models and feedback have not diminished in importance in any way whatever. Therefore, the most important objective of this book is to help the student develop competence in the use of these enduring fundamentals. It is important, however, that these fundamentals be presented in terms of current electronic devices and current engineering practice, and this is a secondary objective of the book. Thus the specific circuits discussed in the book have been chosen, as far as possible, to illustrate basic principles and analytic techniques in terms of current practice.

The first half of the book is concerned mainly with elementary electronic circuits and with the internal physics of the electronic devices used in them. The second half of the book goes on to more elaborate linear amplifier circuits with emphasis on analytical techniques that provide insight and understanding and with considerable attention given to the design of these circuits. The device physics in the first half of the book provides a basis for understanding the dynamics of the devices in both linear amplifiers and switching circuits. In addition, the charge-control characterization of the devices provides a simple but effective means of communication between circuit designers and device designers. However, those users who have only a limited amount of time to devote to device physics will find the book arranged so that the more detailed discussions of the physics can be omitted without serious difficulty. The chapter on power amplifiers is located at the end of the book simply because it does not seem appropriate at any interior point. This state of affairs results largely from the fact that the study of the class B transistor amplifier with complementary symmetry requires familiarity with the emitter follower. In fact, however, power amplifiers as presented in this book can be studied at any point after the emitter follower has been covered.

An effort has been made throughout the book to give a treatment that is well correlated with the more advanced treatment given in the SEEC series on semiconductor electronics. The transistor models and the symbols used are in most respects the same as those used in the SEEC series, and the text contains numerous references to the series for the reader who wants to dig deeper.

Integrated circuits are not treated in a separate chapter set aside for that purpose; they are discussed throughout the text at points where occasion to do so arises naturally. These discussions are concerned with the basic properties and the potentialities of integrated analog circuits rather than with specific technologies. Considerable attention is given to the difference amplifier and the operational microamplifier because these configurations are expected to survive technological changes.

The author wishes to acknowledge with gratitude the many contributions made by his colleagues and students to the preparation of this book. He is also grateful to the Radio Corporation of America, and particularly to Dr. R. W. Ahrons of RCA, for having made it possible for him to work at first hand in the design and use of MOS integrated circuits.

E. James Angelo, Jr.

This book was prepared for courses the author taught while a member of the faculty of the Polytechnic Institute of Brooklyn. The author is now a Member of the Technical Staff at the Bell Telephone Laboratories, Holmdel, New Jersey.

Contents

1

An Overall View
of Electronic Engineering

In this chapter an attempt is made to give an overall picture of the field of electronic engineering and to give some idea of the kinds of things in which electronic engineers are involved. It should give the student some understanding of where the study of electronic circuits typically leads. The range of activities of electronic engineers is so great that it is not possible to be all-inclusive in this discussion without having the forest obscure the trees. However, it is possible with the aid of a few specific examples to give a good picture of the basic activities of the profession. It is also hoped that the student will sense some of the excitement of electronic circuits and systems that have kept the author fascinated for more than thirty years.

1-1 SIGNALS AND SIGNAL PROCESSING

A signal is a measurable quantity, such as voltage, current, length, angular position, and pressure, that varies with time and that, as a result of its variation, contains useful information. The variation of the length of the mercury column in a thermometer is such a signal, as is the variation in the strength of the electromagnetic radiation from a TV transmitting antenna. However, a number of operations must be performed on the TV signal before it can be converted into an image on the screen of a TV receiver. The electronic engineer is concerned with a wide variety of signal-processing operations that convert signals into more useful or more appropriate forms.

Electrical signals (for example, time-varying voltages and currents) lend themselves especially well to signal-processing operations. Thus, when signals exist in a nonelectrical form, one of the first operations to be performed on them is often to convert them to an electrical form. For example, a microphone converts a sound-pressure signal into an electrical signal that varies with time in almost exactly the same way as the sound pressure at the diaphragm; thus the electrical signal contains essentially the same information as the sound-pressure signal. The microphone in this example is called a transducer, and the operation that it performs on the signal is called transduction. Since the electrical signal varies with time in the same way as the sound-pressure signal, the electrical signal is said to be the analog of the sound-pressure signal, and the microphone is said to be an analog transducer. Other forms of transduction are also possible; they are discussed later.

The AM broadcast radio receiver and the TV receiver are familiar electronic systems that could be used to discuss further a variety of signal-processing operations. However, some readers may feel that home entertainment is a somewhat frivolous matter that is by no means essential to the mainstream of human affairs. Therefore, let us choose an example that appears to be more important. Under the assumption that high-speed transportation is important to the standard of living and the general welfare of man, consider the scheduled, all-weather operation of commercial airlines. This operation would be quite impossible without effective, instantaneous communication between the aircraft and ground. An electronic system, the radiotelephone, provides this essential communication, and it is probably the only system capable of filling this need.

The first signal-processing operation in the air-to-ground radio telephone is the transduction of the pilot's voice signal into an electrical signal with the aid of a microphone. The electrical signal from the microphone is too weak to transmit effectively, so the next signal-processing operation is that of amplification, or making the signal stronger. This operation, as with all the rest except the last, is accomplished electronically.

In principle it is now possible to apply the amplified signal to an antenna from which it will radiate to the ground. In fact, however, in order to be an efficient radiator the dimensions of the antenna must be comparable to the wavelength of the signal being radiated, and since the wavelengths of audio-frequency electrical signals are measured in hundreds of kilometers, the antenna required is impossibly large. Therefore, the next signal-processing operation is to shift the information-bearing signal to a much higher frequency in the radio-frequency band, or, from another point of view, to impose the information on a radio-frequency carrier wave. This process is called modulation. There are several ways in which a carrier wave can be modulated, among which amplitude modulation (AM) and frequency modulation (FM) are the best known to the general public. The process of amplitude modulation is illustrated in Fig. 1-1. In this process the amplitude of a radio-frequency

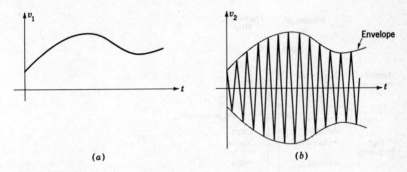

Fig. 1-1 Amplitude modulation. (a) Information-bearing signal; (b) radio-frequency wave carrying the signal.

sine wave, the carrier, is caused to vary instant by instant with the amplitude of the audio-frequency signal. Figure 1-1a shows the audio-frequency signal, and Fig. 1-1b shows the amplitude-modulated carrier wave. The information in the audio signal is contained in the envelope of the AM wave.

Figure 1-2 shows the extent of the commercially useful radio-frequency spectrum, and it indicates a few of the many radio services using the spectrum. At the frequencies employed by AM broadcasting stations a steel tower about 200 m in height is adequate for an antenna. However, this is still a bit large for mobile transmitters such as those in police cars, taxicabs, and airplanes. In the air-to-ground system under discussion a carrier frequency in the band near 120 MHz assigned to aircraft might be used; at this frequency an antenna about 1.5 m long is adequate.

Now the signal in the form of a radio wave with ample strength and appropriate frequency is radiated from the airborne antenna, and when it encounters the ground-based receiving antenna it induces an AM radio-frequency signal on the antenna. Several more signal-processing operations are necessary before the information in the signal can be presented to the control-tower operator in a useful form. First of all, the signal voltage induced on the antenna is likely to be quite small, because the signal radiated by the airplane is radiated in all directions, and only a small amount of the total radiation is intercepted by the receiving antenna. Thus more amplification is needed. Now note further that the radio signal was transmitted from the transmitting antenna to the receiving antenna through open space. But all of the radio transmitters in the world are using the same space, and the receiving antenna picks up many signals simultaneously. Thus an oper-

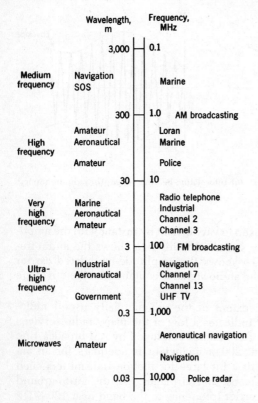

Wavelength, m		Frequency, MHz	
		3,000 — 0.1	
Medium frequency	Navigation SOS		Marine
		300 — 1.0	AM broadcasting
	Amateur		Loran
High frequency	Aeronautical		Marine
	Amateur		Police
		30 — 10	
Very high frequency	Marine Aeronautical Amateur		Radio telephone Industrial Channel 2 Channel 3
		3 — 100	FM broadcasting
Ultra‑high frequency	Industrial Aeronautical		Navigation Channel 7 Channel 13
	Government		UHF TV
		0.3 — 1,000	
Microwaves	Amateur		Aeronautical navigation Navigation
		0.03 — 10,000	Police radar

Fig. 1-2 The radio-frequency spectrum and some of the services using it.

ation to extract the desired signal from the many signals received on the antenna is required; this operation is called filtering. When a radio is being tuned or when a TV channel is selected, the filters are being adjusted to select a particular signal.

Now, with the AM radio-frequency signal suitably amplified and filtered, an operation is needed to extract the information contained in the envelope of the signal. This operation is the inverse of modulation, and it is called demodulation; it is also called detection. The demodulator produces a signal voltage that is a good copy of the original audio signal shown in Fig. 1-1a, although some distortion of the waveform is likely to have occurred in the numerous operations that have been performed on the signal up to this point. This distortion is the reason why voices heard over police radios and taxicab dispatching radios are often hard to understand; those radios are not hi-fi systems. The engineers that design the signal-processing circuits usually have control over the degree of fidelity achieved, but the higher the fidelity the higher the cost.

The signal obtained from the demodulator is usually quite weak, and so additional amplification is usually needed after the demodulator. When the signal has been amplified to a suitable strength, one more operation must be performed. In this operation a loudspeaker or a pair of headphones acts as a transducer to convert the electrical signal back to a sound-pressure signal that is intelligible to the control-tower operator.

Another signal-processing system that many aircraft carry, to permit bad weather ahead to be detected and to aid in landing under conditions of poor visibility, is radar. The radar is an electronic system that utilizes most of the signal-processing operations contained in the radio-telephone described above, and it has several more of its own. In fact, the processes of amplification, modulation, filtering, and demodulation are very basic indeed; they are used in a great variety of systems, including many that do not use radio waves at all and many others that have nothing to do with voice communications.

1-2 FURTHER SIGNAL-PROCESSING OPERATIONS

Several important, basic signal-processing operations are described in Sec. 1-1. However, they are only a small sample of the wide variety of such operations that engage the attention of the design engineer. A few more of these operations are described in this section.

An important process is that of signal generation. For example, the radio transmitter of Sec. 1-1 must contain a circuit to generate the radio-frequency carrier wave. In other applications there are needs for generators of square waves, pulse trains, and ramp waveforms. Radar and TV systems require all three of these waveforms. The cathode-ray oscilloscope requires the ramp (sawtooth) waveform. Digital computers and logic machines require square waves and pulse trains. There are even applications that require generators of pure noise, the sh-h-h sound heard from the TV receiver when the channel to which it is tuned is not transmitting.

Another signal-processing operation is that of waveform shaping. In this operation the waveform of the signal is changed to a greater or lesser degree. When small changes in waveform are introduced, it is often for the purpose of improving the waveform in some way. When large changes are introduced, it is often for the purpose of creating a new signal waveform having some special relation to the original waveform.

Still other signal-processing operations are employed by analog computers. An analog computer is an electronic machine that is capable of solving differential equations. In carrying out its task, it adds signals, subtracts them, multiplies them, integrates them, and it sometimes performs wave-shaping operations on them. All of these operations are performed by electronic circuits.

The idea of an analog signal is developed in Sec. 1-1; there are many systems, including all of broadcast radio and television, that use analog signals. There are also many systems that use digital signals. In digital systems the numerical value of a time-varying quantity at successive instants of time is expressed by a coded set of pulses somewhat similar to the coded set of pulses used in telegraphy to represent the letters of the alphabet. The way in which an analog signal is converted to a digital signal is illustrated in Fig. 1-3. The analog signal is shown in Fig. 1-3a. The first step in the conversion process is to divide the total range of possible values of the analog signal into a number of discrete levels. The value of the signal at any instant is then expressed approximately by the level in which the signal lies at that instant; this process is called quantizing the signal. The signal is then sampled at successive instants of time by an electronic circuit, the level of the signal at each sample is determined, and the appropriate coded set of pulses is generated. The code used may be based on the binary number system. In this case the pulses have amplitudes of 1 or 0, corresponding to the binary digits, and the set of pulses associated with any sample of the signal corresponds to the value of the signal in binary digits. A coded

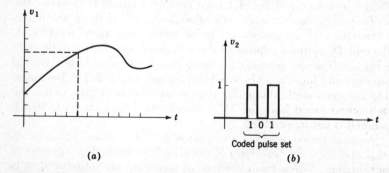

(a)   *(b)*

Fig. 1-3 Digital signals. (*a*) Analog signal; (*b*) digital form of the quantized and sampled analog signal.

set of pulses for one sample of the analog signal in Fig. 1-3a is illustrated in Fig. 1-3b.

A number of telephone systems using digital signals are in operation in the larger cities, and similar systems are being put into use on long-distance circuits. The analog voice signal is converted to a digital signal at a central office. It is then transmitted to another central office near the receiving end of the line where the digital signal is converted back to the analog form. The people speaking over these circuits cannot detect any difference from a purely analog system. One of the advantages of the digital system is the following. The electronic circuits can sample the analog signal and encode the corresponding digital signal in an exceedingly short time. Thus it is possible to sample and encode a number of signals, say five, in rapid succession and return to the first before it has changed very much. As a result it is possible to send all five of the signals over a single pair of wires. Of course it is necessary to separate the signals at the receiving end of the line, but this operation presents no serious problems.

1-3 TRANSISTORS AND DIODES IN SIGNAL-PROCESSING SYSTEMS

In the early days of electronics most of the signal-processing operations were performed by vacuum tubes. The bipolar junction transistor was announced in 1948, and in the decade between 1950 and 1960 it proved itself to be superior to the vacuum tube in almost all respects; this superiority is especially marked in digital-system applications. As a result, most of the new equipment designed since 1960 has been designed with transistors. Transistors and semiconductor diodes together are capable of performing almost all signal-processing operations.

Since 1960 the field-effect transistor has appeared. The field-effect transistor is inferior to the bipolar transistor in most respects, but it has certain special attributes that make it very attractive for a number of special applications. It is a valuable complement to the bipolar transistor.

In order to get the best performance possible from signal-processing circuits using transistors, the circuit designer must know how the transistor works, and he must have a good understanding of the physical processes that go on inside the transistor. In this respect the bipolar transistor is considerably more demanding than the field-effect transistor and the vacuum tube. Moreover, there are many occasions in which it is necessary for the circuit designer to consult with the transistor designer, and for this purpose he must know the language of the transistor designer, that is, the language of the processes that go on inside the transistor. For these reasons a considerable portion of this book is devoted to the study of the physics of transistor action. The rest of the book is devoted to transistors and diodes in various signal-processing applications. The main emphasis has been placed on the process of amplification because amplification is one of the more important processes, because amplifiers are often used to perform or help perform other processes, and because the analytical techniques used in the analysis and design of amplifiers are basic and can be used in connection with other processes.

Although the transistor has replaced the vacuum tube in most electronic equipment designed since 1960, the vacuum tube is by no means extinct. When high power is needed at high signal frequencies, the vacuum tube must be used. Also, much equipment designed with vacuum tubes prior to 1960 continues to perform its assigned function satisfactorily, and it continues to be manufactured; a number of electronic voltmeters used in university and industrial laboratories fall in this category. Furthermore, severe engineering problems have been encountered in getting high-quality performance from transistor circuits in some electronic systems; this has been the case with the color TV receiver. For these reasons it is necessary for the electronic engineer to know something about vacuum tubes and their performance in electronic systems.

1-4 SUMMARY

The electronic engineer is concerned with information-bearing signals and with processing those signals to convert them to a more useful or more appropriate form. A variety of signal-processing operations is described in this chapter. The basic function of the engineer is to design electronic circuits to perform these operations and to invent new signal-processing operations whenever an advantage is to be obtained from such an invention.

The transistor and the semiconductor diode used jointly are capable of performing most of the important signal-processing operations. Consequently this book is concerned mostly with transistors and diodes and how they can be used to process signals. One of the most important processes performed by the transistor is amplification. Accordingly, Chap. 2 introduces the notion of the ideal amplifier and gives some idea of the things that an amplifier can do. Chapter 3 introduces the notion of the ideal diode, and it shows a number of ways in which the diode can be used to process signals. The succeeding chapters show how diodes and transistors and even complete signal-processing circuits can be fabricated in a tiny piece of semiconducting material.

2

Ideal Amplifiers

C*hapter 1 presents a number of signal-processing operations that are employed in electronic systems. The operation that perhaps is the most important and the most often used is that of amplification. The need for signal amplification arises in all sorts of engineering endeavors such as radio, telephony, automatic control of machinery, medical research, instrumentation, and rocket guidance. Consider, for example, a long-distance telephone circuit. The telephone instrument delivers a few milliwatts to the telephone cable, and as the signal is transmitted along the cable, it is attenuated by the losses in the cable. Thus for satisfactory long-distance telephony, it is necessary to locate amplifiers at more or less uniform intervals along the cable. In one system using the digital signals described in Chap. 1, amplifiers are placed at 1-mile intervals along the cable, and each amplifier amplifies the signal by a factor of 100. In a circuit 1,000 miles long, this amounts to a great deal of amplification: 100 raised to the power of 1,000.*

The function of amplification in electronic systems is performed by transistors and vacuum tubes. An amplifier may consist of transistors, resistors, and capacitors connected together in a suitable circuit and supplied with dc power from some source such as a battery. Alternatively, it may consist of transistors and resistors fabricated and interconnected on a tiny chip of silicon no larger than the letter n *as printed here. In the first form the amplifier is said to be made of discrete components; in the second form it is said to be an integrated circuit, and since it is so small, it is conveniently referred to as a microamplifier, also written as μamplifier. The microamplifier also must be supplied with dc power from an external source.*

The purpose of this chapter is to present the notion of an ideal amplifier and to discuss a few of the many functions that it can perform. Real amplifiers of the kinds described above are examined in subsequent chapters. The quality of a real amplifier can be judged by the extent to which its performance approximates that of the ideal amplifier.

2-1 THE IDEAL VOLTAGE AMPLIFIER

One form of ideal electric amplifier is shown in the shaded area in Fig.
2-1a. It consists of an input terminal pair, a voltage source, and an output terminal pair. The voltage at the output terminals is at every instant
directly proportional to the voltage at the input terminals; that is,

$$v_2 = \mu v_1 \tag{2-1}$$

The constant of proportionality μ is called the *voltage-amplification
factor* of the device. If an information-bearing signal voltage is applied
at the input terminals, the waveform of that voltage is repeated exactly
at the output terminals except that its amplitude is magnified by the
amount of the amplification factor. As is implied by Eq. (2-1), the output
voltage is independent of the output current.

The output volt-ampere characteristic for the amplifier is shown in
Fig. 2-1b; it consists of a family of vertical straight lines, one for each
value of the input voltage. If the values of v_1 corresponding to the
curves are separated by 1 volt, then the curves are separated by μ volts
on the output-voltage axis. As is true of all ideal voltage sources, the
ideal voltage amplifier imposes no limit on the output current; the output
current is determined entirely by the external circuit connected to the
output terminals. Since the ideal amplifier consists of a source whose
output voltage is controlled by the input voltage to the amplifier, it is
called a *controlled source*, or a *voltage-controlled source*.

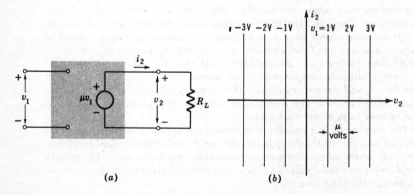

(a)   (b)

Fig. 2-1 The ideal voltage amplifier. (a) Circuit diagram; (b) output volt-ampere characteristic.

The input terminals of the ideal amplifier are isolated from the output terminals in the sense that electrical conditions at the input are completely independent of the electrical conditions at the output terminals and of the external circuit connected to the output terminals. This isolation is an important feature of the amplifier, and there are numerous conditions in which amplifiers are used solely to provide isolation.

If the input voltage v_1 is specified, then the output voltage $v_2 = \mu v_1$ is fixed and is independent of the current drawn by the load resistor. Hence any amount of power can be drawn from the output terminals by proper choice of R_L (assuming, of course, that $v_1 \neq 0$). The input current to the amplifier is zero regardless of the value of v_1; hence the input power is zero under all conditions. From these facts it follows that the power amplification, or power gain, which is the ratio of output power to input power, is infinite. The output power from the amplifier at any instant is

$$p = \frac{(\mu v_1)^2}{R_L} \tag{2-2}$$

Equation (2-2) shows that with a fixed load resistance the output power depends on the input voltage v_1; the greater v_1, the greater the output power. If R_L and μ are both fixed and if the input signal voltage is very small, as it might be at the end of a long telephone cable, then the amplifier may not be able to develop the required signal voltage and power at the load. This difficulty can be remedied by connecting several amplifiers in cascade as illustrated in Fig. 2-2. If the amplification factors are greater than unity, the amplitude of the signal voltage is increased in each successive stage of amplification, and in this way the desired signal voltage and power can be developed at the load even when the input signal is very small. The output voltage of this multistage amplifier is

$$v_4 = \mu_1 \mu_2 \mu_3 v_1 \tag{2-3}$$

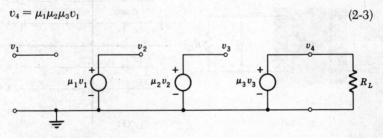

Fig. 2-2 Cascaded voltage amplifiers.

and the output power is

$$p = \frac{(\mu_1\mu_2\mu_3 v_1)^2}{R_L} \tag{2-4}$$

Amplifiers using field-effect transistors and vacuum tubes are conveniently characterized as voltage-controlled amplifiers like those described above, although they fall somewhat short of being ideal. A certain phonograph amplifier using field-effect transistors in integrated-circuit form consists of three stages in cascade. The overall voltage amplification of this circuit is approximately 1,000.

2-2 THE IDEAL CURRENT AMPLIFIER

Another form of ideal electric amplifier is shown in Fig. 2-3a; it is a *current-controlled current source.* The output current of this amplifier is at every instant directly proportional to the input current, and it can be expressed as

$$i_2 = \beta i_1 \tag{2-5}$$

The constant of proportionality β is the *current-amplification factor* of the device. As is implied by Eq. (2-5) the output current is independent of the output voltage. The ideal current amplifier is the dual of the ideal voltage amplifier, and it also provides isolation between the input and output terminals.

Since the output current is independent of the output voltage, any amount of power can be drawn from the output terminals by the proper

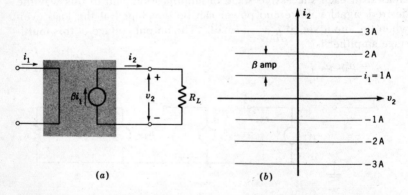

Fig. 2-3 The ideal current amplifier. (*a*) Circuit diagram; (*b*) output volt-ampere characteristic.

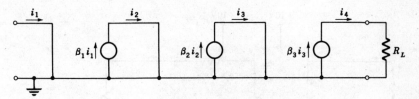

Fig. 2-4 Cascaded current amplifiers.

choice of R_L, provided $i_1 \neq 0$. Moreover, the input voltage is zero regardless of the value of the input current; hence the input power is zero under all operating conditions. It follows from these facts that the power gain of the ideal current amplifier is infinite. The output power from the amplifier at any instant is

$$p = (\beta i_1)^2 R_L \tag{2-6}$$

As in the case of the voltage amplifier, it is often necessary to cascade current amplifiers to obtain the desired signal current and power at the load. A cascade of three current amplifiers is shown in Fig. 2-4. This cascade of current amplifiers is the dual of the cascade of voltage amplifiers shown in Fig. 2-2.

Amplifiers using bipolar transistors are conveniently characterized as current amplifiers, although they fall somewhat short of being ideal. A typical three-stage cascade of bipolar-transistor amplifiers provides a current gain in excess of 30,000.

2-3 A PRACTICAL TRANSISTOR MICROAMPLIFIER

A schematic representation of an integrated-circuit microamplifier using bipolar transistors is shown inside the shaded area in Fig. 2-5. This amplifier can accept two input signals, v_1 and v_2, and the output signal voltage depends on the difference $v_1 - v_2$. Such amplifiers are usually called differential amplifiers. The ± 10-volt terminals are connected to external sources of dc power such as batteries.

Microamplifiers of this kind are fabricated on a rectangular chip of silicon measuring about 1 mm on each side and having a thickness just a little greater than the thickness of the paper on which this page is printed. The circuit diagram shown in Fig. 2-5 is by no means the circuit fabricated on the silicon chip; it is a *circuit model* that represents correctly the relations between the currents and voltages at the terminals of the amplifier. The actual circuit on the silicon chip may contain 10 or 15 transistors and a dozen or more resistors. The circuit model is a

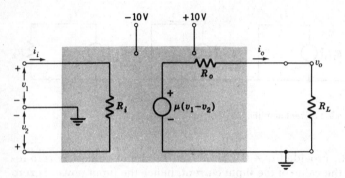

Fig. 2-5 The circuit model for a practical transistor microamplifier.

simplified representation of the amplifier that can be used to calculate, to a good approximation, the performance of the amplifier when it is used in a signal-processing system. This book devotes considerable attention to the techniques for developing circuit models to represent physical devices such as transistors, diodes, and tubes. The ideal amplifier presented in this chapter is the first step in that direction, and the ideal diode of the next chapter is the second step.

The practical amplifier of Fig. 2-5 is not ideal. It consists of an ideal amplifier plus two parasitic elements, the input resistance R_i and the output resistance R_o. As a result of these parasitic elements, the input power is not zero. Moreover, with a given input signal voltage, the output signal voltage depends on the output current, and there is a finite maximum power that can be drawn from the output terminals (with $R_L = R_o$, of course). It follows from these facts that the power gain of the practical amplifier is not infinite. Typical circuit-parameter values for microamplifiers of this type are $\mu = 1,000$, $R_i = 50$ kilohms, and $R_o = 100$ ohms, although these values will vary widely depending on the design of the amplifier.

When the amplifier is used with $v_1 = 0$ and an input signal voltage applied at v_2, the polarity of the output voltage is opposite to the polarity of the input voltage, and the device is said to be a sign-reversing, or an inverting, amplifier. When the signal is applied at v_1 with $v_2 = 0$, the device is called a noninverting amplifier. The existence of these two modes of operation adds a useful flexibility to the amplifier.

When $v_2 = 0$ and a signal is applied at v_1, the voltage amplification, or voltage gain, of the amplifier is defined as the ratio of the output voltage to the input voltage,

$$A_v = \frac{v_o}{v_1}$$

But from Fig. 2-5, the output voltage is

$$v_o = \frac{R_L}{R_o + R_L} \mu v_1$$

Thus the voltage gain is

$$A_v = \frac{R_L}{R_o + R_L} \mu$$

When the signal is applied at v_2 with $v_1 = 0$, the same value of voltage gain is obtained except that it has a minus sign.

The current gain of the amplifier is

$$A_c = \frac{i_o}{i_i}$$

and from Fig. 2-5,

$$i_o = \frac{v_o}{R_L} \qquad \text{and} \qquad i_i = \frac{v_1}{R_i}$$

Thus

$$A_c = \frac{v_o R_i}{R_L v_1} = \frac{R_i}{R_L} A_v$$

When $v_2 = 0$ and a signal voltage is applied at v_1, the power delivered to the input of the amplifier is

$$p_i = \frac{v_1{}^2}{R_i} \tag{2-7}$$

The power delivered by the amplifier to the load is

$$p_o = \frac{v_o{}^2}{R_L} \tag{2-8}$$

The power gain of the amplifier is thus

$$\text{Power gain} = \frac{p_o}{p_i} = \frac{v_o{}^2 R_i}{R_L v_1{}^2} = \frac{v_o}{v_1} \frac{R_i}{R_L} \frac{v_o}{v_1}$$

$$= A_v A_c$$

2-4 GAIN IN DECIBELS

The power gain of an amplifier, calculated above, is a useful measure of the amplifying capability of the amplifier. A related measure that proves to be more useful in some respects is the power gain expressed in decibels (dB); this quantity is defined by

$$G = 10 \log \frac{p_o}{p_i} \tag{2-9}$$

where log is understood to designate the common logarithm. Thus if p_o is 5 watts and p_i is 5 mW, the power gain is 30 dB. The power gain can also be written as

$$G = 10 \log \frac{v_o^2 R_i}{R_L v_1^2} = 20 \log \frac{v_o}{v_1} + 10 \log \frac{R_i}{R_L} \tag{2-10}$$

If $R_i = R_L$, as often happens in communication circuits, then

$$\log \frac{R_i}{R_L} = 0$$

In this case, and *only* in this case, the power gain of the amplifier is

$$G = 20 \log \frac{v_o}{v_1} \qquad \text{when } R_i = R_L \tag{2-11}$$

The logarithm of the voltage amplification proves to be a useful quantity in several respects, regardless of whether $R_i = R_L$ or not. To determine this logarithm, Eq. (2–11) is used; hence the voltage amplification, or voltage gain, in decibels is defined as

$$A = 20 \log \frac{v_o}{v_1} \tag{2-12}$$

Thus if $R_i = R_L$, the voltage gain in decibels is equal to the power gain in decibels. These two gains are not equal, however, when $R_i \neq R_L$; Eq. (2–10) shows that in general the voltage and power gains in decibels differ by the amount 10 log (R_i/R_L). Current gains can also be expressed in decibels by substituting currents for voltages in (2–12).

When the output voltage of an amplifier is greater than the input voltage, the voltage gain in decibels is a positive number. When the output voltage is smaller than the input voltage, the gain in decibels is a negative number. When the output voltage is equal to the input voltage, the gain in decibels is zero. It is possible for the voltage gain in decibels to be positive while the power gain in decibels is negative, and vice versa, if R_i and R_L are not equal.

The decibel is a convenient measure of voltage, current, and power gains for a number of reasons. One of the most important of these is the fact that the frequency characteristics of electric networks, both electronic and otherwise, have especially simple forms when the gain in decibels is plotted against frequency on a logarithmic scale.

The voltage gain of an amplifier can be expressed as the numerical ratio of the output voltage divided by the input voltage, or it can be expressed in decibels according to Eq. (2-12). Since the term "voltage gain" is used for both of these quantities, an ambiguity exists unless a clarifying statement is appended. In this book "voltage gain" means the numerical ratio in all cases unless decibels are specified. The same convention is applied to the terms "current gain" and "power gain."

2-5 OTHER APPLICATIONS FOR AMPLIFIERS

Ideal amplifiers are capable of a number of interesting and useful operations that seem at first glance to bear little relation to the process of signal amplification. Consider, for example, the combination of capacitor and voltage-controlled voltage source shown in Fig. 2-6a. The input current to this circuit is

$$i_1 = C \frac{d}{dt} (v_1 + \mu v_1) = (1 + \mu)C \frac{dv_1}{dt} \qquad (2\text{-}13)$$

Thus the input current is the same as the current that flows into a capacitor of capacitance $(1 + \mu)C$ having the voltage v_1 impressed across its terminals; that is, the circuit of Fig. 2-6b is equivalent to that of Fig. 2-6a insofar as the input terminals are concerned. Circuits of this type are used to obtain the effect of very large capacitors in certain engineering applications.

Figure 2-7 illustrates another application of the ideal amplifier. The gain of this amplifier can be controlled by the voltage v_2; that is,

$$\mu = K v_2 \qquad (2\text{-}14)$$

where the constant K has the dimensions of volt^{-1}. This circuit can perform the function of amplitude modulation illustrated in Fig. 1-1. Amplifiers of this kind are commercially available as integrated circuits,

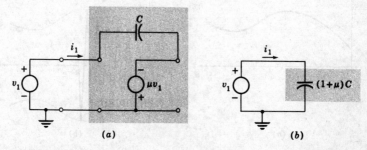

Fig. 2-6 An application for the voltage amplifier. (a) Circuit; (b) an equivalent circuit.

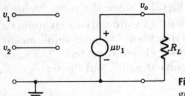

Fig. 2-7 A variable-gain amplifier in which the gain is controlled by the voltage v_2.

and they are used in signal-processing systems for the purpose of amplitude modulation.

Suppose that v_1 is a sinusoidal, radio-frequency voltage. Then it can be expressed as

$$v_1 = V_1 \cos \omega_c t \tag{2-15}$$

Suppose further that v_2 is the sum of a dc component and a time-varying signal voltage. Then v_2 can be expressed as

$$v_2 = V_{dc} + v_s = V_{dc}\left(1 + \frac{v_s}{V_{dc}}\right) \tag{2-16}$$

The output of the amplifier is then

$$v_o = \mu v_1 = KV_{dc}V_1\left(1 + \frac{v_s}{V_{dc}}\right)\cos \omega_c t = V_o \cos \omega_c t \tag{2-17}$$

Typical waveforms for v_2 and v_o are shown in Fig. 2-8. It follows from Eq. (2-17) that the equation of the envelope of the v_o waveform is

$$V_o = KV_{dc}V_1\left(1 + \frac{v_s}{V_{dc}}\right) \tag{2-18}$$

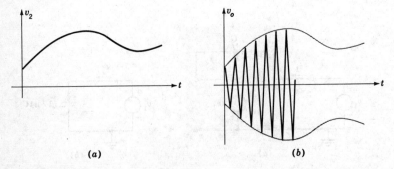

(a)   (b)

Fig. 2-8 Signal waveforms for the amplifier of Fig. 2-7. (a) Input signal v_2; (b) AM output signal.

Thus the variations of the envelope are exactly the same as the variations of v_s, and the envelope contains all of the information contained in v_s. The output voltage is an amplitude-modulated radio-frequency wave; one of the many uses for such waves is described in Chap. 1.

Many different kinds of circuits are used for the purpose of amplitude modulation. The integrated-circuit modulator mentioned above is suitable for use in low-power systems where the output signal power required is no more than a few milliwatts. When large amounts of output power are required, as is the case with radio transmitters, different kinds of modulators must be used.

2-6 SUMMARY

The concept of the ideal amplifier is presented in this chapter. The ideal amplifier is a controlled source; it is a voltage source or a current source whose voltage or current is controlled by an input voltage or current. The concept of the ideal amplifier, or controlled source, is very basic.

The notion of using a circuit model to represent a practical amplifier is also introduced. The model is a simple, approximate representation of a more complicated amplifier structure. In the example used here, the model consists of an ideal voltage amplifier together with two resistors. The concept of the circuit model is also very basic.

In order to analyze and design electronic signal-processing systems, the engineer must have simplifying models for the transistors, diodes, tubes, and other devices that he uses. It turns out that these models use just two new components in addition to the conventional R's, L's, and C's. The new components are the ideal amplifier, or controlled source, and the ideal diode. The ideal amplifier is presented in this chapter; the ideal diode is presented in the next.

PROBLEMS

2-1. *Gain and decibel calculations.* Figure 2-9 shows three amplifiers corresponding roughly to a field-effect transistor, a bipolar transistor, and a high-power vacuum-tube amplifier. (Amplifiers designed for high power often have little voltage gain.) The symbol k designates kilohms, and the symbol M designates megohms.

(a) Calculate the voltage or current gain, as implied in the diagram, for each amplifier. Express the gain in decibels.

(b) Calculate the power gain in decibels for each amplifier. Compare the results with those obtained in part a.

(c) Calculate the voltage gain for the amplifier in Fig. 2-9b.

2-2. *Amplifier analysis.* The amplifier shown in Fig. 2-10 is the same as the practical-microamplifier model shown in Fig. 2-5. The

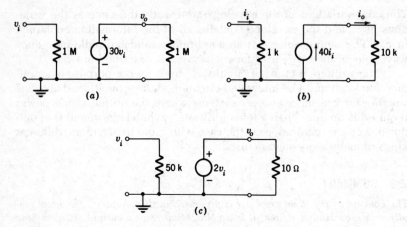

(a)

(b)

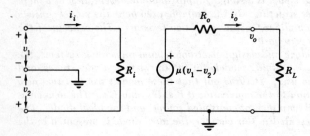

(c)

Fig. 2-9 Amplifiers for Prob. 2-1. *(a)* Field-effect transistor; *(b)* bipolar transistor; *(c)* vacuum-tube amplifier designed for high power.

Fig. 2-10 Amplifier for Probs. 2-2 to 2-5.

parameter values for this circuit are $\mu = 1,000$, $R_i = 50$ kilohms, $R_o = 100$ ohms, and $R_L = 5$ kilohms. A signal voltage is applied at v_1 with $v_2 = 0$. Determine the voltage gain $A_v = v_o/v_1$, the current gain $A_c = i_o/i_i$, and the power gain.

2-3. Maximum power and power gain. The amplifier of Prob. 2-2 is operated with $v_2 = 0$, and a sinusoidal signal having an rms amplitude of 1 mV is applied at v_1. The circuit parameters are $\mu = 1,000$, $R_i = 50$ kilohms, and $R_o = 100$ ohms. The load resistor R_L is adjustable.

(*a*) When R_L is adjusted to draw maximum power from the amplifier, what is the output power? Express the answer in milliwatts.

(*b*) With the adjustment of part *a*, what is the power gain of the amplifier?

(*c*) If the amplifier is held fixed and v_1 and R_L are adjustable, is it possible to achieve more power gain than the value found in part *b*?

2-4. Amplifier analysis. Two signal voltages having the waveforms shown in Fig. 2-11 are applied as input signals to the amplifier of Prob. 2-2. Sketch and dimension the waveform of the output voltage v_o.

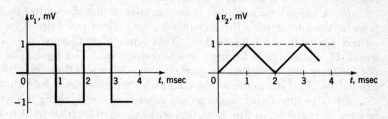

Fig. 2-11 Waveforms for Prob. 2-4.

2-5. Amplifier analysis. A certain microamplifier can be represented by the model shown in Fig. 2-10 with $\mu = 30$, $R_i = 3$ kilohms, and $R_o = 1$ kilohm. Two of these amplifiers are to be connected in cascade with $v_2 = 0$ in each stage. The input resistance of the second amplifier serves as the load resistance for the first amplifier, and the load resistance for the second amplifier is $R_L = 5$ kilohms.

Sketch a diagram of the complete circuit, and determine the overall voltage gain.

2-6. Amplifier design. The 10-mV source and 500-ohm resistor in Fig. 2-12a represent a particular phonograph pickup. The signal voltage from the pickup is to be applied to a loudspeaker having a resistance of

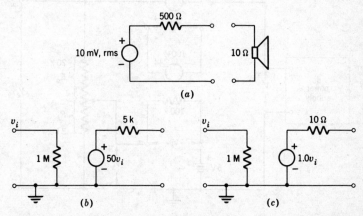

Fig. 2-12 Components for Prob. 2-6. (a) Phonograph pickup and loudspeaker; (b) type 1 amplifier; (c) type 2 amplifier.

10 ohms, and the signal power delivered to the loudspeaker is to be at least 10 watts.

(*a*) If the pickup is connected directly to the loudspeaker, how much power does it deliver to the speaker?

(*b*) A cascade of amplifier stages similar to that shown in Fig. 2-2 is to be used to amplify the signal from the pickup and deliver the required power to the loudspeaker. Two types of amplifiers, shown in Fig. 2-12, are available for the cascade connection. The 1-megohm input resistors can be treated as open circuits. Which of these amplifiers can deliver 10 watts to the loudspeaker with the smallest input voltage?

(*c*) Give the circuit diagram for a cascade of stages, using the types of amplifiers shown in Fig. 2-12, that will deliver at least 10 watts to the loudspeaker with the smallest number of stages. Show the exact number of stages in the diagram. Show how a 1-megohm potentiometer can be used in the first stage as a volume control.

2-7. *Voltage-regulator analysis.* The circuit shown in Fig. 2-13 is the idealized form of a type of voltage regulator that is commonly used in electronic power supplies. The function of the regulator is, among other things, to hold the output voltage constant in the face of changes in the input voltage V_1 and in the load connected across V_2. The amplifier designated by $100V_4$ might be a microamplifier like the one discussed in Sec. 2-3. Any change in the output voltage V_2 is amplified and causes a change in the voltage $2V_5$ that tends to compensate for the initial change

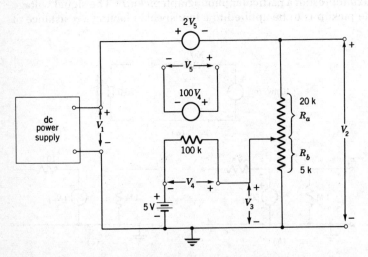

Fig. 2-13   An idealized voltage regulator for Probs. 2-7 and 2-8.

in V_2. The magnitude of the output voltage can be adjusted by adjusting the setting of the potentiometer.

(a) Starting with the relation $V_2 = V_1 - 2V_5$, find V_2 as a function of V_1. The 100-kilohm resistor can be treated as an open circuit.

(b) If $V_1 = 35$ volts dc, what is the value of V_2?

(c) If V_1 increases to 40 volts, what is the new value of V_2?

(d) The ratio $\Delta V_2 / \Delta V_1$ is a measure of the effectiveness of the regulator in stabilizing V_2 against changes in V_1. Evaluate this ratio. *Suggestion:* For best accuracy, differentiate the result obtained in part a.

2-8. Voltage-regulator analysis. The output voltage of the regulator in Prob. 2-7 can be adjusted by adjusting the potentiometer. Let the potentiometer ratio be $K = R_b/(R_a + R_b)$, and let V_1 be constant at 40 volts dc.

(a) Starting with the relation $V_2 = V_1 - 2V_5$, find V_2 as a function of K. The 100-kilohm resistor can be treated as an open circuit.

(b) If $K = 0.2$, what is the value of V_2?

(c) What value of K is required to obtain $V_2 = 20$ volts?

(d) Noting that there is a maximum value that K can have, find the minimum value to which V_2 can be set by means of the potentiometer.

2-9. An electronic integrator. The circuit shown in Fig. 2-14 is an electronic integrator; the output voltage v_3 is the integral of the input voltage v_1 with respect to time. Circuits similar to this one are used in electronic analog computers. Derive the complete expression for v_3 as the integral of v_1.

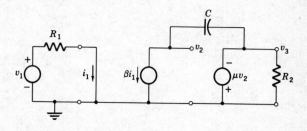

Fig. 2-14 Circuit for Prob. 2-9.

3

The Ideal Diode

*C*hapter *1 describes the role of the electronic engineer as that of a designer of circuits and systems to process electrical signals. Chapter 2 presents the ideal amplifier as a signal-processing device, and it describes several signal-processing operations that can be performed by the ideal amplifier. The diode is another device that can perform useful signal-processing operations. The ideal diode, unlike the ideal amplifier, is a nonlinear device. Circuits containing ideal diodes are therefore nonlinear, and for this reason they can perform operations that cannot be performed by circuits containing linear R's, L's, C's, and ideal amplifiers. In addition to being nonlinear, the ideal diode has the properties of a switch; it can be made to act as an open circuit or a short circuit, depending on the control voltage applied to it. The diode serves many important functions as a switch. Most of the signal-processing operations that engage the attention of electronic engineers can be performed by ideal amplifiers and ideal diodes in various combinations.*

Chapter 2 also points out the desirability of having relatively simple circuit models to represent, within a suitable degree of accuracy, the behavior of the more complicated physical devices such as transistors, tubes, and microamplifiers. It turns out that most of the devices used by electronic engineers can be represented by circuit models consisting of ideal amplifiers and ideal diodes together with parasitic (unwanted) R's, L's, and C's.

3-1 CHARACTERISTICS OF THE IDEAL DIODE

The ideal diode is represented symbolically in Fig. 3-1a, and its volt-ampere characteristic is given by the colored lines in Fig. 3-1b. This characteristic shows that the ideal diode behaves in the following way. When the diode current i_d is positive, the voltage drop v_d across the diode is zero, regardless of the magnitude of the current; that is, the ideal diode behaves as a short circuit to current in the forward direction. The corresponding portion of the volt-ampere characteristic is called

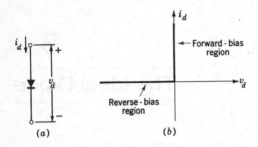

Fig. 3-1 The ideal diode. (*a*) Symbol; (*b*) volt-ampere characteristic.

the forward-bias region. On the other hand, when the voltage drop v_d across the diode is negative, the current through the diode is zero, regardless of the magnitude of the voltage; that is, the ideal diode behaves as an open circuit to voltage in the reverse direction. The corresponding portion of the volt-ampere characteristic is called the reverse-bias region. The diode changes from the open-circuit to the short-circuit condition at the point where both the current through the diode and the voltage drop across it are zero. Of course, no physical device has exactly these characteristics; however, several devices have characteristics that approximate the ideal very closely.

The upper terminal of the diode in Fig. 3-1*a* is called the anode terminal because positive charge flows into the diode at that terminal. The lower terminal is called the cathode terminal because positive charge flows out of the diode at that terminal.

The significant feature of the ideal diode is that its volt-ampere characteristic is not a straight line; it consists of two straight-line segments joined at right angles. The ideal diode is therefore a nonlinear device, and for this reason it can produce results that cannot be obtained with linear R's, L's, C's, and ideal amplifiers. The importance of the diode lies in this fact.

Devices having volt-ampere characteristics consisting of straight-line segments are called piecewise-linear devices because they are linear along each separate piece of the characteristic. If the combinations of current and voltage are restricted to values lying entirely on one piece of the characteristic, the device behaves as a linear device. Its nonlinear nature comes into consideration only if the ranges of current and voltage involved extend across a breakpoint in the characteristic.

3-2 THE HALF-WAVE RECTIFIER

A resistive half-wave-rectifier circuit is shown in Fig. 3-2a. The voltage source supplies a sinusoidal voltage

$$v_s = V_s \sin \omega_s t \qquad (3\text{-}1)$$

to the circuit, R_s represents the source resistance, and R_L represents a resistive load being supplied with power. The function of the diode in this circuit is to supply a unidirectional current to the load. When v_s is positive it produces a current in the positive direction shown in the diagram. Since this is the forward direction for the diode, it acts as a short circuit, and the magnitude of the current is determined by the values of v_s, R_s, and R_L. When v_s is negative, it acts to produce a current in the opposite direction. However, since this is the reverse direction for the diode, it behaves as an open circuit, and no current flows. The voltage developed across the load at each instant is $v_L = R_L i_d$. The waveforms of supply voltage and load voltage are shown in Fig. 3-2b. The load voltage can also be expressed as

$$v_L = \frac{R_L}{R_s + R_L}\, v_s \qquad \text{for } v_s > 0 \qquad (3\text{-}2)$$

$$v_L = 0 \qquad \text{for } v_s < 0 \qquad (3\text{-}3)$$

The load-voltage waveform shown in Fig. 3-2b is periodic, finite, and continuous; therefore it can be represented by a Fourier series. If the peak instantaneous value of v_L is designated by V_L, the series is

$$v_L = \frac{1}{\pi}\, V_L \left(1 + \frac{\pi}{2} \sin \omega_s t - \tfrac{2}{3} \cos 2\omega_s t - \tfrac{2}{15} \cos 4\omega_s t + \cdots \right) \qquad (3\text{-}4)$$

Thus v_L consists of a dc component and sinusoidal components at the fundamental radian frequency ω_s and at integer multiples of ω_s. The

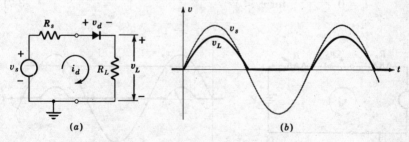

Fig. 3-2 A half-wave diode rectifier. (a) Circuit; (b) waveforms.

voltage across the load therefore contains components at frequencies not present in the voltage applied to the circuit. The appearance of these new frequencies is a consequence of the nonlinearity of the diode. In circuits consisting entirely of linear elements the only frequencies appearing are those present in the applied voltages and currents. One of the primary uses of the diode is the production of these new frequencies. Often, however, the dc component generated by the diode is the quantity of interest. The action by which the diode generates a direct voltage from an alternating supply voltage is called rectification.

The need for rectification is illustrated more clearly by the circuit shown in Fig. 3-3a. This is a circuit that permits a battery to be charged from an ac supply line. In order for charge to accumulate in the battery, the battery current must have a dc component flowing into the positive terminal. The action of the diode in generating a dc component is therefore essential in charging the battery from an ac supply. The action of the circuit can be understood from the waveforms shown in Fig. 3-3b. The ac supply voltage is

$$v_s = V_s \sin \omega_s t \tag{3-5}$$

At those instants when the curve of v_s lies below V (v_s less than V) the net voltage around the loop acts to send current in the reverse direction through the diode. At such instants the diode therefore acts as an open circuit, and no current flows. At those instants when the curve of v_s is above V (v_s greater than V) the net voltage acts to send current in the forward direction through the diode, the diode acts as a short circuit, and current flows in the proper direction to charge the battery. The variable resistance R can be adjusted to give the desired value of charging current. The charging current at each instant is given by

$$i_d = \frac{v_s - V}{R_s + R} \qquad \text{for } v_s > V \tag{3-6}$$

$$i_d = 0 \qquad \text{for } v_s < V \tag{3-7}$$

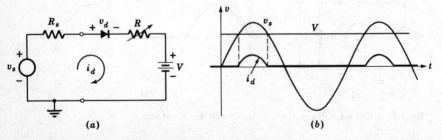

Fig. 3-3 A battery charger. (a) Circuit; (b) waveforms.

An ideal diode can conduct any value of current whatever in the forward direction, and it can withstand any value of inverse voltage. This is not the case, however, with physical diodes. In general there is a limit to the current that can be passed without damage, and there is a limit to inverse voltage that can be applied without causing breakdown. In designing diode circuits it is necessary to ensure that these limits are not exceeded.

No current flows in the circuit of Fig. 3-2 when an inverse voltage exists across the diode; hence there is no voltage drop across the circuit resistances, and the peak inverse voltage that appears across the diode is simply V_s. The peak forward current occurs when v_s has its maximum positive value; the magnitude of this current is $i_{d,\,max} = V_s/(R_s + R_L)$. Similarly, there is no current in the circuit of Fig. 3-3 when an inverse voltage exists across the diode; hence the peak inverse voltage occurs when $v_s = -V_s$, and its value is $PIV = V_s + V$ (see Fig. 3-3b). The peak forward current occurs when $v_s = V_s$, and its value is $i_{d,\,max} = (V_s - V)/(R_s + R)$.

3-3 · THE DIODE LIMITER

The circuit shown in Fig. 3-4a is a diode limiter. It has the property that no matter how wide the range over which v_s varies, the output voltage v_o is restricted to the range between the values of V_1 and $-V_2$. The voltage transfer characteristic of Fig. 3-4b shows how the output voltage varies as a function of the input voltage. If the input voltage makes excursions outside the range between V_1 and $-V_2$, those portions of the input waveform are clipped off. Thus the circuit is often called a diode clipper.

When the input voltage is greater than V_1, diode D_1 conducts, and

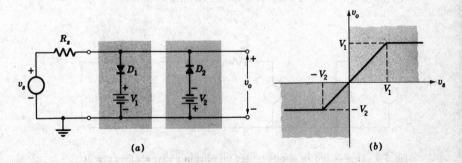

Fig. 3-4 A diode limiter. (a) Circuit; (b) transfer characteristic.

$v_o = V_1$. When the input voltage is more negative than $-V_2$, diode D_2 conducts, and $v_o = -V_2$. For values of v_s between these two limits both diodes are biased in the reverse direction, and therefore both act as open circuits; hence in this range of v_s the output voltage varies with the input voltage, as shown in Fig. 3-4b. If no load is connected across the output terminals, $v_o = v_s$ when v_s is in the range between V_1 and $-V_2$.

The diode limiter can be used to protect a circuit against the application of excessive voltage; for example, it can be used to protect the meter element in an electronic voltmeter against damage from overvoltage. Limiters are also used to protect transistors from overvoltage in some instances. The diode limiter is also used as a signal-processing circuit to alter the waveform of the input signal voltage. For example, if $V_1 = V_2$ and if v_s is a sinusoid with an amplitude much larger than V_1, then the output signal voltage has very nearly a square waveform.

The battery symbols are used in Fig. 3-4a to emphasize the fact that V_1 and V_2 are direct voltages; they do not imply that batteries are necessarily used to obtain these voltages. This practice is followed throughout this book. It is to be understood that battery symbols designate ideal sources of direct voltage.

Figure 3-5 shows a signal-processing circuit consisting of an amplifier followed by a diode limiter. This circuit can accept a small signal v_i at the input, amplify it to a much larger signal μv_i, and clip it to produce a modified wave at the output. The operation performed by the circuit is commonly referred to as "amplify and clip." The circuit can be used as a square-wave generator. It can also be used as a zero-crossing detector; each time the input waveform crosses the zero-voltage axis the output voltage makes a rapid transition from a negative voltage to a positive voltage, or vice versa. There are many circumstances in which the detection of zero crossings is important. For example, one of the most common ways of measuring the frequency of a sinusoidal signal is to count the number of zero crossings in a given interval of time, say 1 sec, with an electronic counter.

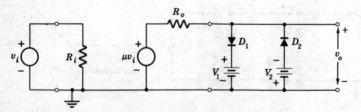

Fig. 3-5 Diodes and an amplifier used together in a wave-shaping circuit.

3-4 THE PEAK RECTIFIER AND DIODE DEMODULATOR

The half-wave rectifier discussed in Sec. 3-2 delivers a pulsating, uni-directional current to the load. Such circuits are satisfactory for many applications such as battery charging and various electrolytic processes. There are many other applications, however, where it is desired to obtain a pure, nonpulsating direct voltage by rectification of the voltage from the standard ac power mains. The peak rectifier shown in Fig. 3-6a is often used for this purpose.

The operation of the circuit [1] † can be understood with the aid of the waveforms shown in Fig. 3-6b. If C is initially uncharged and if the supply voltage $v_s = V_s \sin \omega_s t$ is applied at $t = 0$, then as v_s increases from zero to its maximum positive value, current flows in the forward direction through the diode, and charge is stored in the capacitor. If the source resistance R_s is very small, the voltage drop across it is negligibly small, and v_L, the voltage across C, is essentially equal to v_s at every instant until v_s reaches its maximum positive value. Thus C charges to the voltage V_s. Now if $i_L = 0$, there is no way for C to discharge, for the diode does not conduct current in the reverse direction. Thus, as v_s drops from its maximum positive value, the potential at the anode of the diode becomes less than the potential at the cathode, and a reverse-bias voltage is developed across the diode. The charge accumulated by C during the first quarter cycle is trapped and cannot escape. This trapped charge maintains the voltage across the capacitor at the value V_s. The circuit in Fig. 3-6a is called a peak rectifier because its output voltage is equal to the positive peak value of the input voltage.

If the load on the peak rectifier consists of a large resistor R_L, as shown in Fig. 3-7a, then the capacitor can discharge slowly through R_L while the diode is not conducting. Under these conditions the load

† Superior numbers designate references listed at the end of the chapter.

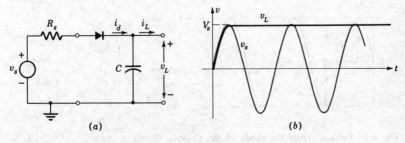

Fig. 3-6 A peak rectifier. (a) Circuit; (b) waveforms.

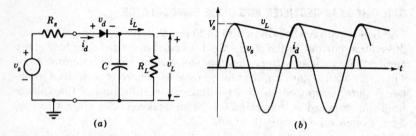

Fig. 3-7 A peak rectifier with a resistive load. (*a*) Circuit; (*b*) waveforms.

voltage consists of a small ripple component superimposed on a large dc component. As R_L is made smaller, the capacitor discharges by a greater amount each cycle, and the ripple becomes larger in magnitude. During a brief interval near the positive peak of v_s in each cycle, a pulse of charging current flows through the diode; this pulse of current restores to the capacitor the charge lost through R_L while the diode is not conducting. The pertinent waveforms are shown in Fig. 3-7*b*.

If the voltage drop across R_s is negligibly small, the voltage across the diode is $v_d = v_s - v_L$. The waveform of this voltage is shown in Fig. 3-8 for the case where the ripple component of v_L is very small. It is clear from this waveform that the peak inverse voltage across the diode is $2V_s$.

A circuit that is capable of generating an amplitude-modulated signal is shown in Fig. 2-7, and the AM signal itself is shown in Fig. 2-8. The AM signal carries information in its envelope. To recover the information it is necessary first to develop a signal voltage that varies in accordance with the envelope; the process by which this is done is called demodulation, or detection. Since the peak rectifier shown in Fig. 3-7

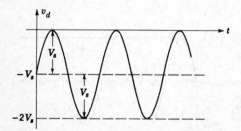

Fig. 3-8 Voltage across the diode in the peak rectifier.

develops an output voltage that is approximately equal to the peak value of the input voltage, it is capable of demodulating the AM signal. When used in this application, the peak rectifier is called a peak detector, or diode detector, or envelope detector. This simple detector is used almost exclusively in AM radio systems.

The waveform of a radio-frequency AM signal v_s applied at the input of a peak detector is shown in Fig. 3-9, and the waveform of the output voltage v_L is also shown. The ripple in the output is a radio-frequency voltage that can be removed by further filtering. The output voltage shown in Fig. 3-9 is the voltage across the capacitor in Fig. 3-7. When the envelope rises, the capacitor voltage rises as a result of charging current through the diode. When the envelope falls, the capacitor voltage falls as a result of discharging current through the resistor R_L; discharging current cannot flow through the diode. It follows that if the time constant CR_L is made too large, the capacitor voltage cannot fall as fast as the envelope falls, and the waveform of the output voltage is a distorted version of the waveform of the envelope. The design of the diode detector involves choosing the time constant small enough to permit the output voltage to follow the waveform of the envelope but not so small that the ripple in the output is excessive. References 1 and 2 give a more extensive discussion of the diode detector.

When the diode detector is used in a signal-processing system, it is usually necessary to know what kind of a load the detector presents to the preceding part of the system. For this purpose an input impedance is commonly specified. However, since the detector is a nonlinear circuit, the notion of an input impedance must be examined with care. In the domain of sinusoidal steady-state circuit analysis the concept of impedance is meaningful only as the ratio of the complex amplitude of a sinusoidal voltage to the complex amplitude of a sinusoidal current having the same frequency. When a sinusoidal voltage is applied to the input of the diode detector, the waveform of the input current is certainly not sinusoidal; it is a train of short pulses as illustrated in Fig.

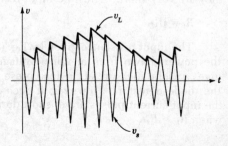

Fig. 3-9 Waveforms for the peak rectifier operating as an AM demodulator.

3-7b. However, it is possible to define a useful quantity, called the input impedance, for the detector in the manner outlined below.

Suppose that the voltage applied to the peak detector is

$$v_s = V_s \cos \omega_s t \qquad (3\text{-}8)$$

The pulses of current that flow through the diode coincide with the positive peaks of the voltage, and if the pulses are very short compared with the interval between pulses, the diode current can be expressed as a Fourier series having the form

$$i_d = I_0 + I_1 \cos \omega_s t + I_2 \cos 2\omega_s t + \cdots \qquad (3\text{-}9)$$

where the I's are the usual Fourier coefficients. The second term in this series is a sinusoidal current having the same frequency and phase as the input voltage. The so-called input impedance of the circuit is defined as the ratio of the amplitude of the input voltage to the amplitude of the component of input current having the same frequency as the input voltage. Since these two sinusoids are in phase, the input impedance is a resistance

$$R_i = \frac{V_s}{I_1} \qquad (3\text{-}10)$$

The dc component of i_d must flow through R_L, and it develops a direct voltage across R_L that, as is indicated in Fig. 3-7b, is very nearly equal to V_s. Thus $V_s = R_L I_0$, and the input resistance can be expressed as

$$R_i = R_L \frac{I_0}{I_1} \qquad (3\text{-}11)$$

But for very short pulses of current the Fourier analysis yields

$$I_1 = 2I_0 \qquad (3\text{-}12)$$

and this relation is independent of the shapes of the pulses provided that they are very short. Thus the input resistance is

$$R_i = \tfrac{1}{2}R_L \qquad (3\text{-}13)$$

The input resistance given by Eq. (3-13) gives the correct value for the power drawn from an ideal voltage source, and in many cases it gives a useful measure of the load presented to the preceding circuits by the diode detector. However, it does depend on the assumption that the input current pulses are of short duration; thus it is the most useful when $R_L \gg R_s$.

3-5 THE DIODE CLAMPER

The circuit of a diode clamper is shown in Fig. 3-10a. This circuit is identical with the peak rectifier shown in Fig. 3-6a except that the positions of the diode and the capacitor are interchanged. Thus the operation of the clamper is like that of the peak rectifier; however, the output voltage is taken from a different pair of terminals. If the input voltage is $v_s = V_s \sin \omega_s t$, and if the voltage drop across R_s is negligibly small, the capacitor charges through the diode to the voltage V_s, as indicated in Fig. 3-10a. The output voltage, which in this case is the voltage across the diode, is $v_d = v_s - V_s$. Thus the waveform of the output voltage is the same as the waveform of the input voltage, but the wave is shifted down by an amount equal to the positive peak value of the input voltage, as illustrated in Fig. 3-10b. Since the output voltage rises just to the value zero when v_s has its positive peak value, the circuit is said to clamp the positive peak of the v_s waveform at 0 volts.

The diode clamper has a number of useful signal-processing applications. For example, in TV receivers it is necessary that the voltages appearing at certain points in the circuit have fixed peak values. Clamping circuits are used to fill this need. As another example, clamping circuits are often used in ac electronic voltmeters; this application is discussed further in Sec. 3-6.

Another form of the diode clamper is shown in Fig. 3-11a. This circuit is like the one in Fig. 3-10 except that the diode is reversed. For variety, a rectangular waveform of input voltage containing a dc component, shown in Fig. 3-11b, is assumed. When v_s is negative, the diode conducts and charges the capacitor to the maximum negative value of v_s with the polarity shown. Thus the voltage across the diode in this circuit is $-v_d = v_s + V_2$, and as shown in Fig. 3-11c, the waveform of the

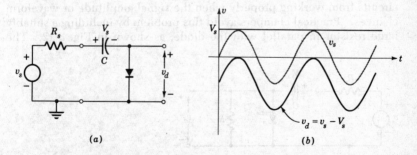

(a)   (b)

Fig. 3-10 A diode clamper. (a) Circuit; (b) waveforms.

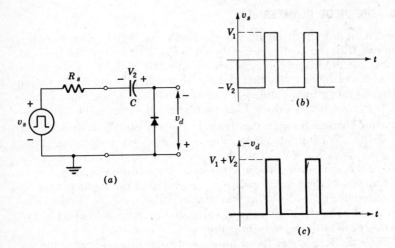

Fig. 3-11 Another diode clamper. (*a*) Circuit; (*b*) input voltage; (*c*) output voltage.

input voltage is shifted up by an amount equal to its negative peak value. This circuit clamps the negative peak of the signal waveform at 0 volts. Note that the dc component in the input signal does not enter into the clamping action; it merely affects the voltage developed across the capacitor.

Signal waveforms can be clamped at levels other than zero by adding a battery in series with the diode in Figs. 3-10 and 3-11.

The capacitors in the clampers of Figs. 3-10 and 3-11 accumulate charge as a result of current through the diodes. This charge is trapped on the capacitors; the capacitors cannot discharge because the diodes cannot conduct current in the reverse direction. This fact prevents these circuits from working properly when the signal amplitude or waveform changes. Practical clampers avoid this problem by including a suitably large resistor in parallel with the diode, as shown in Fig. 3-12. The

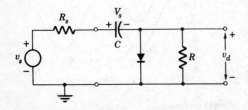

Fig. 3-12 A diode clamper with a resistive load.

capacitor in this circuit discharges slowly through the resistor when a reverse bias exists across the diode. When the resistance is large, it carries negligible current and does not alter the analysis given above.

The practical clamper in Fig. 3-12 is a nonlinear circuit closely related to the peak detector, and it is possible to define an input resistance for the clamper in the same way that the input resistance of the peak detector is defined in Eq. (3-10). To evaluate this resistance, let

$$v_s = V_s \cos \omega_s t \tag{3-14}$$

The fundamental-frequency component of the input current is the sum of the fundamental-frequency components in the diode and shunt resistor R; it is

$$I_{i1} = I_{d1} + I_{R1} \tag{3-15}$$

As in the diode detector, the fundamental-frequency component of diode current is twice the dc component, and thus

$$I_{i1} = 2I_{d0} + I_{R1} \tag{3-16}$$

The direct current in the capacitor is zero; hence the direct current in the resistor is equal to the direct current in the diode. Furthermore, with the positive peak of the signal clamped at zero, the direct voltage across the resistor is V_s, and thus $V_s = RI_{d0}$. The input signal voltage also appears across the resistor, and therefore $V_s = RI_{R1}$. Substituting these relations into (3-16) yields

$$I_{i1} = 2 \frac{V_s}{R} + \frac{V_s}{R} \tag{3-17}$$

from which

$$R_i = \frac{V_s}{I_{i1}} = \tfrac{1}{3}R \tag{3-18}$$

3-6 AN AC ELECTRONIC VOLTMETER

All ac electronic voltmeters except the cathode-ray oscilloscope operate on the principle of converting the ac signal into a dc signal with a related amplitude and measuring the dc signal with a dc meter. Thus to measure a sinusoidal voltage, for example, the peak rectifier of Sec. 3-4 can be used with a high-impedance dc voltmeter to measure the peak value of the voltage. However, electronic voltmeters are often used to measure a small signal voltage, usually sinusoidal, superimposed on a larger direct voltage, and it is the value of the signal voltage alone that is wanted. Since the output of the peak rectifier depends in a major way on any dc component in its input, the peak rectifier is not suitable for this kind of measurement.

The diode clamper of Sec. 3-5, on the other hand, has a capacitor in series with its input, and as is pointed out in Sec. 3-5, its output is independent of any dc component in its input. The clamper also produces a dc component of output voltage that is related to the amplitude of the time-varying signal at the input. Thus the clamper is well suited for use in ac electronic voltmeters, and it is used extensively in that application. The circuit diagram for such an instrument is shown in Fig. 3-13. The high-input-impedance amplifier serves to isolate the low-impedance meter element from the high-impedance clamping circuit; in effect, the amplifier relieves the clamper of the burden of supplying the current needed to deflect the meter. The voltage source v and series resistor R represent the Thévenin equivalent of the circuit to which the instrument is to be connected for measuring purposes.

Waveforms of voltage at various points in the circuit are shown in Fig. 3-13b. Voltage v is the sinusoidal signal to be measured superimposed upon a dc component. The voltage v_d, which appears across the diode, is the clamped waveform; it consists of the signal to be measured minus a dc component V_i that is equal to the peak value of the sinusoid. The combination of R_2 and C_2 forms a filter that removes the

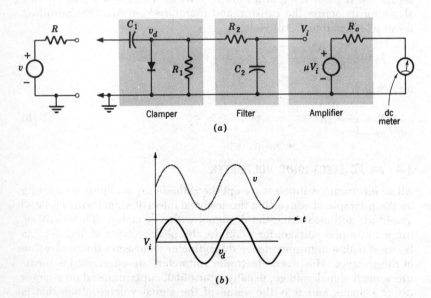

(a)

(b)

Fig. 3-13   An ac electronic voltmeter. (a) Circuit; (b) waveforms.

sinusoidal component of v_d and transmits only the dc component to the amplifier. The filtering is accomplished by designing the RC network so that C_2 acts as a short circuit in comparison with R_2 at all frequencies at which measurements are to be made; of course C_2 acts as an open circuit to the dc component of v_d.

With any input-signal waveform, the positive peak of the waveform is clamped at 0 volts, and the meter indicates the average value of the clamped wave. Thus the meter is said to indicate the *peak-above-average* value of the wave.

This simple circuit performs quite well as an electronic voltmeter, and it is capable of making accurate measurements over the range of frequencies between a few hertz and several hundred megahertz. Instruments designed for use at radio frequencies have the clamper and filter contained in a shielded probe and arranged so that all wires carrying radio-frequency currents are as short as possible; the direct voltage developed at the output of the filter is transmitted to the amplifier over a shielded cable. The principal shortcoming of this instrument is that, owing to the limitations of real diodes, it cannot measure voltages much smaller than 1 volt.

Electronic voltmeters like the one described above are most often used to measure sinusoidal voltages, and they are usually calibrated to read the rms value of sinusoids. However, the deflection of the needle is basically proportional to the peak-above-average value of the signal, and this is true for any signal waveform. Thus it should be understood that the meter scale is calibrated to read the peak-above-average value divided by $\sqrt{2}$. This is the rms value for sinusoids, but it is not the rms value of other waveforms in general. These voltmeters can be used to measure nonsinusoidal waveforms of voltage, but the meter readings must be interpreted with care. When the waveform to be measured is complex, it is usually desirable to use a cathode-ray oscilloscope for the measurement.

In the above discussion it is assumed that the resistance R in series with the voltage being measured is small. If this resistance is large, as sometimes is the case, the current drawn by the voltmeter may cause an appreciable voltage drop across R, with the result that the voltage at the input to the voltmeter is not the desired voltage. In such cases the error in measurement may be appreciable unless a correction is made for the loading effect of the instrument; for this reason it is important that the voltmeter be designed to draw the least possible current from the circuit in which measurements are being made. An input resistance for the instrument can be evaluated by following a procedure similar to the one leading to Eq. (3-18). The input resistance obtained for the circuit in Fig. 3-13 with C_2 treated as a short circuit is the resistance of R_2 in parallel with $R_1/3$; thus it is important that R_1 and R_2 be made as large as possible.

3-7 THE VOLTAGE DOUBLER

The rectifier shown in Fig. 3-14a has the interesting and useful property that it develops a direct voltage at the output equal to the peak-to-peak value of the input voltage.[1] Hence, if the input is a sinusoid $v_s = V_s \sin \omega_s t$, the direct voltage at the output is twice the peak value of the sinusoid. The operation of the circuit can be explained as follows. The peak rectifier consisting of D_2 and C_2 draws negligible current from the clamper consisting of D_1 and C_1, and the clamper operates in the fashion described in Sec. 3-5. Capacitor C_1 charges to the negative-peak value of the input voltage as shown in Fig. 3-14a, and the voltage across D_1 is $v_{d1} = v_s + V_s$. The waveform of this voltage is shown in Fig. 3-14b; it constitutes the input voltage to the peak rectifier. The peak rectifier then charges C_2 to the positive-peak value of v_{d1}. This voltage is the peak-to-peak value of the input voltage to the clamper, $2V_s$ for the conditions illustrated in Fig. 3-14, and it is the output of the voltage doubler.

Since the voltage doubler has a clamper at its input, its output is

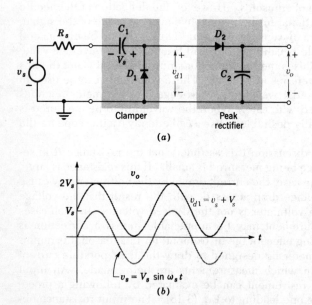

Fig. 3-14   A voltage doubler. (*a*) Circuit; (*b*) waveforms.

independent of any dc component in the input, and it is suitable for use in electronic voltmeters. Voltmeters using the voltage doubler are called *peak-to-peak-reading* voltmeters, and they often have scales labeled "peak-to-peak volts."

By extending the basic ideas involved in the voltage doubler, diode circuits can be devised to act as voltage triplers, quadruplers, . . . , *n*-tuplers. Such circuits are used to develop the very high voltages used in testing high-voltage electrical equipment. Further discussion of these circuits can be found in Ref. 1.

3-8 THE FULL-WAVE RECTIFIER

The circuit of a full-wave rectifier supplying power to a resistive load is shown in Fig. 3-15a. The circuit consists basically of two half-wave rectifiers connected to a single load resistor and supplied with sinusoidal input voltages $v_s = \pm V_s \sin \omega_s t$ that are equal in magnitude but opposite in phase. During the positive half cycle of v_s, D_1 acts as a short circuit, D_2 acts as an open circuit, and $v_L = v_s$. During the negative half cycle of v_s, D_2 acts as a short circuit, D_1 acts as an open circuit, and $v_L = -v_s$. The waveform of v_L is shown in Fig. 3-15b.

The load voltage developed by the full-wave rectifier can be expressed mathematically as

$$v_L = |v_s| = |V_s \sin \omega_s t| \qquad (3\text{-}19)$$

This expression is descriptive of the waveform of v_L, but it is not at all useful in calculating the currents and voltages in a more complicated load configuration. An alternative expression is obtained by recognizing that v_L can be expanded in a Fourier series; this series is

$$v_L = \frac{2}{\pi} V_s (1 - \tfrac{2}{3} \cos 2\omega_s t - \tfrac{2}{15} \cos 4\omega_s t - \cdots) \qquad (3\text{-}20)$$

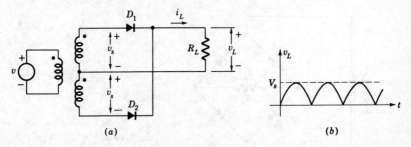

Fig. 3-15 A full-wave rectifier. (*a*) Circuit; (*b*) waveform of output voltage.

The information placed in evidence by this equation is that the load voltage consists of a dc component of magnitude $(2/\pi)V_s$, which is the average value of each half cycle of v_s, and a set of sinusoidal components at radian frequencies that are even multiples of ω_s. Thus, if v_s is a 60-Hz voltage with an rms value of 115 volts, v_L contains a dc component of $(115)(2\sqrt{2}/\pi) = 103.5$ volts plus sinusoidal components at 120, 240, 360, ... Hz.

In the above discussion it is assumed that the voltages across the two halves of the transformer are equal. In practice they are likely to be somewhat unbalanced, with the result that successive peaks in the load-voltage waveform are not equal. In this case the fundamental frequency of the periodic load voltage is 60 Hz, and v_L contains components at 60 Hz and its harmonics.

The peak inverse voltage across the diodes in Fig. 3-15a can be determined in the following way. The sum of the voltages across the two diodes is equal to the full secondary voltage of the transformer, $2v_s$. However, at each instant of time one of the diodes acts as a short circuit and the other acts as an open circuit. Hence the voltage $2v_s$ appears across the diode that acts as an open circuit, and the peak inverse voltage that appears across either diode is $2V_s$.

Another important full-wave-rectifier circuit is shown in Fig. 3-16. This circuit, which is called a bridge rectifier, requires four diodes, but it has several compensating advantages, among which is the fact that it does not require a center-tapped transformer. The circuit is normally operated with a sinusoidal input voltage $v_s = V_s \sin \omega_s t$. During the positive half cycle of v_s current flows through D_1, R_L, and D_3, and during the negative half cycle current flows through D_4, R_L, and D_2. In both half cycles current flows in the same direction through R_L. During the positive half cycle R_L is connected across the transformer by D_1 and D_3; during the negative half cycle it is connected across the transformer by D_2 and D_4 with the terminals reversed. It follows that the waveform of

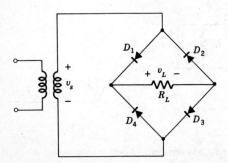

Fig. 3-16 A bridge rectifier.

load voltage is the same as the one shown in Fig. 3-15b and that v_L is given by Eqs. (3-19) and (3-20).

The peak inverse voltage across the diodes can be determined by noting that the sum of the voltages across D_1 and D_4 equals the supply voltage v_s and that, likewise, the sum of the voltages across D_2 and D_3 equals v_s. Thus during the positive half cycle of v_s, when D_1 and D_3 are short circuits, the voltage v_s appears across both D_2 and D_4 in the inverse direction, and the peak inverse voltage across each of these diodes is V_s. During the negative half cycle D_2 and D_4 are short circuits, the voltage v_s appears across both D_1 and D_3 in the inverse direction, and the peak inverse voltage across each of these diodes is V_s. Thus, for a specified dc component of voltage at the load, the diodes in the bridge rectifier need withstand only half as much inverse voltage as those in the rectifier of Fig. 3-15. This is an important consideration when semiconductor diodes are used.

Figure 3-17a shows a full-wave bridge rectifier with a smoothing capacitor connected across its output terminals. The supply voltage for this circuit is $v_s = V_s \sin \omega_s t$. This rectifier is similar to the single-diode peak rectifier discussed in Sec. 3-4; in this case, however, the capacitor receives two charging pulses in each cycle of the input voltage

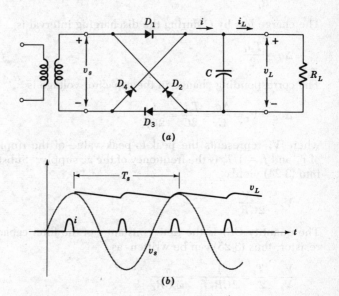

Fig. 3-17 Bridge rectifier with a capacitor filter. (a) Circuit; (b) waveforms.

instead of one.   Consequently, for the same values of R_L and C, the full-wave peak rectifier has less ripple in its output than the half-wave rectifier does.   The circuit in Fig. 3-17a is widely used as the dc power supply in electronic equipment.   Large smoothing capacitors are used in these power supplies in order to keep the ripple in the output voltage small; capacitance values of several hundred microfarads are commonplace.

When the ripple component of the output voltage is small compared with the dc component, the magnitude of the ripple can be estimated easily.   The peak value of v_L is V_s; hence v_L can be written as

$$v_L = V_s - v_r \tag{3-21}$$

where v_r is the sawtooth ripple voltage shown in Fig. 3-17b.   The ripple voltage results from the capacitor discharging through the load resistor during the interval between charging pulses.   The discharging interval is approximately $T_s/2$, half the period of the supply voltage.   When the ripple is small, v_L is nearly constant at the value V_s throughout the discharging interval, and the current discharging C is approximately constant at the value

$$i_L = \frac{V_s}{R_L} \tag{3-22}$$

The charge lost by C during the discharging interval is

$$\Delta q = \frac{T_s i_L}{2} \tag{3-23}$$

The corresponding change in the capacitor voltage is

$$\Delta v_L = V_r = \frac{\Delta q}{C} = \frac{T_s i_L}{2C} = \frac{i_L}{2C f_s} \tag{3-24}$$

where V_r represents the peak-to-peak value of the ripple component of v_L and $f_s = 1/T_s$ is the frequency of the ac supply.   Substituting (3-22) into (3-24) yields

$$V_r = \frac{T_s V_s}{2C R_L} \tag{3-25}$$

The quantity $C R_L$ is the time constant τ of the filter capacitor and load resistor; thus (3-25) can be written as

$$\frac{V_r}{V_s} = \frac{T_s}{2\tau} = \frac{1}{2C R_L f_s} = \frac{\pi}{\omega_s C R_L} \tag{3-26}$$

The capacitor filter described above is the simplest and cheapest of the filters used in electronic power supplies, and it is widely used.

It may be necessary to use more elaborate filters, however, when a high degree of filtering is required. Many power supplies use RC filters like the one shown in Fig. 3-13 to supply low-current portions of the load with a highly filtered direct voltage; these RC filters are used in addition to the filter capacitor shown in Fig. 3-17. Other filters use shunt capacitors and series inductors to achieve a high degree of filtering, although the size and weight of suitable inductors are distinct disadvantages in modern, compact, electronic equipment. Electronic voltage regulators used to stabilize the output voltage of power supplies against variations in ac supply voltage also have the effect of reducing the ripple in the output, and the effect can be very large. Thus when a regulator is needed to stabilize the output voltage, it is also counted on to remove the ripple in the output voltage. The design of power supplies and ripple filters of various kinds are discussed in further detail in Refs. 1 to 5; Refs. 4 and 5 give a variety of charts that are useful in power-supply design.

Example 3-1

A bridge rectifier with a capacitor filter is used to supply the dc power required by an electronic amplifier. The ac supply voltage is sinusoidal with an amplitude of 30 volts and a frequency of 60 Hz; the value of the filter capacitance is 300 μF. It is assumed that the capacitor charges to the peak value of the supply voltage with each pulse of charging current. Under these conditions the amplifier draws 30 mA from the power supply. The ripple voltage is to be determined.

Solution From Eq. (3-24) the ripple is

$$V_r = \frac{i_L}{2Cf_s} = \frac{0.03}{(2)(300)(10^{-6})(60)}$$

$$= 0.83 \text{ volt}$$

Thus the dc component of the output voltage is very nearly 30 volts, and the peak-to-peak value of the ripple is about 3 percent of the output voltage. A ripple voltage of this magnitude is acceptable for some applications, but it is too large for many others.

3-9 THE MONOPULSER

This chapter is brought to a close with the examination of a signal-processing circuit that is somewhat more complex than the circuits presented in the preceding sections. The circuit is shown in Fig. 3-18a; it employs two diodes and a current amplifier along with resistors and a

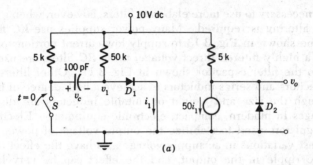

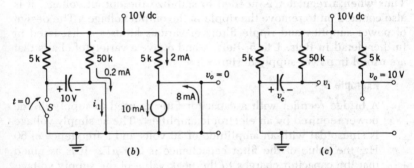

Fig. 3-18 The monopulser. (a) Circuit; (b) circuit details before the switch is closed; (c) circuit details just after the switch is closed.

capacitor. The function of the circuit is to generate a short rectangular pulse of voltage at the output terminals when the switch is closed. It is called a monopulser. The monopulser is used to generate pulses required by logic circuits and systems using digital signals. In these applications the function of the switch is taken over by a bipolar transistor.

Figure 3-18b shows the steady-state conditions in the circuit with the switch open. The 10-volt dc supply turns diode D_1 on, and it delivers 0.2 mA to the input of the amplifier. The output current from the amplifier is thus $(0.2)(50) = 10$ mA. Two milliamperes of this current flow in the 5-kilohm load resistor making the voltage at the output terminal zero. Under this condition diode D_2 acts as a short circuit, and the remaining 8 mA of the amplifier current flow through D_2. The capacitor is charged to 10 volts as shown in the diagram.

When the switch is closed, the voltage on the capacitor cannot change instantaneously; therefore the potential on the right side of the capacitor is forced instantaneously to -10 volts. This condition represents a reverse bias on D_1, and D_1 becomes an open circuit. Thus the input current to the amplifier drops abruptly to zero forcing the output current of the amplifier to zero, and the current source becomes an open circuit. With zero current from the amplifier, there is no voltage drop across the 5-kilohm load resistor, and the output voltage jumps to 10 volts. This output voltage is a reverse bias for D_2, and D_2 becomes an open circuit.

Figure 3-18c shows conditions in the circuit just after the switch is closed. In accordance with the above discussion, both of the diodes and the current source act as open circuits. The abrupt changes in v_1 and v_o that occur when the switch is closed are shown by the waveforms plotted in Fig. 3-19. Immediately after these jumps occur the capacitor starts charging exponentially toward a new steady state in which $v_1 = 10$ volts. As C charges, v_1 is given by

$$v_1 = 10 - 20e^{-t/\tau}$$

where τ is the time constant, (50 kilohms)(100 pF) $= 5$ μsec, with which the capacitor charges. This exponential waveform is shown in Fig. 3-19a. The charging continues until v_1 reaches the value zero. At this point D_1 becomes a short circuit, the input current to the amplifier jumps to 0.2 mA, the output current from the amplifier jumps to 10 mA, and the circuit abruptly reverts to the condition shown in Fig. 3-18b except that the switch is closed. With this action, v_o drops to zero, and the output pulse is ended.

The duration of the output pulse can be calculated from the fact that it ends when the exponentially rising v_1 reaches zero. Substituting $v_1 = 0$ in the equation for v_1 and setting $t = T$, the pulse duration, yields

$$0 = 10 - 20e^{-T/\tau}$$

or

$$T = \tau \ln 2 = (5)(0.693) = 3.47 \ \mu sec$$

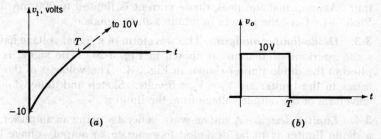

Fig. 3-19 Waveforms of voltage in the monopulser.

It follows from these results that the circuit can be designed to generate a pulse of specified width by proper choice of the RC time constant.

If the switch is now reopened, the capacitor charges rapidly to its original steady state. This action resets the circuit and puts it in condition to generate a new pulse whenever the switch is closed again.

3-10 SUMMARY

This chapter has presented the ideal diode, its characteristics, its circuit properties, and a few of the many applications in which it is used. The diode is a nonlinear device, and it can perform signal-processing operations that cannot be performed by linear circuit components. Two such operations described in this chapter are rectification and demodulation. The ideal diode is also a switch that is turned on or off depending on the polarity of the voltage or current applied to it.

Real diodes are only approximations to the ideal diode, although in many cases they are excellent approximations. Chapter 4 presents the silicon pn junction as a diode, and it examines the physical and electrical properties of the junction.

PROBLEMS

3-1. Battery-charger analysis. A battery charger like the one shown in Fig. 3-3 is used to recharge the 3-volt battery in an electric toothbrush. The supply voltage is $v_s = 6 \sin 377t$.

(*a*) What value of $R + R_s$ is needed to limit the peak diode current to 200 mA?

(*b*) What percentage of the time does the diode conduct?

(*c*) What is the peak inverse voltage that appears across the diode?

3-2. Battery-charger analysis. The electric toothbrush of Prob. 3-1 is used for 1 min, and it draws a current of 200 mA from the battery during that time. How long will it take the battery charger described in Prob. 3-1 to restore the charge drawn from the battery while brushing? Assume that the peak diode current is limited to 200 mA as in Prob. 3-1. Give the answer in minutes and seconds.

3-3. Diode-limiter analysis. The waveform of a signal voltage having some spurious oscillations is shown in Fig. 3-20. The signal is applied to the diode limiter shown in Fig. 3-4. The voltages of the batteries in the limiter are $V_1 = V_2 = 5$ volts. Sketch and dimension the waveform of the output voltage from the limiter.

3-4. Limiter design. A square-wave generator using an amplifier and a diode limiter is to be designed to generate an output voltage having peak positive and negative values of 5 volts. The circuit configuration is shown in Fig. 3-5. The input voltage is to be $v_i = 5 \sin \omega t$, and

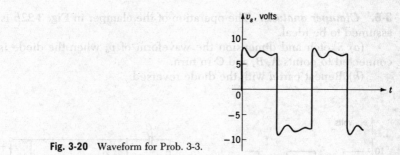

Fig. 3-20 Waveform for Prob. 3-3.

the time required by the output voltage to make the transition from −5 to 5 volts is to be 5 percent of the period of the square wave.

Specify the values of V_1, V_2, and μ that are required.

3-5. Clamper analysis. Three clampers of the type used in TV receivers are shown in Fig. 3-21. The input signal for each circuit, shown in Fig. 3-21d, is $v_s = 3 \sin \omega t$.

(a) Assuming that the operation of each clamper is ideal, sketch and dimension the waveform of output voltage for each circuit. *Suggestion:* Find the voltage across C first.

(b) Give the diagram of a circuit that will clamp the negative peaks of v_s at −2 volts.

(c) If the battery in Fig. 3-21a is reversed, the circuit will not clamp the given signal. Why?

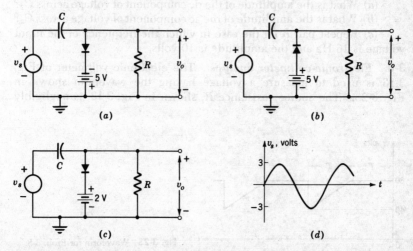

Fig. 3-21 Circuits and waveform for Prob. 3-5.

3-6. *Clamper analysis*. The operation of the clamper in Fig. 3-22*b* is assumed to be ideal.

(*a*) Sketch and dimension the waveform of v_o when the diode is connected to points *A*, *B*, and *C* in turn.

(*b*) Repeat part *a* with the diode reversed.

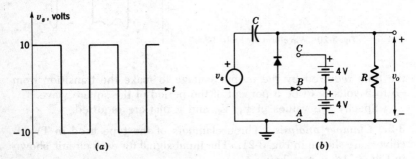

(*a*)   (*b*)

Fig. 3-22 Circuit and waveform for Prob. 3-6.

3-7. *Electronic-voltmeter analysis*. The voltmeter shown in Fig. 3-13 is used to measure a sinusoidal voltage having a peak value of 10 volts and a frequency of 16 kHz. The effectiveness of the *RC* filter is to be examined. The pertinent facts are: $R_2 = 1$ megohm, $C_2 = 0.1\ \mu$F, *R* is negligibly small, and the operation of the clamper is ideal.

(*a*) What is the amplitude of the dc component of voltage across C_2?

(*b*) What is the amplitude of the ac component of voltage across C_2?

(*c*) Repeat part *b* for the case in which the frequency of the input voltage is 16 Hz and the amplitude is 10 volts.

3-8. *Electronic-voltmeter analysis*. The electronic voltmeter of Fig. 3-13 is used to measure a voltage having the waveform shown in Fig. 3-23. The source resistance *R*, shown in Fig. 3-13, is negligibly

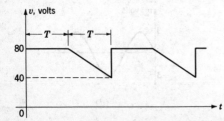

Fig. 3-23 Waveform for Prob. 3-8.

small, and the operation of the clamper is ideal. The voltmeter is calibrated to indicate the rms value of sine waves as discussed in Sec. 3-6.

What is the reading of the voltmeter when it is used to measure the voltage in Fig. 3-23?

3-9. *Full-wave-rectifier analysis.* A full-wave rectifier is shown in Fig. 3-15. Sketch and dimension the waveform of v_L for the case in which both diodes are reversed.

3-10. *CRC ripple-filter analysis.* The power supply shown in Fig. 3-24 supplies the dc power needed to operate a transistor phonograph amplifier. The amplitude of the 60-Hz ac supply voltage is $V_s = 20$ volts. Outputs A and B supply different parts of the circuit.

(*a*) What is the minimum value that C can have if the peak-to-peak ripple voltage at terminal A is not to exceed 2 volts?

(*b*) Assuming for simplicity that the ripple across C is $v_r = \sin 240\pi t$ (2 volts peak-to-peak), calculate the dc component and the ripple component of voltage at B.

(*c*) Why is it impractical to use an RC filter in the 200-mA line?

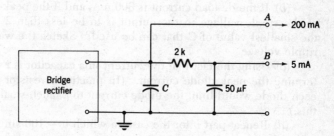

Fig. 3-24 Circuit for Prob. 3-10.

3-11. *Power-supply analysis.* The ac supply voltage for the power supply in Fig. 3-25 has an amplitude $V_s = 40$ volts. Determine the values of V_A and V_B (with respect to ground) under no-load conditions.

If the load current is 400 mA at either terminal, what is the peak-to-peak value of the ripple in the output voltage? Note the ac supply frequency shown in Fig. 3-25.

3-12. *Full-wave rectifier with a trapezoidal input wave.* Peak rectifiers like the one shown in Fig. 3-17 are sometimes supplied with alternating voltages having the waveform shown in Fig. 3-26. The filter capacitor needed with this waveform at the input is much smaller than would be needed with a sinusoidal supply voltage of the same frequency.

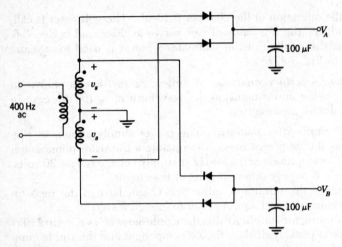

Fig. 3-25   Circuit for Prob. 3-11.

(a) Sketch the waveform of the load voltage when $C = 0$.

(b) If the dc load current is 300 mA, and if the peak-to-peak value
of the ripple voltage in the output is to be less than 2 volts, what is
the smallest value of C that can be used? Sketch the waveform of the
ripple voltage.

(c) Using the fact that the current in a capacitor is $i = C\, dv/dt$, determine the peak diode current. (In practice a resistor in series with
each diode would limit the diode current to a much smaller value than
this.)

(d) Repeat part b for the case in which $v_s = 100 \sin 2{,}000\pi t$.

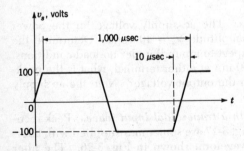

Fig. 3-26   Supply-voltage waveform for the rectifier
of Prob. 3-12.

3-13. *Diode gate.* The circuit shown in Fig. 3-27 has important engineering applications.

(a) Sketch and dimension the waveform of v_o when v_1 and v_2 have the waveforms shown in Fig. 3-27b.

(b) Sketch and dimension the waveform of v_o when v_2 has the waveform shown in Fig. 3-27b and $v_1 = 10 \sin \omega t$.

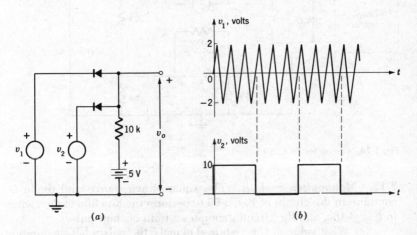

Fig. 3-27 Circuit and waveforms for Prob. 3-13.

3-14. *Electronic ripple filter.* Figure 3-28 shows a bridge rectifier with a capacitor filter and a voltage regulator. The operation of the regulator is described in Prob. 2-7. In this problem the action of the regulator in reducing power-supply ripple is examined.

(a) Treating the 100-kilohm resistor as an open circuit, find V_2 as a function of V_1. *Suggestion:* Start with the relation $V_2 = V_1 - 2V_5$.

(b) Differentiate this result to obtain $\Delta V_2/\Delta V_1$.

(c) The rectifier output V_1 contains a 5-volt ripple component that causes it to vary between 35 and 40 volts. What is the peak-to-peak value of the ripple component in V_2? The best accuracy is obtained by using $\Delta V_2/\Delta V_1$.

(d) If the ripple component of V_2 is to be 5 mV peak to peak, what must be the gain of the differential amplifier (the one whose output is V_5)? *Suggestion:* Let the gain of this stage be A, and repeat parts a and b. Would two amplifiers connected in cascade, each having a gain of 100, be adequate?

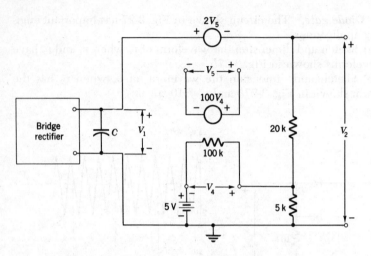

Fig. 3-28 Circuit for Prob. 3-14.

3-15. *Monopulser analysis.* The square-wave source and diode D_3 combine in the circuit of Fig. 3-29 to perform the function of the switch in Fig. 3-18a, and the circuit generates a train of short pulses.

(a) What value of C is required to make the pulses have a duration of 50 μsec?

(b) At which instants on the time scale shown in Fig. 3-29 do the pulses occur?

(c) Is the pulse width affected by an increase in dc supply voltage from 5 to 8 volts?

(d) Sketch and dimension the waveforms of v_1 and v_o during one pulse.

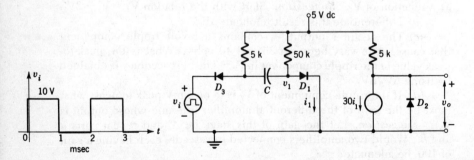

Fig. 3-29 Circuit and waveform for Prob. 3-15.

3-16. *Monopulser analysis.* The form of the circuit in Fig. 3-30 is said to be the *complement* of the form shown in Fig. 3-18a. The circuit is in the steady state with S open for $t < 0$. At $t = 0$, S is closed.

(a) Sketch and dimension the waveforms of v_1 and v_o for $t > 0$.

(b) Show by a circuit diagram how the switch can be replaced by a square-wave generator and diode similar to those in Prob. 3-15. Sketch the waveform of the square wave, and state the requirements on its amplitude.

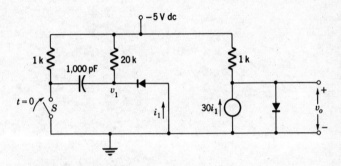

Fig. 3-30 Circuit for Prob. 3-16.

REFERENCES

1. Gray, T. S.: "Applied Electronics," 2d ed., John Wiley and Sons, Inc., New York, 1954.
2. Terman, F. E.: "Electronic and Radio Engineering," 4th ed., McGraw-Hill Book Company, New York, 1955.
3. Arguimbau, L. B., and R. B. Adler: "Vacuum Tube Circuits and Transistors," John Wiley and Sons, Inc., New York, 1956.
4. "RCA Receiving Tube Manual," Radio Corporation of America, Harrison, N.J.
5. "RCA Transistor Manual," Radio Corporation of America, Harrison, N.J.
6. Millman, J., and H. Taub: "Pulse, Digital, and Switching Waveforms," McGraw-Hill Book Company, New York, 1965.

4

Semiconductor Materials and the *pn* Junction

Chapters 2 and 3 present the ideal amplifier and the ideal diode, and they examine some of the useful and interesting signal-processing operations that can be performed by these devices. The next few chapters are concerned with the very important matter of how real amplifiers and real diodes can be made from semiconductor materials. They are also concerned with the equally important matter of the physical processes that take place inside these devices and that make amplifier action and diode action possible. A good understanding of these physical processes is a great asset to the electronic engineer in the design of circuits and systems using amplifiers and diodes.

The pn junction is the heart of almost all semiconductor amplifiers and diodes; therefore, attention is focused initially on the junction. Before the action of the junction can be explained, however, the properties of the semiconductor material must be examined in some detail. Therefore it is helpful at the outset to look ahead a considerable distance to see what the outcome of the study will be. A pn junction is shown pictorially in Fig. 4-1 along with its volt-ampere characteristic. It consists of two semiconductor materials that have different electrical properties and are joined together along a common boundary; this boundary is the junction. The volt-ampere characteristic for the pn junction is very similar to the characteristic of the ideal diode shown in Fig. 3-1. The principal differences are that the pn junction has a small voltage drop across it when it is conducting a forward current and that at a sufficiently large inverse voltage it fails to act as an open circuit. The forward voltage drop is usually less than 1 volt; for a silicon junction it is about 0.7 volt as indicated in Fig. 4-1. The reverse voltage at which the junction breaks down may be less than 5 volts, or it may be greater than 1,000 volts; the breakdown voltage is under the control of the device designer.

The equation for the volt-ampere characteristic, exclusive of the breakdown region, has a relatively simple form; it is

$$I = I_s[\exp (\lambda V) - 1] \qquad (4-1)$$

where I_s and λ are constants and where $\exp x = e^x$.

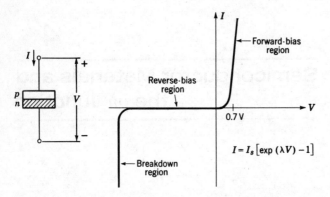

Fig. 4-1 Volt-ampere characteristic for a silicon pn junction.

It follows from the discussion above that the pn junction is, for many purposes, a good approximation to the ideal diode, provided that the peak reverse voltage is not too great. That is, the pn junction is a real diode that approximates the ideal diode. However, the pn junction is much more than a diode. There are applications for the junction in which it operates altogether in the forward-bias region. There are other applications in which it operates altogether in the reverse-bias region, and there are still other applications in which it operates altogether in the breakdown region. These various modes of operation provide the basis for transistors, microamplifiers, and a variety of additional electronic devices. These ideas are developed further in the next few chapters.

4-1 PROPERTIES OF SILICON CRYSTALS

Silicon and germanium are two crystalline solids having physical properties that make them important in semiconductor electronics. In the early days of the transistor, germanium was the more important of the two because at that time it was not possible to make good transistors in silicon. By 1960, however, the technological problems of making silicon transistors were overcome, and since that time, silicon has become by far the more important of the two materials because of its superior properties. Therefore, the discussion that follows is presented in terms of silicon; however, it applies equally well, in a qualitative sense, to germanium and other similar crystalline materials.

A normal silicon atom consists of 14 electrons circulating in orbits around a nucleus made up of 14 protons and 14 neutrons. Since it contains equal amounts of positive and negative charge, it is electrically neutral. Ten of the electrons are very tightly bound to the nucleus; however, the four electrons in the outermost orbits are relatively loosely bound to the atom. These outer four electrons are the valence electrons; they are primarily responsible for the chemical properties of the atom, and it is their presence that permits silicon to form chemical compounds. The valence electrons are required to make the atom electrically neutral, but they represent an excess beyond a preferred chemical state. By way of contrast, the oxygen atom is electrically neutral, but it is deficient by two electrons from a preferred chemical state. If two oxygen atoms and one silicon atom are brought together under favorable conditions, the silicon atom shares two of its excess electrons with each of the oxygen atoms, as illustrated schematically in Fig. 4-2a. These three atoms are bound together in a chemical compound called silicon dioxide; this very stable compound is a form of glass. When joined in this compound and sharing electrons in the manner described, all three of the atoms are closer to a preferred chemical state than when they are separated. The chemical bonds attaching the oxygen atoms to the silicon atom are called electron-pair bonds, or covalent bonds.

If an aggregation of silicon atoms is brought together under favorable conditions, a similar sharing of electrons among the silicon atoms takes place. Each atom shares two electrons in a covalent bond with four of its neighbors in the manner pictured in Fig. 4-2b. For simplicity, a two-dimensional schematic respresentation of the interacting silicon atoms is shown in Fig. 4-2c. Under this condition each silicon atom is closer to a preferred chemical state than when there is no sharing of electrons. The covalent bonds hold the atoms fixed in space relative to each other so that the aggregation forms a regular structure, or lat-

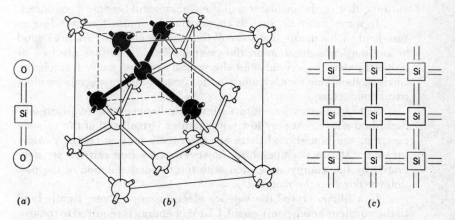

Fig. 4-2 Covalent bonds between atoms. (a) Silicon dioxide molecule; (b) silicon crystal; (c) two-dimensional schematic representation of the silicon crystal.

tice, in space. The entirety of such a regular lattice structure is a crystal. Any particular piece of silicon may consist of a single crystal, or it may consist of many separate crystals oriented in a random manner and joined together in an irregular fashion at their boundaries. In the latter case the silicon is said to have a polycrystalline form. At the present time the best electronic devices are made from single crystals, but polycrystalline materials also have applications in electronics.

The concentration of atoms in a silicon crystal is approximately 5×10^{22} atoms per cubic centimeter. Since each atom is electrically neutral, the crystal itself is electrically neutral on a macroscopic scale.

The single crystal of silicon with a perfectly regular lattice structure as described above is an ideal crystal. Real crystals have imperfections, and these imperfections may have a strong effect on the electrical properties of the crystal. Two-types of imperfections that affect the electrical properties of crystals are dislocations and impurities. A dislocation exists when one or more atoms are missing from the regular sites in the lattice structure or when an atom is located in the space between the regular lattice sites. An impurity is an atom of a different element in the crystal structure; the impurity atom may occupy a regular lattice site, or it may be located in the space between lattice sites.

If the valence electrons of a crystal are tightly bound in covalent bonds, there are no mobile carriers of electric charge in the crystal, and the crystal acts as an insulator. A perfect carbon crystal (diamond) behaves in this manner. If, however, a very strong electric field is applied to such an insulator, the electric forces will tear electrons out of the covalent bonds, thereby setting them free and providing mobile charge carriers; that is, the insulator will break down and become a conductor.

In some materials, such as copper and aluminum, the crystal structure is of such a nature that some of the valence electrons are not bound to any particular location in the crystal. These electrons are free to move through the crystal, and the crystal contains many free charge carriers, about one for each atom in the crystal. Such materials are electrical conductors.

Semiconductors have properties lying between the two extremes described above. At very low temperatures virtually all of the valence electrons are bound, and there are essentially no free charge carriers present. At room temperature, however, many free carriers are generated by the energy associated with the thermal agitation of the particles making up the material.

In a silicon crystal the valence electrons are not very tightly held in the covalent bonds; only about 1.1 eV of energy is required to remove an electron from the bond at room temperature. At room temperature the particles in the crystal are in constant motion, or agitation, by virtue of thermal energy; and as a result of interactions among the particles, energy is continuously interchanged among them. Some electrons acquire energies in excess of 1.1 eV and thereby escape from their bonds. These electrons are mobile carriers of charge, and they endow the crystal with electrical conductivity. Conditions existing in the crystal when an electron escapes from its bond are pictured in Fig. 4-3. The free electron is represented by the minus sign that is not associated with a covalent bond.

An electron is now missing from one of the bonds, and an imperfection, or hole, exists in the regular lattice structure. Associated with this hole, or missing electron, there is an excess of positive charge, indicated in Fig. 4-3 by the plus sign in one of the bonds. It is now a relatively easy matter for an electron in a nearby covalent bond to leave its position and move into the hole left by the thermally ejected electron. When this happens, a hole appears in the nearby bond just vacated; in this manner the hole can move from point to point in the crystal, carrying with it a positive charge equal in magnitude to the electronic charge. Thus, there are two mobile carriers of charge that can move about in the crystal, the free electron with a charge $-q$ and the hole with a charge q. If there are enough of these mobile carriers present, the crystal acts as a fairly good conductor of electricity.

The process by which thermal agitation creates a hole and a free electron is called thermal ionization. The rate at which electron-hole pairs are generated is a strong function of temperature T, and it can be expressed symbolically as

$$\text{Ionization rate} = I = I(T) \tag{4-2}$$

Since holes and electrons are generated in equal numbers by this process, their concentrations are related by

$$n = p = n_i \tag{4-3}$$

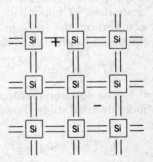

Fig. 4-3 Free electron and hole generated by thermal agitation.

where n is the number of free electrons per unit volume, p is the number of holes per unit volume, and n_i represents the number of holes or electrons per unit volume in an intrinsic (pure) crystal. Also, since holes and electrons are present in equal numbers, the crystal is electrically neutral on a macroscopic scale.

The thermally generated holes and free electrons move in a random manner through the crystal as a result of thermal energy and collisions with other particles. If, in their random wandering, a hole and an electron come close to each other, it is possible that the free electron may move into the empty covalent bond represented by the hole. The hole and the electron cancel each other by this recombination process, and charge carriers disappear from the scene in this way.

In the process of recombination both energy and momentum must be conserved. A more detailed study shows that, under these constraints, the probability of recombination occurring is small in regions where the crystal structure is perfect. It also shows that the probability of recombination is much higher in regions where there are crystal imperfections such as the dislocations and impurities mentioned previously; these imperfections are often referred to as recombination centers. Impurity atoms of gold are especially effective in promoting recombination in silicon crystals, and they are often built into silicon crystals to increase the recombination rate. The surface of a crystal is a surface of imperfection, for there are no atoms outside the surface to form covalent bonds with the layer of atoms on the surface. As a result, the recombination rate at the surface of the crystal is high.

When the concentration of holes and free electrons is not too great, the recombination rate is proportional to the number of carriers available for recombination, and it can be expressed as

$$\text{Recombination rate} = R = \alpha pn = \alpha n_i{}^2 \qquad (4\text{-}4)$$

where α is a constant accounting for the material properties of the crystal and where Eq. (4-3) has been used. Thermal ionization causes the concentrations of holes and free electrons to increase until, at thermal equilibrium, the recombination rate equals the ionization rate. Thus, under equilibrium conditions,

$$I(T) = R = \alpha n_i{}^2 \qquad (4\text{-}5)$$

The equilibrium condition depends strongly on temperature, and a more detailed study shows that

$$n_i{}^2 = KT^3 \exp\left(\frac{-qV_g}{kT}\right) \qquad (4\text{-}6)$$

where T = absolute temperature

q = magnitude of electronic charge

V_g = energy, in eV, required to break covalent bond

k = Boltzmann constant

K = proportionality constant

The quantity q/kT occurs repeatedly in the study of the pn junction, and hence it is convenient to give it a special symbol,

$$\lambda = \frac{q}{kT} \tag{4-7}$$

so that (4-6) becomes

$$n_i^2 = KT^3 \exp(-\lambda V_g) \tag{4-8}$$

At 22°C (71.6°F), $\lambda = 40$ volt^{-1}, and it is convenient to standardize on this value of λ for all room-temperature calculations. Thus at room temperature with $V_g = 1.1$ eV for silicon, Eq. (4-8) yields $n_i \approx 10^{10}$ carriers per cubic centimeter. Since the concentration of atoms in the crystal is approximately 5×10^{22} cm^{-3}, it follows that about one atom in 10^{13} is ionized at room temperature.

The discussion above is concerned with thermal ionization of the atoms in a crystal. There are other processes that produce ionization and that affect the behavior of semiconductor devices. As mentioned earlier, if the crystal is subjected to a strong electric field, the electric forces can ionize atoms by tearing electrons out of covalent bonds. Such electric fields can also accelerate thermally generated carriers to high kinetic energies, and these high-energy carriers can knock electrons out of covalent bonds by collision. If the crystal is irradiated with light photons of suitable energy, the energy of the photons can eject electrons from covalent bonds. All of these ionizing processes play significant roles in semiconductor electronics.

4-2 MOVEMENT OF CHARGE CARRIERS IN CRYSTALS

Charge carriers moving through a crystal are subjected to forces exerted on them by many nearby particles, and the dynamics of their motion is indeed complicated. The problem is further complicated by the fact that, on the atomic scale, the motions of the carriers are subject to laws and restrictions that are not evident on the laboratory scale of measurement and that therefore are not included in the classical (newtonian) mechanics. These important atomic-scale phenomena are accounted for in detail by quantum mechanics, however, and in order to understand in detail the motion of carriers in crystals, it is necessary to employ this branch of physics.

In principle, the quantum theory is capable of providing a complete quantitative description of all material phenomena. In actual fact, however, the mathematical formulations are so complex that it is impossible to solve them for the desired quantities except in the simplest cases. For these reasons an understanding of semiconductor devices is not to be gained by a direct application of either classical physics or quantum physics, but rather by a proper combination of the two. Specifically, quantum physics is used to modify the concepts and visualizations of classical physics in such a way as to bring classical physics into agreement with quantum physics and with physical facts.

For example, it turns out that an electron moving through a crystalline solid has an effective mass that is different from its mass when moving through free space. Another result of quantum-mechanical studies, which came as a surprise to physicists, is the fact that the hole behaves like a particle that has nearly the same effective mass as the electron. Thus, it is very convenient to treat the hole as if it were a charged particle, and, for most purposes, correct results are obtained by taking such a viewpoint. Most of the quantum-mechanical corrections that must be applied to the classical physics are contained in the effective masses of the hole and free electron.

The process of modifying the classical picture to make it yield results that are in agreement with quantum physics requires a deep understanding of both classical and quantum physics, and such an undertaking is not appropriate for this book. Nevertheless, once the necessary modifications are known, the modified classical picture can be used with relative ease to gain valuable understanding of many important semiconductor devices. However, if the examination into the properties of semiconductors is pursued far enough, a point is always reached at which the modified classical picture cannot provide the answer, and recourse to quantum physics becomes necessary.

There are two mechanisms by which holes and electrons move through a crystal. One of these is *diffusion,* associated with the random motion due to thermal energy and illustrated in Fig. 4-4a. The other is *drift,* a motion resulting from an applied electric field, superimposed on the random motion of the particles as shown in Fig. 4-4b. It is helpful in passing to note some of the features of the random motion that exists in the absence of an applied electric field. At room temperature the mean free path of the particles, which is the average distance between collisions, is

mfp ≈ 0.1 μm

$\approx 1,000$ interatomic distances

(a)   (b)

Fig. 4-4   The motion of a charged particle in a crystal.   (a) Random thermal motion; (b) random motion plus drift (dashed line).

(In much existing literature the unit *micrometer* is given the name *micron* and abbreviated μ.)   The rms velocity of the particle at room temperature is

$$v_T \approx 10^7 \text{ cm/sec} \approx 200,000 \text{ mph}$$

A particle moving with a constant velocity equal to v_T would reach the moon in slightly over an hour.   From these values of mean free path and v_T, the mean free time between collisions is found to be

$$\text{mft} \approx 10^{-12} \text{ sec}$$

Thus, on the average, the particles move with a very high velocity and have collisions at a high rate.

When the concentration of free electrons (or holes) is uniform throughout a crystal, there is, on the average, no net flow of electrons in the crystal.   The probability of finding an electron moving from left to right is equal to the probability of finding one moving from right to left. When the concentration of electrons is not uniform, however, random thermal motion results in a net flow from regions of high concentration to regions of low concentration.   This kind of flow, called diffusion, can be observed when a small amount of ink is poured into a tumbler of water.   Figure 4-5a shows a single-crystal silicon bar that is illuminated with a light beam on one end.   The incident light ionizes silicon atoms at the end of the bar, and it thus raises the concentration of free electrons at that end of the bar well above the thermal equilibrium value. It is assumed that the bar is so long that the electron concentration at the other end of the bar is essentially equal to the thermal equilibrium value. Under these conditions the concentration of electrons varies with distance along the bar as shown in Fig. 4-5b.   Now consider a plane cross section at the center of the bar perpendicular to the x direction.   There are more electrons on the left of the cross section than on the right. Therefore, random motion causes more electrons to cross from left to right than from right to left; thus, there is a net flow from the region of

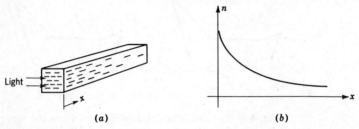

Fig. 4-5 Diffusion of carriers. (*a*) Free electrons generated in a crystal by light; (*b*) concentration of free electrons in the bar.

high concentration to the region of low concentration. If the light is switched off, this diffusion flow continues until the concentration becomes uniform throughout the bar, and since the rms thermal velocity is about 200,000 mph, the condition of uniform concentration is reached very quickly.

The curve shown in Fig. 4-5*b* is called a carrier-distribution curve; the net flow of carriers at any point is related to the gradient, or slope, of the carrier-distribution curve. When the concentration of electrons varies only in the *x* direction and is constant in the *y* and *z* directions, then the current carried by the diffusion of electrons across any plane perpendicular to the *x* direction is proportional to the slope of the curve at that plane, and it is given by

$$I_e = qAD_e \frac{dn}{dx} \tag{4-9}$$

where I_e = current in *x* direction carried by electrons

  q = magnitude of electronic charge

  A = area of cross section

  D_e = proportionality constant, called diffusion constant for electrons

  $n = n(x)$ = concentration of electrons

If dn/dx is positive, electrons move by diffusion in the negative *x* direction; however, since the electrons carry a negative charge, this flow constitutes a conventional current in the positive *x* direction.

Holes obey a similar law and give a diffusion current that is expressed by

$$I_h = -qAD_h \frac{dp}{dx} \tag{4-10}$$

where I_h is the current in the positive *x* direction. The diffusion constants for holes and free electrons in silicon are

$$D_h = 13 \text{ cm}^2/\text{sec} \qquad D_e = 36 \text{ cm}^2/\text{sec} \qquad (4\text{-}11)$$

When an electric field is applied to the crystal, the holes and free electrons continue their violent random motion with about the same mean free path and mean free time as in the field-free case. However, between collisions the particles are accelerated by the field, and they gain a small drift velocity and a small kinetic energy from the field as illustrated in Fig. 4-4b. At each collision the carriers lose some kinetic energy, and they are scattered in random directions. The drift velocity builds up until, on the average, the particles lose just as much energy in collisions as they gain from the field between collisions; thus, a terminal velocity is reached and equilibrium is established with the field applied. Except with very intense applied fields, the equilibrium drift velocity is directly proportional to the strength of the applied field; this fact leads directly to Ohm's law. Moreover, the drift velocity is normally much smaller than the velocity of thermal agitation so that the random motion is not altered appreciably by the superimposed drift velocity.

In particular, the drift velocity for holes in the positive x direction is

$$v_{dh} = -\mu_h \frac{dV}{dx} \qquad (4\text{-}12)$$

where μ_h is a proportionality constant called the mobility of holes, and $V = V(x)$ is the electric potential. The conventional current in the positive x direction due to the drift of holes across a section of area A is

$$I_{dh} = qAp v_{dh} = -qAp\mu_h \frac{dV}{dx} \qquad (4\text{-}13)$$

Similarly, the current in the positive x direction due to the drift of electrons is

$$I_{de} = -qAn\mu_e \frac{dV}{dx} \qquad (4\text{-}14)$$

For silicon at room temperature the mobilities of holes and electrons are

$$\mu_h = 500 \text{ cm}^2/\text{volt-sec} \qquad \mu_e = 1{,}400 \text{ cm}^2/\text{volt-sec} \qquad (4\text{-}15)$$

The total current due to the drift of carriers is thus

$$I = I_{dh} + I_{de} = -qA(p\mu_h + n\mu_e) \frac{dV}{dx} \qquad (4\text{-}16)$$

Equation (4-16) is one form of Ohm's law. Dividing both sides of this expression by $-A \, dV/dx$ yields the definition of the electrical conductivity of the material,

$$\sigma = q(p\mu_h + n\mu_e) \tag{4-17}$$

The resistivity of the material is

$$\rho = \frac{1}{\sigma} = \frac{1}{q(p\mu_h + n\mu_e)} \tag{4-18}$$

Since the resistivity is an easily measured quantity, it is commonly used as an indication of the concentration of mobile carriers in semiconductors.

The intrinsic silicon crystal discussed above provides by thermal ionization equal concentrations of holes and electrons as carriers of electric charge. Since the concentrations are strongly dependent on temperature, such crystals are useful in making resistances that vary with temperature. This is about the limit of their usefulness, however. Semiconductor diodes and transistors require the simultaneous existence of semiconductors in which the carriers are almost all electrons and of other semiconductors in which the carriers are almost all holes. Ideally, each of these semiconductors should contain only one kind of carrier. Means for fabricating such semiconductor materials are presented in the next section.

4-3 DOPED SEMICONDUCTORS

Intrinsic (pure) semiconductors are materials in which holes and free electrons occur in equal numbers. Doped semiconductors are materials in which carriers of one kind predominate. Semiconductors in which the predominant, or majority, carriers are electrons are called *n*-type materials because the charge of the majority carriers is negative; semiconductors in which the majority carriers are holes are called *p*-type materials.

A crystal of *n*-type silicon is represented schematically in Fig. 4-6; the *n* character of this crystal results from the presence of a few impurity

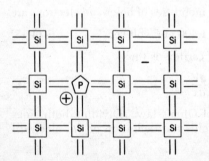

Fig. 4-6 A silicon crystal with phosphorus impurity.

atoms of phosphorus in the crystal lattice. The process of adding this impurity to the crystal is called doping, and the resulting crystal is called a doped silicon crystal. In typical semiconductor devices the concentration of impurity atoms is in the range from 10^{15} to 10^{20} atoms per cubic centimeter, with a value of 10^{17} being representative. Since the concentration of silicon atoms in the crystal is almost 10^{23} atoms per cubic centimeter, it follows that the fraction of impurity atoms in the representative crystal is about 1 part in 10^6. From a metallurgical point of view this is an extremely small amount of impurity, and the metallurgical properties of the crystal are not changed perceptibly by this minute amount of phosphorus. However, the electrical properties of the crystal are greatly changed by the impurity.

One of the ways in which the phosphorus atom is different from the silicon atom is that it has five rather than four valence electrons. When it enters the lattice structure of the silicon crystal, four of the valence electrons form covalent bonds with adjacent silicon atoms. The fifth valence electron is then very lightly attached to the parent atom; only about 0.04 eV of energy is required to separate it from the atom. Hence, at room temperature essentially all of the phosphorus atoms are ionized, and many free electrons are available for the conduction of electric current. Since the normal phosphorus atom is electrically neutral, the loss of the free electron leaves a net positive charge associated with the atomic nucleus. This positive charge cannot move through the crystal because the atomic nucleus is tightly bound in the lattice by the covalent bonds; the circle around the plus sign in Fig. 4-6 signifies that this is a bound charge. However, since the complete crystal contains equal amounts of positive and negative charge, it is electrically neutral on a macroscopic scale. Since the phosphorus impurity in the crystal contributes free electrons for conduction, it is called a *donor* material. Other pentavalent atoms, such as those of arsenic and antimony, can also be used as donors.

It is important to note the difference between the ionization of a silicon atom and the ionization of a donor atom. The ionization of a silicon atom creates two mobile carriers of charge, a hole and a free electron. The ionization of a donor atom creates a free electron and a *bound* positive charge. Thus the donor contributes only electrons for conduction, and this fact gives the crystal its *n* characteristic. The concentration of carriers created by the ionization of silicon atoms depends strongly on temperature, whereas the concentration of carriers created by the ionization of donor atoms, which is essentially equal to the concentration of donor atoms (since virtually all donors are ionized), is independent of temperature, except at very low temperatures.

A crystal of *p*-type silicon is represented schematically in Fig. 4-7;

the *p* character of this crystal results from the presence of a few boron atoms in the crystal lattice. For typical *p*-type silicon, the concentration of impurity atoms lies in the same range as that for the *n* type, 10^{15} to 10^{20} atoms per cubic centimeter, with 10^{17} being a representative value. Thus, the fraction of impurity is again very small, about one part in a million, and metallurgically the crystal is hardly changed by its presence.

One of the ways in which the boron atom is different from the silicon atom is that it has three rather than four valence electrons. When it enters the lattice structure of the silicon crystal, its three valence electrons form covalent bonds with adjacent silicon atoms; however, there is no valence electron to form the fourth bond of the normal lattice structure, and therefore a hole exists in the lattice. Just as in the case of the thermally created hole, this hole is free to move through the lattice carrying with it a positive charge. When the hole moves away from the boron atom, an electron moves into the empty valence bond associated with the boron atom; as a result there is an excess of negative charge associated with the impurity atom. This *bound* negative charge is indicated in Fig. 4-7 by the encircled minus sign at the site of the impurity atom. Since the boron atom captures an electron from the silicon atoms, it is called an *acceptor* material. Other trivalent atoms, such as those of aluminum and gallium, can also be used as acceptors.

As in the case of the donor atoms in the *n*-type silicon, virtually all of the acceptor atoms are ionized at room temperature. And again, the crystal is electrically neutral on a macroscopic scale. In contrast with the ionization of silicon atoms, the ionization of acceptor atoms creates only one mobile carrier, a hole; the counterbalancing negative charge is bound to the acceptor atom. And with virtually all of the acceptor atoms ionized at room temperature, the concentration of carriers contributed by acceptors is equal to the concentration of acceptor atoms and is independent of temperature, except at very low temperatures.

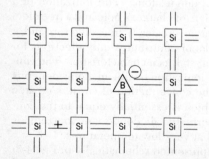

Fig. 4-7 A silicon crystal with boron impurity.

If donor and acceptor impurities are added to a crystal in equal amounts corresponding to the representative concentrations stated above, then the free electrons contributed by the donor atoms fill the holes contributed by the acceptor atoms, and no carriers are present except those generated by thermal ionization. Since the concentration of impurity atoms is still extremely small compared with the concentration of silicon atoms, this crystal has properties almost identical with those of intrinsic silicon. This process, in which the effects of impurity atoms are canceled, is called compensation; it is important in the fabrication of semiconductor devices. By adding an excess of donors or acceptors to the compensated crystal, an n-type or a p-type crystal is obtained. Thus a crystal can be changed back and forth between n-type and p-type by adding successive increments of donor and acceptor atoms.

Thermal ionization takes place in doped silicon crystals just as it does in intrinsic silicon. Since the mechanical properties of the crystal are hardly affected by the minute traces of impurity used, the thermal generation rate is the same as that given by Eq. (4-5) for intrinsic silicon,

$$I(T) = \alpha n_i^2 \tag{4-19}$$

where

$$n_i \approx 10^{10} \text{ cm}^{-3} \tag{4-20}$$

for intrinsic silicon at room temperature. In a typical crystal of n-type silicon, the concentration of donors is

$$N_D = 10^{17} \text{ cm}^{-3} \gg n_i \tag{4-21}$$

Thus, if all of the donors are ionized at room temperature, the concentration of free electrons created by the ionization of silicon atoms is negligible in comparison with the concentration contributed by the donor atoms, and the concentration of free electrons is

$$n_{no} \approx N_D = 10^{17} \text{ cm}^{-3} \tag{4-22}$$

In this expression the symbol n represents the concentration of free electrons, the subscript n signifies n-type material, and the subscript o signifies thermal equilibrium conditions. This concentration depends only on the concentration of impurities and is independent of temperature.

The recombination rate for carriers is given by Eq. (4-4). For n-type material under thermal equilibrium conditions, it takes the form

$$R = \alpha n_{no} p_{no} \tag{4-23}$$

But under thermal equilibrium conditions, the recombination rate must equal the ionization rate given by (4-19); thus

$$R = I(T) = \alpha n_i^2 \tag{4-24}$$

Now equating (4-23) and (4-24) yields

$$n_{no} p_{no} = n_i^2 \tag{4-25}$$

or

$$p_{no} = \frac{n_i^2}{n_{no}} = \frac{n_i^2}{N_D} \tag{4-26}$$

Since N_D is a constant and n_i depends strongly on temperature, the concentration of holes in n-type material depends on temperature. Using the typical values of $n_i = 10^{10}$ and $N_D = 10^{17}$ yields

$$p_{no} = 10^3$$

and

$$n_{no} = 10^{14} p_{no} = N_D$$

The electrons are the majority carriers in n-type material, and holes are the minority carriers; their concentrations are vastly different.

A similar analysis can be applied to p-type silicon with similar results. In this case the majority-carrier concentration is

$$p_{po} = N_A \tag{4-27}$$

and the minority-carrier concentration is

$$n_{po} = \frac{n_i^2}{p_{po}} = \frac{n_i^2}{N_A} \tag{4-28}$$

where N_A is the concentration of acceptor atoms in the p-type material. Again, the majority-carrier concentration is independent of temperature.

Figure 4-8 shows a thin slab of n-type silicon that is illuminated with light capable of ionizing the silicon atoms and generating electron-hole pairs. It is assumed that the slab is so thin that the light passes completely through it and illuminates the whole volume of the slab

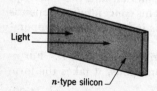

Light

n-type silicon

Fig. 4-8 A thin slab of silicon with uniform illumination.

uniformly. Thus photo-ionization occurs uniformly throughout the slab, and the concentration of minority-carrier holes is increased uniformly above the thermal equilibrium level. Under these conditions the concentration of holes in the slab can be written as

$$p_n = p_{no} + \Delta p_n \qquad (4\text{-}29)$$

where p_{no} is the thermal equilibrium concentration and Δp_n is the excess concentration caused by the light. The recombination rate is $R = \alpha p_n n_n$ as before, and the carrier concentrations build up to the level at which the recombination rate equals the total ionization rate due to thermal and photo ionization. If the light is switched off at $t = 0$, the recombination rate exceeds the ionization rate, and the excess concentration decays exponentially to zero. It can be shown that, for $t > 0$, the concentration of holes is

$$p_n = p_{no} + \Delta p_n \exp \frac{-t}{\tau_h} \qquad (4\text{-}30)$$

where τ_h is known as the lifetime of holes in the silicon slab. The lifetime of minority carriers is an important parameter of semiconductor materials. It depends on the concentration of impurity atoms and on the concentration of recombination centers in the crystal, and it is therefore under the control of the device designer to some extent. Gold atoms incorporated in a silicon crystal do not contribute holes or free electrons for conduction, but they act as highly effective recombination centers. Therefore, silicon crystals are frequently doped with gold to produce material having a short minority-carrier lifetime.

This reasoning can be repeated for a p-type slab of silicon, and it leads to the concept of a lifetime τ_e for minority-carrier electrons in p-type material.

The resistivity of a semiconductor is given by Eq. (4-18) as

$$\rho = \frac{1}{q(p\mu_h + n\mu_e)} \qquad (4\text{-}31)$$

For doped n-type silicon, $n_n \approx N_D \gg p_n$, and the expression reduces to

$$\rho = \frac{1}{qN_D\mu_e} \qquad (4\text{-}32)$$

A similar expression is obtained for p-type material. Since q and μ are known constants, the resistivity is a convenient measure of the impurity concentration in doped crystals. Typical resistivities are of the order of 1 ohm-cm.

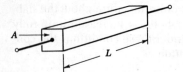

Fig. 4-9   A semiconductor resistor.

Figure 4-9 shows a doped silicon bar that might be used as a re-sistor.   If the bar is made of n-type silicon, its resistance is

$$R = \rho \frac{L}{A} = \frac{L}{qAN_D\mu_e} \tag{4-33}$$

Since the mobilities of carriers decrease with increasing temperature, the resistance of the bar increases with temperature.

4-4   THE pn JUNCTION

With the existence of p-type and n-type silicon crystals, it is possible to make a pn junction simply by pressing crystals of the two types to-gether mechanically.   However, such a process is not suitable for making the very tiny junctions used in semiconductor electronics.   The process used to fabricate these junctions is described in detail in Ref. 2 listed at the end of this chapter; the essential features of the process are described in the paragraphs that follow.

The starting point in the fabrication of pn junctions, diodes, and transistors is a base wafer, called a substrate, of low resistivity (heavily doped), single-crystal silicon shown as n^+-type silicon in Fig. 4-10.   In the first step a layer of n-type silicon is then grown epitaxially on the base wafer as a continuation of the single-crystal lattice of the base wafer. ("Epitaxial" is a coined word based on two Greek roots, *epi* meaning "upon" and *taxis* meaning "arrangement" or "form.")   This growth is accomplished by heating the wafer to about 1200°C in an atmosphere of silicon tetrachloride and hydrogen.   The chloride decomposes at the surface of the wafer, and the silicon atoms attach themselves to the lattice structure of the wafer to form a continuation of the lattice.   The n

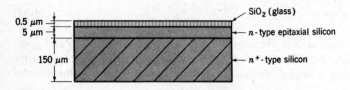

Fig. 4-10   Silicon wafer to be used in making a pn junction.

character of the epitaxial layer is achieved by adding phosphine (PH$_3$) to the atmosphere in which the base wafer is heated. The phosphine also decomposes, and the phosphorus atoms take positions in the crystal lattice where they serve as donor impurities. The impurity concentration in the epitaxial layer is controlled by controlling the amount of phosphine in the atmosphere from which the layer is grown. The layer can be grown to almost any desired thickness; 5 μm, as shown in Fig. 4-10, is typical. (It should be noted that in Fig. 4-10, as in most pictures of semiconductor devices, it is not possible to preserve the same scale in all parts of the drawing.) The *pn* junction is to be fabricated entirely in the epitaxial layer; the base wafer serves only to provide mechanical strength and to provide a low-resistivity electrical connection to the *n*-type material.

When a silicon crystal is exposed to air, oxygen in the air immediately reacts with silicon atoms on the surface of the crystal to form a very thin layer of silicon dioxide, a form of glass, on the surface. Several important advantages are obtained by causing the oxide to grow to a much greater thickness than it normally develops at room temperature. This additional growth is accomplished by heating the wafer in an atmosphere of oxygen or steam. It is customary to grow the oxide to a thickness of about 0.5 μm, as shown in Fig. 4-10. In order to give some degree of calibration to the dimensions shown in Fig. 4-10, note that the thickness of the paper on which this page is printed is about 100 μm, and the wavelength of visible light is in the range between 0.4 and 0.7 μm.

The oxide layer protects the surface of the crystal against chemical and mechanical damage, and it also produces a great improvement in the electrical properties of the crystal by mechanisms that are not well understood. It also turns out that at elevated temperatures boron (acceptor) and phosphorus (donor), along with some other impurities, diffuse fairly rapidly into silicon, whereas they diffuse much more slowly into silicon dioxide. This important fact is the basis for most of the technology used in making *pn* junctions, diodes, transistors, and other semiconductor devices. By removing the oxide from certain portions of the surface of the wafer, donors or acceptors can be diffused into the silicon to change it from one type to the other in selected regions.

Thus, the next step in making a *pn* junction in the wafer of Fig. 4-10 is to use photographic masking and etching techniques to remove the oxide layer from a chosen region on the surface of the wafer. The wafer is then heated in a suitable atmosphere containing boron, and boron atoms diffuse into the silicon where the silicon surface is exposed. These boron atoms enter into the regular lattice structure of the crystal. Thus, they become acceptor impurities, and if their concentration is great enough, they compensate and override the donor

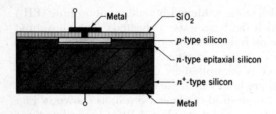

Fig. 4-11 A diffused *pn* junction in silicon.

impurities in the epitaxial layer to form a region of *p*-type silicon. The
resulting structure is illustrated in Fig. 4-11. The boundary between
the *p*-type and *n*-type silicon is the *pn* junction. The depth of the *p*-type
material is governed by the duration of the diffusion; 1 or 2 μm is typical.
The net concentration of impurities in the *p*-type region is governed by
the atmosphere from which the diffusion takes place.

The final step in the fabrication process is to regrow the oxide over
the *p*-type material and to make a metallic connection with the *p*-type
silicon through a hole etched in the regrown oxide. The resulting structure has an epitaxial layer that is a single crystal of silicon with a minute
trace of impurity that changes from donor to acceptor at the *pn* junction.

The base wafer used in the process described above is usually a
disk about 1 in. in diameter. There is no difficulty in making as many as
1,000 *pn* junctions simultaneously on such a wafer. It is also possible
to process as many as 100 wafers simultaneously. Thus 100,000 junctions can be made simultaneously with about the same effort as is required to make a single junction. This fact is of fundamental importance
in lowering the cost of silicon diodes, transistors, microamplifiers, and
so forth.

The step of growing an epitaxial layer in the fabrication process
described above is not essential. The *pn* junction can be formed directly in the base wafer. However, the epitaxial layer gives the device
designer an additional degree of freedom, and it permits him to avoid
several difficult design compromises. Thus, most high-performance
silicon devices are formed in epitaxial layers.

The volt-ampere characteristic for the *pn* junction is given in Fig.
4-1, and it is repeated here in Fig. 4-12. Perhaps it is appropriate to
repeat here some of the comments made in connection with Fig. 4-1.
The volt-ampere law for the junction, exclusive of the breakdown region,
is

$$I = I_s[\exp (\lambda V) - 1] \tag{4-34}$$

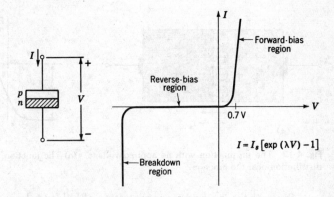

Fig. 4-12 Volt-ampere characteristic for a silicon *pn* junction.

where I_s and $\lambda = q/kT$ are constants and where $\lambda \approx 40$ volt^{-1} at room temperature. This is a theoretical law; real diodes deviate somewhat from this behavior at very small and very large currents. For many purposes the *pn* junction is a good approximation to the ideal diode of Chap. 3; however, the junction is much more than just a diode. There are applications in which the junction operates wholly in the forward-bias region, there are other applications in which it operates wholly in the reverse-bias region, and there are still other applications in which it operates wholly in the breakdown region. Each of these regions is examined in some detail in the chapters that follow this one, and a variety of applications for the various modes of operation is presented.

Conditions existing near a *pn* junction when no external voltage is applied are illustrated in Fig. 4-13. Figure 4-13*a* can be thought of as representing a portion of the *pn* junction in Fig. 4-11, and Fig. 4-13*b* shows the variation in electric potential from one side of the junction to the other. The body of the *p*-type material contains many holes contributed by acceptor atoms. The positive charge of these holes is neutralized by bound negative charges associated with acceptor atoms in the crystal lattice; for simplicity these bound charges are not shown. The body of the *p*-type material also contains a few free electrons created by thermal ionization. Similarly, the body of the *n*-type material contains many free electrons contributed by donor atoms. The negative charge of these electrons is neutralized by bound positive charges associated with the donor atoms; again, for simplicity these bound charges are not shown. The body of the *n*-type material also contains a few holes created by thermal ionization.

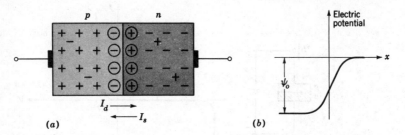

Fig. 4-13 The pn junction with no applied voltage. (*a*) The junction; (*b*) potential distribution near the junction.

Since there is a heavy concentration of holes in the p-type material and only a light concentration of holes in the n-type material, there is a strong tendency for holes to move by diffusion from the p material across the junction into the n material. But, when holes cross the junction, they find themselves in a region heavily populated by free electrons, and they disappear by recombination very quickly. Similarly, there is a strong tendency for free electrons to diffuse across the junction from the n material to the p material, and when they cross the junction, they recombine quickly with holes that are present in abundance in the p material. Thus, there is a steady flow of carriers across the junction with recombination occurring shortly after the junction is crossed; neither carrier penetrates very deeply into the domain of the other. This flow of carriers constitutes the diffusion current I_d indicated in Fig. 4-13a.

As a result of the recombination taking place on each side of the junction, there is a region on each side of the junction in which there are very few carriers present. This region is called the *carrier-depletion region*. The distribution of carriers at a pn junction is illustrated in Fig. 4-14. Carrier concentrations are shown on a logarithmic scale in Fig. 4-14a so that the wide range of concentrations can be seen; the concentrations are also shown on a linear scale in Fig. 4-14b. The carrier-depletion region shows clearly in Fig. 4-14b.

Throughout the body of the p-type and n-type silicon, there are charge carriers present that neutralize the bound charges associated with the ionized impurity atoms. In the carrier-depletion region, however, there are no carriers present to neutralize the bound charges. These charges, shown encircled in Fig. 4-13a, are said to be uncovered. Thus, there is a dipole layer of charge at the pn junction, and the carrier-depletion region is also known as the *space-charge layer*. Since the complete crystal must remain electrically neutral, equal amounts of positive and negative charge are uncovered in the space-charge layer.

As a result of the dipole layer at the junction, an electric field exists

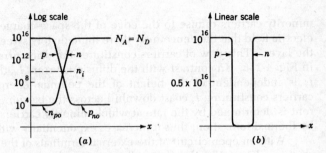

Fig. 4-14 Carrier distributions at a pn junction.

across the junction, and associated with the electric field, there is a potential difference across the junction. The variation of electric potential along a path normal to the junction is shown in Fig. 4-13b. The total potential difference across the space-charge layer is designated ψ_o. Such a potential difference always exists when electrically dissimilar materials are brought into contact with each other; it is called the *contact potential difference,* or simply the *contact potential.*

The electric field across the space-charge layer exerts a force on charged particles such as holes and electrons. The direction of the force is always such as to move positively charged holes toward regions of lower potential and to move negatively charged electrons toward regions of higher potential. Thus the electric field in the space-charge layer opposes the diffusion of holes into the n region and the diffusion of electrons into the p region. If a hole or an electron does succeed in crossing the space-charge layer against the force of the field, it gains in potential energy by the amount $W_o = q\psi_o$. It follows from this fact that only those carriers having kinetic energies greater than this amount in a direction perpendicular to the junction can cross the space-charge layer. The kinetic energy involved is the energy of thermal motion. The larger ψ_o, the smaller the number of carriers having the required kinetic energy. Thus a potential-energy barrier, or a potential hill, exists in the space-charge layer that opposes the diffusion of carriers across the junction. The diffusion current I_d indicated in Fig. 4-13a therefore depends on the height of the barrier. The height of the barrier depends on the strength of the field in the space-charge layer, and this in turn depends on the amount of bound charge uncovered in the layer.

In addition to majority carriers contributed by donor and acceptor atoms, the p and n materials on each side of the junction have relatively small concentrations of minority carriers generated by thermal ionization; these concentrations are shown in Fig. 4-14a. Some of these minority carriers diffuse to the edge of the space-charge layer, and the electric field that they encounter there immediately sweeps them across the layer. This flow of carriers constitutes the drift current I_s indicated in Fig. 4-13a. In contrast with the diffusion current I_d, the drift current I_s is independent of the height of the potential-energy barrier; the carriers constituting I_s coast downhill across the barrier. The drift current is determined by the rate at which minority carriers are generated by thermal ionization; thus it increases exponentially with temperature.

With an open circuit at the external terminals of the *pn* junction as shown in Fig. 4-13a, the current flowing into the device is zero. Therefore, under equilibrium conditions the net current across the junction must be zero; thus the equilibrium condition is $I_s = I_d$. The equilibrium is maintained by ψ_o, the height of the potential-energy barrier. If ψ_o is too small, many carriers are able to diffuse across the barrier, I_d exceeds I_s, and there is a net current across the junction. This current cannot flow out of the open-circuited terminals, so it goes into additional stored charge in the space-charge layer. This additional charge increases ψ_o, and the action continues until the net current across the junction is reduced to zero. If ψ_o is too large, I_s exceeds I_d, and there is a net current across the junction in the direction of I_s. This current reduces the charge stored in the space-charge layer, which reduces ψ_o and increases I_d, and again the action continues until the net current across the junction is reduced to zero.

The voltage across the *pn* junction with no external voltage applied is related to the carrier concentrations on each side of the junction by a simple exponential law,[1-3]

$$p_{no} = p_{vo} \exp(-\lambda \psi_o) \qquad (4\text{-}35)$$

where p_{no} = concentration of holes in n region
p_{vo} = concentration of holes in p region
$\lambda = q/kT = 40$ volt^{-1} at room temperature

The same law holds for the concentrations of free electrons on each side of the junction. It follows from (4-35) that the contact potential is

$$\psi_o = \frac{1}{\lambda} \ln \frac{p_{vo}}{p_{no}} \qquad (4\text{-}36)$$

Substituting Eqs. (4-26) and (4-27) into (4-36) yields

$$\psi_o = \frac{1}{\lambda} \ln \frac{N_A N_D}{n_i^2} \qquad (4\text{-}37)$$

where N_A = concentration of acceptors in p region

N_D = concentration of donors in n region

n_i = thermal equilibrium concentration of carriers in intrinsic material

A typical value for the contact potential across a silicon *pn* junction is 0.9 volt.

Figure 4-13*b* shows the potential distribution in the vicinity of a *pn* junction. It might be inferred from this figure that the junction contact potential difference ψ_o appears as a potential difference between the external terminals connected to the junction. If this were the case, it would be possible to draw energy from the *pn* junction in contradiction to the law of conservation of energy. The potential distribution curve in Fig. 4-13*b* does not show how the potential varies where the metal wires make contact with the semiconductor. A contact potential exists at each of these contacts. The exact nature of the potential variations in the vicinity of the contacts depends on how the contacts are made. However, if the two wires are made of the same material, the contact potentials at the metal-semiconductor contacts just exactly cancel the contact potential across the *pn* junction, and there is no potential difference between the wires. It should be remarked that quantum physics provides a very neat clarification of this point.

A number of important applications for the *pn* junction depend on the width of the space-charge layer, on how the width varies with applied voltage, and on the depth of penetration of the layer into the p and n regions. These matters are examined in detail in Chap. 2 of Ref. 1 and in Chap. 2 of Ref. 2; the important results are summarized below. Since the complete crystal in which the *pn* junction is fabricated must remain electrically neutral, equal amounts of positive and negative charge must be uncovered in the space-charge layer. Therefore, if the concentration of impurity atoms is uniform throughout each of the two regions separated by the junction, then

$$N_A L_p = N_D L_n \tag{4-38}$$

where L_p and L_n are, respectively, the depths of penetration into the p and n regions. The total width of the space-charge layer is

$$L = L_p + L_n \tag{4-39}$$

and it follows from (4-38) and (4-39) that

$$L_p = L \frac{N_D}{N_A + N_D} \tag{4-40}$$

and

$$L_n = L \frac{N_A}{N_A + N_D} \tag{4-41}$$

Many pn junctions are doped very heavily on one side and very lightly on the other in order to achieve desired characteristics. If, for example, $N_A \gg N_D$, then Eqs. (4-40) and (4-41) yield

$$L_p \ll L_n \approx L \tag{4-42}$$

That is, with highly asymmetrical doping the space-charge layer exists almost entirely in the lightly doped material and penetrates only a very short distance into the heavily doped material. This conclusion can also be reached by simple physical reasoning. A slight penetration into the heavily doped region uncovers a large amount of bound charge, and a much deeper penetration into the lightly doped region is required to uncover an equal amount of charge.

It is also shown in the references cited above that the width of the space-charge layer at a pn junction having no externally applied voltage is given by

$$L = \left[\frac{2\epsilon\psi_0}{q} \left(\frac{1}{N_A} + \frac{1}{N_D} \right) \right]^{1/2} \tag{4-43}$$

where ϵ = absolute dielectric permittivity of the semiconductor material
ψ_0 = contact potential across junction
q = magnitude of electronic charge

For the case of unequal doping with $N_A \gg N_D$, (4-43) reduces to

$$L = \left(\frac{2\epsilon\psi_0}{qN_D} \right)^{1/2} \tag{4-44}$$

Typical values of L range from 0.1 to 0.5 μm; this range is just below the range of wavelengths for visible light.

4-5 SUMMARY

This chapter introduces the notion of the doped semiconductor, and it shows how a pn junction can be made using doped p-type and n-type silicon. It also examines the equilibrium conditions at the pn junction under the condition of no externally applied voltage.

The pn junction is the heart of most semiconductor electronic devices. It is an essential component in semiconductor diodes, in transistors of several kinds, and in semiconductor microcircuits. A good understanding of the junction and of the physical processes that determine its electrical properties is of great value to the electronics engineer.

The technique described in this chapter for fabricating pn junctions in silicon, which uses silicon dioxide for masking to permit selective diffusion of impurities into a silicon wafer, is known as the silicon planar technology. This excellent technology is largely responsible for the high performance and low cost of the semiconductor electronic devices that have become available since 1960. It is now used in the fabrication of almost all semiconductor electronic devices, including microcircuits.

The next chapter examines the performance of the pn junction under conditions of reverse bias, and it presents a number of important applications for the junction under such conditions. This study leads directly to the junction field-effect transistor.

PROBLEMS

4-1. Doped silicon resistor. A tiny silicon bar having a square cross section measuring 1 mm on each side is doped with phosphorus at a concentration $N_D = 10^{15}$ cm^{-3}. The mobility of electrons in silicon is 1,400 cm^2/volt-sec, and the magnitude of the electronic charge is 1.6×10^{-19} coul.

(a) Determine the resistivity of the doped semiconductor.

(b) How long must the bar be to provide a total resistance from end to end of 1 kilohm?

4-2. Doped silicon resistor. A silicon bar like the one in Prob. 4-1 is doped with boron at a concentration $N_A = 10^{15}$ cm^{-3}. The mobility of holes in silicon is 500 cm^2/volt-sec, and the electronic charge is given in Prob. 4-1.

Repeat the calculations of Prob. 4-1 for this bar.

4-3. Contact potential. The impurity concentrations on each side of a certain silicon pn junction are $N_A = 10^{19}$ cm^{-3} and $N_D = 10^{16}$ cm^{-3}. Determine the contact potential across this junction at room temperature. The thermal equilibrium concentration of carriers in intrinsic silicon at room temperature is $n_i = 10^{10}$ cm^{-3}.

4-4. Space-charge-layer width. The impurity concentrations on each side of a certain silicon pn junction are $N_A = 10^{18}$ cm^{-3} and $N_D = 10^{16}$ cm^{-3}. The thermal equilibrium concentration of carriers in intrinsic silicon at room temperature is $n_i = 10^{10}$ cm^{-3}, and the dielectric permittivity of silicon is 1.06×10^{-12} farad/cm. The magnitude of the electronic charge is 1.6×10^{-19} coul.

(a) Determine the contact potential across the junction at room temperature.

(b) Determine the width of the space-charge layer.

(c) Determine the depth of penetration of the space-charge layer into the p and n regions on each side of the junction.

REFERENCES

1. Gray, P. E., et al.: "Physical Electronics and Circuit Models of Transistors," John Wiley and Sons, Inc., New York, 1964.
2. Motorola, Inc.: "Integrated Circuits," McGraw-Hill Book Company, New York, 1965.
3. Phillips, A. B.: "Transistor Engineering," McGraw-Hill Book Company, New York, 1962.

5

The *pn* Junction
with Reverse Bias

The objective of this chapter is to examine the physical processes occurring at a pn junction under conditions of reverse bias and to present a number of important applications for the junction operated in this mode. The applications include microcircuit applications as well as applications in discrete-component systems.

5-1 BEHAVIOR OF THE *pn* JUNCTION WITH REVERSE BIAS

Figure 5-1 shows the theoretical volt-ampere characteristic for a silicon *pn* junction with a reverse voltage applied at the terminals. This characteristic shows that for values of reverse voltage smaller in magnitude than the breakdown voltage, a very tiny reverse current flows across the junction. This current results from minority carriers generated in the crystal by thermal ionization, and its magnitude depends on the rate of thermal ionization. Therefore the current is independent of the reverse voltage applied, and it increases exponentially with temperature.

The measured reverse current flowing across a real silicon *pn* junction is much larger than the theoretical value shown in Fig. 5-1; measured values may be as high as 1 nA (nanoampere = 10^{-9} amp) at room temperature. The reverse current flowing across a real silicon junction results largely from leakage currents across the surface of the crystal and from other more complicated surface effects such as generation and recombination of carriers at the surface. These surface phenomena depend on the way in which the silicon surface is prepared, and they cannot readily be included in the theoretical analysis. The component of reverse current resulting from surface effects is usually observed to increase slowly with increasing reverse voltage. However, for almost

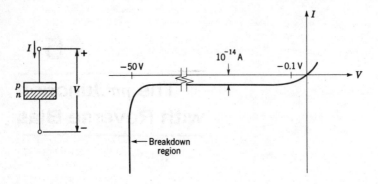

Fig. 5-1 Theoretical volt-ampere characteristic for a silicon *pn* junction with reverse bias.

all engineering purposes a reverse current of 1 nA is completely negligible, and it is customary to treat the *pn* junction as an open circuit under reverse-bias conditions, provided that the breakdown voltage is not exceeded.

When the applied reverse voltage is increased to the breakdown value, the reverse current across the junction increases rapidly as shown in Fig. 5-1. It turns out that in the breakdown region of the junction characteristic, the voltage drop across the junction is very nearly constant and independent of the current across the junction. This constant-voltage-drop property leads to important applications for the junction operating in the breakdown region.

In the study of the *pn* junction it is in many respects more effective to think in terms of an applied current than an applied voltage, for the behavior of the junction is intimately related to the charges stored near the junction. Conditions that exist near the junction when a reverse current $-I$ is applied to the external terminals are pictured in Fig. 5-2. The applied current extracts holes from the *p* region and electrons from the *n* region, and as a result the width of the space-charge layer increases as indicated by the dashed lines in Fig. 5-2a. Thus more bound charges associated with impurity atoms are uncovered, and the height of the potential barrier in the space-charge layer increases as shown in Fig. 5-2b. As a consequence of these changes, fewer carriers have enough kinetic energy to climb the potential hill, and the diffusion current I_d becomes smaller than the drift current I_s. If the applied current is smaller than I_s, an equilibrium is reached with $I_s - I_d = -I$. The increase in height of the potential barrier manifests itself at the external terminals as a negative potential at the *p* terminal relative to the *n* terminal. This polarity is the reverse-bias polarity for the *pn* junction.

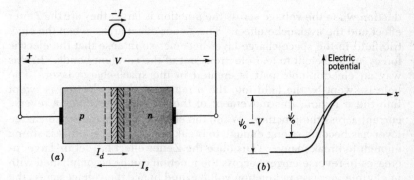

Fig. 5-2 The *pn* junction with a reverse current applied. (*a*) The junction; (*b*) potential distribution at the junction.

Under the conditions described above there is a steady flow of current across the *pn* junction from the *n* side to the *p* side. This flow corresponds to holes flowing from *n* to *p* and electrons flowing from *p* to *n*. Thus the carriers to support this flow are not contributed by the impurity atoms in the crystal. They are generated by thermal ionization in the crystal. Some of the holes generated in the *n* material near the junction diffuse to the edge of the space-charge layer, and they are swept across the layer by the electric field in the layer. Similarly, some of the electrons generated in the *p* material near the junction diffuse to the edge of the space-charge layer and are swept across. But at a fixed temperature carriers are generated by thermal ionization at a fixed, small rate. Hence thermally generated carriers can support only a small reverse current, equal at most to I_s.

Now consider the case in which the applied reverse current is larger than I_s. In this case the height of the potential barrier across the space-charge layer increases, as before, until I_d is reduced to zero, and the only carriers crossing the junction are those generated by thermal ionization. But the thermally generated carriers can support only the current I_s, which is smaller than the applied current by hypothesis. Carriers to support the applied current can be obtained only by a continuing extraction of majority carriers from the *p* and *n* regions. This action results in a continuously widening space-charge layer and a continuously increasing voltage across the junction; the action goes on until, at a sufficiently high junction voltage, a new mechanism enters to supply the carriers needed to support the applied current. This new mechanism comes into action in the breakdown region of the junction characteristic shown in Fig. 5-1.

There are two phenomena that can supply carriers to support conduction when the voltage across the junction is large; they are the Zener effect and the avalanche effect. The Zener effect arises when the electric field in the space-charge layer becomes so intense that the electric forces are sufficient to tear electrons out of the covalent bonds. In this way an electron-hole pair is created in the space-charge layer. The hole is swept by the field into the *p* region, and the electron is swept into the *n* region; this movement of the carriers constitutes a reverse current across the junction. When the electric field in the space-charge layer has become strong enough to break one covalent bond, it is strong enough to break many. Thus once the Zener effect has set in, large increases in reverse current across the junction can be accomplished with negligible increase in junction voltage, and hence the voltage across the junction is nearly constant in the breakdown region. The electric field strength and the junction voltage necessary for Zener breakdown can be calculated.[1] These calculations indicate that the Zener effect is the breakdown mechanism when breakdown occurs at a low voltage, less than about 5 volts in silicon. They indicate that it cannot be the mechanism when breakdown occurs at a high voltage, greater than about 8 volts in silicon.

The second breakdown mechanism is the avalanche effect; it occurs when the breakdown is at high voltage. Thermally generated minority carriers from the *p* and *n* regions crossing the space-charge layer have many collisions with members of the crystal lattice. When the junction voltage is high, these carriers may gain enough energy from the field in the space-charge layer to knock an electron out of a covalent bond. In this way an electron-hole pair is created in the space-charge layer. Again, the hole is swept into the *p* region by the electric field in the layer, and the electron is swept into the *n* region. The movement of these two carriers is equivalent to one carrier moving completely across the space-charge layer, and they too can have ionizing collisions, creating new electron-hole pairs which in turn can have further ionizing collisions. Thus the process builds up in the manner of an avalanche. The ionizing efficiency of the carriers is the average number of ionizing collisions that occur for each primary (thermally generated) carrier that enters the space-charge layer. It increases with the junction voltage, and it may easily equal or exceed 100 percent. When it is 100 percent, each carrier entering the space-charge layer generates the equivalent of one new carrier which in turn generates one more in a process that continues indefinitely. Thus, when the junction voltage is great enough to make the ionizing efficiency 100 percent, any amount of current across the junction can be sustained without further change in junction voltage; it is merely necessary to adjust the concentration of carriers in the space-charge layer to the proper value by a momentary change in voltage.

As mentioned above, the Zener effect is the breakdown mechanism when breakdown occurs at voltages less than about 5 volts in silicon. When breakdown occurs at voltages above about 8 volts in silicon, the avalanche effect is the mechanism. When breakdown occurs at voltages between 5 and 8 volts, the mechanism may be either of these effects, or both effects may operate simultaneously. In any event, the junction is not damaged in any way by operation in the breakdown region, provided that it is not overheated by excessive current and power dissipation.

The breakdown voltage of the *pn* junction can be controlled by controlling the concentration of impurities on each side of the junction. If one side is very heavily doped and the other side is lightly doped, as often is the case, the breakdown voltage depends only on the concentration of impurities in the lightly doped material. The breakdown voltage increases as the concentration is decreased. Thus high-voltage junctions have one side very lightly doped. Breakdown voltages in excess of 1,000 volts are readily achieved.

If the material on both sides of a *pn* junction is heavily doped, the space-charge layer at the junction is very narrow, and the field in the layer is intense. This condition exists in a device known as the tunnel diode, and it leads to some unusual properties, illustrated by the volt-ampere characteristic shown in Fig. 5-3. The electric field in the space-charge layer of the tunnel diode is so intense that the junction is in a state of Zener breakdown with no external voltage applied and even with a small forward voltage applied. The forward voltage applied to the junction reduces the field in the space-charge layer, and at the peak in the characteristic the Zener effect begins to disappear. When the applied forward voltage is made great enough, the tunnel diode characteristic joins the normal diode characteristic.

When the *pn* junction is operated in the breakdown region of its characteristic, it develops a voltage drop that is very nearly constant and independent of the current across the junction. Therefore it is widely used in this operating mode to provide a voltage that is constant and in-

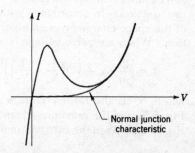

Fig. 5-3 The volt-ampere characteristic for a tunnel diode.

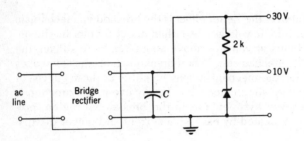

Fig. 5-4   Breakdown diode used to provide a regulated voltage.

dependent of fluctuations in the power-supply voltage.   When it is used
in this application, the junction is referred to as a *breakdown diode,* or,
more commonly, a *Zener diode.*   Zener diodes are commercially available with voltage ratings ranging from 2 to 200 volts.   Figure 5-4 shows
a Zener diode used in conjunction with an electronic power supply to
maintain a constant 10-volt output in the face of variations in the ac
line voltage and variations in the rectifier output caused by varying
loads on the 30-volt output terminal.

The reverse voltage across a Zener diode cannot exceed the breakdown voltage; therefore it can be used as a voltage limiter.   It is sometimes so used to protect delicate equipment from damage by excessive
voltage.

5-2   CHARGE STORED IN THE SPACE-CHARGE LAYER

The space-charge layer at a *pn* junction contains a dipole layer of uncovered bound charges associated with the impurity atoms in the crystal.
Equal charges of opposite sign are stored on each side of the junction,
and the amount of charge stored on each side depends on the concentration of impurity atoms and the depth of penetration of the space-charge
layer on that side.   The depth of penetration is given by Eqs. (4-40),
(4-41), and (4-43) for the case in which no external voltage is applied to
the junction.   In accordance with the discussion of Sec. 5-1, the width
of the space-charge layer changes when a voltage is applied to the junction.   When a reverse voltage is applied, the width is given by [1]

$$L = \left[\frac{2\epsilon}{q} \left(\psi_o + V' \right) \left(\frac{1}{N_A} + \frac{1}{N_D} \right) \right]^{1/2} \tag{5-1}$$

where V' is the *reverse* voltage applied to the junction as defined in
Fig. 5-5a and the remaining quantities are as defined in connection with
Eq. (4-43).   As in the case where no external voltage is applied, the pene-

trations of the space-charge layer on each side of the junction are given by

$$L_p = L \frac{N_D}{N_A + N_D} \tag{5-2}$$

and

$$L_n = L \frac{N_A}{N_A + N_D} \tag{5-3}$$

Again, when one side of the junction is doped much more heavily than the other, the space-charge layer lies almost entirely in the lightly doped region, and the relations given above can be simplified, as is done in Chap. 4.

Since the width of the space-charge layer changes with the applied voltage, it follows that the charge stored on either side of the junction is a function of the applied voltage. When the applied voltage is changed, majority carriers are added to or extracted from the p and n regions to change the width of the space-charge layer and to cover or

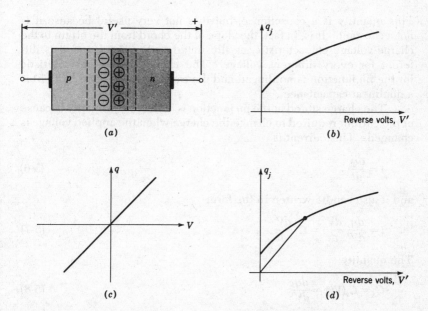

Fig. 5-5 Stored charge and capacitance. (*a*) A *pn* junction; (*b*) charge stored on either side of the junction; (*c*) charge stored in a linear capacitor; (*d*) charge-to-voltage ratio for the *pn* junction.

uncover more charge. Thus, when the applied voltage is changed, a current must flow at the terminals of the device to change the stored charge, and hence there is a capacitive effect associated with the *pn* junction.

Figure 5-5*b* indicates how the charge stored on either side of the junction changes with an applied reverse-bias voltage. Figure 5-5*c* shows how the stored charge varies with voltage for a linear capacitor, such as one made from two parallel metal plates; this capacitor is said to be linear because its charge-voltage characteristic is a straight line passing through the origin of the coordinate axes. The slope of this curve is

$$\frac{q}{V} = C = \text{const} \tag{5-4}$$

This is, of course, the electrical capacitance of the device, and it is a useful quantity because it is a constant, independent of charge and voltage.

The ratio of stored charge to voltage for the *pn* junction is

$$\frac{q_j}{V'} = C_j' \neq \text{const} \tag{5-5}$$

This quantity is a capacitance, but it is not very useful because it is not a constant. It is, in fact, the slope of the chord from the origin to the charge-voltage characteristic, as illustrated in Fig. 5-5*d*, and it is different for every different voltage. The charge-voltage characteristic for the *pn* junction is nonlinear, and the capacitance defined by (5-5) is a nonlinear capacitance.

The charge stored at the *pn* junction is of interest primarily because of the current required to change the charge when the applied voltage is changed. This current is

$$i = \frac{dq_j}{dt} \tag{5-6}$$

and it also can be written in the form

$$i = \frac{dq_j}{dV'} \frac{dV'}{dt} = C_j \frac{dV'}{dt} \tag{5-7}$$

The quantity

$$C_j = C_j(V') = \frac{dq_j}{dV'} \tag{5-8}$$

is called the *incremental capacitance* of the junction. It also is a function of the applied voltage, and for any given voltage it is the slope of the q_j-V' characteristic at that voltage, as shown in Fig. 5-6. For small

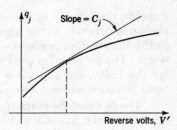

Fig. 5-6 Incremental capacitance of a pn junction.

variations of charge and voltage around a given point on the q_j-V' characteristic, the incremental capacitance is essentially a constant, and it can be used as a constant in Eq. (5-7) to calculate voltages and currents. That is, for small variations in voltage and current, C_j behaves as a linear capacitor. The pn junction is often operated under this *small-signal* condition, and the incremental capacitance is a useful parameter of the device.

Another relation that is sometimes useful is obtained by integrating Eq. (5-7); it is

$$q_j = \int_{V_1'}^{V_2'} C_j(V') \, dV' + q_j(V_1')$$ (5-9)

Thus the charge stored at a pn junction is related to the area under the curve of C_j vs. V'.

The incremental capacitance of the pn junction is examined in detail in Refs. 1 and 2. This study shows that the capacitance is related to the applied voltage by an expression of the form

$$C_j = \frac{K}{(\psi_o + V')^n}$$ (5-10)

where ψ_o = contact potential across junction
   K = constant depending on area of junction and impurity concentrations
   n = constant depending on distribution of impurity atoms near junction

The value of the exponent n ranges from a low of $\frac{1}{3}$ to a high of 3 or 4 for junctions of various types.

When pn junctions are used in the realization of diodes and transistors, the junction capacitance usually amounts to a defect. When a high-frequency voltage is impressed across a junction, the capacitance tends to short-circuit the junction, and this fact usually degrades the performance of diodes and transistors. However, in other applications the junction capacitance can be turned to profit by using the junction as a compact, relatively inexpensive capacitor. Since the value of the capacitance depends on the applied bias voltage in accordance with Eq. (5-10), the junction can be used as a voltage-controlled variable capacitance. The *pn* junction used as a voltage-controlled capacitor is replacing the bulky and expensive mechanical variable capacitors in many radio receivers and radio-frequency oscillators.

The fact that the junction capacitance depends on the applied voltage permits the junction to be used as an amplifier. These amplifiers are special-purpose devices, and they are useful in amplifying signals of very high frequency. Since the amplifying process in these devices depends on the variation of a circuit parameter, the junction capacitance, they are known as parametric amplifiers. Devices designed especially for this application are called *varactors*.

5-3 FURTHER APPLICATIONS FOR THE REVERSE-BIASED *pn* JUNCTION

A *pn* junction and its associated space-charge layer are shown in Fig. 5-7. As stated in Chap. 4, there are very few carriers in the space-charge layer, and it is also known as a carrier-depletion layer. When there is a reverse bias smaller than the breakdown voltage across the junction, the current flowing across the junction is negligibly small. Thus the depletion layer is an insulator, and it provides electrical isolation between the *p* and *n* material on either side of the junction. Isolation of this kind is used extensively in electronic microcircuits. A microcircuit usually consists of diodes, transistors, and resistors, all fabricated in a tiny silicon chip. These components are isolated from one another inside the chip by reverse-biased *pn* junctions, and they are joined together in a circuit by metal connections deposited on the surface of the chip. In order to ensure good isolation, a reverse bias must be maintained across all of the isolation junctions. In accordance with Sec. 5-2, there is a capacitance associated with the isolation junction, and in

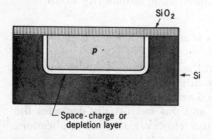

Fig. 5-7 A *pn* junction used to isolate two regions of a silicon crystal.

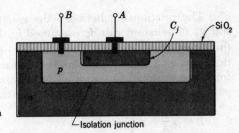

Fig. 5-8 A *pn* junction used as a microcapacitor.

the presence of high-frequency signals it tends to short-circuit the junction. Thus the degree of isolation achieved is impaired at high frequencies.

Figure 5-8 shows a *pn*-junction microcapacitor of the kind sometimes used in microcircuits. This structure contains two junctions, one of which provides isolation and the other provides the desired capacitance. Terminals *A* and *B* provide electrical connection to the microcapacitor. One occasionally troublesome feature of this capacitor is that a reverse bias must be maintained across the capacitor junction so that only capacitive currents can flow. Another disadvantage is that the size limitations imposed on microcircuits are such that it is not possible to realize capacitances greater than a few tens of picofarads with *pn* junctions.

A diffused microresistor using a reverse-biased *pn* junction for isolation is shown in Fig. 5-9. The *p* material serves as the resistor, and terminals *A* and *B* provide electrical connection to the ends of the resistor. The value of the resistance depends on the dimensions of the *p* region and on the resistivity (impurity concentration) of the *p* material. The resistance values that can be obtained range from a minimum of about 100 ohms to a maximum of about 30 kilohms; these limitations are imposed by practical considerations related to the fabrication technique commonly used.

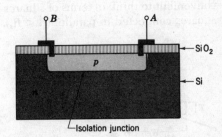

Fig. 5-9 A diffused microresistor using junction isolation.

The relationship between the resistance of the diffused resistor and its dimensions can be examined further with the aid of Fig. 5-10; this figure shows a sheet of conducting material with electrical connections made to opposite faces. If the current flowing through the sheet from one terminal to the other is uniform throughout the volume and parallel to the L direction, then the resistance of the sheet is

$$R = \rho \frac{L}{A} = \rho \frac{L}{TW} \tag{5-11}$$

This expression can be rewritten as

$$R = \frac{\rho}{T} \frac{L}{W} = R_s \frac{L}{W} \tag{5-12}$$

The quantity $R_s = \rho/T$ is called the sheet resistance of the material from which the resistor is made. If $L = W$ so that the sheet is square, then (5-12) yields

$$R = R_s \tag{5-13}$$

Thus the resistance of all square sheets is R_s regardless of the size of the square, and for this reason sheet resistance is specified in units called *ohms per square*, where *square* is a dimensionless quantity. Typical sheet resistances used in diffused microresistors range from 100 to 300 ohms per square. For *p*-type silicon with a uniform distribution of acceptor atoms, the sheet resistance is

$$R_s = \frac{\rho}{T} = \frac{1}{qTN_A\mu_h} \tag{5-14}$$

where q = magnitude of electronic charge
T = thickness of sheet
N_A = concentration of acceptors
μ_h = mobility of holes in silicon

When dealing with resistors having resistances different from R_s, it is convenient to think in terms of squares connected in series ($R > R_s$) and squares connected in parallel ($R < R_s$).

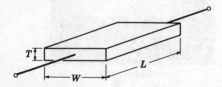

Fig. 5-10 A sheet of conducting material.

5-4 A VOLTAGE-CONTROLLED DIFFUSED RESISTOR

A voltage-controlled capacitor is described in Sec. 5-2, and the usefulness of such a device is pointed out. A voltage-controlled resistor is an
equally useful device, and it can be realized with the aid of two pn junctions as illustrated in Fig. 5-11. The resistor consists of a region of n-
type silicon sandwiched between two regions of p-type material, with
terminals A and B providing electrical contact with each end of the resistor. Provisions are also made for applying a reverse-bias voltage v to
both pn junctions simultaneously. Figure 5-11a shows conditions in
the device when the bias voltage v is zero. Under this condition the
cross-sectional area of the resistor normal to the page is relatively large,
and the resistance appearing between terminals A and B is relatively
small. When a moderate reverse bias is applied to the junctions, the depletion layers at the junctions expand as shown in Fig. 5-11b. As a result, the cross-sectional area of the resistor decreases, and the resistance
between A and B increases. If the reverse-bias voltage is increased still

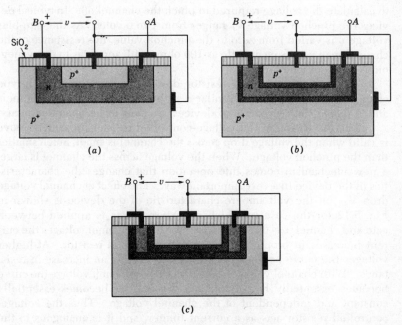

(a)

(b)

(c)

Fig. 5-11 A voltage-controlled resistor. (a) $v = 0$ volts; (b) $v = 2$ volts; (c) $v = 4$ volts.

further, a point is reached at which the depletion layers extend completely across the n region as shown in Fig. 5-11c. Under this condition virtually no current can flow from A to B, and the resistor is virtually an open circuit. When the depletion regions extend completely across the n region, the resistor is said to be pinched off.

The voltage-controlled resistor uses the depletion layers at reverse-biased pn junctions to control the cross-sectional area of the resistor. To optimize the device, it should be designed so that the depletion layers penetrate primarily into the n-type material constituting the resistor. For this purpose the n region, called the channel, is very lightly doped, whereas the p regions, called the gates, are heavily doped. Under these conditions the depletion layers lie almost entirely in the channel, and, if the impurity atoms are uniformly distributed throughout the channel, the width of each depletion layer is given by Eq. (5-1) as

$$L = \left[\frac{2\epsilon}{qN_D} (\psi_o + v) \right]^{1/2} \tag{5-15}$$

where N_D is the concentration of donors in the channel. If the properties of the channel and the junctions are known, Eq. (5-15) can be used to calculate the voltage required to pinch the channel off. In typical devices the pinchoff voltage V_P ranges from 1 to 6 volts. As the gate-bias voltage v is varied from zero to the pinchoff value, the resistance of the channel may vary from as little as 100 ohms with zero bias to many megohms when pinched off.

The voltage-controlled resistor described above uses an n-type channel and p-type gates. Similar resistors are made using p channels and n gates; the two kinds of devices are said to be complementary.

The discussion of the voltage-controlled resistor presented above is valid when the voltage drop across the channel is small, much smaller than the pinchoff voltage. When the voltage across the channel is large, a new mechanism comes into operation that changes the characteristics of the device in a very important way. The effect of channel voltage drop V_{AB} on the volt-ampere characteristic of the device is shown in Fig. 5-12 for the case in which no external bias is applied between gate and channel, $v = 0$ in Fig. 5-11. With low channel voltages the current increases in proportion to the voltage as in a resistor. At higher voltages the curve begins to bend over, indicating an increase in resistance. With channel voltages bigger than the pinchoff voltage the curve becomes essentially horizontal, and the current becomes essentially constant and independent of the channel voltage. Thus the voltage-controlled resistor acts as a current limiter, and it is analogous to the Zener diode, which acts as a voltage limiter. The Zener diode can be used to protect delicate equipment from damage by excessive voltage;

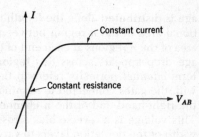

Fig. 5-12 Volt-ampere characteristic for the voltage-controlled resistor with zero gate voltage.

the current limiter can be used for protection from damage by excessive current.

The shape of the volt-ampere characteristic shown in Fig. 5-12 can be explained with the aid of the diagrams in Fig. 5-13. A device with a simplified geometry is shown in this figure so that the important features of the diagrams will stand out clearly. Figure 5-13a shows the electrical connections corresponding to zero external bias between the gates and the channel; these connections are omitted from the remaining parts of the figure for simplicity.

Figure 5-13b shows conditions existing in the device when a small voltage is applied across the channel, making terminal A positive relative to terminal B. A current I flows as indicated, and the applied volt-

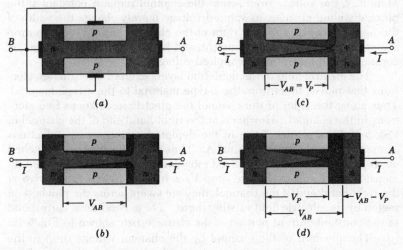

Fig. 5-13 The effect of channel voltage V_{AB} on the voltage-controlled resistor. (a) Circuit connections; (b) small V_{AB}; (c) V_{AB} equal to the pinchoff voltage; (d) V_{AB} larger than the pinchoff voltage.

age is distributed along the length of the channel. Since the cross-sectional area of the *n* region between the gates is much smaller than the area of the *n* regions at each end of the device, essentially all of the voltage drop appears across this region, as indicated in Fig. 5-13*b*. The term *channel* normally refers to the region between the gates. Now, with the gates connected to terminal *B*, the channel voltage drop makes the right-hand end of the *n* channel positive with respect to the gates. This voltage is a reverse bias across the *pn* junctions, and it causes the width of the depletion layers to vary along the channel as shown in Fig. 5-13*b*. The depletion layers restrict the width of the channel somewhat causing the resistance of the channel to increase, and this fact is responsible for the bending of the characteristic shown in Fig. 5-12.

When the channel voltage drop is equal to the pinchoff voltage, it is just great enough to cause the depletion layers to meet at the right-hand end of the channel as shown in Fig. 5-13*c*. Under this condition the channel is pinched off by the channel voltage drop. This condition also corresponds to the beginning of the constant-current portion of the characteristic in Fig. 5-12.

When the applied voltage is greater than the pinchoff voltage, the depletion layers and the pinched-off region expand to the right as shown in Fig. 5-13*d*. The depleted region on the right of the channel isolates the channel from the right-hand end of the device, and changes in applied voltage cannot change the size or shape of the channel. Moreover, the voltage drop across the channel remains constant at the pinchoff value; changes in applied voltage merely change the width of the depletion region on the right of the channel and the voltage drop across it. Thus the channel current, and hence the terminal current, is constant, independent of the applied voltage.

The electric field in the depletion layers exerts a force on free electrons that moves them from the *p*-type material to the *n*-type material. Thus along the sides of the channel the electric force keeps free electrons in the channel. However, at the right-hand end of the channel in Fig. 5-13*d* the electric force in the depleted region sweeps electrons across the region toward terminal *A*. Thus when the channel is pinched off, free electrons move at a fixed rate along a channel having fixed dimensions and a fixed voltage drop V_P across it. When they arrive at the right-hand end of the channel, they are swept across the pinched-off region by the electric field existing there. These conditions correspond to the constant-current portion of the characteristic shown in Fig. 5-12.

The pinchoff of the channel by the channel voltage drop in the voltage-controlled resistor imposes some limitations on the use of the device as a linear resistor. Of far greater importance, however, is the fact that this phenomenon, together with the effects of externally applied gate bias, provides the basis for the field-effect transistor. The field-effect transistor is the principal component in one kind of semiconductor amplifier, and it is the subject of the next chapter.

If the polarity of the voltage V_{AB} applied to the device shown in Fig. 5-13 is reversed, then the voltage drop across the channel constitutes a forward-bias voltage across the pn junctions. The forward bias acts to reduce the width of the depletion layers, and consequently it does not produce pinchoff. With this reversed-polarity voltage, the device acts only as a resistor.

5-5 SUMMARY

This chapter examines the pn junction under reverse-bias conditions, and it presents several applications for the junction operated in this mode. The following important facts about the junction are established:

1. *The width of the space-charge, or depletion, layer depends on the voltage drop across the junction.*
2. *The charge stored in the space-charge layer on either side of the junction depends on the voltage drop across the junction.*
3. *Current must flow into or out of the device to change the stored charge when the voltage drop across the junction is changed.*
4. *The junction has an electrical capacitance.*

These properties of the junction may be either favorable or unfavorable, depending on the application. For example, the capacitance of the junction limits the performance of transistors in response to high-frequency signals, but on the other hand the dependence of the depletion-layer width on the junction voltage makes the junction field-effect transistor possible.

PROBLEMS

5-1. Zener-diode application. The voltage V in Fig. 5-14, supplied by an electronic power supply, is expected to vary between 20 and 25 volts as a result of variations in load current and line voltage. The Zener diode is used to supply a constant output voltage of 10 volts. For

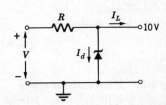

Fig. 5-14 Zener-diode voltage regulator for Prob. 5-1.

good regulation the diode current I_d is to be kept greater than 1 mA, and the power dissipated in the diode must not exceed 250 mW. The diode voltage drop can be assumed to remain constant, although in reality it will vary slightly with variations in I_d.

   (a) Determine the maximum permissible value of I_d.

   (b) What value of R is required to limit I_d to the value determined in part a? Consider the worst case, $V = 25$ volts and $I_L = 0$.

   (c) What is the maximum load current I_L that can be drawn from the output terminals without violating the minimum I_d specification? Consider the worst case, $V = 20$ volts.

5-2. Voltage-regulator design. The voltage regulator of Prob. 5-1 is to be redesigned to provide an output of 30 mA at 10 volts. The range of V and the minimum value of I_d are the same as those given in Prob. 5-1.

   What value of R is required? How much power must the diode be capable of dissipating? Be sure to use worst-case conditions.

5-3. Voltage-controlled capacitance. A certain pn junction is designed to be used as a voltage-controlled capacitor. The incremental junction capacitance is given by Eq. (5-10) with $n = 2$, and the contact potential for the junction is 0.85 volt. The incremental capacitance is 200 pF when a reverse bias of 2 volts is applied to the junction. How much reverse bias is required to reduce the capacitance to 20 pF?

5-4. Voltage-controlled capacitance. The incremental capacitance of the pn junction shown in Fig. 5-15 resonates with the inductor L. The capacitance, given by Eq. (5-10) with $n = 2$, is controlled by the reverse-bias voltage V_c. Derive an expression for the resonant frequency of the circuit as a function of V_c.

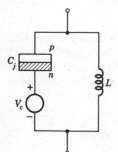

Fig. 5-15 Circuit for Prob. 5-4.

5-5. Microresistor study. A 5-kilohm microresistor is to be fabricated as a p-type layer in a silicon chip with a pn junction for isolation. The

sheet resistance is to be 200 ohms per square, and the thickness of the
p layer is to be 3 μm.

(a) Assuming a uniform distribution of acceptor atoms in the p
layer, what concentration of acceptors is required? The electronic
charge is 1.6×10^{-19} coul, and the mobility of holes in the doped silicon
is 250 cm^2/volt-sec.

(b) How many squares must be connected in series to obtain the
required 5 kilohms?

(c) If the width of the resistor is 25 μm, how long is the resistor?
Give the answer in millimeters.

5-6. *Pinchoff calculation.* A voltage-controlled resistor like the one
shown in Fig. 5-11 has a channel that is 1 μm thick. The donor atoms are
uniformly distributed throughout the channel with a concentration of
10^{16} atoms per cubic centimeter, and the contact potential at the junctions can be taken as 0.85 volt. The electronic charge is 1.6×10^{-19}
coul, and the dielectric permittivity of silicon is approximately 10^{-12}
farad/cm.

Determine the value of the reverse-bias voltage that must be applied to the gate to produce pinchoff. Assume that the depletion layers lie entirely in the channel.

REFERENCES

1. Gray, P. E., et al.: "Physical Electronics and Circuit Models of Transistors," John Wiley
   and Sons, Inc., New York, 1964.
2. Motorola, Inc.: "Integrated Circuits," McGraw-Hill Book Company, New York, 1965.

6

Field-effect Transistors
as Amplifiers

The concept of the ideal amplifier and some of the useful things that can be done with such a device are discussed in Chap. 2. The usefulness of the amplifier having been established, the problem to be confronted is that of realizing an amplifier in the form of a usable physical device. Most electronic amplifiers depend on the existence of pn junctions. The pn junction and some of its properties are presented in Chaps. 4 and 5, and this information provides an adequate basis for understanding the operation of the field-effect transistor (FET) as an amplifier.

The objective of this chapter is to examine the FET as an amplifier and to compare its performance with that of an ideal amplifier in certain respects. The physical laws governing the FET are presented so that the reasons for departure from ideal performance can be perceived and so that the limitations on the use of the FET can be understood. Finally, an introduction to the analysis and design of elementary FET circuits is presented. In this latter phase of the study, the FET is the specific amplifier involved. The methods are quite general, however, and they apply equally well, with minor modifications, to other amplifiers, both electrical and nonelectrical.

6-1 THE BASIC FET AMPLIFIER

Before getting into the details of FET performance, it is helpful first to have a qualitative, overall picture of its behavior as an amplifier. The basic FET amplifier is shown pictorially in Fig. 6-1a, and it is shown schematically in Fig. 6-1b. In order to obtain amplifier action from the FET, the transistor must be used in conjunction with an external power source V_{DD} and series resistor R_d. The voltage source v_1 represents a source of signals, such as a microphone, and R_1 is the internal resistance of the signal source.

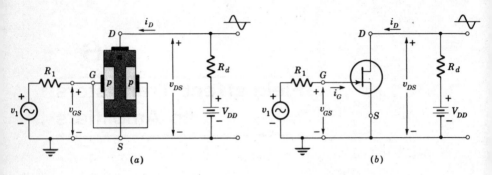

Fig. 6-1 The basic FET amplifier. (*a*) Pictorial representation; (*b*) schematic representation.

Figure 6-1*a* indicates that the physical structure of the FET is the same as the structure of the voltage-controlled resistor presented in Sec. 5-4; in fact, the FET and the voltage-controlled resistor are the same device, and the entire discussion of Sec. 5-4 applies to the FET. The three electrical terminals of the device are called the *drain*, the *gate*, and the *source* as implied by the labels in Fig. 6-1. In normal amplifier operation, majority carriers flow from the source through the channel to the drain; this statement applies to all types of field-effect transistors. The gate junctions normally have a reverse-bias voltage across them, and as a result negligible current flows at the gate terminal. In the schematic representation of Fig. 6-1*b*, the arrowhead at the gate indicates the direction across the junction from the *p* material to the *n* material. Thus in Fig. 6-1*b* it indicates that the channel is made of *n*-type material; this arrowhead is reversed to represent a *p*-channel device. For added clarity in more complex circuits, the gate arrowhead is located at the source end of the channel.

Amplifier action in the FET depends on the fact that the width of the depletion layers at the gate junctions depends on the voltage v_{GS} applied between gate and source. Thus a change in v_{GS} causes a change in the channel dimensions, and this action in turn causes a change in the channel current. The depletion layers act as a valve to control the amount of current in the channel, and this valve is controlled by the gate voltage. If a signal voltage is applied between gate and source as shown in Fig. 6-1, then the signal voltage controls the valve and hence the channel current. The change in channel current is also a change in drain current, and it causes a change in voltage drop across R_d. The change in drop across R_d represents a change in the output voltage v_{DS}. Amplification is obtained because the change in v_{DS} can be many times greater than the change in v_{GS}. In a typical FET amplifier a change in v_{GS} of 0.1 volt may cause a change in i_D of 0.5 mA, which in turn may cause a change in v_{DS} of 1.5 volts. In this case the change in input voltage is amplified by a factor of 15.

The volt-ampere characteristic for a silicon pn junction is shown in Fig. 4-1. This characteristic shows that the junction conducts negligible current for reverse voltages smaller than the breakdown voltage and for positive voltages less than about 0.5 volt. Thus, when the gate voltage for a silicon FET lies between these limits, the current i_G flowing into the gate terminal is negligibly small, usually less than 1 nA at room temperature. It also follows that the power drawn by the FET from the signal source is very small, and the signal-power amplification is very large. With i_G negligible, there is negligible voltage drop across the internal resistance R_1 of the signal source, and the full signal voltage v_1 appears at the input to the transistor. This fact is significant when R_1 is very large, as it is in many cases.

The discussion of the FET given above is in terms of an n-channel transistor. FETs are also made with p-type channels and n-type gates, and the two types of transistors are said to be complementary. The p-channel FET operates in essentially the same manner as the n-channel transistor; the principal difference is that all voltage and current polarities are reversed in the two complementary types. In fact, an n-channel FET in an amplifier circuit can be replaced by a p-channel device of similar ratings provided that the polarity of the power-supply voltage V_{DD} is reversed and provided further that all polarity-sensitive devices in the circuit, such as diodes and electrolytic capacitors, are reversed. All discussions of the FET in this book are in terms of n-channel devices unless it is otherwise stated.

The qualitative, overall picture of the FET amplifier presented above does not give an adequate understanding of the mechanism by which the gate voltage controls the drain current. To gain a better understanding of this mechanism, it is necessary to examine more closely conditions existing inside the transistor. These conditions are illustrated in Fig. 6-2a for the case in which a small negative voltage is applied to the two gates (understood to be connected together electrically) and in which the channel is at the threshold of pinchoff. When the FET is used as an amplifier, it is normally operated with the channel pinched off. The pinchoff mechanism is illustrated in Fig. 5-13 for $v_{GS} = 0$, and it is discussed in detail in Sec. 5-4.

Figure 6-2b shows curves of i_D as a function of v_{DS} for various fixed values of v_{GS}. When v_{DS} is increased from zero with $v_{GS} = 0$, i_D increases with v_{DS} until pinchoff occurs at the drain end of the channel. If v_{DS} is increased beyond the pinchoff point, the pinched-off channel acts as a

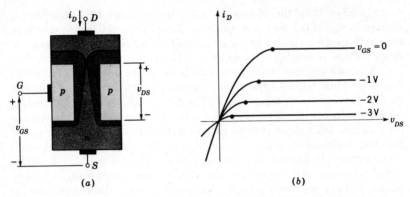

Fig. 6-2   Gate control of the FET.   (*a*) Conditions inside the FET; (*b*) drain current as a function of gate and drain voltages.

current limiter and holds i_D constant at the value existing when pinchoff occurred.   When a small negative voltage is applied to the gates, it causes the depletion layers at the junctions to expand, as indicated at the source ends of the junctions in Fig. 6-2*a*, and v_{GS} aids the channel voltage drop v_{DS} in producing pinchoff.   Thus with v_{GS} negative, pinchoff occurs at a smaller value of v_{DS} and at a correspondingly smaller value of i_D; these facts are illustrated by the curve for $v_{GS} = -1$ volt in Fig. 6-2*b*. Again, if v_{DS} is increased beyond the pinchoff point, the pinched-off channel limits the current to the value existing when pinchoff occurred. The pinchoff point is indicated by a dot on each curve in Fig. 6-2*b*.   On the right of the dots the channel is pinched off, and it limits i_D to a value that depends on v_{GS}.   In this way the gate voltage controls the drain current.

When v_{GS} has a small negative value, the channel is pinched off only at the drain end as shown in Fig. 6-2*a*.   As v_{GS} is made more negative, the pinchoff point remains at the drain end of the channel, but the channel becomes more constricted as a result of the expanding depletion layers.   When v_{GS} is made sufficiently negative, the depletion layers meet along the entire length of the channel, and the channel disappears; under this condition i_D is reduced to zero.   The gate voltage at which i_D is reduced to zero is simply the pinchoff voltage, the voltage required across the junctions to make the depletion layers meet in the middle of the channel.   The definition of the pinchoff voltage is thus

$$V_P = v_{GS} \Big|_{\substack{i_D = \epsilon \\ v_{DS} \neq 0}} \tag{6-1}$$

where ϵ is a very small current and where it is understood that v_{DS} is positive for n-channel transistors and negative for p-channel devices. When V_P is to be measured experimentally, it is often convenient to take $\epsilon = 1\ \mu\text{A}$; many data sheets give V_P at values of ϵ less than 5 nA. It is important to note that

$$V_P < 0 \qquad \text{for } n \text{ channels} \tag{6-2}$$

and

$$V_P > 0 \qquad \text{for } p \text{ channels} \tag{6-3}$$

Figure 6-2a shows an FET in which the channel is just at the threshold of pinchoff at the drain end. The condition corresponding to the pinchoff threshold is that the voltage drop across the depletion layers from gate to channel at the pinchoff point be just equal to the pinchoff voltage. Thus at the pinchoff threshold

$$v_{GS} - v_{DS} = V_P < 0 \tag{6-4}$$

or

$$v_{DS} = v_{GS} - V_P > 0 \tag{6-5}$$

This relation shows quantitatively how the gate voltage affects the value of drain voltage required to produce pinchoff. Since V_P is a negative number for n-channel transistors, negative values of v_{GS} reduce the value of v_{DS} required to produce pinchoff. With a smaller value of v_{DS} appearing across the channel, pinchoff occurs at a smaller drain current. If a small positive voltage is applied to the gate, the value of v_{DS} required to produce pinchoff is increased, and there is a corresponding increase in the drain current at which pinchoff occurs. Normally v_{GS} is not made more positive than about 0.5 volt in order to avoid undesirable current at the gate terminal.

6-2 FET CHARACTERISTICS

The principal volt-ampere characteristics for the n-channel FET are shown in Fig. 6-3. Figure 6-3a is the drain, or output, family of characteristics; these curves are the same as those shown in Fig. 6-2b. The portions of these curves at low values of v_{DS}, where the slopes are not zero, correspond to conditions under which the channel is not pinched off. In this region of the characteristics, the device serves as the voltage-controlled resistor described in Sec. 5-4. The portions of the curves at high values of v_{DS}, where the slopes are zero, correspond to the condition of a pinched-off channel. In this region the transistor is a voltage-controlled constant-current device, and it is in this region that it is normally operated as an amplifier. The sharp breaks in the characteristics

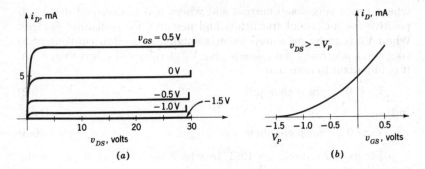

Fig. 6-3 Theoretical static volt-ampere characteristics for the FET. (*a*) Drain, or output, characteristic; (*b*) gate-to-drain transfer characteristic.

at high voltage indicate the onset of avalanche breakdown at the drain ends of the junctions where the reverse bias across the junctions is greatest. Breakdown voltages for typical FETs lie in the range from 20 to 300 volts, with most devices breaking down at less than 50 volts.

Figure 6-3*b* shows the gate-to-drain transfer characteristic for the condition that the channel is pinched off. This curve shows the nature of the control that the gate voltage exerts over the drain current. In particular, it shows that the drain current does not vary linearly with the gate voltage; this fact means that the waveform of the output signal voltage is not an exact copy of the waveform of the input signal voltage. This source of waveform distortion is a departure from the ideal amplifier characteristics, and it is examined in more detail at a later point in this chapter.

Volt-ampere laws for the FET can be derived from a study of the physical processes occurring in the transistor. A detailed derivation is given in Ref. 1 listed at the end of this chapter, and a simplified derivation is given in Appendix 1 at the end of this book. The results of the derivation in Appendix 1 are summarized in the following set of equations:

$$v_{GS} < 0.5 \text{ volt:} \qquad i_G = 0 \tag{6-6}$$

$$v_{GS} - V_P < 0: \qquad i_D = 0 \tag{6-7}$$

$$0 < v_{DS} < v_{GS} - V_P: \qquad i_D = K[2(v_{GS} - V_P)v_{DS} - v_{DS}^2] \tag{6-8}$$

$$0 < v_{GS} - V_P < v_{DS}: \qquad i_D = K(v_{GS} - V_P)^2 \tag{6-9}$$

where K is a constant depending on the geometry and electrical properties of the channel. In particular, K is inversely proportional to the length of the channel. These equations also describe the *p*-channel FET if the inequalities are reversed.

Equation (6-6) states that the gate current is zero when the gate junctions are reverse biased and also when they are forward biased by an amount less than 0.5 volt. Equation (6-7) states that the drain current is zero when the gate voltage is more negative than the pinchoff voltage; this is the condition under which no channel exists in the transistor.

Equation (6-8) applies at low drain voltages when the channel is not pinched off; thus it describes the FET when it acts as a voltage-controlled resistor. For a fixed gate voltage this equation defines a parabola that passes through the origin of the drain-characteristic axes with a positive slope and that reaches a maximum (with zero slope) at the value of v_{DS} corresponding to the pinchoff threshold. The onset of pinchoff changes the physical processes occurring in the transistor, and Eq. (6-8) ceases to describe the transistor when the channel is pinched off.

Equation (6-9) applies when the drain voltage is greater than the value required to produce pinchoff. For a given gate voltage, the drain current through the pinched-off channel is constant—independent of v_{DS}—and it is equal to the drain current existing at the pinchoff threshold. The condition for the pinchoff threshold is given by Eq. (6-5), and thus Eq. (6-9) is obtained by substituting (6-5) for v_{DS} in (6-8). Equation (6-9) describes the constant-current portion of the drain characteristics, and it is the equation of the gate-to-drain transfer characteristic shown in Fig. 6-3b. When the FET is used as an amplifier, it is normally operated in the region of the drain characteristics described by Eq. (6-9). In all discussions of the FET in this book, it is to be understood that operation is in the region described by (6-9) unless otherwise stated.

The boundary between the amplifier region and the voltage-controlled-resistor region of the drain characteristic corresponds to the pinchoff threshold, and the condition for the pinchoff threshold is given by Eq. (6-5). Substituting (6-5) for $v_{GS} - V_P$ in either (6-8) or (6-9) gives the equation of the boundary line on the drain characteristics. The equation obtained is

$$i_D = K v_{DS}^2 \qquad (6\text{-}10)$$

Thus the boundary is a parabola, as shown in Fig. 6-4; Eq. (6-8) applies on the left of the boundary, and (6-9) applies on the right.

Equation (6-9) can be put in a more convenient form for many purposes by factoring $-V_P$ out of the parentheses and noting that the minus sign disappears when the quantity is squared. The result is

$$i_D = K V_P^2 \left(1 - \frac{v_{GS}}{V_P}\right)^2$$

$$= I_{DSS} \left(1 - \frac{v_{GS}}{V_P}\right)^2 \qquad (6\text{-}11)$$

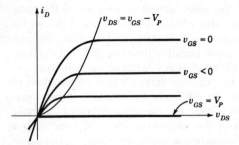

Fig. 6-4 Theoretical static FET drain characteristics.

This relation is subject to the same inequalities limiting Eq. (6-9). The first part of the inequality in (6-9) can be put in a more convenient form for use with (6-11) by dividing by V_P and reversing the inequality because V_P is a negative number for n-channel devices; the result states that (6-11) is valid for $v_{GS}/V_P < 1$.

It is clear from Eq. (6-11) that I_{DSS} is the drain current that flows when $v_{GS} = 0$ and when v_{DS} is large enough to produce pinchoff. This quantity is an important parameter of the transistor, and it is usually given on transistor data sheets along with the pinchoff voltage and other parameters. The relationship between I_{DSS} and the characteristic curves for the transistor is illustrated in Fig. 6-5.

The volt-ampere laws and characteristic curves presented above were derived for a theoretical transistor behaving in the manner described above. The measured characteristics of many real transistors agree reasonably well with the theoretical characteristics, but there are other transistors that have characteristics departing noticeably from the theoretical. These departures result from the geometries and fabrica-

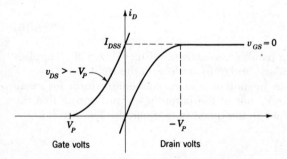

Fig. 6-5 FET characteristics showing the significance of the transistor parameters I_{DSS} and V_P.

tion techniques used to optimize certain aspects of the transistor performance. One feature of many real transistors is that the drain current is not independent of drain voltage when the channel is pinched off; the drain current increases slowly with drain voltage. Another departure of many transistors from the theoretical performance is that they exhibit exponents somewhat different from 2 in the power law of Eq. (6-9). However, for many engineering calculations it is quite satisfactory to treat all transistors as if they obeyed the theoretical laws.

In order to achieve clarity, Figs. 6-1 and 6-2 represent the FET as having a very simple geometry. Two geometries that are more practical are shown in Fig. 6-6. Figure 6-6a gives a cross-sectional view of one geometry, and Fig. 6-6b shows a top view of the same geometry. In fabricating this device, a well of n-type material is diffused into a p-type substrate wafer through a window etched in the surface oxide, and then another p-type layer is formed in the well to serve as the upper gate. The p-type substrate serves as the lower gate. The upper gate extends completely across the n-type well so that there is no path by which current can flow from drain to source without passing under the gate. In

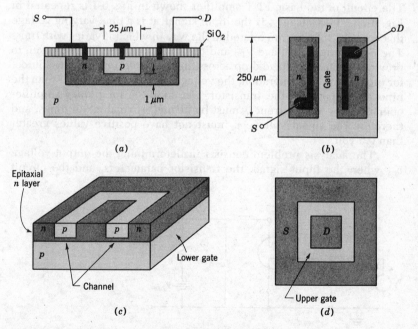

Fig. 6-6 Practical FET geometries. (a) Linear gate; (b) top view of linear gate; (c) rectangular gate; (d) top view of rectangular gate.

this geometry the two gates are connected together internally by the fact that the upper gate extends completely across the well.

Figure 6-6c shows a cross-sectional view of another geometry, and Fig. 6-6d shows a top view of the same device. This transistor is fabricated in an epitaxial layer, and the upper gate, made by diffusion through a window in the surface oxide, forms a hollow rectangle that completely surrounds the drain. Thus again there is no path by which current can flow from drain to source without passing under the gate; however, this result is achieved in this geometry without making an electrical connection between the two gates. Thus, with this geometry, separate terminals can be provided for the gates, and different signals can be applied to the gates to achieve a variety of useful results.

Field-effect transistors are made by a variety of techniques and in many geometric forms, two of which are described above. A detailed discussion of fabrication techniques and geometries is contained in Chap. 8 of Ref. 2.

6-3 ANALYSIS OF THE BASIC FET AMPLIFIER

The circuit of the basic FET amplifier shown in Fig. 6-1 is repeated in Fig. 6-7. The voltage v_1 is the input signal; it is a time-varying voltage that contains information related to the way in which it varies with time. The power-supply voltage V_{DD} and the series resistor R_d are chosen to provide a suitable voltage drop across the transistor from drain to source; for proper amplifier operation, the voltage drop must be greater than the pinchoff voltage for the transistor. In addition, for proper amplifier operation, the gate current i_G must be negligibly small at all times, and therefore the signal voltage v_1 must not have positive values greater than 0.5 volt.

The analysis problem consists in determining the output voltage v_{DS} when the input signal, the transistor parameters, and the circuit

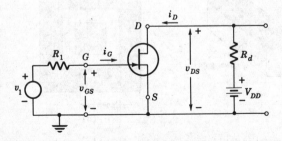

Fig. 6-7 The basic FET amplifier.

parameters are all known. By inspection of the circuit in Fig. 6-7, the output voltage is

$$v_{DS} = V_{DD} - R_d i_D \tag{6-12}$$

when no current is drawn from the output terminals. In normal amplifier operation with the channel pinched off, the drain current i_D is determined solely by the transistor, and it is given by Eq. (6-11) as

$$i_D = I_{DSS} \left(1 - \frac{v_{GS}}{V_P}\right)^2 \tag{6-13}$$

where I_{DSS} and V_P are transistor parameters. As stated in connection with Eq. (6-11), this relation is valid for $v_{GS}/V_P < 1$. Substituting (6-13) into (6-12) yields

$$v_{DS} = V_{DD} - R_d I_{DSS} \left(1 - \frac{v_{GS}}{V_P}\right)^2 \tag{6-14}$$

By inspection of the input circuit in Fig. 6-7, the gate voltage v_{GS} can be written as

$$v_{GS} = v_1 - R_1 i_G \tag{6-15}$$

But i_G is negligibly small in normal amplifier operation, and (6-15) reduces to

$$v_{GS} = v_1 \tag{6-16}$$

Substituting (6-16) into (6-14) yields

$$v_{DS} = V_{DD} - R_d I_{DSS} \left(1 - \frac{v_1}{V_P}\right)^2 \tag{6-17}$$

$$= V_{DD} - R_d I_{DSS} + 2R_d I_{DSS} \frac{v_1}{V_P} - R_d I_{DSS} \left(\frac{v_1}{V_P}\right)^2 \tag{6-18}$$

The first two terms on the right in Eq. (6-18) are constant terms, and the last two terms are time-varying terms. Thus the output voltage of the amplifier consists of a dc component with a time-varying component superimposed upon it. The first of the time-varying terms is directly proportional to the input signal v_1; this is the desired output, a magnified version of the input signal voltage. However, the second time-varying term varies with the square of the input signal. This term is an undesirable distortion term; it causes the waveform of the output voltage to be somewhat different from the waveform of the input voltage. Thus the distortion term alters the information contained in the signal, and if the distortion is severe, the information may be destroyed. All real amplifiers introduce some distortion. Part of the problem in design-

ing amplifier circuits is to design them so that the distortion components
in the output voltage are kept suitably small.

In the amplifier under consideration, if $|v_1| \ll |V_P|$, the term in Eq.
(6-18) depending on $v_1{}^2$ is much smaller than the term depending on v_1.
In this case the time-varying component of v_{DS} is essentially equal to the
linear term, and thus the output voltage is essentially

$$v_{DS} = V_{DD} - R_d I_{DSS} + \frac{2R_d I_{DSS}}{V_P}\, v_1 \tag{6-19}$$

This is a linear relationship between the input and output voltages, and
the time-varying component of the output voltage is directly proportional
to the input voltage.   Under these conditions the input waveform is
preserved in the output signal, and the information contained in the signal is unaltered.

For the n-channel transistor, V_P is a negative number, while R_d and
I_{DSS} are positive numbers.   Thus it is convenient to define

$$-K_v = \frac{2R_d I_{DSS}}{V_P} \tag{6-20}$$

where K_v is a positive number.   Now Eq. (6-19) can be written more
compactly as

$$v_{DS} = V_{DD} - R_d I_{DSS} - K_v v_1 \tag{6-21}$$

As is mentioned above, v_{DS} contains a time-varying component superimposed upon a dc component.   Therefore it is convenient to write

$$v_{DS} = V_{DS} + v_{ds} \tag{6-22}$$

where

$$V_{DS} = V_{DD} - R_d I_{DSS} \tag{6-23}$$

is the dc component and

$$v_{ds} = -K_v v_1 \tag{6-24}$$

is the time-varying component.   The time-varying, or signal, component
of the output voltage is simply the input signal voltage multiplied by
$-K_v$, and thus K_v is the voltage amplification, or voltage gain, of the
amplifier.   The minus sign shows that the amplifier introduces a sign
reversal in the signal; a positive signal at the input produces a negative
signal voltage at the output.   If the input signal is a sinusoidal voltage

$$v_1 = V_1 \cos \omega t \tag{6-25}$$

then

$$v_{ds} = -K_v V_1 \cos \omega t \qquad\qquad (6\text{-}26)$$

and

$$v_{DS} = V_{DS} - K_v V_1 \cos \omega t \qquad\qquad (6\text{-}27)$$

It is appropriate at this point to describe the convention used by most writers in assigning symbols to voltages and currents in transistor circuits. Capital-letter symbols are used to represent voltages and currents that are not functions of time, such as a direct current, or the amplitude of a sinusoidal voltage, or the Laplace transform of a voltage or current. Lowercase letters are used for the instantaneous values of voltages and currents that vary with time. These conventions apply throughout the circuit. For the voltages and currents existing at the terminals of the transistor, additional conventions apply to the subscripts. Symbols with lowercase subscripts represent the time-varying component only of the voltages or current. Thus v_{ds} is the instantaneous value of the time-varying component of drain-to-source voltage, and V_{ds} might represent the Laplace transform of that voltage or the amplitude of a sinusoidally varying component of the voltage. Lowercase symbols with capital-letter subscripts, such as v_{DS}, represent the instantaneous value of the total voltage or current. The dc components of voltage and current are written with capital letters for both symbol and subscripts; an example is V_{DS} appearing in Eq. (6-27).

In summary, capital-letter symbols represent constant quantities; whereas lowercase symbols represent time-varying, or signal, quantities. Capital-letter subscripts denote total voltage or current, and lowercase subscripts denote only signal components. However, the subscript convention applies only to voltages and currents at transistor terminals.

Example 6-1

The performance of an amplifier similar to the one shown in Fig. 6-7 is to be examined. The parameters of the transistor are

$$I_{DSS} = 5 \text{ mA} \qquad V_P = -4 \text{ volts} \qquad BV_{DG} = 40 \text{ volts}$$

where BV_{DG} is the drain-to-gate breakdown voltage. The remaining circuit parameters are

$$V_{DD} = 30 \text{ volts} \qquad R_d = 4 \text{ kilohms} \qquad R_1 = 10 \text{ kilohms}$$

Solution It is assumed initially that the operating conditions are such that Eq. (6-13) applies, and the validity of the assumption is then tested. Consider first the case in which no input signal is applied and $v_1 = 0$; this condition is called the *quiescent* operating condition.   Equation (6-21) then yields

$$v_{DS} = V_{DS} = V_{DD} - R_d I_{DSS} = 30 - (4)(5) = 10 \text{ volts}$$

This voltage is positive and greater than $-V_P = 4$ volts; thus the initial assumption is valid.

Now suppose that

$$v_1 = 0.1 \cos \omega t \qquad \text{volt}$$

The voltage gain of the circuit is

$$K_v = \frac{2 R_d I_{DSS}}{|V_P|} = \frac{(2)(4)(5)}{4} = 10$$

Thus

$$v_{ds} = -K_v v_1 = -1.0 \cos \omega t$$

and

$$v_{DS} = V_{DS} + v_{ds} = 10 - 1.0 \cos \omega t$$

With the value of v_1 used above, there is no problem with current flowing in the gate circuit.   However, it is appropriate to make certain that the channel remains pinched off over the entire cycle of the signal.   The condition that must be satisfied is $v_{DS} > v_{GS} - V_P$, or

$$10 - 1.0 \cos \omega t > 0.1 \cos \omega t + 4$$

$$6 - 1.1 \cos \omega t > 0$$

This last inequality is clearly satisfied for all values of t, and thus the drain remains pinched off at all times.

It is also important to know how much power is dissipated in the transistor under quiescent operating conditions.   With $v_1 = 0$, the power dissipated is

$$P = V_{DS} I_{DSS} = (10)(5) = 50 \text{ mW}$$

A transistor such as this one is normally capable of dissipating several hundred milliwatts without damage in a room-temperature environment.

It is also of interest to examine the amplitude of the distortion term in Eq. (6-18).   This term is

$$R_d I_{DSS} \left(\frac{v_1}{V_P} \right)^2 = (4)(5) \left(\frac{0.1}{-4} \right)^2 \cos^2 \omega t = \frac{1}{80} \cos^2 \omega t$$

Since $\cos^2 \omega t$ never exceeds unity in value, this distortion component of output voltage is small compared to the 1-volt amplitude of the desired signal at the output.

Equation (6-18) shows that the FET introduces some distortion into the signal, especially when the amplitude of the input signal is not small compared with the pinchoff voltage V_P. The distortion term in Eq. (6-18) results directly from the square-law nature of the FET transfer characteristics; it would not exist if the transfer characteristic were linear. There are other sources of signal distortion in the FET amplifier in addition to the square-law transfer characteristic. For example, if the gate voltage becomes more negative than the pinchoff voltage, the drain current is cut off and cannot follow variations in the gate voltage. These sources of distortion can be understood more clearly, and valuable new insights into transistor performance can be obtained, from a graphical analysis of the FET amplifier.

The drain voltage in the amplifier of Fig. 6-7 is given by

$$v_{DS} = V_{DD} - R_d i_D \qquad (6\text{-}28)$$

The drain current is thus

$$i_D = \frac{V_{DD}}{R_d} - \frac{1}{R_d} v_{DS} \qquad (6\text{-}29)$$

$$= I_o - \frac{1}{R_d} v_{DS} \qquad (6\text{-}30)$$

Equation (6-30) is the equation of a straight line on the drain characteristic as shown in Fig. 6-8; this line is called the load line for the amplifier. Equation (6-30) and the load line represent a constraint between i_D and v_{DS} that is imposed by the circuit connected to the transistor; all combinations of voltage and current that satisfy the constraint correspond to points lying on the load line.

An additional constraint between i_D and v_{DS} is imposed by the transistor and is represented by the drain characteristic curves. For a given value of v_{GS}, say -1 volt, all combinations of i_D and v_{DS} that satisfy the transistor constraint correspond to points lying on the drain characteristic for $v_{GS} = -1$ volt. The fact that both of these constraints must be satisfied simultaneously fixes the values of i_D and v_{DS}. It is clear that both constraints are satisfied only at the point where the load line intersects the appropriate drain characteristic; for $v_{GS} = -1$ volt, the intersection is at point P shown in Fig. 6-8.

Point P in Fig. 6-8 is the operating point for the transistor when $v_{GS} = -1$ volt, and the coordinates of the point are the values of i_D and v_{DS} existing when $v_{GS} = -1$ volt. If the value of v_{GS} is changed slowly

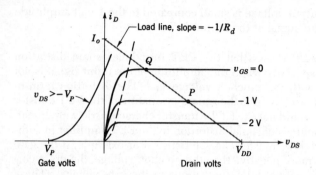

Fig. 6-8 Graphical analysis of the FET amplifier. (Gate and drain voltages are not plotted to the same scale.)

from -1 to -2 volts, the operating point moves along the load line to the drain characteristic for -2 volts; at each instant it lies on the drain characteristic corresponding to the gate voltage at that instant. When no signal is applied at the input of the amplifier in Fig. 6-7 and $v_1 = 0$, the operating point lies at Q in Fig. 6-8. This is the quiescent condition for the amplifier, and point Q is the quiescent operating point.

The graphical construction in Fig. 6-8 contains considerable information about the sources of signal distortion in the FET amplifier. For distortionless amplification, the drain current must vary linearly with the gate voltage, and this occurs only if the drain characteristics for equal increments of gate voltage intersect the load line at equal intervals. The drain characteristics in Fig. 6-8 do not intersect the load line at equal intervals because of the square-law nature of the transfer characteristic, and the distortion term in Eq. (6-18) arises from this fact.

If the gate voltage equals the pinchoff voltage, the drain current is zero, and the operating point lies on the load line at $i_D = 0$, $v_{DS} = V_{DD}$. If the gate voltage is made still more negative, the operating point remains fixed at this location. The view can be taken that, for $v_{GS} < V_P < 0$, all drain characteristics are compressed into one lying on the $i_D = 0$ axis. Clearly this is a region of high distortion, for no signal is transmitted to the output when $v_{GS} < V_P$.

The condition for the pinchoff threshold, $v_{DS} = v_{GS} - V_P$, is shown by the dashed line in Fig. 6-8. On the right of this line the channel is pinched off, and the drain characteristics are straight lines. On the left of this line the channel is not pinched off, and the characteristics are curved. Moreover, the characteristics are crowded together in the region where they are curved; this is a region of high distortion because the characteristics do not intersect the load line at equal intervals. For low-distortion amplification, the operating point must be kept in the constant-current region on the right of the dashed boundary line in Fig. 6-8.

An additional source of distortion arises from the fact that current flows in the gate terminal when v_{GS} is made more positive than about 0.5 volt. To avoid distortion from this source, the operating point must be kept below the drain characteristic for $v_{GS} = 0.5$ volt.

Still another source of distortion is the fact that the gate junctions break down at high values of drain voltage as shown in Fig. 6-3. It is clear from an inspection of the drain characteristics in that figure that the breakdown region is a region of high distortion.

The facts mentioned above represent important limitations on the design of low-distortion amplifiers. When a small signal is applied at the input to the amplifier in Fig. 6-7, the operating point moves along the load line in the vicinity of the quiescent point shown in Fig. 6-8. The voltage amplification of the circuit is given by Eq. (6-20) as

$$K_v = -\frac{2R_d I_{DSS}}{V_P} \qquad (6-31)$$

For a large amplification, R_d should be made as large as possible. However, making R_d large reduces the slope of the load line in Fig. 6-8 and thus moves the quiescent point to the left along the $v_{GS} = 0$ characteristic. When R_d is made large enough, the quiescent point moves across the boundary of the constant-current region into the high-distortion region at the knee of the curve. Thus there is an upper limit to the value of R_d that can be used.

This upper limit can be increased by increasing V_{DD} to shift the whole load line to the right. However, it must be anticipated that large negative gate voltages capable of cutting the drain current off may occur from time to time. With the drain current cut off, the drain voltage is equal to V_{DD}. Therefore, in order to avoid breakdown at the gate junctions under cutoff conditions, it is the usual design practice to limit V_{DD} to a value less than the breakdown voltage for the transistor. Thus the values of R_d and V_{DD} that can be used are limited, and consequently the gain that can be realized is also limited.

The design of the basic amplifier of Fig. 6-7 is limited by the fact that the quiescent operating point must lie on the $v_{GS} = 0$ characteristic. Greater design flexibility can be achieved by adding a direct voltage in series with the input signal in the manner shown in Fig. 6-9. This direct voltage V_{GG} is called the gate-bias voltage. By proper choice of the gate-bias voltage, the quiescent point $(v_1 = 0)$ can be made to lie anywhere along the load line in Fig. 6-8, and this fact often makes improved performance possible.

Equation (6-21) gives the output voltage of the basic amplifier as

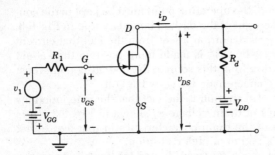

Fig. 6-9 The FET amplifier with a gate-bias voltage V_{GG}.

a function of the input signal voltage and the circuit parameters. It is useful to rederive this equation for the modified amplifier in Fig. 6-9. The gate voltage in Fig. 6-9 can be written as

$$v_{GS} = V_{GS} + v_1 \tag{6-32}$$

where

$$V_{GS} = -V_{GG} \tag{6-33}$$

Substituting (6-32) into (6-14) as before yields

$$v_{DS} = V_{DD} - R_d I_{DSS} \left(1 - \frac{V_{GS}}{V_P} - \frac{v_1}{V_P}\right)^2 \tag{6-34}$$

$$= V_{DD} - R_d I_{DSS} \left(1 - \frac{V_{GS}}{V_P}\right)^2$$

$$+ 2R_d I_{DSS} \left(1 - \frac{V_{GS}}{V_P}\right) \frac{v_1}{V_P}$$

$$- R_d I_{DSS} \left(\frac{v_1}{V_P}\right)^2 \tag{6-35}$$

As before, under small-signal operating conditions with $|v_1| \ll |V_P|$, the distortion term depending on $v_1{}^2$ is negligible in comparison with the desired output signal, and the output voltage is approximately

$$v_{DS} = V_{DS} + \frac{2R_d I_{DSS}}{V_P} \left(1 - \frac{V_{GS}}{V_P}\right) v_1 \tag{6-36}$$

where V_{DS} represents the constant terms in (6-35). The voltage amplification of the modified amplifier is thus

$$K_v = -\frac{2R_d I_{DSS}}{V_P}\left(1 - \frac{V_{GS}}{V_P}\right) \tag{6-37}$$

where $K_v > 0$ and $V_{GS} = -V_{GG}$. The voltage gain of this amplifier depends on the gate bias and thus on the location of the quiescent operating point.

Example 6-2

The parameters of the transistor in an amplifier similar to the one in Fig. 6-9 are $I_{DSS} = 10$ mA and $V_P = -3$ volts. The drain power-supply voltage is to be $V_{DD} = 25$ volts, and the circuit is to be designed so that the quiescent point is at $I_D = 5$ mA and $V_{DS} = 5$ volts. The problem is to specify the required values for V_{GG} and R_d and to calculate the voltage gain of the amplifier.

Solution The quiescent drain current is given by Eq. (6-11) with $v_{GS} = -V_{GG}$; thus

$$I_D = I_{DSS}\left(1 + \frac{V_{GG}}{V_P}\right)^2$$

Substituting the given data in this expression yields

$$5 = 10\left(1 + \frac{V_{GG}}{-3}\right)^2$$

or

$$\frac{V_{GG}}{3} = 1 - \sqrt{0.5} \approx 0.3$$

$$V_{GG} = 0.9 \text{ volt}$$

The quiescent drain voltage is

$$V_{DS} = V_{DD} - R_d I_D$$

$$5 = 25 - 5R_d$$

and

$$R_d = 4 \text{ kilohms}$$

The voltage gain is given by Eq. (6-37); substituting numerical values into this expression yields

$$K_v = -\frac{(2)(4)(10)}{-3}\left(1 - \frac{-0.9}{-3}\right) = \frac{80}{3}(0.7)$$

$$= 18.6$$

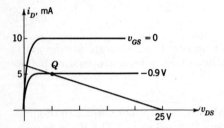

Fig. 6-10 Graphical construction for Example 6-2.

The load line for this amplifier and the quiescent operating point are shown in Fig. 6-10. The boundary between the constant-current region of the drain characteristic and the high-distortion region at low voltages is given by Eq. (6-5) as

$$v_{DS} = v_{GS} - V_P$$

The load line intersects this boundary at

$$v_{DS} \approx -0.9 + 3 = 2.1 \text{ volts}$$

Since the quiescent drain voltage is $V_{DS} = 5$ volts, the signal component of drain voltage cannot be very large without causing the operating point to move into the region of high distortion.

An alternative way of displaying the signal transmission properties of FET amplifiers like the one in Fig. 6-9 is provided by the voltage transfer characteristic shown in Fig. 6-11. This characteristic shows the output voltage v_{DS} as a function of the input signal voltage v_i for a particular circuit. It can be constructed by reading values of v_{DS} from the load line on the drain characteristic for values of v_i covering a suitably wide range. Or alternatively, the portion of the characteristic on the left of point A can be calculated from Eq. (6-34). Point A corresponds to the condition in which the operating point on the drain characteristic lies at the boundary of the constant-current region, and the portion of the curve on the left of A corresponds to operation in the constant-current region of the drain characteristic. The amplifier having the transfer characteristic shown in Fig. 6-11 draws no gate current until v_i has a positive value well to the right of point A. It follows from Eqs. (6-36) and (6-37) that the slope of the transfer characteristic is the voltage gain of the amplifier under small-signal operating conditions.

The construction in Fig. 6-11 shows how the transfer characteristic can be used to determine the waveform of the output voltage when the waveform of the input voltage is known. This construction shows that distortionless amplification requires a linear transfer characteristic.

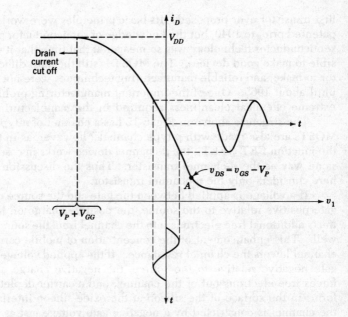

Fig. 6-11 Voltage transfer characteristic for the amplifier of Fig. 6-9 with $I_{DSS} = 10$ mA, $V_P = -4$ volts, $R_d = 6.8$ kilohms, $V_{DD} = 25$ volts, and $V_{GG} = 2.25$ volts.

It is clear that under large-signal operating conditions the curvature of the characteristic in Fig. 6-11 flattens the peaks and exaggerates the valleys in the waveform of the output voltage.

6-4 THE MOST

The structure shown in Fig. 6-12 provides an alternative way to realize a useful field-effect transistor. Two heavily doped wells of n-type silicon are formed in a p-type substrate by diffusion through windows etched in the surface oxide, and these wells are connected by a very shallow channel of lightly doped n-type material lying in the surface of the silicon just under the oxide. The metal electrode deposited on the oxide over the channel serves as the gate for the transistor. This metal-oxide-semiconductor construction gives the device its name: MOS transistor, or MOST. It is also sometimes called a MOS-FET. Since the gate is insulated from the semiconductor by the oxide, the MOST is sometimes called an insulated-gate FET, or IGFET.

As a historical note it is interesting to know that the MOST was the first transistor ever proposed. Its basic principles were worked out and patented prior to 1940, but the knowledge of semiconductor physics and semiconductor technology was so meager at that time that it was impossible to make good devices. The MOST is still the most difficult transistor to make, and reliable manufacturing techniques were not developed until about 1965. One of the important manufacturing problems is the extreme chemical cleanliness required in the manufacturing process.

The structure shown in Fig. 6-12 has a channel of n-type material. MOSTs are also made with p-type channels; however, as in the case of the junction FET (JFET), the p-channel device works in essentially the same way as the n-channel transistor. Thus the discussion presented here considers only the n-channel transistor.

If a voltage is applied between the gate and the source making the gate positive relative to the source, the positive charge on the gate attracts additional free electrons into the channel from the source diffusion well. This enhancement of the concentration of mobile carriers in the channel lowers the channel resistance. If the applied voltage makes the gate negative relative to the source, the negative charge on the gate forces free electrons out of the channel, and a carrier-depletion region forms in the surface of the silicon at the oxide-silicon interface. Thus the channel is constricted by a negative gate voltage just as it is in the FET, and the resistance of the channel increases. It follows from these facts that, like the FET, the MOST can serve as a voltage-controlled resistor. If the gate is made sufficiently negative, the depletion region extends completely across the channel and joins the depletion region at the pn junction on the other side of the channel. Under this condition the channel cannot conduct current between drain and source. The negative gate voltage at which the channel becomes nonconducting is the pinchoff voltage; its value is typically a few volts, as in the FET.

The p-type substrate under the channel in Fig. 6-12 can also serve

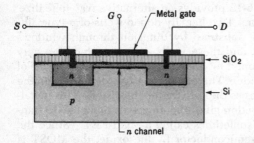

Fig. 6-12 The structure of an n-channel depletion MOS transistor.

as a gate. However, the substrate is usually lightly doped, and it is not a very effective gate. Hence it is the usual practice in MOST amplifiers to connect the substrate to the source.

When an additional voltage is applied between drain and source making the drain positive relative to the source, the behavior of the MOST is essentially the same as the behavior of the FET. There is a voltage drop along the channel, and the drain end of the channel is positive relative to the source end. The positive voltage at the drain end makes the depletion layer at the silicon surface expand just as a negative voltage on the gate does. The depletion layer at the junction between the channel and substrate also widens, but this effect is small because the substrate is a poor gate. Thus, as in the case of the FET, when the gate-to-drain voltage exceeds the pinchoff voltage, the channel becomes pinched off at the drain end, and the transistor becomes a constant-current device. The condition for the pinchoff threshold is thus again

$$v_{GD} = V_P < 0 \tag{6-38}$$

$$v_{GS} - v_{DS} = V_P < 0 \tag{6-39}$$

or

$$v_{DS} = v_{GS} - V_P > 0 \tag{6-40}$$

When the potential of the drain is increased beyond the value required to produce pinchoff, the voltage drop across the channel remains essentially constant at the value given by Eq. (6-40). Now the potential of the drain diffusion well is greater than the potential at the end of the channel, and some mechanism must exist for bridging this potential difference. The positive charge on the drain induces an excess concentration of free electrons in a short length of the channel near the pinched-off end. These excess electrons create a space-charge region at the end of the channel. An electric field exists between this negative space charge and the positive charge on the drain, and the potential difference between the drain and the channel is bridged by the effects of this field.

A more detailed study of the physical processes occurring in the MOST shows that Eqs. (6-7) through (6-9) apply equally well to the MOST. Equation (6-6) also applies, but the inequality limiting (6-6) is not needed, for $i_G = 0$ for all values of v_{GS} provided that the breakdown voltage for the oxide layer is not exceeded. This breakdown voltage typically lies in the range between 30 and 60 volts. Since the FET equations apply to the MOST, it follows that all of the results developed in Sec. 6-3 apply also to the MOST except for the fact that the MOST can also be operated with a positive gate voltage.

Theoretical characteristic curves for the MOST are shown in Fig. 6-13. These curves are the same as the FET characteristics, for they are

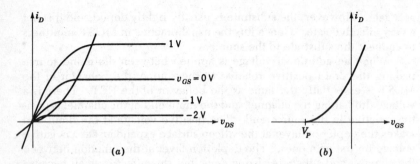

Fig. 6-13 Theoretical static characteristics for the depletion MOST. (*a*) Drain characteristic; (*b*) transfer characteristic.

calculated from the same equations. The one difference is that operation with a large positive gate voltage is indicated for the MOST. If the drain characteristics are extended to sufficiently high values of v_{DS}, drain-to-source breakdown is encountered, and the drain current rises rapidly. The mechanism of this breakdown is the avalanche effect in the space-charge region at the drain end of the channel when the voltage drop across the region is large. Channel breakdown in the MOST may occur at drain voltages as low as 20 volts or as high as 150 volts, depending on the design of the transistor.

Figure 6-14 shows the amplifier circuit of Fig. 6-9 with the FET replaced by a MOST; a symbolic representation of the transistor is used. The arrowhead on the source lead in the transistor symbol indicates the direction of the channel current, and thus it indicates the type of semiconductor in the channel. The arrowhead in Fig. 6-14 indicates an *n*-channel device; the arrowhead is reversed for *p*-channel transistors. The representation of the gate in the transistor symbol is supposed to

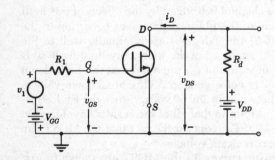

Fig. 6-14 Circuit of a MOST amplifier.

indicate that the gate is insulated from the channel. The connection to the gate is made at the source end of the channel. It should be noted that at the time of this writing, transistor symbols for MOS devices are not standardized, and a variety of symbols is to be found in the literature.

Another structure that is used in making MOS transistors is shown in Fig. 6-15. This structure is identical with the one shown in Fig. 6-12 except that, as shown in Fig. 6-15a, it has no built-in channel between drain and source. The electrical characteristics of this device are similar in most respects to those of the FET and the MOST described above, but there are also significant differences that make this device very attractive for certain applications.

The transistor is normally operated with the p-type substrate connected to the source. If the gate is also connected to the source so that there is no potential difference between the gate and the substrate, no current can flow between drain and source. Regardless of the polarity of the drain-to-source voltage, there is always a reverse-biased junction, either at the drain or the source, blocking the flow of current. With no voltage applied to the gate, the transistor is an open circuit between drain and source.

If v_{GS} in Fig. 6-15b is given a small positive value, say 1 volt, the positive charge on the gate repels holes in the substrate under the gate, and a depletion layer is formed between drain and source as indicated by the dashed line in Fig. 6-15b. If the gate is made still more positive, the charge on the gate attracts free electrons into the depletion layer from the source diffusion well, and an n-type channel is formed at the oxide-silicon interface. This channel extends from drain to source as shown in Fig. 6-15b, and current can now flow through the transistor. The voltage at which the channel forms is called the threshold voltage

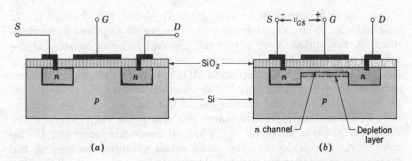

Fig. 6-15 The structure of an n-channel enhancement MOST. (a) No voltage applied; (b) positive voltage applied to the gate.

and designated V_T. The value of the threshold voltage depends on the electrical properties of the substrate and oxide and on the thickness of the oxide; typical values of V_T range from 2 to 4 volts. If the gate voltage is increased above the threshold value, more free electrons are induced into the channel, and the channel resistance decreases. Thus the device provides a voltage-controlled resistance.

The gate voltage of the transistor shown in Fig. 6-12 can be used either to deplete the concentration of carriers in the channel or to enhance the concentration, depending on whether the gate voltage is negative or positive. However, the device is most commonly used in the depletion mode with a negative gate voltage. The gate voltage of the transistor in Fig. 6-15 can be used only to enhance the concentration of carriers in the channel, and the transistor can be used only in the enhancement mode. Consequently, the transistor in Fig. 6-15 is known as an enhancement-mode transistor, whereas the device in Fig. 6-12 is known as a depletion-mode transistor.

When a voltage is applied between drain and source of the enhancement MOST making the drain positive relative to the source, pinchoff occurs at the drain end of the channel just as it does in the depletion MOST and the FET. The minimum potential difference between gate and channel required to maintain the channel is the threshold voltage V_T. Thus pinchoff sets in at the drain end of the channel when

$$v_{GD} = V_T > 0 \tag{6-41}$$

This relation can also be expressed as

$$v_{GS} - v_{DS} = V_T > 0 \tag{6-42}$$

or

$$v_{DS} = v_{GS} - V_T > 0 \tag{6-43}$$

When v_{DS} is less than the value given in (6-43), the channel is not pinched off. When v_{DS} is greater than the value required to produce pinchoff, the voltage drop across the channel remains constant at the value given by (6-43), and the drain current is constant, independent of v_{DS}. As in the case of the depletion MOST, a space-charge region forms at the drain end of the channel to absorb the difference in potential between the drain terminal and the end of the channel.

Enhancement MOS transistors are also made with p-type channels. As always, however, the physical processes occurring in the complementary device are the same in all essential features as the processes occurring in the n-channel transistor, and a separate discussion of the complementary device is not necessary.

Further study of the enhancement MOST shows that it is described

by a set of equations identical in form with Eqs. (6-7) to (6-9) describing
the FET and the depletion MOST. For an n-channel transistor these
equations are:

$$v_{GS} - V_T < 0: \qquad i_D = 0 \tag{6-44}$$

$$0 < v_{DS} < v_{GS} - V_T: \qquad i_D = K[2(v_{GS} - V_T)v_{DS} - v_{DS}^2] \tag{6-45}$$

$$0 < v_{GS} - V_T < v_{DS}: \qquad i_D = K(v_{GS} - V_T)^2 \tag{6-46}$$

where

$$K = \frac{\mu \epsilon W}{2TL} \tag{6-47}$$

The quantities in Eq. (6-47) are

  μ = effective mobility for electrons in channel
  ϵ = absolute permittivity of oxide under gate
  W = width of channel (into the page in Fig. 6-15)
  T = thickness of oxide under gate
  L = length of channel between drain and source

The quantity I_{DSS} defined in Eq. (6-11) has no meaning in connection
with the enhancement MOST because $i_D = 0$ when $v_{GS} = 0$.

Equation (6-44) states that the drain current is zero when the gate
voltage is less than the threshold voltage; no channel exists under this
condition. Equation (6-45) is valid when the channel is not pinched
off; thus it describes the transistor as a voltage-controlled resistor. Equation (6-46) is valid when the channel is pinched off, and it describes the
transistor as a voltage-controlled constant-current device. Since these
equations are identical in form with the equations describing the FET,
the main results derived for the FET apply also to the enhancement
MOST.

Volt-ampere characteristics for the enhancement MOST are shown
in Fig. 6-16; these are the same in form as the FET characteristics. If
the drain characteristics are extended to sufficiently high drain voltages,
channel breakdown occurs. Again, this failure is the result of avalanche
breakdown in the space-charge region at the drain end of the channel.
Most enhancement MOS transistors follow the square-law transfer
characteristic quite closely, and many of them have drain characteristics
agreeing well with the theoretical characteristics. Many other transistors exhibit drain characteristics that rise slowly with drain voltage
in the theoretically constant-current region. This behavior is due, at
least in part, to the fact that the channel length decreases slightly with
increasing v_{DS}, and the channel length affects the drain current through
the constant K given by Eq. (6-47). Transistors optimized for high-fre-

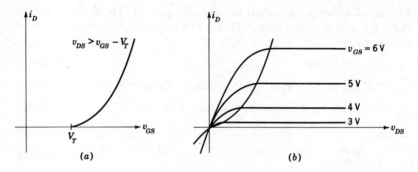

Fig. 6-16 Theoretical static characteristic curves for the enhancement MOST. (*a*) Transfer characteristic; (*b*) drain characteristics.

quency performance have short channels, and they are particularly susceptible to this mechanism.

The enhancement MOS device is quite useful as a microresistor for use in integrated microcircuits, especially when high values of resistance are involved. In Sec. 5-3 it is pointed out that diffused microresistors usually have sheet resistances in the range from 100 to 300 ohms per square and that, owing to the size limitations imposed on microcircuits, diffused resistors are limited in value to about 30 kilohms. Typical MOS resistors have sheet resistances of about 25 kilohms per square, and they can provide resistances up to about 10 megohms for microcircuit applications.

6-5 DISTORTION IN FET AMPLIFIERS

All three of the transistors presented in this chapter are described by equations having the form of Eqs. (6-44) to (6-46). Thus the results derived in this section, which are based on Eq. (6-46), apply equally well to all three types of devices. They also apply qualitatively to other types of electronic amplifiers that are introduced in succeeding chapters.

Figure 6-17 shows an amplifier using an enhancement MOST. (In current practice the same symbol is used for both types of MOS transistors, and thus the type must be identified by other means.) The drain voltage is

$$v_{DS} = V_{DD} - R_d i_D \tag{6-48}$$

In normal amplifier operation the drain current is given by Eq. (6-46), and thus

$$v_{DS} = V_{DD} - K R_d (v_{GS} - V_T)^2 \tag{6-49}$$

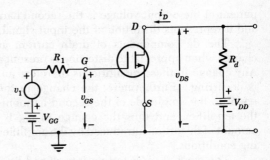

Fig. 6-17 Circuit of an enhancement MOST amplifier.

The gate voltage in the circuit of Fig. 6-17 is

$$v_{GS} = V_{GS} + v_1$$

where V_{GS} is the dc component of v_{GS} and v_1 is the signal component having zero average value. Thus (6-49) can be written as

$$v_{DS} = V_{DD} - KR_d[(V_{GS} - V_T) + v_1]^2 \qquad (6\text{-}50)$$

$$= V_{DD} - KR_d(V_{GS} - V_T)^2$$
$$\quad - 2KR_d(V_{GS} - V_T)v_1 - KR_d v_1^2 \qquad (6\text{-}51)$$

$$= V_{DS} - 2KR_d(V_{GS} - V_T)v_1 - KR_d v_1^2 \qquad (6\text{-}52)$$

where V_{DS} represents the dc component of v_{DS}.

If the input signal is a sinusoid,

$$v_1 = V_1 \cos \omega t \qquad (6\text{-}53)$$

then the output voltage is

$$v_{DS} = V_{DS} - 2KR_d(V_{GS} - V_T)V_1 \cos \omega t$$
$$\quad - KR_d V_1^2 \cos^2 \omega t \qquad (6\text{-}54)$$

$$= V_{DS} - 2KR_d(V_{GS} - V_T)V_1 \cos \omega t$$
$$\quad - \tfrac{1}{2}KR_d V_1^2 - \tfrac{1}{2}KR_d V_1^2 \cos 2\omega t \qquad (6\text{-}55)$$

The first term in this expression is a dc term; it is the value of v_{DS} under quiescent operating conditions, $V_1 = 0$. The second term is an undistorted and amplified reproduction of the input sinusoid; it is the desired output signal. The third term is a dc component of v_{DS} that has a magnitude depending on the amplitude of the input signal and that is related to the distortion introduced by the amplifier. The last term is a sinusoid whose frequency is twice the frequency of the input signal. This component of the output voltage is the second harmonic of the input signal, and it represents distortion of the input signal.

The dc components of drain current and drain voltage always change when appreciable distortion is present as indicated by Eq. (6-55). Any change in these quantities is therefore an indication that distortion is occurring. Furthermore, the change in the dc component of v_{DS} is equal to the amplitude of the second-harmonic voltage at the output of the amplifier, and thus the change in v_{DS} is a direct measure of the amount of distortion introduced by the amplifier under sinusoidal operating conditions.

The amount of distortion introduced by an amplifier under sinusoidal operating conditions can be specified by stating the amplitude of the second-harmonic voltage as a percentage of the fundamental-frequency component. In audio systems used for amplifying speech and music, the distortion commonly ranges from 1 to 5 percent at maximum volume. In high-fidelity systems, however, it is considered desirable to keep the distortion well below 1 percent. Equation (6-55) gives the amplitude of the fundamental-frequency component of v_{DS} as

$$V_f = 2KR_d(V_{GS} - V_T)V_1$$

and it gives the amplitude of the second-harmonic component as

$$V_h = \tfrac{1}{2}KR_dV_1{}^2$$

Thus the percentage of second-harmonic distortion is

$$D = 100\,\frac{V_h}{V_f} = 100\,\frac{V_1}{4(V_{GS} - V_T)} \qquad (6\text{-}56)$$

If the distortion is not to exceed a specified value of D, the input signal amplitude must be kept less than

$$V_1 = \frac{(V_{GS} - V_T)}{25}\,D \qquad (6\text{-}57)$$

where V_{GS} is the dc component of the gate voltage. For small distortion with large signal levels, V_{GS} must be substantially bigger than the threshold voltage V_T, that is, the transistor must operate on the high-current portion of the transfer characteristic where the curvature is small.

All of the results obtained above apply equally well to the JFET and the depletion MOST when V_T is replaced with the pinchoff voltage V_P.

In the practical case where the input signals are not sinusoidal, an additional type of distortion not contained in Eq. (6-55) appears. Suppose, for example, that the input signal consists of the sum of two sinusoids that are not harmonically related,

$$v_1 = V_a \cos \omega_a t + V_b \cos \omega_b t$$

Then the last term in the output voltage given by Eq. (6-52) contains

$$v_1{}^2 = V_a{}^2 \cos^2 \omega_a t + 2V_a V_b \cos \omega_a t \cos \omega_b t$$
$$+ V_b{}^2 \cos^2 \omega_b t \qquad (6\text{-}58)$$

The terms on the right can be expanded as follows:

$$V_a{}^2 \cos^2 \omega_a t = \tfrac{1}{2} V_a{}^2 (1 + \cos 2\omega_a t) \qquad (6\text{-}59)$$

$$V_b{}^2 \cos^2 \omega_b t = \tfrac{1}{2} V_b{}^2 (1 + \cos 2\omega_b t) \qquad (6\text{-}60)$$

and

$$2V_a V_b \cos \omega_a t \cos \omega_b t = V_a V_b \cos (\omega_a + \omega_b)t$$
$$+ V_a V_b \cos (\omega_a - \omega_b)t \quad (6\text{-}61)$$

All of these components of voltage appear at the output of the amplifier. Thus the distortion produces not only harmonics of the input sinusoids but also voltage components at the sum and difference frequencies. The mechanism by which the sum and difference frequencies are produced is called intermodulation. Since it results in voltage components that are not harmonically related to the input signal, it is one of the most objectionable types of distortion in audio amplifiers. On the other hand, however, there are many useful engineering applications for this process of generating sum and difference frequencies.

6-6 POWER RELATIONS IN ELECTRONIC AMPLIFIERS

In principle the life of a transistor is unlimited, for there is nothing in a transistor that is used up or worn out in normal operation. In practice, however, chemical reactions having nothing to do with the operation of the transistor take place, and these reactions may limit the useful life of the device. Such reactions take place between the silicon crystal and residual contaminants enclosed in the capsule with the transistor. There are also reactions between the metals used to make electrical contact with the transistor and between these metals and the silicon. When these reactions proceed far enough, the electrical characteristics of the transistor may be seriously degraded, or the transistor may fail completely. Thus, although the failure rate for transistors is quite small, it is not zero. To achieve low failure rates, the highest possible degree of chemical cleanliness is required in the manufacturing process.

The chemical reactions mentioned above proceed at a rate that increases more or less exponentially with temperature. Thus the higher the operating temperature of the transistor, the shorter its life expectancy. The silicon transistor and the leads connected to it can withstand temperatures up to 200°C without excessive reduction in life expectancy, and transistors mounted in metal cans are usually rated for operation and storage at temperatures up to 175 or 200°C. Transistors that are encapsulated in plastic for low manufacturing cost often have lower temperature limits, as low as 125°C in some cases. High operating temperatures can arise from two sources, high ambient temperatures and self-heating. Self-heating is caused by the current flowing through the transistor between drain and source. The rate at which electrical energy is converted to thermal energy in the transistor at any instant is

$$p_T = v_{DS}i_D \qquad\qquad\qquad\qquad (6\text{-}62)$$

This quantity is the instantaneous power dissipation in the transistor. Under quiescent operating conditions the power dissipation is a constant,

$$P_T = V_{DS}I_D \qquad\qquad\qquad\qquad (6\text{-}63)$$

The power dissipated in the transistor causes its temperature to rise above the ambient temperature, and thermal energy flows out of the transistor to the surrounding environment. The rate at which thermal energy flows out of the transistor depends on the thermal conductivity of the structure and on the temperature rise above the ambient temperature. Thermal equilibrium is reached when the temperature rise is such that thermal energy flows out of the transistor just as fast as it is generated. Typical small transistors reach equilibrium with a 150°C temperature rise when the power dissipation is about 300 mW; thus, if the ambient temperature is 25°C, the temperature of the transistor is 175°C when it is dissipating 300 mW. The manufacturer of the transistor usually specifies a maximum permissible continuous device dissipation at an ambient temperature of 25°C.

If the quiescent power dissipation given by Eq. (6-63) is set equal to the maximum permissible device dissipation, which is a constant, the resulting relation describes a hyperbola on the drain characteristics. The general form of this maximum-permissible-dissipation hyperbola is shown in Fig. 6-18. Amplifiers intended for use in ambient temperatures of 25°C or less must be designed so that the quiescent operating point lies below this hyperbola. Operating points lying above the hyperbola result in an excessive temperature rise and a likelihood of early transistor failure. Operating points lying well above the hyperbola may produce instant, catastrophic failure. However, problems of excessive dissipation ordinarily do not arise in the design of small-signal amplifiers of the kind discussed in this chapter.

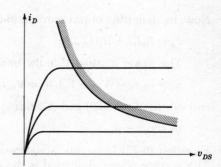

Fig. 6-18 Hyperbola of constant power dissipation on the drain characteristics.

When an input signal is applied to an amplifier such as the one shown in Fig. 6-17, the drain current can be expressed as

$$i_D = I_D + i_d \tag{6-64}$$

where I_D is the average value of the current, and i_d is the signal component having zero average value. Similarly, the drain voltage is

$$v_{DS} = V_{DS} + v_{ds} \tag{6-65}$$

where V_{DS} is the average value of the voltage, and v_{ds} is the signal component having zero average value. The power drawn from the drain power supply at each instant is thus

$$p_{DD} = V_{DD}i_D = V_{DD}I_D + V_{DD}i_d \tag{6-66}$$

Since V_{DD} is constant and i_d has zero average value, the average value of the last term in (6-66) is zero; hence the average power drawn from the power supply is

$$P_{DD} = V_{DD}I_D \tag{6-67}$$

If the distortion introduced by the amplifier is negligibly small and if the input signal has zero average value, then I_D is equal to the quiescent drain current and is independent of the signal amplitude. Under these conditions the power drawn from the drain power supply is also independent of the signal level.

The power absorbed by the drain load resistor R_d at each instant is

$$p_R = R_d i_D^2 = R_d I_D^2 + 2R_d I_D i_d + R_d i_d^2 \tag{6-68}$$

Since i_d has zero average value, the next-to-last term in (6-68) has zero average value, and the average power absorbed by R_d is

$$P_R = R_d I_D^2 + R_d (i_d^2)_{av} \tag{6-69}$$

Now, by definition of rms current, (6-69) can be written as

$$P_R = R_d I_D{}^2 + R_d (I_{d,\,\mathrm{rms}})^2 \tag{6-70}$$

The power dissipated in the transistor at each instant is

$$p_T = v_{DS} i_D = (V_{DD} - R_d i_D) i_D = V_{DD} i_D - R_d i_D{}^2 \tag{6-71}$$

and substituting (6-66) and (6-68) into (6-71) yields

$$p_T = p_{DD} - p_R \tag{6-72}$$

Equation (6-72) is merely a statement of the conservation of energy. The average power dissipated in the transistor is

$$P_T = P_{DD} - P_R \tag{6-73}$$

Since P_{DD} and P_R are always positive and since P_R has its minimum value when no signal is applied, it follows from (6-73) that the power dissipated in the transistor is maximum under quiescent operating conditions. When signal is applied, P_R increases, and P_T decreases by exactly the same amount; the power drawn from the drain power supply remains constant as long as the distortion introduced by the amplifier is negligibly small.

It follows from these relations that in designing a small-signal amplifier the circuit parameters must be chosen so that the quiescent operating point on the drain characteristics lies on or below the hyperbola of maximum permissible device dissipation shown in Fig. 6-18. Then, for signals of any magnitude or waveform, it is ensured that the power dissipated in the transistor will not exceed the permissible value, provided that distortion does not cause a change in the average value of the drain current.

The currents and voltages in electronic circuits usually consist of a dc component plus a time-varying signal component, and the signal component may have a wide variety of waveforms. Therefore computations of average power must be made with care. If v is the potential difference between any pair of terminals and if i is the current from the high-potential to the low-potential terminal, then the power delivered to the terminals at any instant is

$$p = vi \tag{6-74}$$

This relation is always true; it results directly from the definitions of voltage and current. The average power delivered to the terminals, which is often the quantity of most interest, is accordingly

$$P = (vi)_{\mathrm{av}} \tag{6-75}$$

The average of the product vi must be found by proper means. In general,

$$P \neq v_{av}i_{av} \tag{6-76}$$

The average power equals the product $v_{av}i_{av}$ only in the special cases where either v or i, or both, do not vary with time. If $v = V + v_s$ and $i = I + i_s$, where V and I are dc components and v_s and i_s are signal components with zero average value, the instantaneous power is

$$p = (V + v_s)(I + i_s) = VI + Vi_s + v_sI + v_si_s \tag{6-77}$$

Since v_s and i_s have zero average value and V and I do not vary with time, the average power is

$$P = VI + (v_si_s)_{av} \tag{6-78}$$

Thus the dc power and the signal power can be calculated separately, and it is often desirable to do so. If the current i flows through a resistance R, the instantaneous power absorbed by the resistor is

$$p = R(I + i_s)^2 = R(I^2 + 2Ii_s + i_s^2) \tag{6-79}$$

and the average power absorbed is

$$P = RI^2 + R(i_s^2)_{av} = RI^2 + R(I_{s,\,rms})^2 \tag{6-80}$$

Example 6-3

An amplifier using an enhancement MOS transistor is shown in Fig. 6-19. The threshold voltage for the MOST is $V_T = 3.0$ volts, and the input signal is $v_1 = 0.5 \cos \omega t$. With this signal applied, the operating point for the transistor remains always in the constant-current region of the drain characteristics, and the drain current in milliamperes is given by Eq. (6-46) with $K = 0.3$. Power relations and distortion in this amplifier are to be examined.

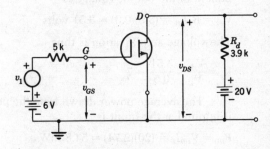

Fig. 6-19 Amplifier for Example 6-3.

Solution   The quiescent operating conditions in the circuit are the following:

$$I_D = K(V_{GS} - V_T)^2 = (0.3)(6-3)^2 = 2.7 \text{ mA}$$

$$V_{DS} = V_{DD} - R_d I_D = 20 - (3.9)(2.7) = 9.5 \text{ volts}$$

The power drawn from the drain power supply is

$$P_{DD} = V_{DD}I_D = (20)(2.7) = 54 \text{ mW}$$

The quiescent power dissipated in the transistor is

$$P_T = V_{DS}I_D = (9.5)(2.7) = 25.6 \text{ mW}$$

and the power dissipated in R_d is

$$P_R = R_d I_D{}^2 = (3.9)(2.7)^2 = 28.4 \text{ mW}$$

When the specified input signal is applied, the drain current is

$$i_D = 0.3[(V_{GS} - V_T) + v_1]^2 = 0.3(3 + v_1)^2$$

$$= 2.7 + 1.8v_1 + 0.3v_1{}^2$$

$$= 2.7 + 0.9 \cos \omega t + 0.075 \cos^2 \omega t$$

$$= 2.74 + 0.9 \cos \omega t + 0.038 \cos 2\omega t$$

The percentage of second-harmonic distortion in i_D, and hence in the output voltage, is

$$D = 100 \, \frac{0.038}{0.9} = 4.2 \text{ percent}$$

The amplitude of the fundamental-frequency component of i_D is

$$I_{d1} = 0.9 \text{ mA}$$

Neglecting the small distortion component of i_D, the signal component of the output voltage is

$$V_{ds} = R_d I_{d1} = (3.9)(0.9) = 3.51 \text{ volts}$$

The voltage amplification is thus

$$K_v = \frac{V_{ds}}{V_1} = \frac{3.51}{0.5} \approx 7$$

The average power drawn from the power supply when signal is applied at the input is

$$P_{DD} = V_{DD}I_D = (20)(2.74) = 54.8 \text{ mW}$$

This value is slightly greater than the quiescent value because of the small distortion introduced by the amplifier. The average power dissipated in R_d is

$$P_R = R_d I_D{}^2 + R_d(I_{d,\text{rms}})^2$$

and, if the small second-harmonic component of drain current is not neglected,

$$(I_{d,\text{rms}})^2 = I_1{}^2 + I_2{}^2$$

where I_1 and I_2 are the rms values of the two sinusoidal components of i_D. Thus

$$(I_{d,\text{rms}})^2 = \tfrac{1}{2}(0.9)^2 + \tfrac{1}{2}(0.038)^2$$

$$= 0.405 + 0.0007 \approx 0.405$$

Thus the second harmonic makes a completely negligible contribution to $I_{d,\text{rms}}$ and to the power dissipated in R_d. Finally,

$$P_R = (3.9)(2.74)^2 + (3.9)(0.405)$$

$$= 29.2 + 1.58 = 30.8 \text{ mW}$$

(Note that using current in milliamperes and resistance in kilohms in these calculations yields power directly in milliwatts. In fact, volts, milliamperes, kilohms, and milliwatts form a consistent set of units that is very convenient in the analysis and design of transistor circuits.) The average power dissipated by the transistor with signal applied is

$$P_T = P_{DD} - P_R = 54.8 - 30.8 = 24 \text{ mW}$$

6-7 SUMMARY

This chapter presents three different types of field-effect transistors, the JFET, the depletion MOST, and the enhancement MOST. All three of these devices can be used to realize useful electronic amplifiers, and they are also used for other kinds of signal-processing operations. These transistors are all described by equations having the same algebraic form, and thus they behave alike in many respects. However, the enhancement MOST has features that make it more attractive than the depletion type, and it is expected to displace the depletion type eventually. It is also expected to displace the JFET, especially in high-frequency applications.

The basic amplifier circuit is also presented in this chapter. This circuit configuration is capable of satisfactory performance, but other configurations are more practical. These more practical circuit configurations are presented in Chap. 7. The next chapter also develops further the techniques for the analysis and design of transistor amplifiers.

PROBLEMS

6-1. *FET voltage-controlled resistor.* An n-channel junction FET is used as a voltage-controlled resistor under small-signal operating conditions. The drain current is given by Eq. (6-8) with $V_P = -5$ volts. With any fixed gate voltage, the small-signal conductance between drain and source is the slope of the corresponding drain characteristic at $v_{DS} = 0$.

   (*a*) Derive an expression for the small-signal drain-to-source resistance as a function of v_{GS} with $v_{DS} = 0$.

   (*b*) If the resistance is 100 ohms when $v_{GS} = 0$, what value of v_{GS} is required for a resistance of 1 kilohm?

   (*c*) Sketch and dimension a curve of drain-to-source conductance G vs. v_{GS} for values of v_{GS} between 0 and -5 volts.

6-2. *FET breakdown.* The gate junctions of a certain n-channel FET go into avalanche breakdown when the drain is 30 volts positive with respect to the gate. What drain-to-source voltage v_{DS} causes the junctions to break down when $v_{GS} = 0$, -2, and -4 volts?

6-3. *FET amplifier design.* An n-channel FET is used in the amplifier shown in Fig. 6-20. The transistor parameters are $I_{DSS} = 5$ mA and

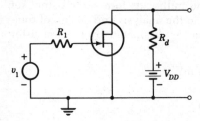

Fig. 6-20   Amplifier for Prob. 6-3.

$V_P = -4$ volts. The amplifier is to be designed so that $v_{DS} = 8$ volts when $v_1 = v_{GS} = 0$, and the power-supply voltage is to be $V_{DD} = 25$ volts.

   (*a*) Determine the value of R_d required.

   (*b*) Determine the voltage amplification of the circuit under small-signal operating conditions, $|v_1| \ll |V_P|$.

6-4. *p-channel FET.* A basic FET amplifier using a p-channel transistor is shown in Fig. 6-21. The transistor parameters are $I_{DSS} = -4$ mA and $V_P = +4$ volts. The power-supply voltage is 25 volts, and the circuit is to be designed so that $v_{DS} = -7$ volts when $v_1 = v_{GS} = 0$.

   (*a*) Determine the value of R_d required.

   (*b*) Determine the voltage amplification of the circuit under small-signal operating conditions, $|v_1| \ll |V_P|$.

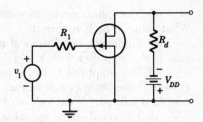

Fig. 6-21 Amplifier for Prob. 6-4.

6-5. Graphical analysis. The parameters of the n-channel transistor in the amplifier of Fig. 6-20 are $I_{DSS} = 5$ mA and $V_P = -4$ volts. The circuit parameters are $R_d = 3.3$ kilohms and $R_1 = 5$ kilohms, and the power-supply voltage is $V_{DD} = 25$ volts.

(a) Sketch and dimension a reasonably accurate drain characteristic for $v_{GS} = 0$. (Calculate a few points using the transistor equations, if necessary.)

(b) Construct the load line for the transistor on the sketch of part a. Give the values of voltage and current at which the load line intersects the axes.

(c) Give the values of v_{DS} and i_D at the quiescent operating point.

6-6. FET amplifier analysis. The parameters of the n-channel FET in the amplifier of Fig. 6-22 are $I_{DSS} = 12$ mA and $V_P = -5$ volts. The cir-

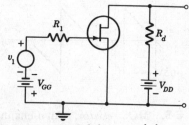

Fig. 6-22 Amplifier for Prob. 6-6.

cuit parameters are $R_d = 3.3$ kilohms and $R_1 = 50$ kilohms, and the applied voltages are $V_{GG} = 1.75$ volts and $V_{DD} = 25$ volts.

(a) Determine the quiescent values of v_{DS} and i_D.

(b) Determine the small-signal voltage gain of the amplifier.

6-7 FET amplifier design. The FET used in an amplifier similar to the one shown in Fig. 6-22 has the parameters $I_{DSS} = 10$ mA and $V_P = -4$ volts. The drain power-supply voltage is $V_{DD} = 25$ volts. The effect of the gate bias on the voltage gain of the amplifier is to be examined.

(a) The quiescent drain voltage V_{DS} is to be 7 volts with $V_{GG} = 0$. What value of R_d is required?

(b) Under the conditions of part a, what is the small-signal voltage gain of the amplifier with $V_{GG} = 0$?

(c) The amplifier is to be redesigned so that the quiescent drain voltage is the same as that in part a and so that the quiescent drain current I_D is 2.5 mA. What values of V_{GG} and R_d are required?

(d) For small-signal operation under the conditions of part c, what is the voltage gain?

(e) Which of these two designs is the more susceptible to nonlinear distortion?

6-8. *Voltage transfer characteristic.* An n-channel depletion MOS transistor is used in an amplifier like the one shown in Fig. 6-22. The parameters of the transistor are $I_{DSS} = 10$ mA and $V_P = -4$ volts. The circuit parameters are $R_1 = 100$ kilohms and $R_d = 6.8$ kilohms. The supply voltages are $V_{GG} = 2.25$ volts and $V_{DD} = 25$ volts.

(a) Plot an accurate voltage transfer characteristic, v_{DS} vs. v_1, for the amplifier corresponding to operation in the constant-current region of the drain characteristics. Cover the range of v_1 between -2.5 and 0.5 volt. *Suggestion:* Use Eq. (6-34).

(b) Using the curve of part a, construct the waveform of the output voltage v_{DS} when v_1 has the waveform shown in Fig. 6-23. If carefully constructed, the output waveform should show a slight distortion.

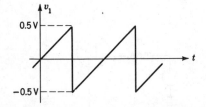

Fig. 6-23   Signal waveform for Prob. 6-8.

6-9. *MOS resistor.* An n-channel enhancement MOS device is used as a resistor in an integrated microcircuit. Under small-signal operating conditions, the drain current is given by Eq. (6-45) with $V_T = 3$ volts. With any fixed value of gate voltage, the small-signal conductance between drain and source is the slope of the corresponding drain characteristic at $v_{DS} = 0$.

(a) Derive an expression for the small-signal drain-to-source resistance as a function of the gate voltage with $v_{DS} = 0$.

(b) The sheet resistance of the channel is 25 kilohms per square when $v_{GS} = 10$ volts. How many squares must be connected in series to realize a drain-to-source resistance of 500 kilohms with this gate bias?

(c) If the channel width is 20 μm, how long is the resistor of part b? Give the answer in millimeters.

(d) What value of v_{GS} is required to make the drain-to-source resistance be 1 megohm?

6-10. Enhancement MOST amplifier. An n-channel enhancement MOS transistor is used in the amplifier shown in Fig. 6-24. The output voltage for this amplifier is given by Eq. (6-52).

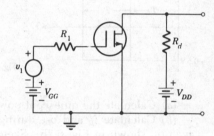

Fig. 6-24 Circuit diagram for Prob. 6-10.

(a) What is the voltage gain of the amplifier, in terms of the circuit parameters, under small-signal operating conditions, $|v_1| \ll |V_{GS} - V_T|$?

(b) The transistor parameters are $K = 0.3$ mA/volt² and $V_T = 3$ volts, and the drain power-supply voltage is 25 volts. Specify the values of V_{GG} and R_d that will place the quiescent operating point at $V_{DS} = 6$ volts and $I_D = 2.7$ mA.

(c) What is the value of the small-signal voltage gain under the conditions of part b?

6-11. Distortion study. An n-channel enhancement MOS transistor is used in an amplifier like the one shown in Fig. 6-24. The transistor parameters are $K = 1.0$ mA/volt² and $V_T = 3$ volts, and the drain power-supply voltage is to be $V_{DD} = 30$ volts. This amplifier is to be designed so that the second-harmonic distortion is 2 percent when $v_1 = 0.5 \cos \omega t$, and the operating point is to remain in the constant-current region of the drain characteristics throughout the entire cycle of the signal.

(a) Determine the required value of gate-bias voltage V_{GG}.

(b) The quiescent drain voltage is to be $V_{DS} = 12$ volts. Determine the value of R_d required.

(c) Calculate the voltage gain and the amplitude V_{ds} of the sinusoidal component of the output voltage.

(d) Is the inequality associated with Eq. (6-46) satisfied at every instant? The worst case is when v_1 has its peak positive value.

(e) What is the power dissipated by the transistor under quiescent operating conditions?

6-12. *Power dissipation.* An *n*-channel JFET is used in an amplifier similar to the one shown in Fig. 6-22. The transistor parameters are $I_{DSS} = 10$ mA and $V_P = -4$ volts. The applied voltages are $V_{GG} = 2.25$ volts and $V_{DD} = 25$ volts, and the circuit parameters are $R_1 = 1$ kilohm and $R_d = 6.8$ kilohms.

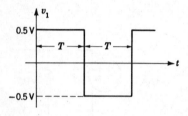

Fig. 6-25 Signal waveform for Prob. 6-12.

(*a*) Calculate the quiescent power dissipated by the transistor.

(*b*) Calculate i_D and v_{DS} during each half cycle when v_1 has the waveform shown in Fig. 6-25. Sketch and dimension the waveform of v_{DS}.

(*c*) Determine the average power dissipated by the transistor under the conditions of part *b*. Compare the result with the quiescent power dissipation. *Comment:* For the case of a square-wave signal, the average power can be determined very simply from the instantaneous power.

6-13. *Distortion and power relations.* An *n*-channel JFET is used in the amplifier described in Prob. 6-12 with a sinusoidal input signal $v_1 = V_1 \cos \omega t$. The transistor and circuit parameters have the values given in Prob. 6-12.

(*a*) What is the largest value that V_1 can have if the second-harmonic distortion is not to exceed 2 percent?

(*b*) Determine the small-signal voltage gain of the amplifier.

(*c*) Determine the average power drawn from the drain power supply, the average power dissipated in R_d, and the average power dissipated in the transistor under quiescent operating conditions.

(*d*) Repeat part *c* for the condition in which the signal found in part *a* is applied at the input. The small effects of distortion can be neglected.

6-14. *Nonlinear distortion.* A certain nonlinear device has a transfer characteristic given by

$$i = a_0 + a_1 v + a_2 v^2 + a_3 v^3$$

where the *a*'s are constants. The input voltage is $v = V \cos 2\pi ft$, where

$f = 1$ kHz. Find the frequencies of all the sinusoidal components of the current i.

6-15. *MOST signal-processing circuit.* Figure 6-26 shows two n-channel enhancement MOS transistors, Q_1 and Q_2, in a signal-processing circuit. This circuit has a number of useful applications that depend

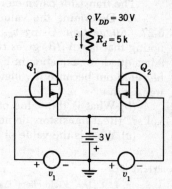

Fig. 6-26 A signal-processing circuit for Prob. 6-15.

on the special nature of the transfer characteristic i vs. v_1. The transistors are identical with parameters $V_T = 3$ volts and $K = 0.3$ mA/volt^2.

(*a*) By simple physical reasoning make a sketch showing the general shape of the transfer characteristic i vs. v_1 for v_1 in the range between -3 and $+3$ volts.

(*b*) Derive the equation for the curve sketched in part *a*. It can be assumed that $v_{DS} > v_{GS} - V_T$.

6-16. *Elementary MOST voltage regulator.* Figure 6-27 shows an n-channel enhancement MOS transistor used as a voltage regulator with an electronic power supply. Diode D is a 40-volt Zener diode; it can be treated as a 40-volt battery. If V_L tends to decrease, v_{GS} increases, and this change in v_{GS} acts to increase i_D. The increase in i_D tends to restore

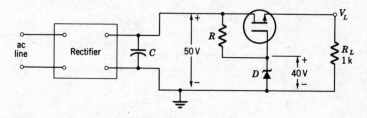

Fig. 6-27 Voltage regulator for Prob. 6-16.

V_L to its original value. If V_L tends to increase, the circuit again reacts in a way that tends to restore V_L to its original value. Thus the function of the regulator is, among other things, to hold V_L constant in the face of changes in line voltage and to reduce the ripple component of the output voltage. As an additional feature, V_L can be adjusted over a range of values by adjusting the voltage applied to the gate of the MOST.

The transistor parameters are $K = 0.7$ mA/volt^2 and $V_T = 3$ volts.

(a) Determine the value of V_L for the conditions shown in Fig. 6-27. *Suggestion:* Using $v_{GS} = 40 - V_L$ in the transistor equation and noting that $i_D = V_L/R_L$ gives two equations that can be combined to obtain a quadratic equation in V_L. One of the two roots of this equation can be ruled out because it violates one of the physical requirements of the transistor.

(b) What is the value of V_L if the rectifier output increases to 60 volts? (Real transistors do not regulate this well.)

(c) What is the value of V_L if R_L is increased to 2 kilohms?

REFERENCES

1. Sevin, L. J., Jr.: "Field-effect Transistors," McGraw-Hill Book Company, New York, 1965.
2. Motorola, Inc.: "Integrated Circuits," McGraw-Hill Book Company, New York, 1965.

7

Practical FET Amplifiers and Small-signal Models

The basic amplifier circuits presented in Chap. 6 are perfectly good amplifiers; however, from a practical point of view they leave something to be desired. In particular, the gate-bias voltage V_{GG} can usually be obtained in a more satisfactory way by a slight circuit modification. These modified circuits often require the use of capacitors, and when capacitors are present, the circuit behaves in one way for the dc components of current and voltage and in a different way for the time-varying signal components. Practical amplifiers are sometimes required to drive reactive loads, and again the amplifier reacts differently to the dc and the signal components of voltage and current. One objective of this chapter is to present a few important practical circuits and to examine their behavior by extending the graphical methods of analysis introduced in Chap. 6.

The behavior of the FET amplifier can also be calculated algebraically, although the calculations are complicated by the fact that the transfer characteristic for the FET is nonlinear. As is shown in Chap. 6, however, under small-signal operating conditions the nonlinear effects become negligibly small, the amplifier becomes essentially a linear device, and the algebraic analysis becomes simple. This fact is formalized in this chapter by the development of the small-signal model to represent the FET as a linear device under small-signal operating conditions. The concept of the small-signal model is very important, for it permits the powerful and extensive theory of linear circuits to be brought to bear on the problems of analyzing and designing electronic amplifiers.

7-1 A PRACTICAL ENHANCEMENT MOST AMPLIFIER

The enhancement MOS transistor provides the simplest of all practical amplifiers; this is one of several features that make this device especially attractive. The circuit of this amplifier is shown in Fig. 7-1; it differs from the basic circuit of Fig. 6-17 only in the way that the gate-bias voltage is supplied. For proper operation as an amplifier, the quiescent operating point for the transistor must lie in the constant-current region of the drain characteristics. This condition is guaranteed if

$$V_{DS} > V_{GS} - V_T > 0 \tag{7-1}$$

where V_{DS} and V_{GS} represent quiescent values. If the gate is connected to the drain through a large resistor R_g as shown in Fig. 7-1a, no current flows in R_g, and

$$V_{GS} = V_{GG} = V_{DS} \tag{7-2}$$

Clearly the inequality (7-1) is satisfied with this connection provided only that $V_{DS} > V_T$, and this requirement is satisfied if $V_{DD} > V_T$.

If a low-impedance source of signal voltage is now connected to the input of the amplifier in Fig. 7-1a, the bias conditions may be disturbed, and the quiescent operating point may be shifted to an unsatisfactory location. If there is a dc path through the signal source, current flows through R_g, and V_{GS} is reduced, possibly to a value less than the threshold voltage V_T. If the output of the signal source contains a dc component, as it often does, this component will also affect the bias condition. Therefore it is usually necessary to add a dc blocking capacitor (or coupling capacitor) to the circuit as shown in Fig. 7-1b. This capacitor is an open circuit for dc components of voltage, and it ensures that $V_{GS} = V_{DS}$ regardless of the source connected to the input. The blocking capacitor is also chosen large enough to act as a short circuit to the time-varying signal component of the input voltage.

No direct current can flow through the biasing resistor R_g; however, with C acting as a short circuit for signal components of voltage, signal components of current do flow through R_g. In most cases it is possible to make this current negligibly small by making R_g very large, 10

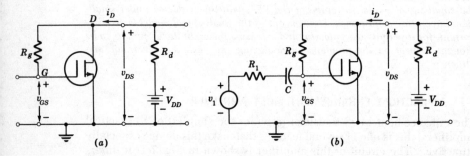

Fig. 7-1 Practical enhancement MOST amplifier. (a) Basic circuit; (b) circuit with a dc blocking capacitor.

megohms or more. Under these conditions, R_g provides bias voltage for
the gate of the transistor, but it does not affect the operation of the circuit
in any other way.

The simple circuit configuration of Fig. 7-1b eliminates the need
for a separate battery or power supply to provide gate bias; it derives the
gate-bias voltage from the main power supply V_{DD}. Of equal or greater
value is the fact that it permits the signal source v_i and the power supply
V_{DD} to have a common terminal. A common terminal for voltage sources
is generally a necessity in electronic systems in order to avoid problems
associated with stray currents flowing through ground paths.

Neglecting the current in R_g, the drain voltage in the amplifier of
Fig. 7-1b is

$$v_{DS} = V_{DD} - R_d i_D \qquad (7\text{-}3)$$

or

$$i_D = \frac{V_{DD}}{R_d} - \frac{1}{R_d} v_{DS} \qquad (7\text{-}4)$$

This is the equation of the load line shown on the drain characteristics
in Fig. 7-2. A second constraint, imposed on the circuit by the bias
connection under quiescent conditions, is

$$v_{DS} = v_{GS} \qquad (7\text{-}5)$$

The dashed line in Fig. 7-2 is the parabola bounding the constant-current region of the drain characteristics; points on this parabola correspond to $v_{DS} = v_{GS} - V_T$. It follows that points on the drain characteristic that satisfy (7-5) lie on the boundary parabola shifted to the right

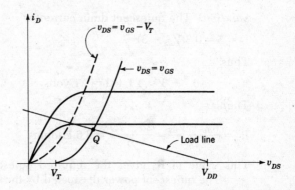

Fig. 7-2 Graphical analysis of the MOST amplifier of Fig.
7-1.

by the amount V_T as shown in Fig. 7-2. Equations (7-4) and (7-5) are satisfied simultaneously at the intersection of this shifted parabola with the load line at the point Q shown in Fig. 7-2; this is the quiescent operating point.

When a small signal voltage v_1 is applied at the input to the amplifier, v_{GS} contains this signal superimposed on the bias voltage, and Eq. (7-5) does not apply. Under this condition the operating point moves along the load line with the signal in accordance with the load-line analysis given in Chap. 6.

If the amplifier is completely specified and it is desired to find the quiescent operating point, there are two ways of proceeding. One way is by a graphical analysis like the one shown in Fig. 7-2. The other is an algebraic solution. The equation for the drain current in the transistor

$$i_D = K(v_{GS} - V_T)^2 \tag{7-6}$$

can be solved simultaneously with Eqs. (7-3) and (7-5) to obtain a quadratic equation in v_{DS} which can then be factored to obtain v_{DS}.

The more usual problem is that of designing an amplifier for a specified quiescent point. If the transistor parameters are known and if V_{DD} and the quiescent drain current I_D are specified, then v_{DS} can be found from (7-6) and (7-5). With v_{DS} known, R_d can be found from (7-3).

Example 7-1

The transistor in an amplifier like the one in Fig. 7-1b has the parameters $V_T = 3$ volts and $K = 0.3$ mA/volt^2. The amplifier is to be designed so that the quiescent drain current is $I_D = 5$ mA when $V_{DD} = 25$ volts. The power dissipation and the small-signal voltage gain are to be calculated.

Solution   The quiescent drain current is

$$I_D = 5 = 0.3(V_{DS} - 3)^2$$

Thus

$$V_{DS} = \sqrt{16.7} + 3 \approx 4.1 + 3 = 7.1 \text{ volts}$$

Then

$$R_d = \frac{V_{DD} - V_{DS}}{I_D} = \frac{25 - 7.1}{5} = 3.6 \text{ kilohms}$$

This value of R_d gives the required quiescent operating point.

The quiescent power dissipated by the transistor is

$$P_T = V_{DS}I_D = (7.1)(5) = 35.5 \text{ mW}$$

If the blocking capacitor acts as a short circuit for the signal and if R_g is large and carries negligible current, then the small-signal voltage gain can be obtained from Eq. (6-52); it is

$$K_v = 2KR_d(V_{GS} - V_T)$$

$$= (2)(0.3)(3.6)(7.1 - 3) = 8.85$$

7-2 THE JFET AMPLIFIER WITH SELF-BIAS

The circuit of a JFET amplifier with self-bias is shown in Fig. 7-3. The JFET differs from the enhancement MOS transistor in that it requires that the gate be biased negatively with respect to the source. This negative gate bias is achieved in the circuit of Fig. 7-3 by grounding the gate and then biasing the source positively with respect to ground by means of the voltage drop across R_s. This circuit is also used with the depletion MOST amplifier.

The blocking capacitor C performs the same function as its counterpart in Fig. 7-1b. Resistor R_g connects the gate to ground and at the same time provides a high-resistance load for the input signal source v_1. With a reverse bias across the gate junctions, the direct current in R_g is the very small reverse-bias leakage current for the junctions. Thus at room temperature with values of R_g up to a few megohms, the dc voltage across R_g is negligibly small. If R_g is made too large, however, the voltage drop across it may become big enough to affect the quiescent operating point appreciably, especially at elevated temperatures. Resistor R_s is the bias resistor. The drain current flowing through R_s develops a voltage drop that biases the source positively with respect to ground and thus with respect to the gate.

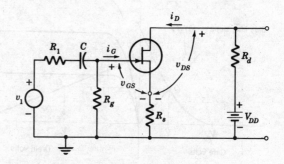

Fig. 7-3 JFET amplifier with self-bias.

Under quiescent operating conditions with $v_1 = 0$ and with i_G negligibly small,

$$V_{GS} = -R_s I_D \tag{7-7}$$

or

$$I_D = -\frac{V_{GS}}{R_s} \tag{7-8}$$

Also

$$V_{DD} = V_{DS} + (R_d + R_s)I_D \tag{7-9}$$

and thus

$$I_D = \frac{V_{DD}}{R_d + R_s} - \frac{1}{R_d + R_s} V_{DS} \tag{7-10}$$

Equation (7-10) describes the load line on the drain characteristics, and Eq. (7-8) describes the bias line on the transfer characteristic; these two lines are shown in Fig. 7-4. All combinations of V_{GS} and I_D that satisfy Eq. (7-8) lie on the bias line, and all combinations that satisfy the law of the transistor lie on the transfer characteristic. The intersection of these two curves at Q satisfies both constraints simultaneously, and it is the quiescent operating point for the amplifier. The quiescent point determined in this way can then be projected onto the load line as shown in Fig. 7-4.

A problem that must be solved frequently is that of designing an amplifier to have a specified quiescent operating point. If the transistor

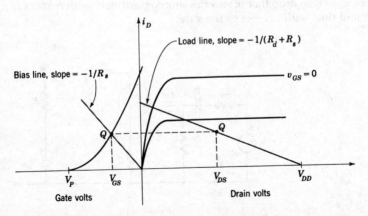

Fig. 7-4 Graphical analysis of the JFET amplifier of Fig. 7-3. Drain volts and gate volts are not plotted to the same scale.

parameters and V_{DD} are known for an amplifier like the one in Fig. 7-3, and if the coordinates V_{DS} and I_D of the quiescent point are specified, then the value of V_{GS} is obtained from

$$I_D = I_{DSS} \left(1 - \frac{V_{GS}}{V_P} \right)^2 \tag{7-11}$$

Then R_s is determined by Eq. (7-7), and R_d is determined by (7-9).

When a signal voltage v_1 is applied to the amplifier in Fig. 7-3, a signal voltage appears across R_g. If $R_g \gg R_1$ and if C is large enough to act as a short circuit to the signal voltage, then the voltage across R_g is essentially v_1. Thus with a signal voltage applied at the input, the gate-to-source voltage is

$$v_{GS} = v_1 - R_s i_D \tag{7-12}$$

A positive increment in v_1 gives a positive increment in v_{GS}, which in turn gives a positive increment in i_D. The increase in i_D results in an increase in the voltage drop across R_s which, as indicated by Eq. (7-12), subtracts from the change in v_{GS}. Thus the positive increment in v_{GS} is not as great as the increment in v_1, and the transistor behaves as if the input signal were smaller than v_1. The result is a smaller change in the output voltage and a smaller voltage gain. The resistor R_s that provides gate-to-source bias also provides a degeneration that reduces the voltage gain.

The degeneration introduced by the bias resistor can be removed, insofar as the signal is concerned, by connecting in parallel with R_s a capacitor that acts as a short circuit for the signal component of i_D. This capacitor is shown in Fig. 7-5; it bypasses the signal component of i_D around R_s and hence is known as a bypass capacitor. In the steady state, the average value of the current through the capacitor must be zero; otherwise the charge on the capacitor and the voltage across it would in-

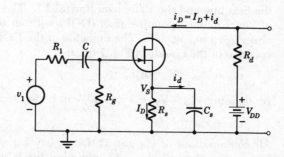

Fig. 7-5 The FET amplifier with a bypassed bias resistor.

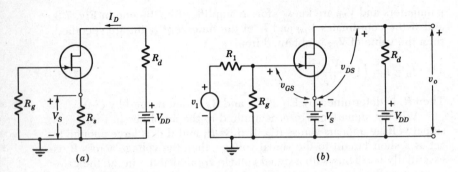

Fig. 7-6 Representation of a JFET amplifier with a bypassed bias resistor. (a) Circuit for dc components of voltage and current; (b) circuit for total voltages and currents.

crease indefinitely with time. Also, since C_s acts as a short circuit to the signal component of i_D, there is no signal component of current through R_s. Hence the current in R_s is I_D, and the current in C_s is i_d. The voltage drop across the parallel combination is a direct voltage, $V_S = R_s I_D$.

With the bypass capacitor C_s added to the circuit, its behavior for the dc components of current and voltage is different from its behavior for the signal components. Insofar as the dc components are concerned, the circuit is that shown in Fig. 7-6a. For total currents and voltages (dc plus the signal component) the circuit behaves as the one shown in Fig. 7-6b. The battery V_S accounts for the voltage drop across C_s; its voltage is $V_S = R_s I_D$. The circuit of Fig. 7-6a can be solved to find the quiescent point and the value of V_S, and then the circuit of Fig. 7-6b can be solved to find the output voltage.

The circuit in Fig. 7-6a is the same as the circuit in Fig. 7-3 under quiescent operating conditions, and hence the graphical analysis shown in Fig. 7-4 applies. The graphical construction is given in Fig. 7-7 by the bias line and the static load line (SLL). The graphical construction of the dynamic operating path (DOP) with an input signal applied is also shown in Fig. 7-7. The equation of the DOP can be written by inspection of the circuit in Fig. 7-6b; it is

$$v_{DS} = V_{DD} - V_S - R_d i_D \tag{7-13}$$

By defining $V'_{DD} = V_{DD} - V_S$, (7-13) can be written in the standard form

$$v_{DS} = V'_{DD} - R_d i_D \tag{7-14}$$

All combinations of v_{DS} and i_D that satisfy Eq. (7-13) lie on the DOP. Thus when a small signal is applied at the input of the amplifier, v_{GS}

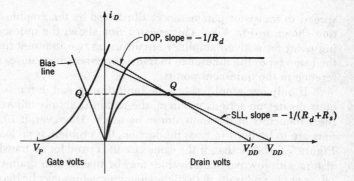

Fig. 7-7 Graphical analysis of the JFET amplifier with bypassed bias resistor. Gate and drain voltages are not plotted to the same scale.

varies accordingly, and the operating point moves along the DOP in the vicinity of the quiescent operating point.

At any instant when i_D equals the quiescent value I_D with an input signal applied, Eq. (7-13) gives

$$v_{DS} = V_{DD} - V_S - R_d I_D = V_{DD} - (R_d + R_s)I_D \qquad (7\text{-}15)$$

But according to Eq. (7-9), the right-hand side of (7-15) is the quiescent drain voltage V_{DS}. Thus at any instant when $i_D = I_D$,

$$v_{DS} = V_{DS} \qquad (7\text{-}16)$$

and it follows from this fact that the DOP passes through the quiescent operating point as shown in Fig. 7-7. Thus the easiest way to construct the DOP is to pass a line through the quiescent point with a slope $-1/R_d$. (Under large-signal operating conditions the nonlinearity of the transistor causes the dc component of i_D to change from its quiescent value, as is demonstrated in Sec. 6-5. This change in I_D causes a small change in V_S, and as a consequence the DOP does not pass through the quiescent point under large-signal conditions. In reasonably linear amplifiers such as the ones under discussion here, this effect is of no consequence and it can be ignored.)

The manufacturing process by which junction FETs are made is of such a nature that it does not provide good control of the transistor parameters V_P and I_{DSS}; as a result, these parameters cannot be held to close tolerances. Transistors of the same type made by the same manufacturer using the same process may exhibit spreads of 5 to 1 or more in both of these parameters. One of the consequences of this production-line spread in transistor parameters is illustrated by the graphical construction shown in Fig. 7-8. This construction shows the quiescent operating point for a given amplifier circuit using two different transistors of the same type; the difference in transistor parameters causes a wide difference in the quiescent points.

If only one amplifier is to be built, the spread in transistor parameters creates no serious problem; the circuit is simply tailored to match the parameters of the transistor to be used. However, if 10,000 amplifiers are to be built in mass production, the problem is no longer trivial. Figure 7-8 shows that, if the same circuit is used for all transistors, transistors with low parameter values may be biased near drain-current cutoff, as at Q_1, and units with high parameter values may be biased outside the constant-current region of the drain characteristic, as at Q_2. In either case excessive distortion may result. To tailor each circuit to match its particular transistor is undesirable from a cost standpoint, for it nullifies the advantages of mass production. Thus it is necessary to seek a modified circuit configuration that is less sensitive to the production-line spread in transistor parameters.

The problem described above would be eliminated if the bias line in Fig. 7-8 were arranged so that all transistors drew nearly the same drain current regardless of the values of V_P and I_{DSS}. Thus a bias line with a smaller slope is needed. However, if the slope of the bias line is reduced simply by increasing R_s, all transistors are biased in the high-distortion region near drain-current cutoff. This problem is avoided with the circuit configuration shown in Fig. 7-9a by applying a positive

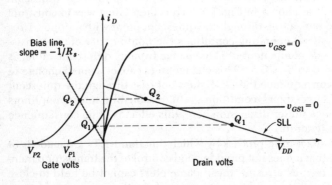

Fig. 7-8 Graphical construction showing the effect of the spread in transistor parameters on the quiescent point. Gate and drain voltages are not plotted to the same scale.

bias voltage to the gate of the transistor when the value of R_s is increased. Thus, as illustrated by the graphical construction in Fig. 7-9b, a reasonably large drain current is maintained when R_s is increased. The gate-to-source voltage in the circuit of Fig. 7-9a under quiescent conditions is

$$v_{GS} = V_{GG} - R_s i_D \tag{7-17}$$

Thus

$$i_D = \frac{V_{GG}}{R_s} - \frac{1}{R_s} v_{GS} \tag{7-18}$$

This is the equation of the bias line in Fig. 7-9b.

The quiescent operating point for JFET amplifiers also depends to some extent on temperature. The reverse-bias leakage current across the gate junctions increases exponentially with temperature, and at elevated temperatures this current may cause an appreciable voltage across R_1 and R_2 in Fig. 7-9a or across R_g in Fig. 7-5. Thus the gate-leakage current may cause a shift in the quiescent point at high temperatures if the gate-circuit resistances are too high. In addition, the mobility of the carriers in the channel of the transistor and the contact potential across the gate junctions are both functions of temperature, and they can cause the quiescent point to shift with temperature. However, the last two of these temperature effects are opposed to each other, and they can be made to cancel by proper choice of the quiescent point. This matter is discussed in detail in Chap. 2 of Ref. 1.

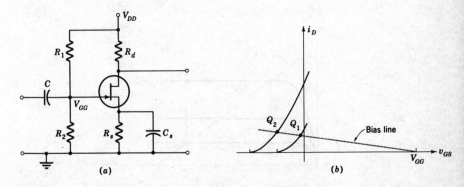

Fig. 7-9 A modified bias circuit to stabilize the quiescent drain current. (a) Circuit; (b) graphical analysis.

7-3 A JFET AMPLIFIER WITH A REACTIVE LOAD

Electronic amplifiers are often required to drive reactive loads; Fig.
7-10*b* shows such an amplifier loaded with a capacitor. Again, with a
reactive component present the circuit responds differently to the dc
and the time-varying signal components of current and voltage. Useful insight into the properties of this circuit can be gained by examining
the very simple case in which the input signal is the square wave shown
in Fig. 7-10*a*. The half-period T of the input signal is much longer than
the time constant $R_d C$, and the circuit reaches the steady state during
each half cycle of the input wave. Thus at the end of each half cycle the
operating point is on the static load line shown in Fig. 7-10*c*; for example, at the end of the time interval between 0 and T, the operating point
is at *a* in Fig. 7-10*c*. During the interval between T and $2T$, the gate
voltage is −2 volts, and at the end of the interval the operating point is
at *c* in Fig. 7-10*c*; however, it does not move along the static load line
from *a* to *c*. The voltage across the capacitor cannot change instantane-

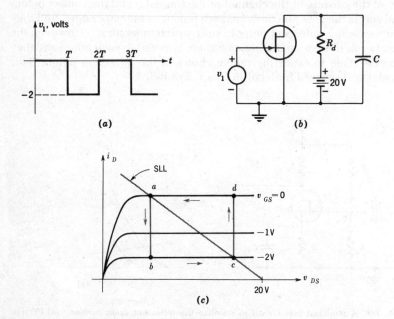

Fig. 7-10 Graphical analysis of an amplifier with a reactive load. (*a*) Input signal
waveform; (*b*) circuit; (*c*) graphical construction.

ously with finite currents flowing. Therefore, when v_1 changes abruptly from 0 to -2 volts at the instant $t=T$, the operating point moves instantaneously along a line of constant v_{DS} to point b on the drain characteristic for $v_{GS} = -2$ volts. The capacitor then begins to charge, and the drain voltage builds up to its new steady-state value with the operating point moving along the line of constant $v_{GS} = -2$ volts to point c. Similarly, when the gate voltage changes abruptly to 0 volts at $t = 2T$, the operating point moves abruptly along a line of constant v_{DS} to point d on the $v_{GS} = 0$ characteristic. Then the capacitor discharges, and the operating point moves along the line of constant $v_{GS} = 0$ to point a. Thus the DOP for a full cycle of the signal is the contour $abcd$ shown in Fig. 7-10c.

7-4 A MOST AMPLIFIER FOR INTEGRATED CIRCUITS

Figure 7-11 shows the circuit diagram of the standard amplifier used in integrated MOS circuits. A signal-processing operation realized in MOS-integrated-circuit form might use 10 or more of these amplifiers along with a number of MOS resistors, and the entire circuit might exist in a square silicon chip measuring a little more than 1 mm on each side. As in the fabrication of the pn junction discussed in Sec. 4-4, these integrated circuits are made on a silicon wafer having a diameter of about 1 in., and since the circuits are so small, it is possible to make several hundred of them simultaneously on a single wafer. It is also possible to process many wafers simultaneously. Thus if each wafer contains

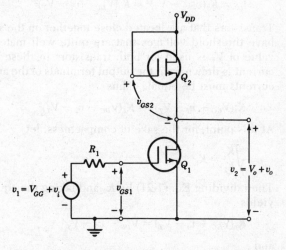

Fig. 7-11 A MOST amplifier for integrated circuits.

400 circuits and if 50 wafers are processed simultaneously, then 20,000 circuits are produced simultaneously. The wiring interconnecting the circuit components is also formed during the process by depositing strips of metal film on the surface of the oxide covering the wafer. The wafers are then cut apart into 20,000 individual circuits, and the separate circuits are mounted in suitable packages. This process is capable of making very tiny, highly reliable circuits at a low cost. However, the low cost is realized only if there is a large market capable of absorbing a large output volume; thus some signal-processing operations are suitable for integration, but many others are not.

The amplifier shown in Fig. 7-11 has several interesting and unique properties, some of which are examined in the following discussion. It is assumed for simplicity that the input voltage v_i contains a dc component that provides a suitable bias for the transistor, a condition that often exists. If this condition does not exist, then a self-bias network like the one shown in Fig. 7-1b can be added to the circuit. Transistor Q_1 is the amplifying transistor, and Q_2 serves as the drain resistor R_d for Q_1. Each transistor is biased so that the channel is pinched off and the operating point is in the constant-current region of the drain characteristic. Thus, using subscripts 1 and 2 to designate transistors Q_1 and Q_2, the drain currents are

$$i_{D1} = K_1(v_{GS1} - V_T)^2 = K_1(V_{GG} + v_i - V_T)^2 \qquad (7\text{-}19)$$

and

$$i_{D2} = K_2(v_{GS2} - V_T)^2 = K_2(V_{DD} - v_2 - V_T)^2 \qquad (7\text{-}20)$$

Transistors that are located close together on the silicon wafer normally have threshold voltages that are quite well matched; hence the same value of V_T is used for both transistors in these equations. When no current is drawn from the output terminals of the amplifier, the two drain currents must be equal. Thus

$$K_1(V_{GG} + v_i - V_T)^2 = K_2(V_{DD} - v_2 - V_T)^2 \qquad (7\text{-}21)$$

At this point, for the sake of compactness, let

$$\sqrt{\frac{K_1}{K_2}} = K_v \qquad (7\text{-}22)$$

Then dividing Eq. (7-21) by K_2 and taking the square root of both sides yields

$$K_v(V_{GG} + v_i - V_T) = V_{DD} - v_2 - V_T \qquad (7\text{-}23)$$

and

$$v_2 = V_{DD} + (K_v - 1)V_T - K_v V_{GG} - K_v v_i \tag{7-24}$$

The first three terms on the right in Eq. (7-24) are constants, independent of the input signal v_i; the sum of these constants is designated by V_o in Fig. 7-11. Thus

$$v_2 = V_o - K_v v_i \tag{7-25}$$

and the signal component of the output voltage is

$$v_2 - V_o = v_o = -K_v v_i \tag{7-26}$$

Equation (7-25) shows that the output voltage is a linear function of the input voltage. This fact is important, for it means that the amplifier provides distortionless amplification for moderately large signals as well as for small signals. In contrast with FET amplifiers using conventional resistors for R_d, the voltage-transfer equation (7-25) does not contain a term depending on v_i^2. The nonlinear resistance characteristic of Q_2 in Fig. 7-11 exactly cancels the nonlinear transfer characteristic of Q_1, and distortionless amplification results, provided that no current is drawn from the output terminals and that the operating points for Q_1 and Q_2 remain in the constant-current regions of the drain characteristic.

Equation (7-26) relates the signal component of the output voltage to the signal component of the input voltage, and it follows from this relation that the voltage gain is

$$K_v = \sqrt{\frac{K_1}{K_2}} \tag{7-27}$$

But K_1 and K_2 are given by Eq. (6-47) as

$$K = \frac{\mu \epsilon W}{2TL} \tag{7-28}$$

and for transistors located close together on the silicon wafer, μ, ϵ, and T are normally closely matched. Thus Eq. (7-27) for the voltage gain becomes

$$K_v = \left(\frac{W_1 L_2}{W_2 L_1}\right)^{1/2} \tag{7-29}$$

where L_1, L_2 = channel lengths from drain to source
    W_1, W_2 = channel widths

Thus the gain of the amplifier is determined by the geometries of the transistors. Gains of 10 or more are achieved in practical amplifiers.

The voltage transfer characteristic for the amplifier in Fig. 7-11 is shown in Fig. 7-12. The portion of the curve between points A and

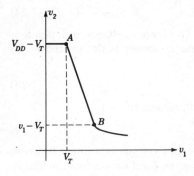

Fig. 7-12   Voltage transfer characteristic for the amplifier in Fig. 7-11.

B is the linear portion; it corresponds to the condition in which the operating points for both transistors are in the constant-current region of the drain characteristic.   On the right of point B, the operating point for Q_1 is in the voltage-controlled resistance region of the drain characteristic. On the left of point A, Q_1 is cut off, and the voltage drop across Q_2 is just slightly greater than V_T to turn Q_2 on very slightly so that it can supply the tiny leakage currents that invariably flow from the output terminal to ground.   (Q_2 must also supply the current drawn by the voltmeter used to measure the output voltage v_2.)

The quiescent operating condition for the amplifier of Fig. 7-11 is given by Eq. (7-24) with $v_i = 0$; it is

$$v_2 = V_o = V_{DD} + (K_v - 1)V_T - K_v V_{GG} \qquad (7\text{-}30)$$

In designing the amplifier, V_{GG} is chosen so as to place the quiescent operating point at a suitable position on the linear portion of the voltage transfer characteristic.   The self-bias network of Fig. 7-1 is often used with this amplifier, and in this case

$$V_{GG} = V_o \qquad (7\text{-}31)$$

Substituting this value for V_{GG} in Eq. (7-30) yields

$$V_o = \frac{V_{DD}}{K_v + 1} + \frac{K_v - 1}{K_v + 1} V_T \qquad (7\text{-}32)$$

7-5 SMALL-SIGNAL LINEAR MODELS FOR FETs

It is shown in Chap. 6 that the FET amplifier using a conventional resistor for R_d is basically a nonlinear device.   The nonlinearity is a direct consequence of the square-law nature of the FET transfer characteristic. It is also shown in Chap. 6 that the nonlinearity has a negligible effect on the performance of the amplifier under small-signal operating con-

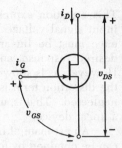

Fig. 7-13 Symbolic representation of the JFET.

ditions and that under these conditions the amplifier can be treated as a linear device. These facts are formalized by the small-signal model for the transistor. The concept of the small-signal linear model is applicable to all nonlinear devices, and it is an extremely important concept. Its importance lies in the fact that it makes possible the systematic application of the powerful theory of linear circuits to the problems of analyzing and designing electronic signal-processing circuits.

Figure 7-13 shows the symbolic representation of a junction FET. A small-signal model for this transistor is to be developed by a simple algebraic process based on the volt-ampere laws of the device. Since all three of the FETs presented in Chap. 6 are described by volt-ampere laws having the same algebraic form, the results obtained are equally applicable to all three devices.

Normal amplifier operating conditions are assumed with $i_G = 0$ and with the operating point for the transistor in the constant-current region of the drain characteristics. The drain current is then given by

$$i_D = I_{DSS}\left(1 - \frac{v_{GS}}{V_P}\right)^2 \tag{7-33}$$

Voltages and currents in small-signal amplifiers always have the form of a signal component superimposed upon a dc component. Therefore it is convenient to let

$$v_{GS} = V_{GS} + v_{gs} \qquad \text{and} \qquad i_D = I_D + i_d \tag{7-34}$$

where v_{gs} and i_d are the signal components of v_{GS} and i_D. Then Eq. (7-33) becomes

$$i_D = I_{DSS}\left[\left(1 - \frac{V_{GS}}{V_P}\right) - \frac{v_{gs}}{V_P}\right]^2 \tag{7-35}$$

$$= I_{DSS}\left(1 - \frac{V_{GS}}{V_P}\right)^2 - 2I_{DSS}\left(1 - \frac{V_{GS}}{V_P}\right)\frac{v_{gs}}{V_P}$$

$$+ I_{DSS}\left(\frac{v_{gs}}{V_P}\right)^2 \tag{7-36}$$

This equation expresses the dependence of the drain current on the input signal voltage v_{gs}. For distortionless amplification this dependence must be linear; therefore the second term in Eq. (7-36) is the desired response, and the third term represents undesired distortion. However, as is shown in Chap. 6, if the input signal v_{gs} is small enough, the distortion term is much smaller than the linear term and it can be neglected. That is, under small-signal conditions the FET is essentially a linear device.

An estimate of how small v_{gs} must be to qualify as a "small signal" can be obtained by forming the ratio of the distortion term to the linear term in (7-36),

$$\frac{\text{Distortion term}}{\text{Linear term}} = \frac{v_{gs}}{2(V_P - V_{GS})} \tag{7-37}$$

The ratio in (7-37) is closely related to the percentage of second-harmonic distortion given by Eq. (6-56); there is a trivial difference in sign due to the fact that (6-56) is derived for an enhancement device. Equation (7-37) states that the distortion term can be made arbitrarily smaller than the linear term by making v_{gs} sufficiently small. In a qualitative sense, small-signal operation exists when

$$|v_{gs}| \ll |V_P - V_{GS}| \tag{7-38}$$

There are many uses for amplifiers in which this condition is satisfied.

Then, by neglecting the distortion term under small-signal conditions, Eq. (7-36) becomes

$$i_D = I_D + i_d = I_{DSS}\left(1 - \frac{V_{GS}}{V_P}\right)^2 - 2I_{DSS}\left(1 - \frac{V_{GS}}{V_P}\right)\frac{v_{gs}}{V_P} \tag{7-39}$$

There is no signal component of i_D when no signal voltage is applied to the gate; thus $i_d = 0$ when $v_{gs} = 0$, and

$$I_D = I_{DSS}\left(1 - \frac{V_{GS}}{V_P}\right)^2 \tag{7-40}$$

This is the quiescent drain current. Equation (7-39) can now be written more compactly as

$$i_D = I_D - \frac{2I_{DSS}}{V_P}\left(1 - \frac{V_{GS}}{V_P}\right)v_{gs} \tag{7-41}$$

The coefficient on v_{gs} in Eq. (7-41) has the dimensions of conductance, and thus it is convenient to define a new symbol

$$g_m = -\frac{2I_{DSS}}{V_P}\left(1 - \frac{V_{GS}}{V_P}\right) > 0 \tag{7-42}$$

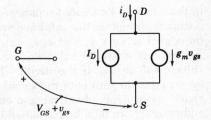

Fig. 7-14 A circuit model for the FET under small-signal conditions.

With this definition g_m is a positive number because V_P is a negative number for the n-channel transistor. This conductance relates the small change in drain current to the small change in voltage applied to the gate; thus it is called the gate-to-drain transfer conductance, or, more simply, the mutual conductance. Equation (7-41) now takes the simple form

$$i_D = I_D + g_m v_{gs} \tag{7-43}$$

This equation, together with the fact that $i_G = 0$, implies that the voltages and currents in the transistor are correctly represented by the circuit model shown in Fig. 7-14. The current I_D in this model is the quiescent drain current given by Eq. (7-40); it is constant when the gate-bias voltage V_{GS} is constant.

Figure 7-15 shows how the circuit model of Fig. 7-14 is used to represent the transistor in an amplifier. The current source I_D accounts for the quiescent drain current; it is a constant current. The response of the transistor to the applied signal voltage v_1 is accounted for entirely by the current source $g_m v_{gs}$. One of the important features of the model in Fig. 7-15b is the fact that it permits the current and voltage sources

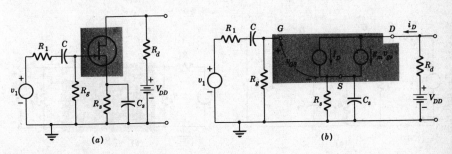

Fig. 7-15 Circuit model for a JFET amplifier. (a) Symbolic representation; (b) circuit model for small-signal operating conditions.

in the circuit to be treated separately by superposition, for it is a *linear* circuit representing the amplifier under small-signal conditions.   Thus the signal sources v_1 and $g_m v_{gs}$ can be set equal to zero with the dc sources left active to calculate the dc, or quiescent, components of voltage and current in the circuit.   Then the dc sources I_D and V_{DD} can be set equal to zero with the signal sources left active to calculate the signal components of voltage and current.   These two sets of voltages and currents can then be combined by superposition to obtain the total voltages and currents.   However, interest is centered primarily on the signal components, for these are the components that describe the signal-processing action of the circuit.

Figure 7-16a shows the circuit of Fig. 7-15b with the dc sources set equal to zero; by the principle of superposition, this circuit is used to calculate the signal components of voltage and current in the amplifier under small-signal operating conditions.   The capacitors C and C_s in this circuit are normally chosen large enough to act as short circuits to the signal components of voltage and current, and thus the circuit of Fig. 7-16a can be reduced to the very simple form shown in Fig. 7-16b. The small-signal voltage gain of the amplifier can be determined from this last circuit by the following simple calculation:

$$v_o = -R_d g_m v_{gs}$$

$$= -g_m R_d \frac{R_y}{R_1 + R_y} v_1 \tag{7-44}$$

The voltage gain is thus

$$\frac{v_o}{v_1} = -g_m R_d \frac{R_y}{R_1 + R_y} = -K_v \tag{7-45}$$

If $R_y \gg R_1$, the expression for the gain becomes

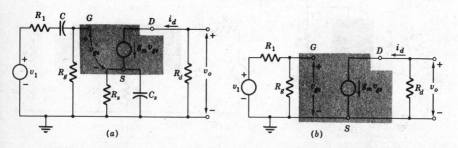

Fig. 7-16 Small-signal model for the amplifier of Fig. 7-15. (*a*) Model including the capacitors; (*b*) model treating the capacitors as short circuits for signal components of current.

$$K_v \approx g_m R_d \qquad\qquad (7\text{-}46)$$

Substituting Eq. (7-42) for g_m in (7-46) yields the value of K_v given by Eq. (6-37).

Figure 7-16 shows the small-signal model for an FET amplifier. The portion of that model representing the transistor only is shown in Fig. 7-17. This is the small-signal model for the FET; it produces the small change in drain current that results from a small change in voltage applied at the gate of the transistor. This model pertains to the transistor only, and it is independent of the circuit in which the transistor is used. In fact, it can be obtained directly from Fig. 7-14 by setting $I_D = 0$. Thus the small-signal model for the transistor can be used in any amplifier circuit configuration provided that the circuit supplies appropriate bias voltages and that the small-signal condition is met. This model is derived for an ideal, theoretical transistor. Second-order effects existing in real transistors sometimes require that additional circuit elements be included in the model; these second-order effects are examined in Sec. 7-6.

The small-signal model for the FET implies that the transconductance is the important parameter of the transistor when it is used in a small-signal linear amplifier, and it is appropriate to examine this parameter in more detail. Since the transconductance gives the change in drain current resulting from a change in gate voltage, it is related to the derivative of i_D with respect to v_{GS}. Differentiating Eq. (7-33) with respect to v_{GS} yields

$$\frac{di_D}{dv_{GS}} = -\frac{2I_{DSS}}{V_P}\left(1 - \frac{v_{GS}}{V_P}\right) \qquad\qquad (7\text{-}47)$$

Evaluating this derivative at a fixed value of $v_{GS} = V_{GS}$ and comparing the result with Eq. (7-42) leads to

$$\frac{di_D}{dv_{GS}} = -\frac{2I_{DSS}}{V_P}\left(1 - \frac{V_{GS}}{V_P}\right) = g_m \qquad\qquad (7\text{-}48)$$

Thus the transconductance is equal to this derivative, and this fact means that the transconductance is equal to the slope of the transfer characteristic for the transistor at the operating point. This derivative relationship is the basic definition of the transconductance. It is much more

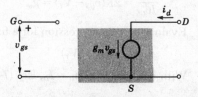

Fig. 7-17 Small-signal model for the theoretical FET.

general than the definition given by Eq. (7-42), and it can be applied to any electronic amplifier in which a voltage applied at one pair of terminals controls the current at another pair.

When $V_{GS} = 0$,

$$g_m = g_{mo} = -\frac{2I_{DSS}}{V_P} \qquad (7\text{-}49)$$

This quantity is important because it is usually given on the manufacturer's data sheets for junction FETs. On data sheets it is usually designated y_{fs}, where y denotes admittance, f denotes forward direction, and s indicates that the source terminal is grounded. The transconductance can now be written as

$$g_m = g_{mo}\left(1 - \frac{V_{GS}}{V_P}\right) \qquad (7\text{-}50)$$

An alternative form for Eq. (7-50) can be obtained by solving Eq. (7-40) to obtain

$$1 - \frac{V_{GS}}{V_P} = \sqrt{\frac{I_D}{I_{DSS}}} \qquad (7\text{-}51)$$

whence

$$g_m = g_{mo}\sqrt{\frac{I_D}{I_{DSS}}} \qquad (7\text{-}52)$$

Thus g_m depends on the quiescent drain current I_D, and since I_D is less than or at most equal to I_{DSS} in junction FETs, g_m is less than or equal to g_{mo}. Typical values of g_{mo} range from 1 to 20 mmhos.

As stated at the beginning of this section, all of the results obtained in this section apply to MOS transistors as well as to the junction FET. However, the parameter I_{DSS} has no meaning in relation to the enhancement MOST, and a slightly different formulation for the transconductance is needed. The most direct way to evaluate the transconductance of the MOS transistor is to differentiate the equation for the drain current. Thus

$$i_D = K(v_{GS} - V_T)^2 \qquad (7\text{-}53)$$

$$\frac{di_D}{dv_{GS}} = 2K(v_{GS} - V_T) = g_m \qquad (7\text{-}54)$$

Evaluating this expression for a fixed value of $v_{GS} = V_{GS}$ yields

$$g_m = 2K(V_{GS} - V_T) \qquad (7\text{-}55)$$

With $v_{GS} = V_{GS}$ and $i_D = I_D$, Eq. (7-53) gives

$$V_{GS} - V_T = \sqrt{\frac{I_D}{K}}$$ (7-56)

and thus (7-55) can be written as

$$g_m = 2\sqrt{KI_D}$$ (7-57)

If the transconductance is known for a particular reference value of drain current I_{DR}, then its value at any other drain current is

$$g_m = g_{mR}\sqrt{\frac{I_D}{I_{DR}}}$$ (7-58)

Example 7-2

Figure 7-18a shows the circuit of an FET amplifier, and Fig. 7-18b shows the small-signal model for the amplifier under the assumption that the coupling and bypass capacitors act as short circuits to the signal components of voltage and current. The manufacturer's data sheet gives the following typical parameter values for the transistor: $I_{DSS} = 10$ mA, $V_P = -4$ volts, and $g_{mo} = 5$ mmhos. The values of R_d and R_s are to be chosen to give a quiescent operation point at $I_D = 2.5$ mA and $V_{DS} = 8$ volts, and the small-signal performance of the amplifier is to be examined.

Solution The quiescent drain current given by Eq. (7-40) is

$$I_D = 2.5 = 10\left(1 - \frac{V_{GS}}{-4}\right)^2$$

Thus

$$1 + \frac{V_{GS}}{4} = \sqrt{0.25} = 0.5$$

$$V_{GS} = 4(0.5 - 1) = -2 \text{ volts} = -V_S$$

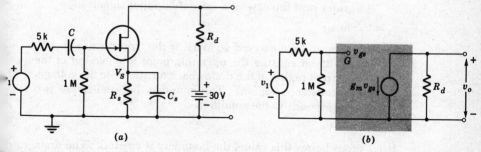

(a)   (b)

ig. 7-18 Amplifier for Example 7-2. (a) Circuit; (b) small-signal model.

The bias resistor needed to provide this voltage is

$$R_s = \frac{V_S}{I_D} = \frac{2}{2.5} = 0.8 \text{ kilohm}$$

The drain-circuit resistance needed to establish the specified value of V_{DS} is

$$R_d = \frac{V_{DD} - V_S - V_{DS}}{I_D} = \frac{30 - 2 - 8}{2.5} = 8 \text{ kilohms}$$

Thus the specified quiescent point is established, and the next step is to examine the performance of the circuit as a small-signal amplifier.

The transconductance of the transistor is given by Eq. (7-52) as

$$g_m = 5\sqrt{\frac{2.5}{10}} = 2.5 \text{ mmhos}$$

The voltages in the small-signal model of Fig. 7-18b are

$$v_{gs} = v_1 \frac{1,000}{1,000 + 5} \approx v_1$$

and

$$v_o = -g_m R_d v_{gs} = -g_m R_d v_1$$
$$= -(2.5)(8)v_1 = -20v_1$$

Thus

$$\frac{v_o}{v_1} = -20$$

and the voltage gain is

$$K_v = 20$$

Consider next the case of a sinusoidal input signal

$$v_1 = V_1 \cos \omega t$$

The question to be answered is: What is the largest value that V_1 can have without causing the operating point to move out of the constant-current region of the drain characteristics into the voltage-controlled resistance region? The boundary between these two regions corresponds to the condition

$$v_{DS} = v_{GS} - V_P$$

If v_{DS} drops below this value, the boundary is crossed. The drain voltage is

$$v_{DS} = V_{DS} + v_{ds} = 8 - 20V_1 \cos \omega t$$

and the gate voltage is

$$v_{GS} = V_{GS} + v_1 = -2 + V_1 \cos \omega t$$

Substituting these values into the condition expressed above and setting $V_P = -4$ volts yields

$$8 - 20V_1 \cos \omega t = -2 + V_1 \cos \omega t + 4$$

$$6 - 21V_1 \cos \omega t = 0$$

Thus the operating point just reaches the boundary of the constant-current region when

$$21V_1 = 6$$

$$V_1 = 0.286 \text{ volt}$$

The corresponding sinusoidal component of the output voltage has an amplitude

$$V_o = 20V_1 = 5.7 \text{ volts}$$

With this signal level, the percentage of second-harmonic distortion, given by Eq. (6-56), is

$$D = \frac{25V_1}{V_{GS} - V_P} = \frac{25(0.286)}{2} \approx 3.6 \text{ percent}$$

For a 1 percent distortion level, the output voltage must be restricted to an amplitude of about 1.5 volts.

The amplifier shown in Fig. 7-18a is about the simplest form in which a practical, general-purpose, small-signal FET amplifier can be realized. It is therefore appropriate to compare this amplifier with the ideal amplifier of Chap. 2. Figure 7-19a shows the small-signal model of the practical amplifier under the assumption that the coupling and bypass capacitors act as short circuits for the signal. Figure 7-19b shows the small-signal model after the current source in parallel with R_d has been converted into a voltage source in series with R_d. The conversion process yields $K_v = g_m R_d$, and thus K_v is the small-signal voltage gain of the amplifier. The ideal voltage amplifier of Fig. 2-1 is shown in Fig. 7-19c for comparison.

The small-signal model of Fig. 7-19b falls short of the ideal in two ways. First, the input resistance is not infinite; thus the amplifier loads the source of signals, and the voltage-divider action of R_g and R_1 may introduce appreciable attenuation of the signal. Second, the output

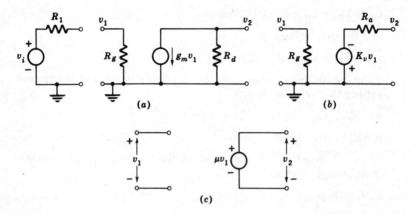

Fig. 7-19 Comparison of the FET amplifier with the ideal voltage amplifier. (a) Small-signal model for the FET amplifier; (b) the model of part a after a source conversion; (c) the ideal voltage amplifier.

resistance of the amplifier is not zero, and hence if a load resistor is connected across the output terminals, additional signal attenuation is introduced.   However, the isolation between the input and output terminals of the ideal amplifier is preserved in the practical amplifier, and this fact is important. (Unfortunately, the isolation is lost when the signal frequencies are high, for reasons developed in Sec. 7-6.) The voltage gain provided by typical practical amplifiers is less than 20.

The comparison made above is between the small-signal model of the practical amplifier and the ideal voltage amplifier. In the actual amplifier, the output voltage contains a dc component in addition to the signal component; this fact is another departure from the ideal, and it raises some problems when amplifiers are to be connected in cascade, especially when the amplifiers are integrated circuits on a single silicon chip. A further deviation from the ideal is the fact that the practical amplifier introduces signal distortion when the small-signal condition is violated. The information carried by the signal is contained in its waveform, and if the waveform is distorted, the information is altered. And finally, the performance of the practical amplifier changes with changes in temperature and with changes in power-supply voltage.

Means are available for minimizing the effects of all these nonideal properties of practical amplifiers. The analysis of amplifier circuits is concerned largely with determining the effects of these (and other) nonideal features of the circuits. The design of amplifiers is concerned largely with developing means for obtaining good performance in spite of the nonideal features.

7-6 SECOND-ORDER EFFECTS IN FETs

The FET drain characteristics given in the preceding sections are calculated curves for theoretical transistors. A family of measured characteristics for a real transistor is shown in Fig. 7-20a; the principal departure from the theory exhibited by these curves lies in the fact that the drain current is not constant in the region corresponding to channel pinchoff. The gradual increase in drain current with increasing drain voltage is due, at least in part, to the fact that the channel length does not remain constant under pinched-off conditions as assumed in the theoretical analysis. The channel length decreases slightly with increasing drain voltage, and this fact results in a slight increase in drain current. The magnitude of this effect depends on the design of the transistor. In general, transistors designed for good performance with high-frequency signals have short channels, and in this case the effect is relatively large. On the other hand, some transistors designed for audio and supersonic frequencies exhibit virtually constant-current characteristics as predicted by the theory.

The nonzero slope of the drain characteristics does affect circuit performance in some cases, and it is therefore desirable to include its effect in the small-signal model for the transistor. The graphical construction in Fig. 7-20b shows the increment in drain current Δi_D that results from an increment in drain voltage. It follows from this construction that

$$\Delta i_D = m \, \Delta v_{DS} \qquad\qquad (7\text{-}59)$$

where m is the slope of the drain characteristic. The slope m has the dimensions of conductance, and it is helpful to designate it as such; thus a new symbol $g_d = m$ is defined and Eq. (7-59) is written as

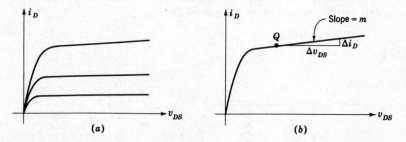

Fig. 7-20 Nonideal FET characteristics. (a) Drain characteristics for a real transistor; (b) graphical construction relating Δi_D and Δv_{DS}.

$$\Delta i_D = g_d\, \Delta v_{DS} = \frac{1}{r_d}\, \Delta v_{DS} \tag{7-60}$$

If the increments of voltage and current in Eq. (7-60) represent small-signal components, then (7-60) can be written as

$$i_d = g_d v_{ds} = \frac{1}{r_d}\, v_{ds} \tag{7-61}$$

The change in drain current caused by a small change in gate voltage has already been shown to be $i_d = g_m v_{gs}$. Thus when both the drain and the gate voltage change, the total change in drain current is

$$i_d = g_m v_{gs} + \frac{1}{r_d}\, v_{ds} \tag{7-62}$$

This equation implies the small-signal model shown in Fig. 7-21. For typical low-power transistors, the incremental drain resistance r_d lies in the range between 20 and 100 kilohms, with some units having resistances above this range.   The value of r_d depends on the quiescent drain current I_D; it decreases with increasing I_D.

The plausibility argument presented in the preceding paragraph can be made rigorous and general in the following way. The family of drain characteristics represents in graphical form a functional relation between the drain current and the drain and gate voltages,

$$i_D = f(v_{GS}, v_{DS}) \tag{7-63}$$

If small increments are given to both voltages, the resulting small increment in the current can be expressed symbolically as

$$\Delta i_D \approx di_D = \frac{\partial i_D}{\partial v_{GS}}\, dv_{GS} + \frac{\partial i_D}{\partial v_{DS}}\, dv_{DS} \tag{7-64}$$

The partial derivatives have the dimensions of conductance, and the differentials are small-signal components of voltage and current; thus (7-64) can be written more compactly as

$$i_d = g_m v_{gs} + g_d v_{ds} \tag{7-65}$$

or

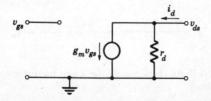

Fig. 7-21 Small-signal model for the FET including the effect of drain voltage on drain current.

$$i_d = g_m v_{gs} + \frac{1}{r_d} v_{ds} \qquad (7\text{-}66)$$

This equation, obtained rigorously, is identical with Eq. (7-62), and it corresponds to the model shown in Fig. 7-21.

If the functional relation in Eq. (7-63) is known in analytic form, the parameters g_m and g_d can be evaluated as the partial derivatives in Eq. (7-64). Alternatively, the partial derivatives show that g_m is equal to the slope of the transfer characteristic for the transistor at the quiescent operating point and that g_d is equal to the slope of the drain characteristic at the operating point. Thus the parameters can be determined from measured characteristic curves. In practice, however, it is more common to measure the parameters directly with the transistor operating under small-signal conditions.

Another second-order effect, which takes on first-order importance when the transistor is used to amplify radio-frequency signals, is associated with the charge stored in the space-charge layers at the gate junctions. The space-charge layers are shown as shaded regions in Fig. 7-22; this figure illustrates conditions inside the transistor when the channel is pinched off. Any change in the voltage drop across the junctions causes a change in the thickness of the space-charge layers, and this change in turn is associated with a change in the unneutralized space charge stored on each side of the junction. To change the thickness of the space-charge layers, majority carriers must flow into or out of the transistor at the transistor terminals, a fact that is discussed in detail in Sec. 5-2.

The charge stored in the space-charge layers on the left of the pinchoff point in Fig. 7-22 must change when the gate voltage changes, and to effect this change in stored charge, current must flow through the transistor between the gate and source terminals. The charge stored in the space-charge layers on the right of the pinchoff point must change

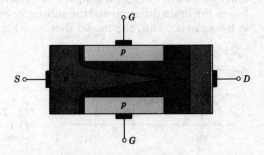

Fig. 7-22 The geometry of a junction FET.

when either the gate or the drain voltage is changed, and to effect this change, current must flow through the transistor between the gate and the drain terminals. The current that flows as a result of a change in terminal voltage is given by Eq. (5-7) as

$$i = C_j \frac{dV'}{dt} \tag{7-67}$$

where V' = reverse voltage across the junction
C = incremental junction capacitance

The incremental capacitance is a constant for small variations of voltage about a fixed operating point. These considerations imply that incremental capacitances must be added to the small-signal model for the FET as shown in Fig. 7-23 to account for the currents that flow at the terminals when the thickness of the space-charge layers changes. Capacitance C_{gd} accounts for the current required to charge the portion of the layers on the right of the pinchoff point in Fig. 7-22, and C_{gs} accounts for the current that charges the portion of the layers on the left of the pinchoff point.

When the transistor is used to amplify radio-frequency signals, the reactances of these parasitic junction capacitances become small, and they tend to short-circuit the input signal voltage. Thus the voltage gain is reduced, and at sufficiently high frequencies there is no gain at all. In small transistors designed for high-frequency applications, C_{gs} and C_{gd} are usually about equal, and their values range from about 1 to 5 pF. To keep these capacitances small, the volume of the space-charge layers to be charged must be kept small. This fact means that the physical dimensions of the transistor must be kept as small as possible.

MOS transistors also have space-charge layers that change with applied voltage, and similar capacitive effects are associated with them. In addition, MOS transistors have a direct capacitance C_{ds} between drain and source; this capacitance is associated with the junction between the drain diffusion and the substrate, which is normally connected to the source. Thus the model shown in Fig. 7-23 applies also to MOS

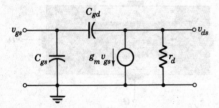

Fig. 7-23 Small-signal model for the FET including the effects of charge stored in the space-change layer.

devices when an additional capacitance C_{ds} is connected across the output terminals. MOS transistors can be made somewhat smaller than junction FETs, and the parasitic capacitances are correspondingly smaller. Typical values of C_{gd} in MOS transistors intended for high-frequency applications range from about 0.5 to 1.0 pF; C_{gs} and C_{ds} are typically 1 to 3 pF.

It is clear from the model in Fig. 7-23 that the input terminals are not isolated from the output terminals at high frequencies because of the presence of the gate-to-drain capacitance C_{gd}. This capacitance raises problems that are considerably more serious when the transistor is used with radio-frequency signals than is indicated by the model in Fig. 7-23, and every effort is made to keep C_{gd} as small as possible.

The currents that are required to charge and discharge the space-charge layers in Fig. 7-22 must flow for some distance through bulk semiconductor material, and this material does not have zero resistance. Thus there should be a resistor in series with each capacitor in the model of Fig. 7-23. However, these resistances are usually less than 100 ohms, and consequently they are negligible in comparison with the capacitive reactances except at very high frequencies. The model shown in Fig. 7-23 is usually valid at frequencies up to about 100 MHz for transistors designed to be used with high-frequency signals.

Another second-order effect in transistors is associated with the fact that the flow of drain current is not a smooth, continuous flow like the flow of a viscous fluid. The current is carried by charged particles drifting along the channel from source to drain, and these particles have random thermal velocities that on the average are far greater than the drift component of velocity (see Sec. 4-2). Thus the arrival of carriers at the drain has an element of randomness about it; it is somewhat like raindrops falling on a sidewalk. As a consequence of this fact, the drain current contains a very small, randomly varying component that is called noise. Noise currents of this kind are responsible for the sh-h-h- sound heard when a television receiver is tuned to a channel that has no station assigned to it. It is clear that this noise places a lower limit on the amplitude of signal that can be amplified successfully; if the signal amplitude is less than the noise amplitude, the signal is lost in the noise. The very faint signals returning to earth from space probes traveling to Venus and Mars present special problems in this regard, and their detection requires special low-noise amplifiers. In most applications the junction FET, whose noise is generated by the thermal-agitation process described above, generates less noise than any other transistor or vacuum tube.

7-7   THE DUAL-GATE MOS TRANSISTOR

It is pointed out in Sec. 7-6 that the drain-to-gate capacitance in the FET destroys the isolation between the input and output of the transistor at high frequencies and that it raises some serious design problems when transistors are called upon to amplify radio-frequency signals. The dual-gate MOS transistor shown in Fig. 7-24 has a geometry that is designed to minimize this capacitance. This transistor has the normal n-type drain and source diffusions, and it has an additional n-type diffusion which forms an island in the channel between the source and the drain. Thus the channel is broken into two portions covered by separate gates as shown in Fig. 7-24. Gate 1 is the signal gate, and it is used in the same way as the signal gate in the basic MOS transistor. Gate 2 is biased with a fixed positive voltage, usually about half of the quiescent drain voltage. As a result of the separation between the drain and the signal gate and of the electrostatic shielding provided by gate 2, a substantial reduction in C_{g1d} is realized. Values of C_{g1d} as low as 0.01 pF have been achieved, a fact which makes the dual-gate transistor very attractive for use in radio-frequency amplifiers.

The dual-gate transistor consists simply of two MOS transistors connected in series. Thus the quiescent operating conditions in this device are very similar to those in the two-transistor amplifier shown in Fig. 7-11. There are four different modes of operation, depending on whether or not each channel is pinched off. In normal operation, both channels are pinched off. The current in each channel depends on the potential difference between its gate and the source end of that channel. Since the channels are connected in series, they must carry the same current, and if the channels are identical, this fact requires

$$V_{G2I} = V_{G1S} \tag{7-68}$$

where V_{G2I} is the potential difference between gate 2 and the channel island. Charge accumulates on the island in the proper amount to make

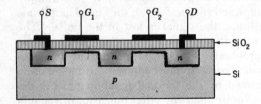

Fig. 7-24   The structure of a dual-gate MOS transistor.

the potential of the island have the value required to satisfy Eq. (7-68). If gate 1 is biased at −2 volts and gate 2 is biased at 6 volts, for example, then the potential of the island must be 8 volts. If an input signal applies a small positive increment of voltage to gate 1, there must be a negative increment of voltage at the island having the same magnitude.

It follows from these facts that the drain current in the dual-gate transistor is controlled by the signal gate G_1 and that the drain current obeys the same law as the drain current of the single-gate transistor. Thus the small-signal model for the dual-gate structure has the same form as the model in Fig. 7-23 with an additional capacitor C_{ds} connected from drain to source to account for the output capacitance of the MOS structure.

7-8 SUMMARY

The concept of enduring importance in this chapter is that of the small-signal model for the transistor. This model provides a linear representation for the nonlinear transistor operating under small-signal conditions, and it thereby makes possible the application of the powerful and extensive theory of linear circuits to the analysis and design of transistor amplifiers.

The concept of the model is of far greater importance than can be indicated by this chapter. Whether they are aware of it or not, engineers represent the entire physical world by models of one sort or another. In electrical engineering, every resistor, every transformer, and every transistor, along with the rest of the system components, is represented by a model for the purpose of circuit analysis and design. These models are of varying degrees of complexity, and they represent the physical device with varying degrees of accuracy. The models usually represent the device only for a restricted set of operating conditions. For example, a physical resistor burns up and perhaps even explodes if too much voltage is applied to it, but the model—Ohm's law or the zigzag line used to symbolize it—does not indicate this fact. Thus the models commonly used represent the physical devices approximately under a limited range of operating conditions. In the early days of electronics the models for electronic devices (vacuum tubes) were called equivalent circuits, but in recent years the term models has gained preference, to shift the emphasis from the equivalence to the approximate nature of the representation.

Chapters 4 through 7 are concerned with the pn junction under reverse-bias conditions, and they present a number of valuable applications for the pn junction operated in this way. Chapter 8 begins a study of the junction under forward-bias conditions. This study develops additional important applications for the junction, and it leads to the bipolar transistor, the most important of all transistors.

PROBLEMS

7-1. *MOST amplifier design.* The parameters of the n-channel enhancement MOS transistor in the amplifier of Fig. 7-25 are $V_T = 3$ volts and $K = 0.1$ mA/volt^2. The circuit is to be designed so that the quiescent drain current is $I_D = 2$ mA.

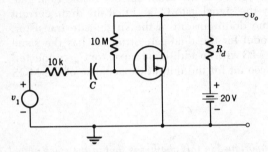

Fig. 7-25 Circuit for Prob. 7-1.

(a) Determine the value of R_d required.

(b) Determine the quiescent power drawn from the 20-volt power supply and the quiescent power dissipated in the transistor.

(c) Assuming that the signal current in the 10-megohm bias resistor is negligibly small and that C acts as a short circuit for the signal voltage, calculate the small-signal voltage gain.

7-2. *Two-stage-amplifier study.* The enhancement transistors in the amplifier of Fig. 7-26 are identical, and their parameters are $V_T = 3$ volts and $K = 0.1$ mA/volt^2.

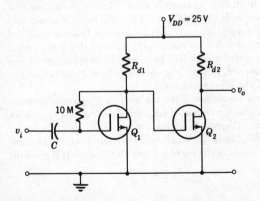

Fig. 7-26 Two-stage amplifier for Prob. 7-2.

(a) Determine the value of R_d required to make $I_{D1} = 2.5$ mA.

(b) If $R_{d2} = R_{d1}$, what are the values of V_{GS2}, V_{DS2}, and I_{D2}? (These values can be determined without making any calculations.)

(c) Under the conditions of parts a and b, what is the overall small-signal voltage gain v_o/v_i?

7-3. Amplifier-design considerations. The small-signal voltage gain of the amplifier in Fig. 7-25 depends on the value chosen for the quiescent drain current I_D and on the power-supply voltage V_{DD}. The consequences of this dependence are to be examined.

(a) Sketch and dimension a reasonably accurate curve of gain K_v as a function of I_D. Use the power-supply voltage and the transistor parameters given in Prob. 7-1. *Suggestion:* Calculate K_v for 1-mA increments in I_D between 1 and 4 mA.

(b) Comment on the compromise between gain and distortion that the design engineer must face.

(c) Comment on the relations among gain, distortion, and power-supply voltage.

7-4. JFET-amplifier analysis and design. The components R_g and V_{GG} in the circuit of Fig. 7-27 represent the Thévenin equivalent circuit for R_1, R_2, and V_{DD} in Fig. 7-9; with $V_{GG} = 0$, the circuit reduces to the one in Fig. 7-5. The parameters of the transistor are $I_{DSS} = 8$ mA and $V_P = -4$ volts. The circuit is to be designed to have its quiescent operating point at $V_{DS} = 6$ volts and $I_D = 2$ mA with $V_{GG} = 0$, $V_{DD} = 20$ volts, and $R_g = 1$ megohm.

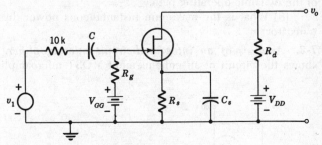

Fig. 7-27 Amplifier for Prob. 7-4.

(a) Specify the values of R_s and R_d required.

(b) Sketch and dimension a reasonably accurate drain characteristic for $v_{GS} = 0$. Construct the static load line and the dynamic operating path for the amplifier on this sketch. Give the values of voltage and current at which these lines intersect the coordinate axes.

(c) Assuming that C and C_s act as short circuits for signal components of current, determine the small-signal voltage gain of the amplifier. (The results obtained in Chap. 6 can be used for this calculation.)

7-5. Bias-circuit design. The components R_g and V_{GG} in Fig. 7-27 represent the Thévenin equivalent circuit for R_1, R_2, and V_{DD} in Fig. 7-9. The circuit in Fig. 7-9 is to be designed so that $R_g = 1$ megohm and $V_{GG} = 2.5$ volts when $V_{DD} = 20$ volts. Determine the required values of R_1 and R_2.

7-6. JFET-amplifier design. The parameters of the transistor in an amplifier like the one shown in Fig. 7-27 are $I_{DSS} = 6$ mA and $V_P = -4$ volts. The bias resistor is chosen to be $R_s = 3.3$ kilohms to provide suitable stability of the quiescent operating point, and the power-supply voltage is $V_{DD} = 20$ volts. The circuit is to be designed for a quiescent point at $I_D = 1.5$ mA and $V_{DS} = 6$ volts.

(a) Determine the required values of V_{GG} and R_d.

(b) If $R_g = 1$ megohm and if C and C_s act as short circuits to the signal components of current, what is the small-signal voltage gain of the amplifier? (The results of Chap. 6 can be used for this calculation.)

7-7. Amplifier with a reactive load. The transistor in the amplifier of Fig. 7-10b has the parameters $I_{DSS} = 5$ mA and $V_P = -3$ volts. The drain-circuit resistance is $R_d = 3$ kilohms. The input signal voltage has the waveform shown in Fig. 7-10a, and the period of this wave is long enough to permit the steady state to be reached during each half cycle.

(a) Determine the coordinates (voltage and current) of each corner of the dynamic operating path.

(b) What is the maximum instantaneous power dissipated in the transistor?

7-8. Analysis of an integrated-circuit microamplifier. Figure 7-28 shows the circuit of an enhancement MOST microamplifier with self-

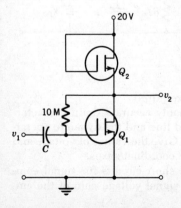

Fig. 7-28 Microamplifier for Prob. 7-8.

bias. (The capacitor C is a discrete component separate from the amplifier.) The threshold voltage for both transistors is $V_T = 3$ volts, and Q_1 is characterized by $K_1 = 0.2$ mA/volt². The dimensions of the transistors in micrometers are $W_1 = 250$, $L_1 = 10$, $W_2 = 25$, and $L_2 = 75$.

(a) Determine the small-signal voltage gain of the amplifier. The signal current flowing in the 10-megohm bias resistor can be neglected.

(b) Determine the quiescent drain current and drain-to-source voltage for each transistor.

7-9. *Analysis of a microamplifier.* The circuit configuration shown in Fig. 7-29 is sometimes used in MOST integrated circuits. The enhancement transistors have the same threshold voltages V_T, and they are char-

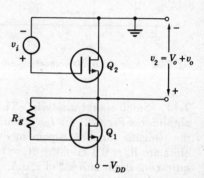

Fig. 7-29 Microamplifier for Prob. 7-9.

acterized by parameters K_1 and K_2. Let $K_1/K_2 = C^2$ and determine the small-signal voltage gain v_o/v_i for the circuit in terms of C. A procedure similar to the one leading to Eq. (7-24) can be used. *Note:* The voltage gain of this circuit is less than unity.

7-10. *Electronic-voltmeter study.* Figure 7-30 shows the circuit of an elementary dc electronic voltmeter. The resistance of the dc microammeter M is negligibly small, and the parameters of the transistor are $I_{DSS} = 4$ mA and $V_P = -3$ volts.

(a) The meter reads zero when $i_D = 1$ mA. What value of R_s is needed to make the meter read zero when the input voltage is $v_x = 0$?

(b) The meter gives full-scale deflection when $i_m = 200$ μA, and this value of i_m corresponds to $i_D = 1.3$ mA when the calibrate potentiometer is set for 0 ohms. What value of v_x is required to produce full-scale deflection of the meter when R_s has the value found in part a and the calibrate potentiometer is set for 0 ohms?

(c) How much current does the instrument draw from the source of v_x when it is indicating full scale?

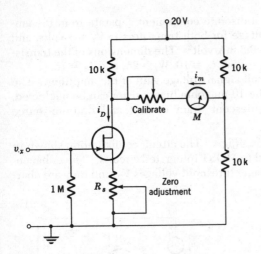

Fig. 7-30 A dc electronic voltmeter for Prob. 7-10.

7-11. *Small-signal analysis.* The parameters of the transistor in the amplifier of Fig. 7-27 are $I_{DSS}=8$ mA, $g_{mo}=4$ mmhos, and $r_d=30$ kilohms; the parasitic junction capacitances are negligible. The circuit parameters are $R_d=6.8$ kilohms, $R_g=1$ megohm, and R_s is chosen to give a quiescent drain current of 2 mA. The coupling and bypass capacitors act as short circuits to signal components of current.

    (*a*) Give a small signal model for the amplifier.
    (*b*) Determine the value of g_m at the stated quiescent drain current.
    (*c*) Determine the small-signal voltage gain of the circuit.

7-12. *Small-signal analysis of a MOST amplifier.* The transistors used in an amplifier like the one shown in Fig. 7-26 are identical, and they have the parameters $K=0.2$ mA/volt², $V_T=3$ volts, and $r_d=25$ kilohms. The parasitic transistor capacitances are negligibly small. The drain-circuit resistances are $R_{d1}=R_{d2}=8.6$ kilohms, and V_{DD} is adjusted to make $I_{D1}=I_{D2}=2$ mA. The input coupling capacitor acts as a short circuit and the 10-megohm bias resistor acts as an open circuit for signal components of voltage.

    (*a*) Give a small-signal model for the circuit.
    (*b*) Determine the value of g_m at the stated quiescent drain current, and calculate the overall small-signal voltage gain v_o/v_i.

7-13. *Small-signal models.* There are important applications for each of the circuit configurations shown in Fig. 7-31. The transistors are biased so that there is no gate current, and they are characterized by the small-signal parameters g_m and r_d. The parasitic junction capacitances

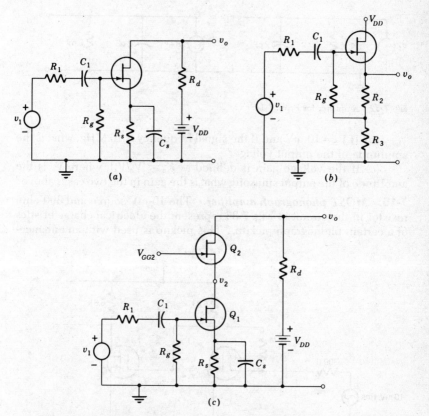

Fig. 7-31 Amplifier configurations for Prob. 7-13. (*a*) Grounded-source amplifier; (*b*) source follower; (*c*) cascode amplifier.

can be neglected. The voltage V_{GG2} in Fig. 7-31*c* is a dc bias voltage.

Give a small-signal model for each circuit. Show clearly the controlling voltage for each controlled source. Do not assume that the bypass and coupling capacitors are short circuits.

7-14. *High-frequency effects.* Figure 7-32 shows the small-signal model of an amplifier using a dual-gate MOS transistor. The 2-pF capacitors represent the parasitic transistor capacitances C_{gs} and C_{ds}; it is assumed that C_{gd} is negligibly small for the dual-gate device. The transconductance is $g_m = 1.5$ mmhos. The input signal is a sinusoidal voltage $v_1 = V_1 \cos \omega t$, and under small-signal conditions, the output voltage is also sinusoidal.

(*a*) If $V_1 = 10$ mV and if the signal frequency is 1 kHz, what is the amplitude of the sinusoidal output voltage?

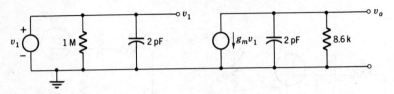

Fig. 7-32 Amplifier for Prob. 7-14.

(b) If $V_1 = 10$ mV and if the signal frequency is 10 MHz, what is the amplitude of the output voltage?

(c) If the voltage gain is defined as $K_v = |V_o/V_1|$, where V_o is the amplitude of the output sinusoid, what is the gain in the two cases above?

7-15. *MOST phonograph amplifier.* The 10-mV source and 300-ohm resistor in the circuit of Fig. 7-33 represent the electrical characteristics of a certain phonograph pickup. This pickup is used with an enhance-

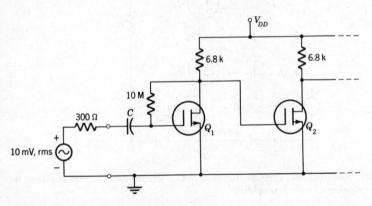

Fig. 7-33 Phonograph amplifier for Prob. 7-15.

ment MOST amplifier to obtain a signal voltage of at least 1.0 volt, rms. The transistor parameters are $g_m = 1.2$ mmhos and $r_d = 30$ kilohms; the parasitic transistor capacitances are negligible. The coupling capacitor C is a short circuit for signal components of current, and the 10-megohm bias resistor is an open circuit for signal currents.

How many identical stages like those shown in Fig. 7-33 must be connected in cascade to provide at least 1.0 volt, rms, at the output of the amplifier? Small-signal operation can be assumed.

REFERENCES

1. Sevin, L. J., Jr.: "Field-effect Transistors," McGraw-Hill Book Company, New York, 1965.

The pn Junction with Forward Bias

C*hapters 5 to 7 present a number of important applications for the pn junction operating under reverse-bias conditions. This chapter considers the junction operating with a forward bias, and it leads to further uses of the junction in a new variety of electronic devices. When a forward bias is applied to the junction, new physical processes are initiated in the vicinity of the junction. These processes are involved with the flow of carriers across the junction and with the associated injection of minority carriers into the regions on each side of the junctions. The primary objective of this chapter is to examine these new physical processes and to develop the volt-ampere law for the junction under forward-bias conditions. Another objective is to show that the junction serves as a real diode having characteristics that approximate those of the ideal diode quite well for many purposes. This study also provides the foundation for the presentation of the bipolar junction transistor in Chap. 9.*

8-1 A QUALITATIVE VIEW OF THE pn JUNCTION WITH FORWARD BIAS

Figure 8-1a shows a pn junction with a forward-bias current I applied at its terminals. Figure 8-1b shows the volt-ampere characteristic for the junction. The equation for this characteristic, exclusive of the breakdown region, is

$$I = I_s(e^{\lambda V} - 1) \tag{8-1}$$

where λ is given by Eqs. (4-6) and (4-7) and has the value 40 volt^{-1} at room temperature. This equation is developed in Sec. 8-2 from a consideration of the physical processes occurring at the junction. These processes can be described with the aid of Fig. 8-1c and d.

Figure 8-1c shows the potential distribution in the vicinity of the junction with the potential of the n-type material chosen as the zero-voltage reference. The black curve shows the potential when no bias current is applied, $I = 0$, and it is the same as the curve shown in Fig.

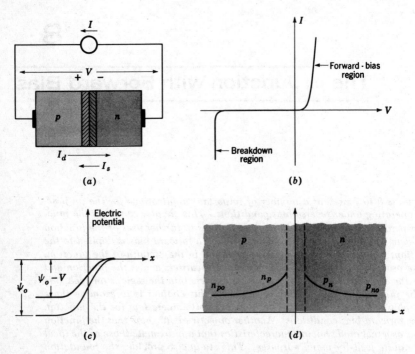

Fig. 8-1 Conditions at the pn junction with forward bias. (a) The junction; (b) the volt-ampere characteristic; (c) the potential distribution; (d) the minority-carrier distributions.

4-13.   When the applied current I is positive, representing a forward bias, electrons flow in the external circuit from the p terminal to the n terminal.   This flow constitutes a flow of majority-carrier electrons into the n region and a flow of majority-carrier holes into the p region.   This increase in majority-carrier population neutralizes some of the bound charge in the space-charge layer at the junction.   As a result the width of the layer decreases, the charge and the electric field in the layer decrease, and the height of the potential barrier at the junction decreases as shown by the colored curve in Fig. 8-1c.   Thus the potential of the p region increases, and the change in height of the potential barrier is manifest as a potential difference V at the terminals of the device as shown in Fig. 8-1a.

With the height of the potential barrier decreased, many more holes in the p region have enough kinetic energy to climb the barrier and enter the n region, and similarly, many more free electrons in the n region are able to cross the junction and enter the p region.   These carriers that cross the junction become minority carriers, and they are said to have been injected into the p and n regions by the bias applied to the junction. The zero-bias equilibrium condition, $I_d = I_s$, is destroyed by this action, as is indicated in Fig. 8-1a, and there is a net current across the junction from the p region to the n region. The equilibrium condition is now $I_d - I_s = I$, where I is the terminal current.

The injection of minority carriers into the regions on each side of the junction causes the minority-carrier concentrations to exceed the thermal-equilibrium values in the regions near the junction as shown in Fig. 8-1d. These excess carriers diffuse away from the junction toward regions of lower concentration, and in the process they recombine with majority carriers that are present in abundance. A short distance away from the junction all of the excess carriers have disappeared by recombination, and the minority-carrier concentration equals the thermal-equilibrium value.

In regions where the minority-carrier concentration equals the thermal-equilibrium value, the recombination rate equals the rate at which carriers are created by thermal ionization. In regions where the concentration exceeds the thermal-equilibrium value, the recombination rate exceeds the thermal generation rate; equilibrium exists when the excess minority carriers recombine at a rate equal to the rate at which they are injected into the region. Thus in equilibrium all of the excess carriers crossing the junction disappear by recombination, and hence the recombination process accounts for current flowing across the junction. Majority carriers must flow toward the junction to feed this recombination and to supply the carriers that are injected as minority carriers into the region across the junction. The overall picture of the carrier flow is thus the following: Majority carriers are forced into the semiconductor at the terminals by the external bias source. These carriers move toward the junction; some of them recombine with excess minority carriers as they approach the junction, and others cross the junction and recombine on the other side. The total current flow is accounted for by the recombination that takes place on each side of the junction.

The excess concentration of minority carriers near the junction represents an excess charge stored in the device as a result of the bias applied at the terminals. To meet the requirement of macroscopic space-charge neutrality, a similar excess concentration of majority carriers forms near the junction. When the bias applied at the terminals is changed, terminal current must flow to change the amount of charge stored in these excess concentrations. Thus under forward-bias conditions the pn junction exhibits a new capacitive effect in addition to the one associated with the charge stored in the space-charge layer.

The discussion presented above provides a basis for deriving the

volt-ampere law for the pn junction given by Eq. (8-1). The derivation
consists of the following three parts:

1. The junction current is expressed in terms of the rate of recombina-
    tion of excess minority carriers on both sides of the junction, and
    this rate is related to the excess charge stored outside the space-
    charge layer.
2. The excess stored charge is then related to the excess concentration
    of minority carriers at the edge of the space-charge layer.
3. The excess concentration of step 2 is then related to the voltage
    drop across the terminals of the device.

These three steps are combined to yield Eq. (8-1). With respect to understanding the properties of the forward-biased pn junction, step 1 is
the most important of the three. Steps 2 and 3 are needed to express the
excess stored charge in terms of a measurable quantity, the terminal
voltage V.

8-2 THE STATIC VOLT-AMPERE LAW OF THE JUNCTION

The first step in developing the volt-ampere law of the junction is to
relate the current across the junction, which is also the terminal current,
to the excess charge associated with the excess of minority carriers near
the junction. The concentration of minority carriers in the n region
shown in Fig. 8-1d can be written as

$$p_n = p_{no} + p_n' \tag{8-2}$$

where p_{no} = thermal equilibrium concentration of holes
      p_n' = excess concentration of injected holes

The recombination rate is proportional to the concentration of minority
carriers, and it can be expressed as

$$R = \frac{p_{no} + p_n'}{\tau_h} \tag{8-3}$$

where τ_h is the lifetime for holes in the n-type material, presented in
connection with Fig. 4-8 and Eq. (4-30). In fact, Eq. (8-3) can be taken
as the definition of the minority-carrier lifetime. The term p_{no}/τ_h on the
right-hand side of Eq. (8-3) is the recombination rate under thermal
equilibrium, and it is counterbalanced by an equal thermal-generation
rate. Thus the net recombination rate is

$$R' = \frac{p_n'}{\tau_h} \quad \text{holes/cm}^3\text{-sec} \tag{8-4}$$

Since each carrier has a charge q equal to the magnitude of the electronic charge, charge vanishes from the scene at a rate of

$$qR' = q\,\frac{p'_n}{\tau_h} \qquad \text{coul/cm}^3\text{-sec} \tag{8-5}$$

The quantity qp'_n is the charge per cubic centimeter, and thus Eq. (8-5) can be written as

$$qR' = \frac{\rho'_h}{\tau_h} \tag{8-6}$$

where ρ'_h is the excess charge density associated with the excess holes injected into the n-type material.

The current carried across the junction by holes equals the total rate at which excess charge disappears in the entire volume of the n-type material; thus the hole current is obtained by integrating Eq. (8-6) over the volume,

$$i_h = \int_{\text{vol}} \frac{\rho'_h}{\tau_h}\,dV = \frac{1}{\tau_h}\int_{\text{vol}} \rho'_h\,dV \tag{8-7}$$

But the integral of the charge density over the volume is simply the total charge contained in the volume, and thus Eq. (8-7) becomes

$$i_h = \frac{q'_h}{\tau_h} \tag{8-8}$$

where q'_h is the total excess charge associated with the holes injected into the n-type material. By a similar analysis, the current carried across the junction by electrons injected into the p-type material is

$$i_e = \frac{q'_e}{\tau_e} \tag{8-9}$$

where q'_e = magnitude of excess charge stored in p-type material
τ_e = lifetime of electrons in p-type material

The total current crossing the junction is equal to the current at the terminals, and it is given by

$$i = i_h + i_e = \frac{q'_h}{\tau_h} + \frac{q'_e}{\tau_e} \tag{8-10}$$

Three important conclusions about the pn junction can be drawn from this simple analysis:

1. The current across the junction is a linear function of the excess charge stored if the minority-carrier lifetimes are constant.

2. To change the junction current, the excess stored charge must be changed. This charge is located outside the space-charge layer and is in addition to the charge stored in the layer.
3. To make rapid changes in junction current possible, the stored charge must be as small as possible. Thus the volume of the device and the minority-carrier lifetimes must be made small.

The device designer has control over the lifetimes, and he uses this fact to control the charge stored in the device. As is mentioned in connection with Eq. (4-30), doping the silicon crystal with gold atoms is an effective way of reducing the lifetimes without altering the other properties of the crystal appreciably; thus gold doping is often used in the fabrication of high-speed junctions.

The junction current is given by Eq. (8-10) in terms of the stored charges. Although this equation is very informative, the charges are generally unknown, and (8-10) cannot be used to calculate the current. The charges can be determined by evaluating the integral in Eq. (8-7) if an equation for the minority-carrier distribution shown in Fig. 8-1*d* is known. If the length of the semiconductor material from the junction to the terminals is long enough for all the excess minority carriers to disappear by recombination before they reach the terminal, this equation is [1]

$$p'_n = p'_n(0) \exp \frac{-x}{L_h} \tag{8-11}$$

where $p'_n(0)$ = excess minority-carrier concentration at edge of space-charge layer, $x = 0$

$\quad\quad L_h$ = constant called diffusion length for holes

A similar equation holds for electrons in the *p*-type material. Multiplying (8-11) by the magnitude of the electronic charge q yields the charge density which is then substituted into Eq. (8-7) and integrated to obtain

$$q'_h = qAL_h p'_n(0) \tag{8-12}$$

where A is the area of the junction. Similarly, the magnitude of the excess charge stored in the *p*-type material is

$$q'_e = qAL_e n'_p(0) \tag{8-13}$$

where $n'_p(0)$ is the excess concentration of electrons at the edge of the space-charge layer in the *p*-type material.

Equations (8-12) and (8-13) give the excess stored charges in terms of $p'_n(0)$ and $n'_p(0)$, and it is now necessary to relate these quantities to the voltage applied at the terminals. For the condition of no voltage applied to the terminals, Eq. (4-35) gives

$$p_{no} = p_{po} \exp(-\lambda \psi_o) \qquad (8\text{-}14)$$

where ψ_o is the height of the potential barrier at the junction. When a voltage is applied to the terminals, either as a forward bias or as a reverse bias, this relation becomes [1]

$$p_n(0) = p_{po} \exp[-\lambda(\psi_o - v)] \qquad (8\text{-}15)$$

where $\psi_o - v$ is the height of the potential barrier at the junction with a voltage v applied at the terminals. Rearranging this expression and making use of Eq. (8-14) yields

$$p_n(0) = p_{no}e^{\lambda v} \qquad (8\text{-}16)$$

and

$$p_n'(0) = p_n(0) - p_{no} = p_{no}(e^{\lambda v} - 1) \qquad (8\text{-}17)$$

Similarly,

$$n_p'(0) = n_{po}(e^{\lambda v} - 1) \qquad (8\text{-}18)$$

Equations (8-17) and (8-18) can be combined to form the law of the junction as follows:

$$\frac{p_n'(0)}{p_{no}} = e^{\lambda v} - 1 = \frac{n_p'(0)}{n_{po}} \qquad (8\text{-}19)$$

The carrier concentrations involved in this law are illustrated in Fig. 8-2.

Substituting the law of the junction into Eqs. (8-12) and (8-13) yields

$$q_h' = qAL_h p_{no}(e^{\lambda v} - 1) = q_{ho}(e^{\lambda v} - 1) \qquad (8\text{-}20)$$

and

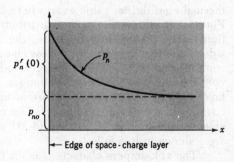

Fig. 8-2 Diagram illustrating the quantities in Eq. (8-19).

$$q'_e = qAL_e n_{\nu o}(e^{\lambda v} - 1) = q_{eo}(e^{\lambda v} - 1) \tag{8-21}$$

Thus the junction current given by Eq. (8-10) becomes

$$i = \left(\frac{q_{ho}}{\tau_h} + \frac{q_{eo}}{\tau_e}\right)(e^{\lambda v} - 1) \tag{8-22}$$

$$= I_s(e^{\lambda v} - 1) \tag{8-23}$$

When v is more negative than about -0.1 volt, the exponential term in Eq. (8-23) is much smaller than unity, and the theoretical reverse-bias current across the junction is $i = -I_s$. The value of I_s for small silicon junctions is of the order of 10^{-15} amp. In real silicon junctions under reverse-bias conditions, however, there are additional currents resulting from leakage paths across the surface of the crystal and from other surface effects. Thus the junction current is more accurately written as

$$i = I_s e^{\lambda v} - I_s - I_L \tag{8-24}$$

where the leakage current I_L depends on the applied voltage and is typically of the order of 10^{-9} amp under moderate reverse-bias voltages. It follows that the term $-I_s$ in (8-24) is negligible under all operating conditions. Under reverse-bias conditions the exponential term is negligible, and the current is $-I_L$. Under forward-bias conditions with i greater than about 0.1 μA, the exponential term dominates, and the current is given by

$$i = I_s e^{\lambda v} \tag{8-25}$$

For most engineering purposes Eq. (8-25) gives the junction current with sufficient accuracy under all operating conditions.

Minority-carrier distributions near a *pn* junction are shown in Fig. 8-3*a* under forward-bias conditions. There is an excess of carriers near the junction, and there are excess charges q'_e and q'_h associated with the excess carriers. Figure 8-3*b* shows the carrier distributions when no bias is applied to the junction; in this case the concentrations equal the thermal equilibrium value everywhere outside the space-charge layer. Figure 8-3*c* shows the carrier distributions under reverse-bias conditions. The reverse bias depresses the carrier concentrations on each side of the junction, and the excess charge stored near the junction becomes a deficiency. When the reverse-bias voltage exceeds about 0.1 volt, the minority-carrier concentrations at the edge of the space-charge layer are essentially zero, and further increases in reverse bias do not change the carrier distributions. The charges q_{ho} and q_{eo} introduced as constants in Eqs. (8-20) and (8-21) are proportional to the shaded areas shown in Fig. 8-3*c*.

The volt-ampere characteristic for the *pn* junction, corresponding

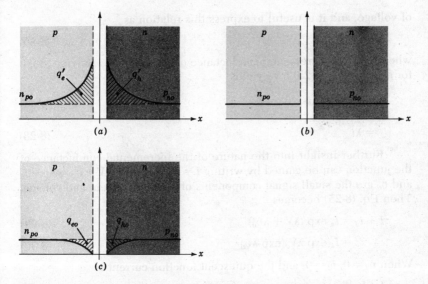

Fig. 8-3 Minority-carrier distributions near a pn junction. (a) With forward bias; (b) in thermal equilibrium with no bias applied; (c) with reverse bias.

to Eq. (8-25), is shown in Fig. 8-4. The ratio of junction current to junction voltage is a conductance, and it is equal to the slope of the chord drawn from the origin of the coordinates to the characteristic as shown in Fig. 8-4. However, as the applied voltage is changed the value of this conductance varies over such a wide range that it is not a very useful concept. On the other hand, as in the case of the transistors in Chap. 7, the junction is often operated with small variations of voltage and current superimposed on constant dc components. If the variations are sufficiently small, the increment of current is proportional to the increment

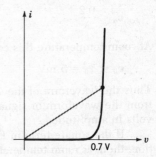

Fig. 8-4 A silicon-junction characteristic.

of voltage, and it is useful to express this relation as

$$di = g \, dv \tag{8-26}$$

where g is the incremental conductance of the junction. Solving (8-26) for g and using (8-25) for i yields

$$g = \frac{di}{dv} = \lambda I_s e^{\lambda v} \tag{8-27}$$

$$= \lambda i \tag{8-28}$$

Further insight into the nature of the incremental conductance of the junction can be gained by writing $i = I + i_d$ and $v = V + v_d$, where i_d and v_d are the small signal components of junction current and voltage. Then Eq. (8-25) becomes

$$I + i_d = I_s \exp (\lambda V + \lambda v_d) \tag{8-29}$$

$$= I_s(\exp \lambda V)(\exp \lambda v_d) \tag{8-30}$$

When $v_d = 0$, $i_d = 0$, and the quiescent junction current is

$$I = I_s e^{\lambda V} \tag{8-31}$$

Substituting this relation into (8-30) and expanding the second exponential in its power series yields

$$I + i_d = I + I[\lambda v_d + \tfrac{1}{2}(\lambda v_d)^2 + \cdots] \tag{8-32}$$

Thus

$$i_d = I[\lambda v_d + \tfrac{1}{2}(\lambda v_d)^2 + \cdots] \tag{8-33}$$

For small-signal operation, the terms in the power series beyond the square term are negligible, and if the square term in (8-33) is to be less than 10 percent of the linear term, it is necessary that

$$\tfrac{1}{2}(\lambda v_d)^2 < 0.1(\lambda v_d) \tag{8-34}$$

or

$$v_d < \frac{0.2}{\lambda} \tag{8-35}$$

At room temperature this condition corresponds to

$$v_d < \tfrac{0.2}{40} = 5 \text{ mV} \tag{8-36}$$

Thus the waveform of the signal current may be considerably different from the waveform of signal voltage if the voltage exceeds a few millivolts in amplitude.

If the square term in Eq. (8-33) is neglected in small-signal operation, then, at room temperature,

$$\frac{v_d}{i_d} = r = \frac{1}{\lambda I} = \frac{25}{(I)_{\text{mA}}} \text{ ohms} \tag{8-37}$$

Thus for small-signal operation with a quiescent current I of 10 mA, the incremental resistance of the junction is 2.5 ohms. It follows from Eq. (8-37) that the junction can be used as a current-controlled or voltage-controlled resistor. The resistance range that can be covered by changing the applied bias is from about 1 ohm to several megohms in the case of small junctions. The lower limit is set by the bulk resistance of the silicon crystal, and the upper limit is set by leakage currents. Junctions are used in this way to implement automatic-gain-control features in communication systems; however, when they are used in such applications, they must be located at a point in the system where the signal level is less than 1 mV to avoid waveform distortion.

8-3   DYNAMIC BEHAVIOR OF THE *pn* JUNCTION

When a bias voltage is applied to the *pn* junction, excess charges are stored in the semiconductor material on each side of the junction. The behavior of the junction is intimately related to these charges; for example, Eq. (8-10) shows that the current across the junction is directly proportional to the stored charge. In addition, the dynamic behavior of the junction is affected by the stored charge. The total excess charge stored near the junction is

$$q = q_h' + q_e' \tag{8-38}$$

and substituting Eqs. (8-20) and (8-21) into (8-38) yields

$$q = (q_{ho} + q_{eo})(e^{\lambda v} - 1) \tag{8-39}$$

$$= q_o(e^{\lambda v} - 1) \tag{8-40}$$

Thus the stored charge is an exponential function of the voltage applied to the junction, and a graph of this relation is shown in Fig. 8-5. Any change in applied voltage requires a change in stored charge which in turn requires a current flowing at the terminals of the device. In addition, a change in applied voltage causes a change in the thickness of the space-charge layer at the junction, and it thus causes a change in the charge stored in the layer. Therefore, under dynamic conditions Eq. (8-25) for the terminal current of the device becomes

$$i = I_s e^{\lambda v} + \frac{dq}{dt} + \frac{dq_j}{dt} \tag{8-41}$$

where q_j = charge stored in space-charge layer
q = charge stored outside space-charge layer

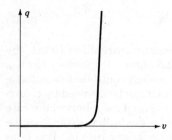

Fig. 8-5 Charge-voltage characteristic for charge stored near the *pn* junction in the form of excess minority carriers.

Both q and q_j are nonlinear functions of the junction voltage v. The nonlinear relation between q_j and v is discussed in some detail in Sec. 5-2. In order to gain additional insight, the relation between q and v is approached in a slightly different way. The charge q can be written as

$$q = Cv \tag{8-42}$$

where C is a nonlinear capacitance that is a strong function of the voltage v. Thus the second term in Eq. (8-41) becomes

$$\frac{dq}{dt} = C\frac{dv}{dt} + v\frac{dC}{dt} \tag{8-43}$$

An alternative and more useful approach is to write

$$\frac{dq}{dt} = \frac{dq}{dv}\frac{dv}{dt} \tag{8-44}$$

The factor dq/dv has the dimensions of capacitance, and it is helpful to identify it as such by writing

$$\frac{dq}{dt} = C_d\frac{dv}{dt} \tag{8-45}$$

The capacitance C_d is an incremental capacitance associated with the charge stored outside the space-charge layer at the junction, and it is often called the diffusion capacitance of the junction. It is a function of the junction voltage and is equal to the slope of the charge-voltage characteristic shown in Fig. 8-5. However, *pn* junctions are often operated with small signal variations of voltage and current superimposed on constant dc components. If the signal components are small enough, the diffusion capacitance C_d can be treated as a constant. Since Eq. (8-40) has the same form as Eq. (8-23), the small-signal conditions established in Sec. 8-2 for the incremental junction resistance apply also to the incremental diffusion capacitance.

The total junction capacitance is

$$C_t = C_d + C_j \tag{8-46}$$

where C_j is developed in Sec. 5-2 and accounts for the third term in Eq. (8-41). The diffusion capacitance is

$$C_d = \frac{dq}{dv} = \lambda q_o e^{\lambda v} = \lambda (q + q_o) \tag{8-47}$$

in which the derivative is evaluated with the aid of Eq. (8-40). Under forward-bias conditions q is large and C_d is large, often much larger than C_j. Under reverse-bias conditions, $q = -q_o$, and $C_d = 0$.

Under large-signal conditions involving forward bias, the charge stored at the junction is a highly nonlinear function of the junction voltage. In this case the concept of junction capacitance can be quite misleading, and its use is not recommended under these conditions.

Some interesting and important features of the *pn* junction under large-signal conditions can be developed with the aid of Eq. (8-40). Solving this equation for the junction voltage yields

$$v = \frac{1}{\lambda} \ln \left(\frac{q}{q_o} + 1 \right) \tag{8-48}$$

This relation yields immediately the following important fact: After the application of forward bias, a reverse-bias voltage ($v < 0$) cannot be built up until all of the stored charge q has been removed from the device and q becomes negative. When the junction is used as a diode or as a switch, this fact limits the speed at which it can be changed from the conducting to the nonconducting state. The dynamic behavior of the junction in a switching application is illustrated in Fig. 8-6. Initially the circuit

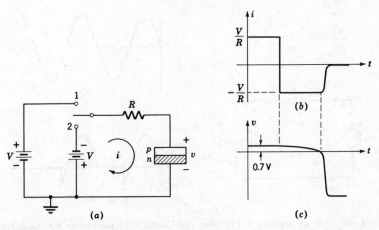

Fig. 8-6 Dynamics of the *pn* junction. (*a*) Circuit; (*b*) junction current; (*c*) junction voltage.

is in the steady state with the switch in position 1. Thus a forward bias is applied to the junction, and a small forward voltage, about 0.7 volt for silicon, exists across the junction. If the applied voltage V is much larger than 0.7 volt, then the current is approximately V/R, as shown in Fig. 8-6b. Then the switch is thrown abruptly to position 2. The charge stored at the junction cannot change abruptly, and Eq. (8-48) states that therefore the voltage across the junction cannot change abruptly. Thus just after the switch is thrown, a reverse current $-V/R$ flows through the device, as shown in Fig. 8-6b. This reverse current extracts the stored charge from the junction, and according to Eq. (8-48), $v = 0$ when $q = 0$. Reverse current continues to flow for a brief period after $q = 0$ to give q a small negative value and to charge the space-charge layer up to a voltage equal to the applied reverse-bias voltage. Once q has been reduced to zero, the buildup of reverse bias can be extremely fast, a fact that has important engineering applications. More details on the dynamics of this switching action can be found in Chap. 5 of Ref. 1.

Figure 8-7 shows the *pn* junction used as a diode rectifier with a sinusoidal supply voltage. As the supply voltage goes through its positive values, the diode conducts and accumulates stored charge. When the supply voltage goes negative, a negative current flows through the diode until all the stored charge is removed as shown in Fig. 8-7c. When the stored charge is removed, the current drops quickly to zero, and a reverse-bias voltage builds up quickly across the diode. The reverse current that flows through the diode reduces its rectification ef-

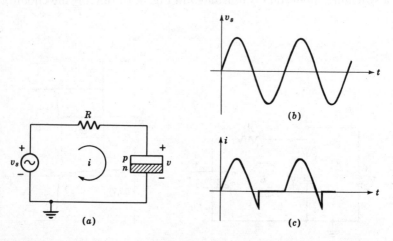

Fig. 8-7   The *pn* junction as a rectifier.   (*a*) Circuit; (*b*) input-voltage waveform; (*c*) junction-current waveform.

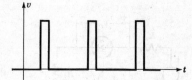

Fig. 8-8 Waveform of the output of a pulse
generator.

ficiency. If the frequency of the supply voltage is high, the reverse current may flow for an appreciable fraction of the negative half cycle, and
the performance of the rectifier may be quite poor.

On the other hand, the sharp drop in diode current that occurs
each cycle when the diode cuts off has valuable engineering applications. If this current (or the voltage across the junction) is applied to
a differentiating circuit, a train of very short voltage pulses like the one
shown in Fig. 8-8 can be generated. Junctions designed for optimum
behavior in this kind of operation are called step-recovery diodes or
snap-off diodes. One commercial pulse generator using this principle
generates pulses that are almost 1 volt in amplitude and that are less
than 70 psec in duration. Such pulse trains are important in signal-
processing operations that involve high-speed switching. Moreover,
the Fourier series for such a pulse train is very rich in harmonics; in
many cases, useful harmonics out to the hundredth are generated.
There are also important applications for these harmonic voltages.

In order for a *pn* junction to be capable of fast switching and good
rectification efficiency, it must have a small stored charge. This condition is achieved by keeping the volume of the device as small as possible and, as is indicated by Eq. (8-10), by making the lifetimes of the
minority carriers small. The use of gold doping to control the lifetimes
is discussed in connection with Eq. (8-10).

8-4 THE *pn* JUNCTION AS A DIODE

Under reverse-bias conditions the silicon *pn* junction is an open circuit
for most engineering purposes. Under forward-bias conditions it is not
a short circuit, but the voltage drop across it is less than 1 volt in all
normal operating conditions. Thus the junction is a useful approximation to the ideal diode, and it is used extensively in all of the diode circuits presented in Chap. 3. Figure 8-9*a* shows a silicon diode used in a
simple half-wave rectifier with a supply voltage

$$v_s = V_s \sin \omega t \qquad\qquad (8\text{-}49)$$

The symbol used for the diode is intended to indicate that it is a real
diode, not an ideal diode. It should be noted that in general this prac-

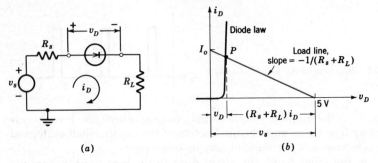

Fig. 8-9 Graphical analysis of a diode circuit. (*a*) Circuit; (*b*) graphical construction.

tice is not followed in the engineering literature; the ideal-diode symbol of Chap. 3 is also used to represent real diodes. However, this fact causes no serious problems because the meaning of the symbol is normally clear from the context in which it is used.

The behavior of the half-wave-rectifier circuit is governed by the diode law

$$i_D = I_s \exp \lambda v_D \tag{8-50}$$

and by the loop equation

$$v_s = v_D + (R_s + R_L)i_D \tag{8-51}$$

Equation (8-50) is a relation between i_D and v_D imposed by the diode; (8-51) is a relation imposed by the circuit connected to the diode; both relations must be satisfied simultaneously. However, since (8-50) is a nonlinear transcendental equation, graphical methods must be used to obtain the simultaneous solution. Graphical methods of analysis also frequently provide valuable insights into circuit properties that are difficult to obtain by other means. The diode law is given graphically by the diode characteristic curve shown in Fig. 8-9*b*. The loop equation given by (8-51) can be rewritten in the form

$$i_D = \frac{v_s}{R_s + R_L} - \frac{1}{R_s + R_L} v_D \tag{8-52}$$

$$= I_o - \frac{1}{R_s + R_L} v_D \tag{8-53}$$

For any fixed value of v_s, I_o is fixed, and (8-53) describes a straight line on the i_D-v_D coordinates. This line, called the load line, is shown in

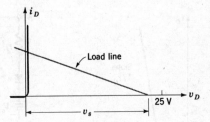

Fig. 8-10 Graphical analysis of the circuit in Fig. 8-9 when the input voltage is large.

Fig. 8-9b for one value of v_s. All combinations of i_D and v_D that satisfy the diode law lie on the diode characteristic curve, and all combinations that satisfy the loop equation lie on the load line. The two equations are satisfied simultaneously at point P where the two curves intersect. Point P is called the operating point for the circuit.

As time passes and v_s goes through its cycle of values, the load line shifts horizontally in accordance with the value of v_s at each instant, and the operating point moves along the diode characteristic as the load line shifts. The corresponding values of i_D can be read from the scale of ordinates, and the waveform of i_D can be plotted if desired. The load voltage at each instant can be calculated from the corresponding value of i_D.

The graphical construction shown in Fig. 8-9 is appropriate for conditions in which the supply voltage v_s does not exceed 5 volts. For substantially larger supply voltages a different voltage scale must be used, and the graphical construction takes the form illustrated in Fig. 8-10. Since the voltage drop across the diode is less than 1 volt, it appears to be virtually zero in this case, and it follows that the silicon diode can be treated as an ideal diode with good accuracy insofar as load voltage and current are concerned. That is, the ideal diode is a model that represents the silicon diode under these operating conditions.

Simple circuit models that represent the silicon diode to a good approximation with supply voltages of only a few volts also exist. The procedure used in developing these models is shown in Fig. 8-11. The basic step in the procedure is the approximation of the diode characteristic by two straight-line segments as shown in Figs. 8-11a and c. Characteristics consisting of a series of straight-line segments are called piecewise-linear characteristics; they are used in approximating the characteristics of a variety of nonlinear devices. The value of the piecewise-linear approximation is the fact that it greatly simplifies the analysis of many electronic circuits and gives results that are in good agreement with the actual circuit performance.

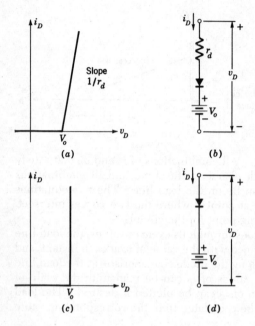

Fig. 8-11 Piecewise-linear models for the *pn* junction.

The circuit in Fig. 8-11*b* consists of an ideal diode in series with a resistor and a voltage source. It is readily seen that the volt-ampere characteristic of this circuit corresponds exactly to the piecewise-linear characteristic in Fig. 8-11*a*; thus the circuit is a piecewise-linear model for the silicon diode and, more generally, for the silicon *pn* junction. The parameters V_o and r_d in Fig. 8-11*a* are chosen to give the best fit between the diode characteristic and the piecewise-linear approximation; these parameters are then used in the piecewise-linear model for the diode.

Figure 8-11*c* and *d* shows a slightly coarser model for the silicon diode; it is the same as the model in Fig. 8-11*b* with $r_d = 0$. This simple model is entirely adequate for most diode applications, and it is very widely used to represent silicon junctions in general. The value of V_o is usually chosen to be 0.7 volt for silicon, but the value can be adjusted to meet the requirements of any particular application. When silicon diodes are used in the circuits of Chap. 3, it is very convenient for purposes of analysis to represent the diode by this model.

When the silicon diode is forward biased, it accumulates a store of

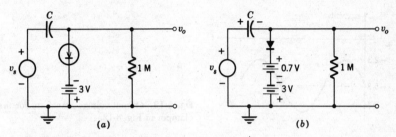

Fig. 8-12 Diode clamper for Example 8-1. (*a*) Circuit; (*b*) piecewise-linear model for the circuit.

excess charge, and this excess charge affects the dynamic behavior of the diode in the manner described in Sec. 8-3. When the diode is used under the small-signal operating conditions described in Sec. 8-3, the stored charge can be accounted for by a capacitance, given by Eq. (8-46), connected across the terminals of the model in Fig. 8-11*b*. Under large-signal conditions such as exist in a rectifier, however, the concept of capacitance is not useful; the behavior of the diode under these conditions is discussed in connection with Figs. 8-6 and 8-7.

Example 8-1

Figure 8-12*a* shows the circuit of a clamper using a silicon diode, and Fig. 8-12*b* shows a model for the circuit in which the diode is represented by the model in Fig. 8-11*d* with $V_o = 0.7$ volt. The problem is to find the output voltage when the input voltage is

$$v_s = 3 \sin \omega t$$

Solution The voltage drop across the diode when it is conducting subtracts from the voltage of the 3-volt battery; thus it affects the clamping level. The ideal diode in the model of Fig. 8-12*b* does not let the output voltage become more positive than −2.3 volts, and therefore the positive peaks of the input signal are clamped at −2.3 volts.

In more detail, the capacitor C charges up to a voltage

$$V = 3 + 3 - 0.7 = 5.3 \text{ volts}$$

with the polarity shown in the diagram. If C is large enough, this voltage remains essentially constant throughout the cycle of the input voltage. Thus the output voltage is

$$v_o = v_s - V = 3 \sin \omega t - 5.3 \qquad \text{volts}$$

The waveform of this voltage is shown in Fig. 8-13.

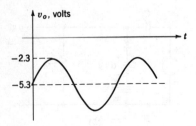

Fig. 8-13 Output-voltage waveform for the clamper in Fig. 8-12.

8-5 SUMMARY

This chapter develops the volt-ampere law for the pn junction. The results show that, among its many important applications, the junction provides a useful approximation to the ideal diode of Chap. 3.

One of the central ideas developed in this chapter is the notion of the excess minority-carrier charge stored at the junction and the relation of this charge to the junction current. Since the bipolar junction transistor of Chap. 9 uses a forward-biased junction, this idea is important to the understanding of that device. The idea is not important in the case of the FETs of Chap. 6 because minority carriers do not play a significant role in those devices.

A second important concept developed in this chapter is that of the piecewise-linear model for nonlinear devices. This model is used widely to simplify the analysis and design of large-signal nonlinear circuits.

PROBLEMS

8-1. *Junction volt-ampere law.* A certain silicon *pn* junction is described by Eq. (8-23) with $I_s = 10^{-15}$ amp and $\lambda = 40$ volt^{-1}. Calculate and plot on graph paper the volt-ampere characteristic. Do not use broken scales for *v* or *i*. *Suggestion:* Calculate *i* for $v = 0.6, 0.65, 0.7, 0.725,$ and 0.75 volt. The purpose of this problem is to emphasize the great rate at which the current increases with increases in forward bias and to illustrate the fact that a turn-on threshold voltage must be exceeded before appreciable current flows.

8-2. *Junction resistance.* A certain silicon junction is operated with small signal variations in voltage and current superimposed on dc components. The incremental resistance of the junction is to be varied between 10 ohms and 100 kilohms by varying the dc component of junction current. The junction is characterized by $I_s = 10^{-15}$ amp and $\lambda = 40$ volt^{-1}.

(*a*) Over what range must the dc component of current be varied to obtain the specified range of resistance?

(*b*) What is the corresponding range of the dc component of junction voltage?

8-3. *Junction current density.* A certain pn junction is rectangular in shape with sides measuring 7×10^{-4} cm and 30×10^{-4} cm. What is the current density flowing across this junction when the junction current is 10 mA? Give the answer in amperes per square centimeter.

8-4. *Stored charge.* A pn junction used in the circuit of Fig. 8-6 produces the current waveform shown in Fig. 8-14 when the switch is thrown from position 1 to position 2. The n side of the junction is doped

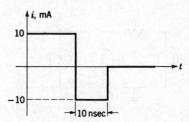

Fig. 8-14 Waveform for Prob. 8-4.

much more heavily than the p side. Thus the junction current is carried almost entirely by electrons, and it is given by Eq. (8-10) with $q'_h = 0$. For the purposes of this problem, the charge stored in the space-charge layer can be neglected, and $q_{eo} \ll q'_e$.

(*a*) Assuming that negligible charge is lost by recombination during the 10-nsec reverse-recovery interval, how much charge was stored at the junction just before the switch was thrown?

(*b*) Determine τ_e, the lifetime for electrons on the p side of the junction.

(*c*) From the results of part b, what can be said about the assumption made in part a?

8-5. *Piecewise-linear model.* A certain pn junction having the volt-ampere characteristic shown in Fig. 8-15 is to be represented by the piecewise-linear model in Fig. 8-11b.

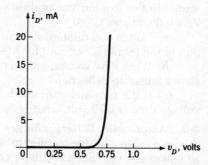

Fig. 8-15 Diode characteristic for Prob. 8-5.

(a) Choose values of V_o and r_d to be used in the model for a good representation at currents up to $i_D = 20$ mA.

(b) Repeat part *a* for diode currents up to 4 mA.

8-6. *Diode circuit analysis.* The silicon diodes in the circuit of Fig. 8-16 can be represented by a piecewise-linear model with $V_o = 0.7$ volt and $r_d = 0$. The input voltage is $v_s = 10 \sin \omega t$ volts. Sketch and dimension the waveform of the output voltage v_o.

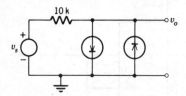

Fig. 8-16   Diode circuit for Prob. 8-6.

8-7. *Microammeter protection circuit.* Figure 8-17 shows two silicon diodes used to protect a delicate microammeter from damage by excessive current. The meter shows full-scale deflection when $i_m = 200$ μA,

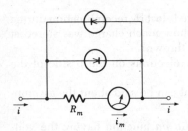

Fig. 8-17   Circuit for Prob. 8-7.

and the meter resistance R_m is 1.5 kilohms. The diodes can be represented in this low-current application by a piecewise-linear model with $V_o = 0.6$ volt and $r_d = 0$.

(a) Sketch and dimension a curve of the meter current i_m as a function of the terminal current i for i between ± 1 mA.

(b) What bias voltage appears across the diodes when the meter shows full-scale deflection?

(c) If the terminal current i becomes very large so that the diode voltage drop is 0.75 volt, what is the meter current i_m?

8-8. *Automobile battery charger.* Figure 8-18 shows four silicon diodes used in a bridge rectifier to charge an automobile battery from an ac power source. Such rectifiers are used with automobile alternators

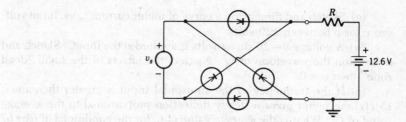

Fig. 8-18 Battery charger for Prob. 8-8.

(ac generators) to charge the battery, but the automobile system is complicated somewhat by the action of the automatic voltage regulator which controls the charging rate. The diodes can be represented by a piecewise-linear model with $V_o = 0.7$ volt and $r_d = 0$; the input voltage is $v_s = 50 \sin \omega t$ volts.

(a) What value of R is needed to limit the peak instantaneous battery current to 30 amp?

(b) With this value of R, sketch and dimension the waveform of battery current. What percentage of the time does charging current flow through the battery?

(c) What is the peak inverse voltage that appears across the diodes?

8-9. Voltmeter rectifier. Figure 8-19 shows a circuit using two silicon diodes that permits alternating voltages to be measured with a dc microammeter. The resistance of the meter is negligibly small, and the diodes can be represented by a piecewise-linear model with $V_o = 0.6$ volt and $r_d = 0$.

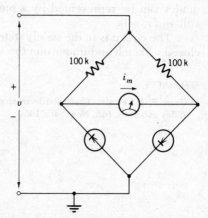

Fig. 8-19 Circuit for Prob. 8-9.

(*a*) Sketch and dimension a curve of meter current i_m vs. input voltage v for v between ± 20 volts.

(*b*) A voltage $v = 20 \sin \omega t$ volts is applied at the input. Sketch and dimension the waveform of i_m. Neglect the effects of the small "dead zone" near $v = 0$.

(*c*) If the frequency of the sinusoidal input is greater than about 15 Hz, the meter gives a steady deflection proportional to the average value of i_m. What is the average value of i_m for the conditions of part *b*?

(*d*) Sketch the waveform of i_m when the input is $v = 5 + 10 \sin \omega t$ volts. Dimensions are not required. *Note:* The 100-kilohm resistors are often replaced with capacitors to remove any dc component that exists in the input voltage.

8-10. *Monopulser analysis.* Figure 8-20 shows the monopulser of Fig. 3-18 with silicon diodes in place of the ideal diodes. The silicon

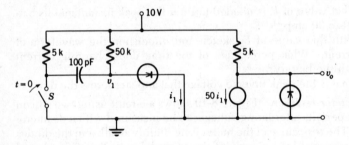

Fig. 8-20 Monopulser for Prob. 8-10.

diodes can be represented by a piecewise-linear model with $V_o = 0.7$ volt and $r_d = 0$.

The circuit is in the steady state with S open for $t < 0$. At $t = 0$, S is closed. Sketch and dimension the waveforms of v_1 and v_o for $t > 0$.

REFERENCES

1. Gray, P. E., et al.: "Physical Electronics and Circuit Models of Transistors," John Wiley and Sons, Inc., New York, 1964.

9

The Bipolar Junction Transistor as an Amplifier

*T*his chapter presents the bipolar junction transistor (BJT). The BJT is the most important of all the electronic amplifiers. As an illustration of its importance it is probably fair to say that today's giant high-speed computers would not be possible without the BJT, and without the computers the program to explore outer space could not be undertaken. In this application alone the BJT has had a major impact on the economy of the nation. What the future holds with BJTs, along with FETs, in integrated circuits is largely unknown at the time this is being written, but the prospects are exciting and the engineers participating in this development will have stimulating and rewarding opportunities.

The physical processes occurring inside the BJT involve both minority and majority carriers, whereas only majority carriers play a significant role in the FETs of Chap. 6. This fact is responsible for the word bipolar in the name of the device, and for the same reason the FET was known at one time as a unipolar transistor. In comparison with the FET, the BJT can provide substantially greater gain, and its performance at very high frequencies is somewhat better than that of the FET. On the other hand, the BJT generates more waveform distortion than the FET in many applications, and it has a low input impedance which, for practical reasons, is very troublesome in many cases. Thus each type of transistor has applications in which it excels, and optimum design often results in both types being used in one piece of equipment.

The bipolar transistor has been developed to the point where its performance is near the theoretical limits. The field-effect transistor has not reached this state of development, and its performance is limited by the technology used in its design and fabrication. Thus the FET has the potential for further improvement, but whether it will be able to match, or exceed, the performance of the BJT cannot be foreseen at this time. Engineers participating in the further development of the FET may have some exciting discoveries in store for them.

9-1 INTRODUCTION TO THE BJT

Figure 9-1a shows a pictorial representation of the BJT in the basic amplifier circuit, and Fig. 9-1b shows the symbolic representation of the same circuit. The npn transistor consists of a thin slice of p-type silicon sandwiched between two pieces of n-type silicon, and the three regions are separated by two pn junctions. The behavior of the transistor depends on the flow of current across these junctions.

In normal amplifier operation the input voltage v_s in Fig. 9-1a is positive, and the lower junction is biased in the forward direction. Hence a net flow of holes is injected across the junction from p to n, and a net flow of electrons is injected across the junction from n to p. The electron component of this current is the important component, and hence the lower region of n material is called the emitter. The central slice of p material is called the base of the transistor. The electrons injected from the emitter into the base are minority carriers in the base, and they behave much as electrons generated in the base by thermal ionization. They diffuse through the base away from the emitter junction in a random manner, and some of them recombine with holes that are present in great quantity in the base. In normal amplifier operation the upper junction in Fig. 9-1a is biased in the reverse direction; hence

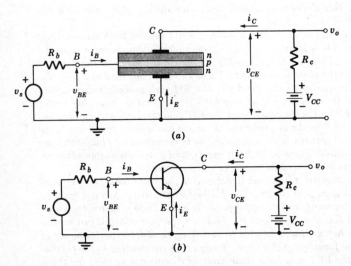

Fig. 9-1 The basic BJT amplifier. (*a*) Pictorial representation; (*b*) symbolic representation.

any free electrons in the base that diffuse to the edge of the upper junction are swept across it into the upper piece of n material, called the
collector, creating a reverse current across the upper junction. If the
width of the base is very small, almost all the electrons injected into the
base by the emitter are collected by the collector, and the collector current i_C is approximately equal to the electron current flowing across the
emitter junction. The base current i_B supplies the hole current that
flows across the emitter junction from base to emitter, and it also supplies the holes that are required to feed the recombination of carriers
that takes place in the base. The hole current across the junction can
be made much smaller than the electron current by doping the emitter
much more heavily than the base, and the base current needed to feed
recombination can be made small by making the base width small. Thus
it is possible for a small input current i_B to control a much larger output
current i_C, and current amplification is obtained.

These relations are illustrated in more detail in Fig. 9-2. The current i_{CE} is the current associated with electrons flowing across the base
from the emitter to the collector; this current constitutes the output current of the transistor, and it is the desired component of current. The
current i_{BB} is the current associated with the recombination of injected
electrons in the base; it is an undesired current, and in a sense it represents a defect in the device. The current i_{BE} is associated with the holes
injected into the emitter by the forward bias across the emitter junction;
it also is an undesired, or defect, current. The terminal currents can be
expressed as

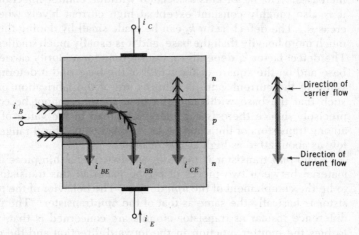

Fig. 9-2 Components of current in the BJT.

$$i_C = i_{CE} \tag{9-1}$$

and

$$i_B = i_{BB} + i_{BE} \tag{9-2}$$

It also turns out, as is shown in Sec. 9-2, that the three components of current shown in Fig. 9-2 are proportional to each other, and thus

$$i_{BB} = k_1 i_{CE} = k_1 i_C \tag{9-3}$$

and

$$i_{BE} = k_2 i_{CE} = k_2 i_C \tag{9-4}$$

Substituting these relations into Eq. (9-2) yields

$$i_B = (k_1 + k_2) i_C \tag{9-5}$$

or

$$i_C = \frac{1}{k_1 + k_2} i_B = \beta i_B \tag{9-6}$$

where

$$\beta = \frac{1}{k_1 + k_2} \tag{9-7}$$

is the current amplification factor for the transistor.

The factor k_1 is associated with the base-recombination defect of the transistor; it is roughly constant except at low current levels where it increases. The factor k_2 is associated with the emitter-injection defect; it is also roughly constant except at high current levels where it increases. The defect factor k_2 can be made small by doping the emitter much more heavily than the base, and it is usually much smaller than k_1. The defect factor k_1 depends on the lifetime for minority carriers in the base and on the square of the width of the base, and it determines the value of the current gain β. The nature of the fabrication process is such that the base width and the base lifetime cannot be controlled precisely; hence there is considerable spread in the value of β, even among transistors of the same type. Typical values for β range from as low as about 20 to as high as 500 or more.

The *pnp* transistor is made by sandwiching a thin piece of *n*-type material between two pieces of *p*-type material; this transistor is said to be the complement of the *npn* device. The behavior of the *pnp* transistor is basically the same as that of the *npn* transistor. The principal difference insofar as transistor circuits are concerned is that, in order to bias the emitter junction in the forward direction and the collector junction in the reverse direction, the voltages applied to the *pnp* transistor must be opposite in polarity from those shown in Fig. 9-1. However, the behavior of the two transistors is essentially identical insofar as the signal components of voltage and current are concerned. In fact, an *npn* transistor in an amplifier circuit can be replaced by a *pnp* transistor of similar ratings provided that the polarities of all bias batteries are reversed and that the connections to all polarity-sensitive devices such as diodes and electrolytic capacitors are reversed.

The arrowhead at the emitter in the symbolic representation of the transistor in Fig. 9-1*b* indicates the direction of forward current for the emitter junction. Thus it always points from *p* to *n*, and it indicates whether the transistor is an *npn* or a *pnp* device.

The volt-ampere characteristics for a real transistor are shown in Fig. 9-3. The input characteristic shown in Fig. 9-3*a* is simply the exponential volt-ampere characteristic of a *pn* junction. The output, or collector, characteristic is shown in Fig. 9-3*b*; since these curves are nearly uniformly spaced for uniform increments of i_B, it follows that the current gain β is nearly constant. However, these curves deviate somewhat from the simple theory presented above in that the collector current is not constant for constant i_B but increases slowly with increasing collector-to-emitter voltage. This and other second-order effects are discussed in Sec. 9-3.

In normal amplifier operation the emitter junction is forward biased, and, as shown in Fig. 9-3, v_{BE} is nearly constant at the value V_o. Then in the amplifier circuit of Fig. 9-1,

$$v_s = V_o + R_s i_B \tag{9-8}$$

and from Eq. (9-6),

$$i_C = \beta i_B \tag{9-9}$$

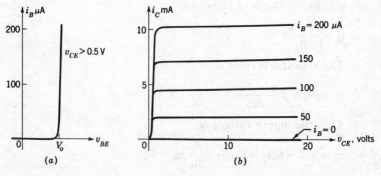

Fig. 9-3 BJT characteristics. (*a*) Input characteristic; (*b*) output, or collector, characteristic.

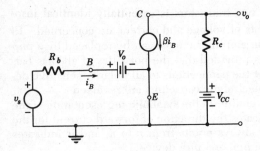

Fig. 9-4 A circuit model for the amplifier of Fig. 9-1.

These two equations imply that in normal amplifier operation the circuit in Fig. 9-1 can be represented by the model shown in Fig. 9-4. This simple model is an adequate representation of the BJT for many purposes, and it is extensively used by design engineers, especially in the initial stages of the design.

It is instructive to compare this model for the transistor with the ideal current amplifier of Fig. 2-3. The principal differences are that the input voltage to the transistor is not zero and that the input current i_B cannot be negative. Apart from these differences, the BJT is a good approximation to the ideal current amplifier.

Example 9-1

The performance of an amplifier having a model like the one shown in Fig. 9-4 is to be examined under the condition that $R_c = 15$ kilohms, $R_b = 50$ kilohms, $\beta = 50$, $V_{CC} = 25$ volts, and $v_s = 2$ volts.

Solution   The base current is

$$i_B = \frac{v_s - V_o}{R_b}$$

The collector current is

$$i_C = \beta i_B = \beta \frac{v_s - V_o}{R_b}$$

Thus the output voltage is

$$v_o = V_{CC} - R_c i_C = V_{CC} - \beta \frac{R_c}{R_b}(v_s - V_o)$$

$$= V_{CC} + \beta \frac{R_c}{R_b} V_o - \beta \frac{R_c}{R_b} v_s$$

The output voltage is therefore a linear function of the input signal voltage v_s, and the small change in output voltage caused by a small change in input voltage is

$$\Delta v_o = -\beta \frac{R_c}{R_b} \Delta v_s$$

The voltage gain of the amplifier is thus

$$K_v = \beta \frac{R_c}{R_b}$$

Substituting the given numerical values into these relations yields

$$i_B = \frac{2 - 0.7}{50} = 26 \ \mu A$$

$$i_C = 50 i_B = 1.3 \text{ mA}$$

$$v_o = V_{CC} - R_c i_C = 25 - (15)(1.3) = 5.5 \text{ volts}$$

and

$$K_r = 50 \ \tfrac{15}{50} = 15$$

As an aid to understanding the requirements that must be met in fabricating a good transistor, the conditions existing inside a certain *npn* device are illustrated in Fig. 9-5. Figure 9-5b shows the electric potential distribution, and Fig. 9-5c shows the concentration of minority carriers in the device. In silicon transistors under normal amplifier operating conditions, the thermal equilibrium concentrations of minority carriers are orders of magnitude smaller than the excess concentrations resulting from the injection of carriers across the emitter junction; thus the thermal equilibrium concentrations do not show in Fig. 9-5c, and the total concentrations are essentially equal to the excess concentrations. The dashed curve in Fig. 9-5b shows the potential distribution when no bias is applied at the terminals of the device, and the solid curve shows the potential when normal amplifier bias is applied. The forward bias applied to the emitter junction reduces the height of the potential barrier at this junction, with the result that excess holes are injected into the emitter and excess electrons are injected into the base as shown in Fig. 9-5c. The electrons injected into the base move across the base toward the collector junction, and many of them recombine with holes on the way. In fact, the minority-carrier distribution curve for the base of this device shows that all of the injected carriers disappear by recombination before they reach the collector junction. Thus the collector current is not influenced by the forward bias at the emitter junction, and

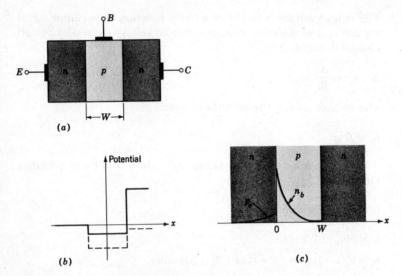

Fig. 9-5 Potential and minority-carrier distributions in an *npn* device. (*a*) The device; (*b*) electric-potential distribution; (*c*) minority-carrier distribution.

there is no transistor action. The base-recombination defect factor k_1 in Eq. (9-7) is very large for this device, and the current gain β is essentially zero.

To obtain efficient transistor action, the recombination of minority carriers in the base must be kept small by making the base width W much smaller than that indicated in Fig. 9-5. If all of the carriers injected into the base recombine before reaching the collector, and if the base is doped with a uniform concentration of impurity atoms, then the minority-carrier concentration in the base is given by Eq. (8-11) as

$$n_b(x) = n_b(0) \exp \frac{-x}{L_{eb}} \tag{9-10}$$

where $n_b(0) \approx n_b'(0)$ = excess concentration of minority carriers in base at edge of emitter junction

L_{eb} = diffusion length for electrons in base

Thus in order to have a small loss of carriers by recombination in the base, the base width W must be much smaller than L_{eb} so that the entire base extends over only a tiny portion at the beginning of the exponential distribution curve shown in Fig. 9-5c. Typical values of W for transistors designed to operate at low power levels range from 0.3 to 1.0 μm. For a comparison, the wavelength of visible light is in the range from 0.4 to

0.72 μm, and the thickness of the paper on which this page is printed is roughly 100 μm. The transverse dimensions of such a transistor are usually on the order of 100 μm. The fabrication of transistors is thus concerned with the precise control of the geometry of a device having dimensions on the micrometer scale, and the problems encountered are by no means trivial.

Figure 9-6a shows a cross-sectional view of a silicon epitaxial planar diffused BJT, and Fig. 9-6b shows a top view of the same device. The collector junction is formed by etching a window in the oxide layer and diffusing a suitable impurity, usually boron, into the crystal to form the p-type base. The oxide is then regrown, a new window is cut, and the n-type emitter is formed by a second diffusion. The depths of the junctions below the crystal surface, and hence the base width, are controlled by the temperature and duration of the diffusions. A typical transistor of this type has a rectangular emitter junction with sides measuring 7.0 and 30 μm, and it has a base width, measured under the emitter junction, of 0.5 μm.

Transistors can be fabricated directly in a suitable substrate wafer without the preliminary step of growing an epitaxial layer. However, the high-resistivity epitaxial layer provides a high breakdown voltage and a small capacitance for the collector junction, whereas the low-resistivity substrate under the layer provides a low-resistance path to the collector terminal. The various processes for making planar diffused BJTs are discussed in detail in Ref. 1 listed at the end of this chapter.

The diffusion process used in forming the base of the transistor in Fig. 9-6 results in a base that is not uniformly doped; the base is more heavily doped on the emitter side and less heavily doped on the collector

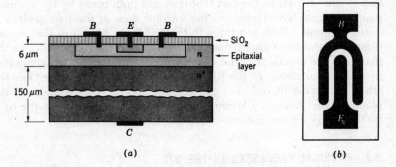

Fig. 9-6 Structure of an epitaxial planar diffused BJT. (a) Cross-sectional view; (b) top view.

side. Such transistors are referred to as graded-base transistors. Transistors having uniformly doped bases are made by starting with a suitable p-type wafer and forming the two junctions by simultaneous diffusions from opposite faces of the wafer. Such transistors, called hometaxial-base transistors, have collector-junction breakdown characteristics that make them well suited for high-power applications.

The npn sandwich constituting the transistor in Fig. 9-1 is still an npn sandwich if the emitter and collector connections are interchanged, and in principle it should also be a transistor in this inverted connection. However, it is not a very good transistor when used in this manner. Figure 9-6a shows that the collector and emitter junctions are quite different in area, and, moreover, in order to optimize the performance of the device in the normal mode, the emitter is the most heavily doped region and the collector is the most lightly doped region. As a result of these facts the silicon planar transistor usually has a current gain less than unity in the inverted mode. It should be noted, however, that the inverted mode has useful applications in some switching circuits, and a few transistors are designed specifically to be operated in this mode.

The silicon wafer on which the transistor is made usually has room for about 1,000 transistors. Thus if 100 wafers are processed simultaneously, 100,000 transistors are made simultaneously. As a result of this batch-processing, the cost of a transistor on the wafer amounts to 2 or 3 cents. If the transistors are to be cut apart and sold as discrete units, the sale price and the life expectancy of the devices are determined by the testing and packaging operations. If, on the other hand, the transistor is to be left on the silicon chip as part of an integrated circuit, its cost is very small.

The geometry of the BJT shown in Fig. 9-6a is very similar in many respects to the geometry of the junction FET shown in Fig. 6-6a, and this is due in part to the fact that they are both made by the diffused planar technology. However, the current flow in the two devices is quite different. Both junctions in the JFET are biased in the reverse direction, and the current flow through the channel is parallel to the planes of the junctions. The current in the JFET is carried altogether by majority carriers. In the BJT one junction is forward biased and the other is reverse biased, and the current flow is perpendicular to the planes of the junction. Minority carriers play a major role in the BJT, for they carry the current across the base of the transistor.

9-2 PHYSICAL PROCESSES IN THE BJT

In Chap. 8 it is shown that the current across a pn junction is directly proportional to the excess charge stored near the junction. A similar analysis applied to the BJT shows that the transistor currents also are proportional to the excess charge stored in the device, and this fact leads to valuable new insights into the nature of the transistor. The transistor is shown in Fig. 9-7a; this illustration is intended to represent a small portion of the active region of the transistor under the emitter junction in Fig. 9-6a, and it neglects the small end effects in the regions on either side of the emitter junction. Figure 9-7b shows the minority-carrier distributions under normal amplifier conditions in a transistor with a uniformly doped base. The forward bias at the emitter junction causes excess concentrations of carriers to appear on each side of the junction in accordance with the law of the junction, Eq. (8-19), and the reverse bias at the collector junction forces the minority-carrier concentration to zero at the collector junction. With a very narrow, uniformly doped base, the carrier-distribution curve for the base region is essentially a straight line as shown in Fig. 9-7b.

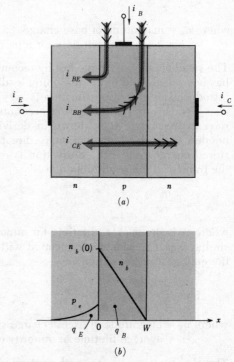

Fig. 9-7 Current flow in the BJT. (a) Components of current; (b) minority-carrier distributions in a homogeneous-base transistor. (The thermal equilibrium concentrations are too small to show.)

The total excess charge stored in the base of the transistor as a result of injection from the emitter is designated q_B in Fig. 9-7b, and the excess charge stored in the emitter is designated q_E. The requirement of space-charge neutrality is satisfied by majority carriers flowing into the transistor at the base and emitter terminals, and thus any change in stored charge must be accompanied by a charging current at the base and emitter terminals. The charge stored in the base is a central quantity in determining the behavior of the transistor. This charge is associated with minority carriers in transit across the base carrying the current i_{CE} shown in Fig. 9-7a. The time required for a carrier to cross the entire base from the emitter junction to the collector junction is called the transit time. In a time interval equal to the transit time, the total base charge q_B moves across the base, and the associated current is

$$i_{CE} = \frac{q_B}{\tau} \tag{9-11}$$

where q_B = magnitude of base charge
  τ = average transit time for minority carriers crossing base

The small amount of charge lost by recombination of carriers in the base has been neglected in the reasoning leading to Eq. (9-11), but the error introduced by this approximation is very small.

The current i_{BB} in Fig. 9-7a accounts for the recombination of carriers in the base. It is shown in deriving Eq. (8-8) that the current needed to feed recombination is directly proportional to the excess stored charge, and if the derivation is repeated for the base region of the transistor, the result obtained is

$$i_{BB} = \frac{q_B}{\tau_b} \tag{9-12}$$

where τ_b is the average lifetime for minority carriers in the base. In a similar way i_{BE}, which is associated with recombination of carriers in the emitter, can be written as

$$i_{BE} = \frac{q_E}{\tau_e} \tag{9-13}$$

where q_E = magnitude of excess charge stored in emitter
  τ_e = average lifetime of minority carriers in emitter

The charges q_B and q_E are both controlled by the same junction voltage, v_{BE}, and it follows from this fact that for a given transistor structure these charges are proportional to each other. A detailed calculation shows that

$$q_E = Kq_B = \frac{2L_{he}}{W}\frac{p_{eo}}{n_{bo}}q_B \qquad (9\text{-}14)$$

where L_{he} = diffusion length for holes in emitter
$\quad\quad W$ = base width
$\quad\; p_{eo}, n_{bo}$ = thermal equilibrium concentrations of minority carriers in emitter and base, respectively

Thus Eq. (9-13) can be rewritten as

$$i_{BE} = \frac{q_B}{\tau_e'} \qquad (9\text{-}15)$$

where τ_e' is a modified lifetime including the factor K in Eq. (9-14).

The terminal currents for the transistor can now be expressed in terms of the charge stored in the base as

$$i_C = i_{CE} = \frac{q_B}{\tau} \qquad (9\text{-}16)$$

and

$$i_B = i_{BB} + i_{BE} = \frac{q_B}{\tau_b} + \frac{q_B}{\tau_e'} \qquad (9\text{-}17)$$

$$= \frac{q_B}{\tau_B} \qquad (9\text{-}18)$$

where τ_B is an effective lifetime for the base. Thus the terminal currents are directly proportional to the charge stored in the base, and to change the currents, the stored charge must be changed. The necessity of adding or subtracting stored charge from the base limits the speed with which the terminal currents can be changed. The currents needed to change the stored charge are not included in Eqs. (9-16) and (9-18); hence these equations are valid only for static conditions. The static current gain of the transistor is given by (9-16) and (9-18) as

$$\frac{i_C}{i_B} = \beta = \frac{\tau_B}{\tau} = \frac{\text{effective lifetime}}{\text{average transit time}} \qquad (9\text{-}19)$$

Thus for a large current gain the lifetime must be much larger than the transit time; one way to achieve a small transit time is by making the base width small.

In order to get more information from Eqs. (9-16) and (9-18), it is necessary to evaluate the magnitude of the charge stored in the base of the transistor. For the conditions pictured in Fig. 9-7b, the average concentration of minority carriers in the base is $\frac{1}{2}n_b(0)$, where $n_b(0)$ is the concentration of carriers at the edge of the emitter junction. It then follows that

$$q_B = \tfrac{1}{2}AWqn_b(0) \tag{9-20}$$

where A = area of emitter junction

$\quad W$ = width of base

$\quad AW$ = volume of active base region

$\quad q$ = magnitude of electronic charge

Then from the law of the junction, Eq. (8-19), with $n_p'(0) = n_b(0)$ and $v = v_{BE}$,

$$q_B = \tfrac{1}{2}AWqn_{bo}[\exp(\lambda v_{BE}) - 1] \tag{9-21}$$

$$= q_{Bo}[\exp(\lambda v_{BE}) - 1] \tag{9-22}$$

Thus the stored charge is an exponential function of the base-to-emitter voltage. When the silicon transistor is used in normal amplifier operation, the exponential term in Eq. (9-22) is orders of magnitude greater than unity, and hence the unity term can be dropped. Making this simplification and substituting the result into Eqs. (9-16) and (9-18) yields

$$i_C = \frac{q_{Bo}}{\tau} \exp(\lambda v_{BE}) \tag{9-23}$$

$$= I_{Cs} \exp(\lambda v_{BE}) \tag{9-24}$$

and

$$i_B = \frac{q_{Bo}}{\tau_B} \exp(\lambda v_{BE}) \tag{9-25}$$

$$= I_{Bs} \exp(\lambda v_{BE}) \tag{9-26}$$

Also, Eq. (9-19) gives

$$i_C = \beta i_B \tag{9-27}$$

and from Fig. 9-7a,

$$i_E = -i_C - i_B \tag{9-28}$$

These equations describe the BJT under static conditions in normal amplifier operation, and since Eq. (9-26) is simply the volt-ampere law for a pn junction, they imply that the transistor can be represented by the circuit model shown in Fig. 9-8a. The junction diode shown in this model represents the emitter junction of the transistor. The input volt-ampere characteristic corresponding to Eq. (9-26) is shown in Fig. 9-8b. This model is a nonlinear model that represents the transistor with considerable accuracy.

The current amplification factor for the BJT is $\beta = \tau_B/\tau$, and it is worthwhile to examine this important parameter in more detail. Con-

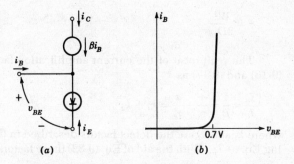

Fig. 9-8 Circuit model for the BJT with reverse bias at the collector junction. (a) Model; (b) input volt-ampere characteristic.

sider first a transistor with a uniformly doped base. In this case there is space-charge neutrality in the base, there is no electric field in the base, and the minority carriers flow across the base by diffusion. Therefore the current i_{CE} in Fig. 9-7a, which is associated with these carriers, is given by Eq. (4-9) as

$$i_{CE} = i_C = -qAD_{eb} \frac{dn_b}{dx} \qquad (9\text{-}29)$$

where the minus sign is required because the positive reference direction for i_{CE} is in the negative x direction and where

A = area of emitter junction
n_b = concentration of electrons in base

The concentration of electrons in the base is given by the linear distribution curve in Fig. 9-7b, and thus the concentration gradient is $dn_b/dx = -n_b(0)/W$. Equation (9-29) can therefore be written as

$$\begin{aligned} i_C &= -qAD_{eb}\left[-\frac{n_b(0)}{W}\right] \\ &= \frac{AD_{eb}}{W} qn_b(0) \end{aligned} \qquad (9\text{-}30)$$

Now solving Eq. (9-20) for $qn_b(0)$ and substituting the result into (9-30) yields

$$i_C = \frac{2D_{eb}}{W^2} q_B = \frac{1}{\tau} q_B \qquad (9\text{-}31)$$

Thus the average transit time for carriers crossing a uniformly doped base is

$$\tau = \frac{W^2}{2D_{eb}} \tag{9-32}$$

The reciprocal of the current amplification factor is given by Eqs. (9-16) and (9-17) as

$$\frac{i_B}{i_C} = \frac{1}{\beta} = \frac{\tau}{\tau_b} + \frac{\tau}{\tau_e'} = k_1 + k_2 \tag{9-33}$$

where k_1 and k_2 are the defect factors described in the paragraph following Eq. (9-7). With the aid of Eq. (9-32) these factors can be expressed as

$$k_1 = \frac{W^2}{2D_{eb}\tau_b} \tag{9-34}$$

and

$$k_2 = \frac{W^2}{2D_{eb}\tau_e'} \tag{9-35}$$

After a considerable amount of symbol manipulation, (9-35) can be put in the form

$$k_2 = \frac{W\rho_e}{L_{he}\rho_b} \tag{9-36}$$

where W = base width
$\quad L_{he}$ = diffusion length for holes in emitter
$\quad \rho_e, \rho_b$ = resistivity of emitter and base, respectively

For the transistor to have a high current gain, these factors must both be small, and thus it is clear that a small base width is essential for a high gain. The resistivities in Eq. (9-36) are inversely proportional to the associated doping concentrations, and they are thus under the control of the device designer; it is usually possible to make k_2 negligibly small in comparison with k_1.

Many of the results derived in this section depend on the existence of uniform doping in the base of the transistor. When the base region is formed by diffusion, as it is in most silicon transistors, a nonuniform-base, or graded-base, transistor results. The nonuniform distribution of impurities in the base has two important effects: It causes a built-in electric field to exist in the base that aids the transit of minority carriers across the base, and it causes the minority-carrier distribution in the base to depart from the straight-line distribution shown in Fig. 9-7b. Thus the graded base affects the value of the transit time given by Eq. (9-32), and it affects the value of the stored charge given by Eq. (9-20). However, it does not affect the basic current-charge relations given by Eqs. (9-16) to (9-18), nor does it affect the exponential form of the basic charge-voltage relation given by Eq. (9-22). Thus the graded base does not affect the form of the volt-ampere relations given by Eqs. (9-23) to (9-27); it merely alters the coefficients in these equations. The effect of the graded base on these coefficients turns out to be quite minor.

The important results of this section can be summarized as follows:

$$i_B = \frac{q_B}{\tau_B} \qquad (9\text{-}37)$$

$$i_C = \frac{q_B}{\tau} \qquad (9\text{-}38)$$

$$i_B = I_{Bs} \exp(\lambda v_{BE}) \qquad (9\text{-}39)$$

$$i_C = I_{Cs} \exp(\lambda v_{BE}) \qquad (9\text{-}40)$$

and

$$i_C = \beta i_B \qquad (9\text{-}41)$$

These equations do not include the components of current necessary to change the charge stored in the transistor, and thus, strictly speaking, they are valid only for static operation. In practice, however, the charging currents are negligible with sinusoidal signals having frequencies up 100 kHz or more.

The equations given above correspond to the nonlinear transistor model shown in Fig. 9-8a. The analysis and design of BJT circuits are greatly facilitated if the model is linearized in the way that the junction diode is linearized in Fig. 8-11. For this purpose the input volt-ampere characteristic is approximated by a piecewise-linear characteristic as shown in Fig. 9-9a. This characteristic implies that the transistor can be represented approximately by the piecewise-linear model in Fig.

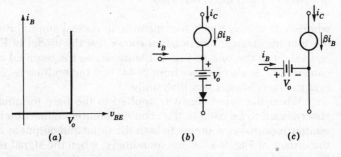

Fig. 9-9 Linearized model for the BJT with normal amplifier bias. (a) Piecewise-linear input characteristic; (b) piecewise-linear model; (c) linearized model.

9-9b; the ideal diode and battery represent an approximation for the emitter junction. This model does not impose any limitation on the input current i_B when the ideal diode is forward biased; thus it is a good representation for the transistor only when i_B comes from a suitably high-resistance source that limits the current. A source resistance greater than 5 kilohms is usually suitably high. However, the model is not such a bad representation as it might seem, for, if a 1-volt battery is connected between the base and emitter terminals of a real transistor, the transistor will be destroyed at once by excessive currents.

With normal amplifier bias applied to the transistor, the ideal diode in Fig. 9-9b acts as a short circuit, and thus it can be eliminated from the model. The voltage drop across the ideal current source βi_B does not affect the current delivered by the source; hence the lower terminal of the current source can be shifted to the other side of the battery to obtain the alternative model shown in Fig. 9-9c. This model is identical with the one developed by elementary reasoning and shown in Fig. 9-4.

There are some applications for the BJT in which the input signal is applied to the emitter terminal instead of the base terminal, and in such cases it is advantageous to have the current-controlled current source in the model expressed as a function of the emitter current. Figure 9-10a shows the model for the transistor with the input signal applied to the emitter terminal. The collector current is

$$i_C = \beta i_B \qquad\qquad (9\text{-}42)$$

and by using Kirchhoff's current law, it can be expressed as

$$i_C = \beta(-i_C - i_E) \qquad\qquad (9\text{-}43)$$

Thus

$$i_C = -\frac{\beta}{1+\beta}\, i_E = -\alpha i_E = \beta i_B \qquad\qquad (9\text{-}44)$$

where i_E is always a negative quantity in normal amplifier operation of the npn transistor. This relation shows that the model in Fig. 9-10b is equivalent to the one in Fig. 9-10a insofar as the terminal currents are concerned. It also follows from (9-44) that the emitter-to-collector current gain α is always less than unity.

When the input signal is applied to the base terminal, the transistor is said to be used in the common-emitter connection because the emitter terminal is common to both the input and output as indicated in the circuit of Fig. 9-4. Correspondingly, when the signal is applied to the emitter terminal, the transistor is used in the common-base connection. A meaningful comparison between these two connections cannot be made until some of the second-order effects in the transistor have

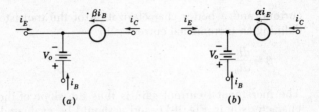

Fig. 9-10 Common-base model for the BJT. (a) Current source controlled by i_B; (b) current source controlled by i_E.

been examined. It can be stated, however, that, although the current gain in the common-base connection is always less than unity, the voltage and power gains can be substantial.

9-3 SECOND-ORDER EFFECTS IN THE BJT

The collector current in the theoretical transistor of Sec. 9-2 is directly proportional to the base current, and thus the theoretical transistor provides distortionless current amplification. The curve of collector current as a function of base current in Fig. 9-11 shows that this is not exactly the case with real transistors; the characteristic displays some curvature at low currents and again at high currents. Thus the current amplification factor $\beta = i_C/i_B$ is not a constant for real transistors. The ratio i_C/i_B is the slope of the chord drawn from the origin of coordinates in Fig. 9-11 to the characteristic, and it varies with the current level.

Transistors are often operated with small signal variations of current superimposed on fixed dc components. In such cases the operating point for the transistor makes small excursions along the characteristic in Fig. 9-11 in the vicinity of the point fixed by the dc components of

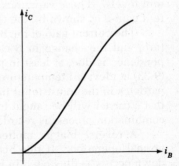

Fig. 9-11 Collector current as a function of base current for a real transistor.

current, and a better characterization of the transistor is obtained by defining an incremental current gain

$$\beta = \frac{di_C}{di_B} \tag{9-45}$$

The incremental current gain is thus the slope of the current transfer characteristic in Fig. 9-11, and it should be evaluated at the operating point established by the dc components of current. The incremental current gain is usually called the ac current gain, although the notion of alternating current has nothing to do with it. The current gain defined as

$$\beta = \frac{i_C}{i_B} \tag{9-46}$$

also has areas of application, and it is usually called the dc current gain, although again the notion of direct current is not directly pertinent. As a matter of simplicity, these two current gains are assumed to be equal in the remainder of this book.

The variations in β with current level are due mainly to variations in the lifetimes τ_b and τ_e' in Eq. (9-33). The recombination of minority carriers in the base takes place for the most part at recombination centers. At moderate and high current levels these centers are mostly occupied, and their effectiveness in promoting recombination is reduced. At low current levels many recombination centers are vacant and thus available to aid recombination; hence the effectiveness of the centers is high, and the defect factor k_1 in Eq. (9-33) increases at low current levels. At high current levels the concentration of minority carriers in the base near the emitter junction is high. The requirement of space-charge neutrality results in a corresponding increase in majority carriers at the edge of the emitter junction, and the increased majority-carrier concentration results in an increased hole current injected from the base into the emitter. Thus the defect factor k_2 in Eq. (9-33) increases at high current levels. These variations in k_1 and k_2 result in variations in the current gain β as shown by the characteristic in Fig. 9-11.

The current gain of the BJT also increases with increasing temperature, and the change in β can be substantial. This temperature dependence is due, at least in part, to an increase in the lifetimes in Eq. (9-33) at elevated temperatures. The increased thermal agitation of the particles in the transistor at increased temperatures makes it less likely that a carrier will be caught in a recombination center, and thus the recombination process is retarded.

A reverse bias is applied to the collector junction of the BJT in normal operation. If this bias is made large enough, collector breakdown occurs as illustrated in Fig. 9-12; this figure shows the breakdown characteristic for three different modes of operation. Curve 1 is the breakdown characteristic for the transistor operated in the common-base connection shown in Fig. 9-12b. In this case the emitter terminal is open circuited, and the applied voltage appears across the collector junction. Thus the breakdown that occurs is simply avalanche breakdown of the collector junction as described in Sec. 5-1. The breakdown voltage under this condition is designated BV_{CBO}.

Curve 2 is the breakdown characteristic for the transistor operated in the common-emitter connection with a constant current applied at the base as shown in Fig. 9-12c. The breakdown process is somewhat complicated in this case by the fact that both junctions are involved, and breakdown occurs at a voltage that may be substantially lower than BV_{CBO}. The currents flowing in the transistor are shown in Fig. 9-13. When the voltage across the collector junction is large enough, the electrons carrying the current i_{CE} generate electron-hole pairs in the space-charge layer by high-energy collisions with the crystal lattice. The holes

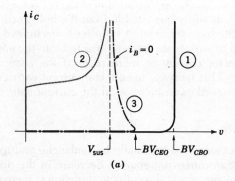

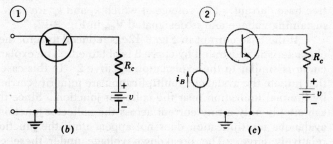

Fig. 9-12 Collector breakdown in the BJT. (a) Collector characteristics; (b) common-base connection; (c) common-emitter connection.

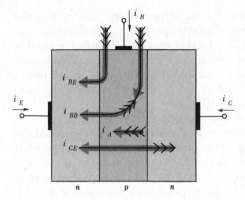

Fig. 9-13 Currents in the common-emitter BJT.

so generated are swept into the base region, and they constitute a hole current i_A flowing from the collector junction into the base. This process is called avalanche multiplication of the collector current. Now the hole current i_A is indistinguishable from the hole current i_B flowing into the base at the base terminal; it supplies holes to feed recombination in the base and to supply the current injected into the emitter. Thus it causes the collector current to increase *as if* the base current had been increased. This increase in collector current with a constant base current can be viewed as an increase in the current gain,

$$\beta = \frac{1}{k_1 + k_2}$$

The holes supplied internally by avalanche multiplication at the collector junction cause an apparent decrease in the defect factors k_1 and k_2. If the voltage across the collector junction is increased further, k_1 and k_2 can be reduced to zero and even made negative, corresponding to a negative base current. The voltage at which k_1 and k_2 are zero is called the sustaining voltage and is designated V_{sus} in Fig. 9-12.

If the base current in Fig. 9-12c is reduced to zero, the breakdown characteristic illustrated by curve 3 is obtained. The explanation of this curve is similar to the explanation of curve 2. In this case the carriers that initiate the avalanche multiplication are minority carriers generated by thermal ionization near the collector junction. Since these carriers can supply only a tiny current across the collector junction, significant avalanche multiplication does not appear until the junction voltage is relatively large. The breakdown voltage under these conditions is designated by BV_{CEO} in Fig. 9-12; this voltage is usually specified on transistor data sheets.

Values of BV_{CEO} for various types of BJTs range from as low as 10 volts to as high as 300 volts. Most transistors designed for small-signal amplifiers have ratings between 20 and 50 volts. Large-signal, high-power transistors have higher ratings.

Another second-order effect in the BJT is the fact that the base width does not remain constant; it varies somewhat with variations in the junction voltages. The junctions and their associated space-charge layers are pictured in Fig. 9-14. The widths of the space-charge layers change in accordance with Eq. (5-1) when the junction voltages are changed, and the result is a change in the base width W. This change, called base-width modulation, is entirely similar to the change in channel thickness caused by the gate voltage in the junction FET. The doping concentrations near the emitter junction are relatively large, and the changes in v_{BE} are quite small under forward-bias conditions; thus v_{BE} has a negligible effect on the base width. On the other hand, the doping concentrations at the collector junction are relatively small, and the reverse bias across the collector junction can change over a large range; hence the collector voltage can produce significant changes in the base width. With the aid of Eqs. (9-33) and (9-34) the current gain can be expressed as

$$\beta = \frac{1}{k_1 + k_2} \approx \frac{1}{k_1} = \frac{2D_{eb}\tau_b}{W^2}$$

As v_{CE} increases, the base width W decreases, and β increases. Thus with constant i_B, i_C increases with increasing v_{CE} as shown by the collector characteristic curves in Fig. 9-3b.

Base-width modulation also causes small changes in voltage and current at the base terminal, and hence the input to the BJT is not completely isolated from the output. This effect is small, however, and it is almost always negligible. In fact, for many purposes the total effect of base-width modulation can be neglected.

Another phenomenon in the BJT that has a counterpart in the junc-

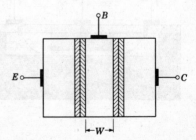

Fig. 9-14 The space-charge layers at the BJT junctions.

tion FET is the pinchout effect. Base current to feed recombination in the transistor of Fig. 9-15 must flow in a transverse direction into the active base region under the emitter junction. Since the base is extremely thin, it presents an appreciable resistance to this flow, and a significant transverse voltage drop is developed. This voltage drop makes the region of the base under the center of the emitter junction less positive than the region under the edge of the junction, and thus the perimeter of the emitter junction is forward biased more heavily than the central region. The transverse voltage drop results in a nonuniform junction bias just as the channel voltage drop does in the junction FET. The nonuniform bias causes the current across the emitter junction to concentrate at the perimeter of the junction, especially at high current levels, and the central portion of the junction may contribute little to the operation of the transistor. This fact has led transistor designers to seek emitter geometries that maximize the ratio of perimeter to area. The geometry illustrated in Fig. 9-6, which uses a narrow rectangular emitter almost completely surrounded by the metallic base connection, is widely used. For high-current devices, several emitter stripes similarly surrounded by base metallization are connected in parallel; such designs are called interdigitated (between fingers) geometries.

The resistance causing the pinchout effect described above acts somewhat as a resistance in series with the base connection. This resistance, called the base resistance or the base-spreading resistance, ranges from about 15 to 100 ohms for small-signal transistors. The base resistance and the pinchout effect which it causes are discussed in detail in Chap. 8 of Ref. 2.

The output current that flows at the collector of the BJT is not a smooth, continuous flow like the flow of a viscous fluid. The current is carried by charged particles that climb the potential barrier at the emitter junction by virtue of their random thermal velocities and that move across the base with random thermal velocities to the collector

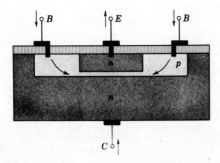

Fig. 9-15 Planar BJT geometry.

junction. Thus the arrival of carriers at the collector has an element of randomness about it, and as a consequence the collector current contains a very small, randomly varying component called noise. The consequences of this noise component of collector current are discussed in connection with the FET at the end of Sec. 7-6; they impose limitations on the capabilities of the transistor, and they sometimes present challenging problems to device and circuit designers. A detailed study of noise in general and of BJT noise in particular is presented in Chap. 4 of Ref. 3.

9-4 GRAPHICAL ANALYSIS OF THE BASIC BJT AMPLIFIER

Graphical methods for the analysis of FET amplifiers are presented in Chap. 6; similar graphical techniques can be used with BJT amplifiers, and they provide helpful insights into the large-signal performance of the amplifiers. The circuit diagram for a BJT amplifier is shown in Fig. 9-16a. Since current flows in the BJT only when the emitter junction is forward biased, a bias battery V_{BB} is included in the circuit to permit negative excursions of the signal voltage v_s to be amplified as well as positive excursions.

The collector current of the theoretical BJT is directly proportional to the base current, and this relation is very nearly true for many real transistors. Thus in many respects it is convenient to think of the BJT as a current amplifier. However, when it comes to building and testing BJT amplifiers, there are many excellent instruments available for measuring signal voltages, whereas there are no instruments that are completely satisfactory for measuring signal currents. Therefore, since the engineer is largely concerned with building and testing, it is appropriate for him to center his attention on signal voltages rather than signal currents.

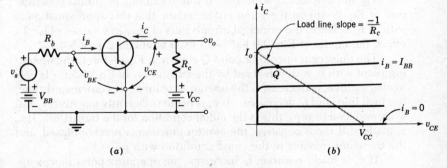

Fig. 9-16 Graphical analysis of a BJT amplifier. (a) Circuit; (b) graphical construction.

The collector voltage in the amplifier of Fig. 9-16a, which is also the output voltage, is given by

$$v_{CE} = V_{CC} - R_c i_C \qquad (9\text{-}47)$$

or

$$i_C = \frac{V_{CC}}{R_c} - \frac{1}{R_c} v_{CE} = I_o - \frac{1}{R_c} v_{CE} \qquad (9\text{-}48)$$

This is the equation of the load line on the collector characteristic; the load line is shown in Fig. 9-16b. For any given value of base current i_B, the operating point for the transistor lies at the intersection of the load line with the characteristic curve for that value of i_B.

The base current can be determined by a graphical construction on the input characteristic (see Fig. 9-3a) similar to the one presented in Fig. 8-9b. However, the nature of the BJT is such that i_B can be calculated directly from the circuit in Fig. 9-16 with sufficient accuracy when R_b is large, several kilohms or more. The base current is given by

$$i_B = \frac{V_{BB} + v_s - v_{BE}}{R_b} \qquad (9\text{-}49)$$

But, as shown in Fig. 9-3a, when normal currents flow in the transistor, v_{BE} is approximately constant at a value designated V_o, where V_o is about 0.7 volt for silicon. Thus Eq. (9-49) can be expressed approximately as

$$i_B = \frac{V_{BB} + v_s - V_o}{R_b} = I_{BB} + \frac{v_s}{R_b} \qquad (9\text{-}50)$$

where $I_{BB} = (V_{BB} - V_o)/R_b$ is the quiescent (no-signal) value of the base current. Equation (9-50) is a linear relation between i_B and v_s, whereas (9-49), containing the variable voltage v_{BE}, is a nonlinear relation. The linearization is effected by treating v_{BE} as a constant, that is, by neglecting Δv_{BE} in comparison with Δv_s. It follows from a graphical construction of Eq. (9-49) on the input characteristic that this approximation is valid if R_b is large; the principal error occurs at very low values of i_B. In effect, the model of Fig. 9-9 has been used to obtain Eq. (9-50).

The quiescent operating point Q shown in Fig. 9-16b is the operating point with $v_s = 0$; it is fixed by the current I_{BB} in Eq. (9-50). If v_s is made negative, i_B decreases, the operating point moves downward along the load line, and i_C decreases. If v_s is made sufficiently negative, i_B and i_C are reduced to zero; this is the cutoff condition for the transistor. If v_s is made still more negative, the emitter junction is reverse biased, and the transistor remains in the cutoff condition with $v_{CE} = V_{CC}$.

If v_s is made positive, i_B increases, the operating point moves upward along the load line, and i_C increases. If v_s is made sufficiently positive, the operating point moves to the point where the collector characteristics all converge into virtually a single line. This is the saturation region of the characteristics; when the transistor is in saturation, the collector junction is forward biased, and amplifier action ceases.

The portion of the load line between saturation and cutoff is called the active region for the amplifier; in this region the transistor amplifies the input signal v_s. Since the collector characteristics for equal increments of i_B intersect the load line at nearly equal intervals, the waveforms of collector current and output voltage are reasonably good reproductions of the input-signal waveform.

The voltage transfer characteristic for the amplifier of Fig. 9-16 is shown in Fig. 9-17. This characteristic can be constructed graphically by assuming various values of v_s, calculating i_B from Eq. (9-50), and reading the corresponding values of output voltage $v_o = v_{CE}$ from the load line. The incremental voltage gain of the amplifier is defined as $K_v = dv_o/dv_s$, and it is equal to the slope of the voltage transfer characteristic. The gain is thus zero in the cutoff and saturation regions, and, for the characteristic shown in Fig. 9-17, it is nearly constant in the active region. The output voltage in the saturation region is typically about 0.2 volt for silicon transistors. The use of the voltage transfer characteristic in the graphical analysis of amplifiers is illustrated in Fig. 6-11.

It is important to note that the transfer characteristics for BJT amplifiers are not always as linear as the one shown in Fig. 9-17. The characteristic in Fig. 9-17 represents an amplifier having a large value of R_b, and hence the nonlinear input characteristic has a small effect that has been neglected. In many BJT amplifiers the value of R_b is not very large, and the input characteristic can introduce considerable curvature in the transfer characteristic. This curvature can be a very troublesome source of waveform distortion.

The coordinates of the points at the ends of the active region of the transfer characteristic can be calculated approximately in the following

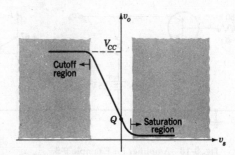

Fig. 9-17 Voltage transfer characteristic for the BJT amplifier.

way. At the threshold of the cutoff region, $i_B = 0$, $i_C = 0$, and $v_o = v_{CE} = V_{CC}$. Also, with $i_B = 0$, Eq. (9-50) gives

$$V_{BB} + v_s - V_o = 0$$

or

$$v_s = -V_{BB} + V_o \qquad (9\text{-}51)$$

Thus the values of v_s and v_o at the cutoff threshold are established. At the threshold of the saturation region, $v_o \approx 0.2$ volt $\ll V_{CC}$; thus

$$V_{CC} = R_c i_C$$

$$i_C = \frac{V_{CC}}{R_c} = \beta i_B$$

Substituting Eq. (9-50) for i_B in this expression yields

$$v_s = \frac{R_b}{\beta R_c} V_{CC} - V_{BB} + V_o \qquad (9\text{-}52)$$

and the values of v_s and v_o at the saturation threshold are established.

The graphical techniques presented above are seldom used for the quantitative analysis of BJT amplifiers; the simple models of Fig. 9-9 provide numerical answers with much less work. Moreover, characteristic curves for transistors are seldom available because the model is adequate for engineering needs. However, the graphical analysis provides an additional kind of insight into the large-signal behavior of the amplifier, and engineers frequently sketch load lines on i_C-v_{CE} coordinates even when there are no characteristic curves on the coordinates.

Example 9-2

The BJT in the amplifier shown in Fig. 9-18 can be characterized by the parameters $\beta = 50$ and $V_o = 0.7$ volt. Certain aspects of the

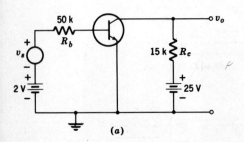

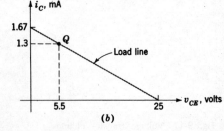

Fig. 9-18 Amplifier for Example 9-2.

performance of this amplifier are to be examined by graphical analysis.

Solution First, the load line is to be sketched on the i_C-v_{CE}. coordinates. The load line intersects the v_{CE} axis at $V_{CC} = 25$ volts as shown in Fig. 9-18, and it intersects the i_C axis at

$$I_o = \frac{V_{CC}}{R_c} = \frac{25}{15} = 1.67 \text{ mA}$$

as is also shown in the diagram.

The quiescent base current in Eq. (9-50) is

$$I_{BB} = \frac{V_{BB} - V_o}{R_b} = \frac{1.3}{50} = 0.026 \text{ mA} = 26 \text{ } \mu\text{A}$$

and the quiescent collector current is

$$I_C = \beta I_{BB} = (50)(0.026) = 1.3 \text{ mA}$$

The quiescent collector voltage is then

$$V_{CE} = V_{CC} - R_c I_C = 25 - (15)(1.3) = 5.5 \text{ volts}$$

and this is also the quiescent output voltage. The quiescent operating point is shown on the load line in Fig. 9-18.

In the absence of any special knowledge about the input signal v_s, it must be assumed that it causes the operating point to make equal excursions on either side of the quiescent point. As the amplitude of the signal is increased, the magnitude of the excursions increases, and with a sufficiently large signal, the amplifier is driven into saturation. Thus, as a low-distortion amplifier, this circuit is said to be saturation limited. It is important to know the signal level at which saturation is reached, and the value of v_s that brings the amplifier just to the threshold of saturation is given by Eq. (9-52) as

$$v_s = \frac{(50)(25)}{(50)(15)} - 2 + 0.7 = 0.37 \text{ volt}$$

If the input voltage is a sinusoid with the maximum amplitude that does not cause serious distortion, then

$$v_s = 0.37 \sin \omega t \qquad \text{volts}$$

From a consideration of the motion of the operating point caused by this input signal, the output voltage is estimated to be

$$v_o = 5.5 - 5.5 \sin \omega t$$

The amplitude of the output sinusoid is

$$K_v = \frac{5.5}{0.37} = 14.9$$

times as big as the input sinusoid, and K_v is the voltage gain.

9-5 DYNAMIC BEHAVIOR OF THE BJT

The collector current in the BJT is given by Eq. (9-16) as

$$i_C = \frac{q_B}{\tau} \qquad\qquad (9\text{-}53)$$

and the charge stored in the base is given by (9-22) as

$$q_B = q_{Bo}[\exp{(\lambda v_{BE})} - 1] \qquad\qquad (9\text{-}54)$$

Thus if $i_C > 0$, $q_B > 0$, and in order to change i_C, it is necessary to change the stored charge q_B. (It is also usually necessary to change the charge stored in the junctions, but these small charges can be neglected without affecting the qualitative discussion that follows in any major way.) In particular, the collector current cannot be turned off until all of the stored charge is removed from the base and $q_B = 0$. Also, from Eq. (9-54)

$$v_{BE} = \frac{1}{\lambda} \ln\left(\frac{q_B}{q_{Bo}} + 1\right) \qquad\qquad (9\text{-}55)$$

Thus, as in the case of the isolated pn junction considered in Chap. 8, reverse bias cannot be built up across the emitter junction until q_B has been reduced to zero. These factors limit the speed at which digital computers and logic machines can operate, for in such machines the transistors are simply switched back and forth from the turned-on to the cutoff condition.

The turnoff characteristic of the BJT is illustrated in Fig. 9-19. With the switch in position 1, the transistor is turned on, and it is assumed that $V \gg V_o$ so that the base current is V/R. It is also assumed that R is large enough to prevent saturation of the transistor. When the switch is thrown to position 2 to turn the transistor off, v_{BE} cannot change much until q_B is reduced to zero, and thus the base current is approximately $-V/R$ as shown in Fig. 9-19b. As q_B is decreased by the reverse base current, i_C drops as shown in Fig. 9-19c. When q_B is reduced to zero, $i_C = 0$, i_B drops rapidly to zero, and a reverse bias builds up across the emitter junction.

If the switch is now thrown back to position 1, there is a similar delay in the buildup of collector current while the required charge is being delivered to the base. These delays, called the rise and fall times,

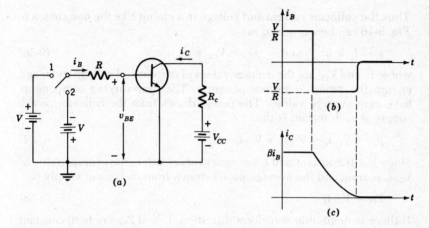

Fig. 9-19 Turnoff behavior of the BJT. (*a*) Circuit; (*b*) waveform of base current; (*c*) waveform of collector current.

depend on the transistor and on the turnon and turnoff voltages applied at the input; typical values for small transistors range from about 10 to 100 nsec. Thus the delays are negligible at switching rates up to at least 100 kHz.

In order to make a transistor that can be switched rapidly from the conducting state to the nonconducting state, the transistor must be designed for the smallest possible charge stored in the base. Small stored charge is achieved by making the volume of the base small and, as is implied by Eq. (9-37), by making the lifetime for minority carriers in the base small. As has been mentioned earlier, gold doping is often used to achieve small lifetimes in high-speed transistors. It is worthwhile to note that the current gain, given by Eq. (9-19), suffers when the lifetime is made small.

More details on the dynamic behavior of the BJT can be found in Chap. 10 of Ref. 2.

9-6 POWER RELATIONS IN THE BJT AMPLIFIER

The effects of power dissipation and elevated temperature on the life expectancy of the BJT are the same as those described in Sec. 6-6 for the FET, and the power relations that exist in the BJT amplifier are essentially the same as those developed in Sec. 6-6. Also, the convention followed in assigning symbols for voltage and current in BJT circuits is the same as the one described in the paragraph following Eq. (6-27).

Thus the collector current and voltage in a circuit like the one shown in Fig. 9-16 can be expressed as

$$i_C = I_C + i_c \quad \text{and} \quad v_{CE} = V_{CE} + v_{ce} \tag{9-56}$$

where I_C and V_{CE} are the average values of current and voltage and i_c and v_{ce} are the time-varying components. The time-varying components have zero average value. The power drawn from the collector power supply at each instant is thus

$$p_{CC} = V_{CC}i_C = V_{CC}I_C + V_{CC}i_c \tag{9-57}$$

Since V_{CC} is constant and i_c has zero average value, the average value of $V_{CC}i_c$ is zero, and the average power drawn from the power supply is

$$P_{CC} = V_{CC}I_C \tag{9-58}$$

If there is negligible waveform distortion, I_C and P_{CC} are both constant and independent of the signal amplitude.

The power absorbed by the resistor R_c at each instant is

$$p_R = R_c i_C{}^2 = R_c(I_C + i_c)^2 \tag{9-59}$$

$$= R_c I_C{}^2 + 2R_c I_C i_c + R_c i_c{}^2 \tag{9-60}$$

The quantity $2R_c I_C$ is constant, and the average value of i_c is zero; hence the average value of the second term on the right-hand side of Eq. (9-60) is zero, and the average power absorbed by R_c is

$$P_R = R_c I_C{}^2 + R_c(i_c{}^2)_{av} = R_c I_C{}^2 + R_c(i_{c,\,rms})^2 \tag{9-61}$$

The power dissipated by the transistor at each instant is

$$p_T = v_{CE}i_C + v_{BE}i_B \tag{9-62}$$

Since $i_B = i_C/\beta$ and $v_{BE} \ll v_{CE}$ in normal amplifier operation, the second term on the right in (9-62) normally makes a negligible contribution to p_T, and the transistor dissipation can be written as

$$p_T = v_{CE}i_C = (V_{CC} - R_c i_C)i_C = V_{CC}i_C - R_c i_C{}^2 \tag{9-63}$$

Now, using Eqs. (9-57) and (9-59), this expression can be put in the form

$$p_T = p_{CC} - p_R$$

The average power dissipated by the transistor is therefore

$$P_T = P_{CC} - P_R \tag{9-64}$$

Thus in distortionless operation, the transistor dissipation is maximum under quiescent operating conditions, and it decreases as the signal level is increased.

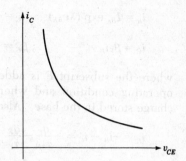

Fig. 9-20 Hyperbola of maximum permissible power dissipation for the BJT.

The power dissipated by the transistor

$$p_T = v_{CE}i_C \tag{9-65}$$

causes the temperature of the device to rise with consequences discussed at the beginning of Sec. 6-6. To avoid excessive temperature rise, the manufacturer of the device usually specifies a maximum permissible continuous device dissipation at an ambient temperature of 25°C. If the power dissipation given by Eq. (9-65) is set equal to the maximum permissible device dissipation, which is a constant, the resulting relation describes a hyperbola on the collector characteristics as shown in Fig. 9-20. Amplifiers intended for use in ambient temperatures of 25°C or less must be designed so that the quiescent operating point lies on or below this hyperbola. Operating points lying above the hyperbola result in excessive temperature rise and a likelihood of early transistor failure. Operating points lying well above the hyperbola may produce instant, catastrophic failure. However, problems of excessive power dissipation ordinarily do not arise in the design of small-signal amplifiers of the kind discussed in this chapter.

9-7 THE EBERS-MOLL MODEL FOR THE BJT

The minority-carrier concentrations in a BJT under normal amplifier conditions with a forward bias at the emitter junction and a reverse bias at the collector junction are given in Fig. 9-7, and they are repeated in Fig. 9-21a. The charge associated with the excess minority-carrier concentration in the base is designated q_F in Fig. 9-21a to indicate the forward, or normal, operating condition. The relations existing under this condition, given by Eqs. (9-37) through (9-41), are

$$i_B = \frac{q_F}{\tau_{BF}} \qquad\qquad i_C = \frac{q_F}{\tau_F} \tag{9-66}$$

$$i_B = I_{Bs} \exp(\lambda v_{BE}) \qquad i_C = I_{Cs} \exp(\lambda v_{BE}) \tag{9-67}$$

$$i_C = \beta_F i_B \qquad \beta_F = \frac{\tau_{BF}}{\tau_F} \tag{9-68}$$

where the subscript F is added to designate the forward, or normal, operating condition and where q_F represents the magnitude of the charge stored in the base. Also, the emitter current is given by

$$i_E = -i_B - i_C = -\frac{q_F}{\tau_{BF}} - \frac{q_F}{\tau_F} \tag{9-69}$$

When the BJT is used in the reverse, or inverted, mode with a forward bias at the collector junction and a reverse bias at the emitter junction, the minority carriers have the distribution shown in Fig. 9-21b. The reasoning used in establishing Eqs. (9-66) to (9-69) can be applied to the inverted mode of operation, and results having a similar algebraic form are obtained. They are

$$i_B = \frac{q_R}{\tau_{BR}} \qquad i_E = \frac{q_R}{\tau_R} \tag{9-70}$$

$$i_E = \beta_R i_B \qquad \beta_R = \frac{\tau_{BR}}{\tau_R} \tag{9-71}$$

and

$$i_C = -i_B - i_E = -\frac{q_R}{\tau_{BR}} - \frac{q_R}{\tau_R} \tag{9-72}$$

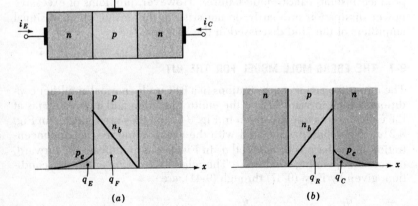

Fig. 9-21 Minority-carrier concentrations in the BJT. (a) Normal operation, $v_{BE} > 0$, $v_{CE} > v_{BE}$; (b) inverted operation, $v_{BC} > 0$, $v_{EC} > v_{BC}$.

where the subscript R designates the reverse, or inverted, mode of operation and where q_R represents the magnitude of the charge stored in the base. The inverted-mode parameters τ_R and τ_{BR} would be equal to the corresponding normal-mode parameters τ_F and τ_{BF} if the transistor were completely symmetrical with respect to geometry and doping. However, as is pointed out in connection with Fig. 9-6, real transistors are not symmetrical with respect to either geometry or doping, and in addition, graded-base transistors have an electric field in the base that aids the flow of carriers in the normal mode and impedes the flow in the inverted mode. Thus the inverted-mode parameters are not equal to the normal-mode parameters. These matters are discussed in more detail in Sec. 10.1 of Ref. 2.

The charges q_F and q_R depend on the emitter and collector voltages, respectively, and they are independent of each other; thus Eqs. (9-66), (9-69), (9-70), and (9-72) can be added to obtain expressions for the transistor currents when either or both of the junctions are forward biased. The results obtained are

$$i_B = \frac{q_F}{\tau_{BF}} + \frac{q_R}{\tau_{BR}} \tag{9-73}$$

$$i_C = \frac{q_F}{\tau_F} - \frac{q_R}{\tau_{BR}} - \frac{q_R}{\tau_R} \tag{9-74}$$

and

$$i_E = -\frac{q_F}{\tau_{BF}} - \frac{q_F}{\tau_F} + \frac{q_R}{\tau_R} \tag{9-75}$$

Solving Eq. (9-73) for q_F yields

$$q_F = \tau_{BF} i_B - \frac{\tau_{BF}}{\tau_{BR}} q_R$$

and substituting this relation into (9-74) gives

$$i_C = \beta_F i_B - \left(\frac{1}{\tau_{BR}} + \frac{1}{\tau_R} + \frac{\tau_{BF}}{\tau_F \tau_{BR}} \right) q_R \tag{9-76}$$

Using Eq. (9-73) in a similar manner to eliminate q_R from (9-75) yields

$$i_E = \beta_R i_B - \left(\frac{1}{\tau_{BF}} + \frac{1}{\tau_F} + \frac{\tau_{BR}}{\tau_R \tau_{BF}} \right) q_F \tag{9-77}$$

The charge q_F, which is designated as q_B in Sec. 9-2, is given by Eq. (9-22) as an exponential function of the base-to-emitter voltage. Similarly, the inverted-mode stored charge q_R is an exponential function of the base-to-collector voltage v_{BC}. Using these charge-voltage relations permits Eqs. (9-76) and (9-77) to be expressed as

$$i_C = \beta_F i_B - I_{CEO}[\exp(\lambda v_{BC}) - 1] \qquad (9\text{-}78)$$

and

$$i_E = \beta_R i_B - I_{ECO}[\exp(\lambda v_{BE}) - 1] \qquad (9\text{-}79)$$

where I_{CEO} and I_{ECO} are constants. These equations are one form of the Ebers-Moll equations describing the behavior of the theoretical transistor under all conditions of applied voltage. Since the equations do not include the components of terminal current required to change the charge stored in the transistor, they are valid only for static and low-frequency operation.

The current I_{CEO} is the theoretical collector current that flows when $i_B = 0$ with a reverse bias applied at the collector junction, and I_{ECO} has a similar significance in relation to the emitter current. In silicon transistors these currents are of the order of 10^{-14} to 10^{-15} amp. In real transistors, however, reverse-bias leakage currents associated with surface effects are typically of the order of 10^{-9} amp, and they dominate the current that flows with $i_B = 0$. It also follows from these facts that the unity term inside the brackets in the Ebers-Moll equations can be neglected as is done in Eqs. (9-67).

A more detailed study of the transistor shows that the four parameters in the Ebers-Moll equations are related by the equation

$$\beta_F I_{ECO} = \beta_R I_{CEO} \qquad (9\text{-}80)$$

Thus only three of the parameters are independent. It also turns out that β_R is usually less than unity for diffused planar transistors.

The exponential terms in Eqs. (9-78) and (9-79) are simply volt-ampere laws for pn junctions; hence the equations imply that the transistor can be represented by the circuit model shown in Fig. 9-22a. The diodes in this model represent pn junctions, and the symbol beside each diode is the characteristic current for the diode. This model is a nonlinear model that represents the transistor with good accuracy for many applications. An approximation that makes the model easier to work with is shown in Fig. 9-22b; in this model the pn junctions are approximated by ideal diodes in series with batteries, and it is assumed that the conducting voltage drops are the same for the two junctions. Now the two identical batteries in Fig. 9-22b can be replaced by a single battery as shown in Fig. 9-22c. And finally, since the behavior of a current source is not affected by the voltage across its terminals, the current sources can be moved to the other side of the battery as shown in Fig. 9-22d without changing the terminal voltages and currents.

The models in Fig. 9-22 represent the npn transistor in all modes of

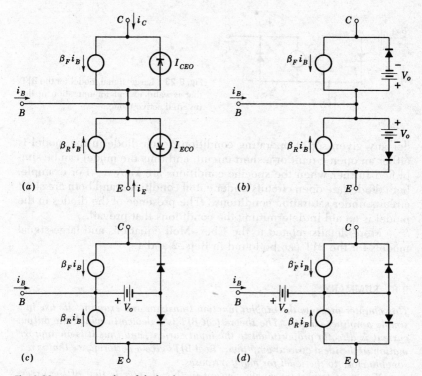

Fig. 9-22   Large-signal models for the BJT.

operation except for junction-breakdown conditions. They are also pictorial representations in a way, for the diodes correspond to the collector and emitter junctions, and each current source represents current across one junction resulting from the injection of carriers into the base at the other junction. These models also describe the *pnp* transistor if the diodes and the batteries are reversed.

The various operating modes for the transistor can be identified as normal active, cut off, saturated, and inverted active. The inverted active mode is rarely encountered. In the remaining three modes, either the emitter junction is forward biased so that the emitter diode in Fig. 9-22*d* is a short circuit, or the transistor is cut off with $i_C = i_B = 0$. It follows from this fact that, except for the inverted active mode, the model shown in Fig. 9-23 is equivalent to the one in Fig. 9-22*d*. The subscript is omitted from the current gain in this model because it is not needed. The model in Fig. 9-23 does not retain the pictorial features of the models in Fig. 9-22, but it is somewhat easier to use in circuit analysis. Un-

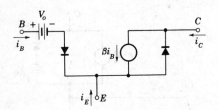

Fig. 9-23 Large-signal model for the BJT that is valid except for operation in the inverted active mode.

der any given set of operating conditions, each diode in the model is either an open circuit or a short circuit, and thus the model can be simplified further when the specific conditions are known. For example, both diodes are open circuits under cutoff conditions, and both are short circuits under saturation conditions. The presence of the diodes in the model is an aid in determining the conditions that prevail.

More details related to the Ebers-Moll equations and large-signal models for the BJT can be found in Refs. 2 and 4.

9-8 SUMMARY

This chapter presents the bipolar junction transistor and examines its use in a simple amplifier circuit. The theoretical BJT is a device in which the output current is directly proportional to the input current, and thus it is an approximation to the ideal current amplifier. Real BJTs give a performance that is reasonably close to the ideal for many purposes.

The most important single concept in this chapter is that of the excess charge stored in the base of the transistor and its relation to the currents flowing at the terminals of the transistor. A clear understanding of the central role played by this stored charge provides valuable insights into the nature of the BJT, and it aids the engineer in designing circuits for optimum performance. It also provides the circuit designer with a point of view and a language that permits him to communicate effectively with the transistor designer.

PROBLEMS

9-1. Basic amplifier analysis. The silicon transistor in the amplifier of Fig. 9-24 can be represented by the model shown in Fig. 9-4 with $\beta = 100$ and $V_o = 0.7$ volt. The circuit parameters and voltages are $R_b = 100$ kilohms, $R_c = 20$ kilohms, $V_{BB} = 1.7$ volts, and $V_{CC} = 25$ volts.

(a) Sketch the diagram of the model for the circuit, and determine the quiescent values ($v_s = 0$) of V_{CE}, I_C, and I_B.

(b) Determine the voltage gain $K_v = |\Delta v_o / \Delta v_s|$.

(c) If $v_s = 0.01 \sin \omega t$ volts, what is the output voltage v_o as a function of time?

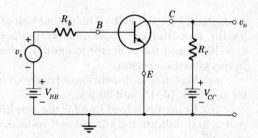

Fig. 9-24 Circuit for Prob. 9-1.

9-2. *Germanium BJT amplifier.* An *npn* germanium BJT is used in the amplifier of Fig. 9-24. The transistor can be represented by, the model shown in Fig. 9-4 with $\beta = 50$ and $V_o = 0.2$ volt. The circuit parameters and voltages are the same as those given in Prob. 9-1.

Carry out the analysis specified by parts *a*, *b*, and *c* of Prob. 9-1.

9-3. *Quiescent-point design.* An *npn* silicon transistor is used in an amplifier like the one shown in Fig. 9-24. The transistor can be represented by the model given in Fig. 9-4 with $\beta = 50$ and $V_o = 0.7$ volt, and the supply voltages are $V_{BB} = 1.7$ volts and $V_{CC} = 25$ volts.

(*a*) The circuit is to be designed to have a quiescent operating point $(v_s = 0)$ at $I_C = 2$ mA and $V_{CE} = 5$ volts. Sketch the diagram of the model for the circuit, and determine the values of R_b and R_c required.

(*b*) What is the voltage gain $K_v = |\Delta v_o / \Delta v_s|$ for this amplifier?

9-4. *Amplifier analysis and design.* When the transistor in the amplifier of Fig. 9-24 is represented by the model shown in Fig. 9-4, the voltage gain of the circuit $K_v = |\Delta v_o / \Delta v_s|$ is completely determined by the two quantities $V_{CC} - V_{CE}$ and $V_{BB} - V_o$, where V_{CE} is the quiescent collector voltage. These quantities are the quiescent voltage drops across R_c and R_b.

(*a*) Sketch the diagram of the model for the circuit, and derive an expression for the voltage gain in terms of the two quantities stated above.

(*b*) An amplifier of this kind using a silicon transistor ($V_o = 0.7$ volt) is to be designed for a voltage gain of 15. The collector supply voltage available is $V_{CC} = 20$ volts, and the quiescent point can be chosen at $I_C = 1$ mA and $V_{CE} = 2$ volts. Specify the values of V_{BB} and R_c required. Specify the value of R_b as a function of the transistor current gain β.

9-5. *Graphical analysis.* An *npn* silicon transistor is used in the amplifier of Fig. 9-24 with $V_{CC} = 25$ volts, $V_{BB} = 1.7$ volts, $R_b = 100$ kilohms, and $R_c = 20$ kilohms. The base-to-emitter voltage drop can be treated as a constant, $V_o = 0.7$ volt. The output volt-ampere characteristic curves for the transistor are approximately horizontal lines corresponding to $i_C = 100i_B$, and the collector-to-emitter voltage drop is essentially zero in the saturation region.

(a) Sketch and dimension a family of output characteristic curves for $i_B = 0, 5, 10, 15,$ and 20 μA.

(b) Construct a load line for the amplifier on the characteristics of part a, and indicate the current and voltage at which the load line intersects the axes.

(c) Show the quiescent operating point on the load line of part b, and give its coordinates.

(d) Sketch and dimension the voltage transfer characteristic, v_o vs. v_s, for the amplifier.

9-6. *Amplifier design.* An *npn* silicon transistor is used in the amplifier shown in Fig. 9-24 with $V_{CC} = 20$ volts and $V_{BB} = 2$ volts. The transistor is characterized by $\beta = 100$ and $V_o = 0.7$ volt, and v_{CE} is essentially zero when the transistor is saturated. The maximum permissible power dissipation for the transistor is 360 mW.

(a) Specify the values of R_b and R_c that will locate the quiescent operating point at $V_{CE} = 5$ volts and $I_C = 1$ mA.

(b) What is the quiescent power dissipated by the transistor?

(c) What is the voltage gain of the amplifier?

(d) If the input signal v_s is a sinusoidal voltage, approximately what is the greatest amplitude that the sinusoidal component of the output voltage v_o can have without driving the operating point into saturation or cut off? *Suggestion:* Sketch a load line on the i_C-v_{CE} coordinates and locate the quiescent operating point on this line.

9-7. *Electronic-voltmeter analysis.* An elementary ac electronic-voltmeter circuit is shown in Fig. 9-25. The circuit has a high input im-

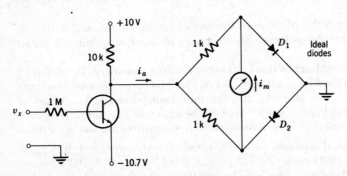

Fig. 9-25 Electronic voltmeter for Prob. 9-7.

pedance so that it does not load the circuit being measured, and it permits an alternating voltage to be measured with a dc microammeter. Diode D_2 conducts when $v_x > 0$, D_1 conducts when $v_x < 0$, and the meter resistance is negligibly small. It follows from these facts that $|i_a| = 2i_m$.

The transistor can be represented by the model in Fig. 9-9c with $\beta = 100$ and $V_o = 0.7$ volt.

(*a*) Sketch the diagram of the model for the circuit, and show that $i_a = Kv_x$. Give the value of the constant K.

(*b*) Sketch and dimension a curve of i_m vs. v_x for v_x between ± 5 volts.

(*c*) Suppose that $v_x = 5 \sin \omega t$ volts. If the frequency of v_x is greater than about 15 Hz, the meter gives a steady deflection proportional to the average value of i_m. Determine the average value of i_m in microamperes.

9-8. *Sawtooth generator analysis.* The *pnp* transistor in the circuit of Fig. 9-26 can be represented by the common-base model of Fig. 9-10*b* if the battery in the model is reversed. The transistor parameters are

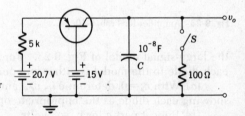

Fig. 9-26 Sawtooth generator for Prob. 9-8.

$V_o = 0.7$ volt and $\alpha \approx 1.0$. The switch is open for 100 μsec, and then it is closed for 100 μsec; this cycle is repeated periodically.

Sketch and dimension the waveform of the output voltage v_o.

9-9. *Stored charge.* A BJT used in the circuit of Fig. 9-19 produces the base-current waveform shown in Fig. 9-27 when the switch is thrown from position 1 to position 2.

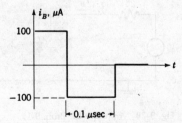

Fig. 9-27 Waveform for Prob. 9-9.

(a) Assuming that negligible charge is lost by recombination during the 0.1-μsec turnoff interval, estimate the amount of charge that was stored in the transistor just before the switch was thrown.

(b) From a consideration of the base current flowing before the switch was thrown, estimate the effective lifetime τ_B for minority carriers in the base.

(c) From the results of part b, what can be said about the assumption in part a?

9-10. BJT-model study. Figure 9-28 shows the circuit model for an amplifier using an *npn* silicon transistor; the transistor is represented by

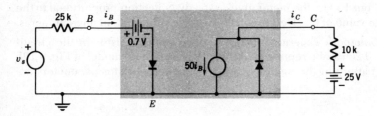

Fig. 9-28 Amplifier for Prob. 9-10.

the large-signal model of Fig. 9-23.   For any given operating condition, each diode in the model is either an open circuit or a short circuit.

(a) With $v_s = 0$, determine i_C, v_{CE}, and v_{BE}.   Sketch the circuit model showing each diode as the appropriate open circuit or short circuit.

(b) Repeat part a for $v_s = 3$ volts.

(c) Repeat part a for $v_s = 1.7$ volts.

(d) For each of the three parts above, specify whether the transistor is cut off, active, or saturated.

9-11. Germanium pnp transistor model. Figure 9-29 shows the circuit model for an amplifier using a germanium *pnp* transistor; the transistor is represented by the large-signal model of Fig. 9-23 with the

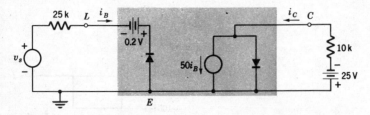

Fig. 9-29 Amplifier for Prob. 9-11.

diodes and the battery reversed. The value of 0.2 volt is commonly used for V_o with germanium transistors. For any given operating condition, each diode in the model is either an open circuit or a short circuit.

(a) With $v_s = 0$, determine i_C, v_{CE}, and v_{BE}. Sketch the circuit model showing each diode as the appropriate open circuit or short circuit.

(b) Repeat part a for $v_s = -3$ volts.

(c) Repeat part a for $v_s = -1.2$ volts.

(d) For each of the three parts above, specify whether the transistor is cut off, active, or saturated.

9-12. BJT monopulser. Figure 9-30 shows a circuit in which the *npn* transistor is represented by the large-signal model of Fig. 9-23. Except for the 0.7-volt battery, this circuit is identical with the monopulser cir-

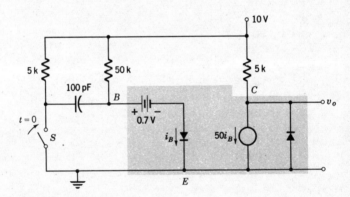

Fig. 9-30 Circuit for Prob. 9-12.

cuit shown in Fig. 3-18; however, in the practical circuit of Fig. 9-30 the current amplifier and both diodes are provided by a single BJT.

The circuit is in the steady state with S open for $t < 0$. At $t = 0$, S is closed. Sketch and dimension the waveforms of v_{BE} and $v_o = v_{CE}$ for $t > 0$.

REFERENCES

1. Motorola Inc.: "Integrated Circuits," McGraw-Hill Book Company, New York, 1965.
2. Gray, P. E., et al.: "Physical Electronics and Circuit Models of Transistors," John Wiley and Sons, Inc., New York, 1964.
3. Thornton, R. D., et al.: "Characteristics and Limitations of Transistors," John Wiley and Sons, Inc., New York, 1966.
4. Ebers, J. J., and J. L. Moll: Large-signal Behavior of Junction Transistors, *Proc. IRE*, vol. 42, pp. 1761–1772, December, 1954.

diodes, and the battery resistor etc. The value of 0.2 watt is commonly used for V_{γ} with normal transistors. For silicon semiconductors this diode in the model is either an open circuit or a short circuit.

(c) With $v_o > v_i$ determine v_o, v_{os}, and v_{ds}, sketch the corresponding anomalous each diode as the approximate open circuit or short circuit.

(d) Repeat part a for negative voltage.

(e) Repeat part b for $v_i = 4$ and 8 volts.

(f) Repeat part c. Rearrange parts above to describe either the transition or cut-off of the saturated region.

8.17. (a) The resistance, Figure 8-18 denotes a circuit in which the v_o are alternatives given by the corresponding region for 3-V. For the v_i to be 0 such figure. The circuit is indicated, which is shown below etc.

Fig. 8-18 Circuit for Problem 8.17.

Problem for Fig. 8-24, below of the result illustrated, above of the current amplifier, and both diodes are modeled as a resistor R.

(a) With $R = 1k$ and a 2-V silicon V_{γ} and on $v_i = 5$ V, find V_o, V_s just and all gain, determine the voltage-gain as constant g_m each input.

REFERENCES

1. G. Sedra, Adel S., and Kenneth C. Smith, Microelectronic Circuits, New York, 1987.
2. Millman, Jacob, and C. Halkias, Integrated Electronics, New York, McGraw-Hill, 1972.
3. Boylestad, R., and L. Nashelsky, Electronic Devices and Circuit Theory, Prentice-Hall, 1987.
4. Sedra, A. S., and K. C. Smith, Microelectronic Circuits, Holt, Rinehart and Winston, New York, 1987.

10

Practical BJT Amplifiers and Small-signal Models

The basic BJT amplifier circuit presented in Chap. 9 is the prototype amplifier. Practical amplifiers usually consist of this circuit with certain modifications and additions that simplify the physical realization and improve the performance of the amplifier. For example, it is usually possible and desirable to derive the bias voltages for both the collector and the base from a single dc power supply. Furthermore, it is almost always necessary to incorporate some feature in the circuit to compensate for the fact that the current gain β may vary widely from one transistor to another and that the transistor parameters vary with temperature. Circuit configurations used to achieve these ends are presented in this chapter.

The BJT, like the FET, is often used with small-signal voltages superimposed on large dc components of voltage. The analysis and design of transistor circuits for this mode of operation is greatly aided by the use of small-signal models to represent the transistor. These small-signal models are developed in this chapter.

10-1 THE BJT AMPLIFIER WITH A SINGLE BATTERY

When an *npn* BJT is used in normal amplifier operation, both the collector terminal and the base terminal are at positive potentials relative to the emitter terminal, and therefore both bias voltages can be obtained from a single dc power supply in the manner illustrated in Fig. 10-1a. The voltage V_{CC} is a positive direct voltage commonly in the range between 10 and 30 volts. The quiescent base current I_B is determined primarily by V_{CC} and R_b, the quiescent collector current is $I_C = \beta I_B$, and the quiescent collector voltage is determined by V_{CC}, I_C, and R_c. The dc blocking capacitor C prevents the flow of direct current through the signal source v_1, and it thereby renders the dc components of voltage

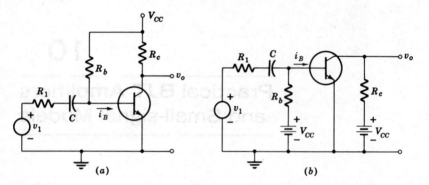

Fig. 10-1 BJT amplifier with a single battery. (a) Circuit; (b) an equivalent circuit.

and current in the amplifier independent of the signal source. The circuit of Fig. 10-1a is equally useful with *pnp* transistors, the only difference being that V_{CC} is a negative voltage for *pnp* transistors.

   Although only one physical battery is used in the amplifier of Fig. 10-1a, it is clear that the circuit shown in Fig. 10-1b is entirely equivalent to the actual circuit. This equivalent form of the circuit is often more convenient for analytical studies.

10-2 QUIESCENT-POINT PROBLEMS IN BJT AMPLIFIERS

A large-signal model for the amplifier in Fig. 10-1 is shown in Fig. 10-2. The transistor is represented in this circuit by a large-signal model similar to the one developed in Fig. 9-9. In addition to the components shown in Fig. 9-9, the model in Fig. 10-2 contains a current source I_{CEO} to account for the current flowing across the reverse-biased collector junction as a result of thermal generation of carriers near the junction. This current is negligibly small in silicon transistors, but it is a significant current in germanium transistors; it is included here so that the discussion that follows will contain this important feature of the germanium transistor. From this model, the quiescent base current is

$$I_B = \frac{V_{CC} - V_o}{R_b} \tag{10-1}$$

and the quiescent collector current is

$$I_C = \beta I_B + I_{CEO} = \beta \frac{V_{CC}}{R_b} - \beta \frac{V_o}{R_b} + I_{CEO} \tag{10-2}$$

Thus the quiescent operating conditions in the amplifier depend on the

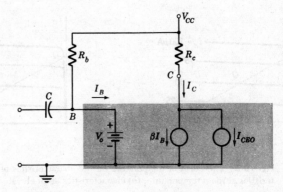

Fig. 10-2 Large-signal model for the BJT amplifier of Fig. 10-1.

three transistor parameters β, V_o, and I_{CEO}, and variations in these parameters can cause undesirable shifts in the quiescent operating point. Variations in the parameters can cause the quiescent point to shift to a region of high distortion near saturation or near cutoff, and in some cases they can cause the quiescent point to move into a region of excessive power dissipation.

It is pointed out in Chap. 9 that the current gain β cannot be closely controlled in mass production. As a result, the production-line spread in β for transistors of a given type made in the same batch-processing operation may be as great as 10 to 1, covering a range from 30 to 300 for example. Thus with I_B constant as indicated by Eq. (10-1), the quiescent collector current may also vary over a range of 10 to 1 when various transistors of a given type are used in the circuit of Fig. 10-2. This fact would raise serious problems in the mass production of the circuit of Fig. 10-2 and in the replacement of transistors in existing amplifiers of this type.

The current gain of the transistor also depends on the temperature of the device. For example, a temperature rise of 100°C above room temperature may cause the current gain to double. Thus an amplifier having a satisfactory quiescent operating point at room temperature may be in saturation at an elevated temperature. This fact is illustrated in Fig. 10-3. The production-line spread in β has an effect similar to the one shown in Fig. 10-3.

The base-to-emitter voltage drop V_o is also a function of temperature. When I_B is constant, this dependence can be determined from Eq. (9-39) by making use of the temperature dependence of I_{Bs} and λ, and the result is that V_o decreases with increasing temperature at a rate

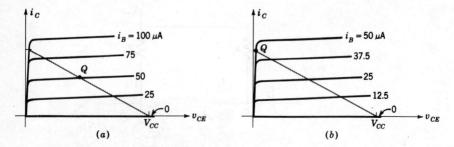

Fig. 10-3 The effect of temperature on the characteristics of a silicon BJT. (a) Characteristics at room temperature; (b) characteristics at an elevated temperature.

of about 2.5 mV/°C. This variation in V_o with temperature affects the quiescent operating conditions in the amplifier of Fig. 10-2 as indicated by Eqs. (10-1) and (10-2).

The thermally generated current I_{CEO} in the circuit of Fig. 10-2 is an exponential function of temperature, and it doubles, approximately, for each 10°C rise in temperature. It affects the quiescent collector current as indicated by Eq. (10-2). In silicon transistors this current is very small, and its variation with temperature is masked by the variation in β with temperature. In germanium transistors, however, the temperature dependence of I_{CEO} may be the dominant effect.

The temperature of a transistor is determined by two factors, the ambient temperature and self-heating in the transistor. Self-heating results from the power dissipated in the transistor, and it causes the temperature of the transistor to rise above the ambient temperature; it can be a serious matter. It turns out that to a good approximation the steady-state temperature rise above ambient is proportional to the power dissipated in the transistor. Thus, letting T_T and T_A designate the transistor and ambient temperatures, respectively, the steady-state temperature rise can be expressed as

$$\Delta T = T_T - T_A = \theta P_T = \theta V_{CE} I_C \tag{10-3}$$

By analogy with Ohm's law,

$$\theta = \frac{\Delta T}{P_T} \tag{10-4}$$

is called the thermal resistance of the transistor; its value depends in a large measure on the way in which the transistor is mounted in its environment. Typical values of θ for transistors of various designs range between about 1 and 1000°C/watt. The lower values of thermal resistance are achieved by mounting the transistor on an aluminum block, called a heat sink, that is equipped with large fins to aid the transfer of heat to the surrounding atmosphere.

The effects of self-heating in transistors can be examined further with the aid of the diagrams in Fig. 10-4. Figure 10-4a shows a static load line and three hyperbolas of constant transistor dissipation on the i_C-v_{CE} plane. It is clear from this diagram that any load line drawn on these coordinates is tangent to just one hyperbola of constant dissipation, although not necessarily one of the three shown in the figure. It also follows from Fig. 10-4a that, for any given load line, the transistor dissipation is greatest when the operating point is at the point of tangency. When the operating point is below the point of tangency, increasing the collector current moves the operating point to a region of greater dissipation; thus with such operating points, self-heating has a cumulative effect. When the operating point is above the point of tangency, increasing the collector current moves the operating point to a region of less dissipation; in this case, self-heating has a self-correcting effect.

Figure 10-4b shows a single hyperbola of constant dissipation P_T and a static load line BC; the load line is tangent to the hyperbola at point A. It is useful here and in other studies to know that, for any such load line, the point of tangency A bisects the line segment BC. This fact can be established with the aid of the construction shown in Fig. 10-4b. The equation of the hyperbola is

$$i_C = \frac{P_T}{v_{CE}} \tag{10-5}$$

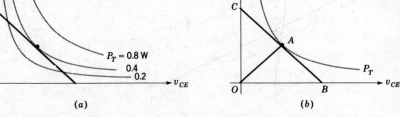

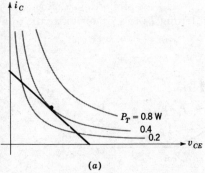

(a)   (b)

Fig. 10-4. Analysis of self-heating in transistors.

and its slope at any point is

$$\frac{di_C}{dv_{CE}} = -\frac{P_T}{v_{CE}{}^2} = -\frac{i_C}{v_{CE}} \tag{10-6}$$

But i_C/v_{CE} is just the slope of the line OA in Fig. 10-4b; thus the slope of the hyperbola at any point is the negative of the slope of the chord drawn from the origin to that point. It follows directly that the triangle OAB is isosceles and that therefore the tangent point A bisects the line segment BC.

It follows from the foregoing argument that, if a transistor is biased so that the quiescent collector voltage V_{CE} is less than $\frac{1}{2}V_{CC}$, the quiescent point lies in a region where the self-heating is self-correcting, and self-heating has a relatively small effect on the quiescent point. This condition is pictured in Fig. 10-5 with V_{CC1} and R_{c1}. On the other hand, if V_{CE} is greater than $\frac{1}{2}V_{CC}$, as shown in Fig. 10-5 with V_{CC2} and R_{c2}, the quiescent point lies in a region where self-heating is cumulative, and the effect on the quiescent point may be large. As shown in Fig. 10-5, it is quite possible for the transistor dissipation to become excessive and for the transistor to be destroyed by high temperature. The phenomenon of cumulative self-heating is often referred to as thermal runaway. The possibility of thermal runaway is almost always a problem when the collector voltage is applied through an inductor, such as a transformer primary, instead of through a resistor.

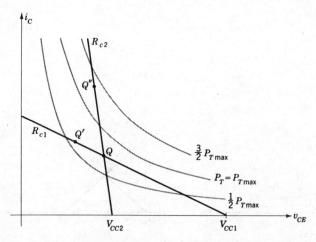

Fig. 10-5 The effect of self-heating on the quiescent operating point. $P_{T\,max}$ = maximum permissible value of P_T.

10-3 STABILIZATION OF THE QUIESCENT OPERATING POINT WITH EMITTER DEGENERATION

The quiescent-point problems described in Sec. 10-2 make it imperative that steps be taken to stabilize the location of the quiescent point in a suitable region of the collector characteristic. The objective in stabilizing is to make the quiescent point largely independent of the transistor parameters and to make it depend instead on the stable circuit elements external to the transistor. There are a number of ways in which stabilization can be accomplished; the method most commonly used with discrete components (as opposed to integrated circuits) is by the use of the circuit arrangement shown in Fig. 10-6a. Stabilization results primarily from the presence of the resistor R_e connected in series with the emitter lead. The base bias is provided through the voltage divider consisting of R_1 and R_2, rather than through a series resistor as in Fig. 10-1a, for the reason that the series resistance required for a suitable quiescent point usually does not permit sufficient stabilization. The use of two resistors in the voltage-divider arrangement permits the resistance in the base circuit and the quiescent base current to be adjusted separately, thus providing both a suitable quiescent point and good stabilization.

A model for the circuit with no signal applied at the input terminals is shown in Fig. 10-6b. This model has been simplified by replacing V_{CC} and the voltage divider R_1-R_2 by a Thévenin equivalent circuit consisting of V_{BB} and R_b; thus

$$V_{BB} = \frac{R_2}{R_1 + R_2} V_{CC} = \frac{V_{CC}}{1 + R_1/R_2} \tag{10-7}$$

and

$$R_b = \frac{R_1 R_2}{R_1 + R_2} = \frac{R_1}{1 + R_1/R_2} = R_1 \frac{V_{BB}}{V_{CC}} \tag{10-8}$$

This model is actually much simpler than it looks because of the fact that the collector is represented by ideal current sources. It is clear from this model that any positive increment in I_{CEO} divides between R_b and R_e; hence it causes a negative increment in I_B and a corresponding negative increment in βI_B. This negative increment in βI_B tends to compensate for the original increment in I_{CEO}, and the net change in I_C is smaller than the increment in I_{CEO}. Changes in β and V_o are compensated in a similar manner, and the shift in the quiescent point resulting from these changes is thereby reduced. By proper design of the circuit, the quiescent point can be made quite stable, even in the pres-

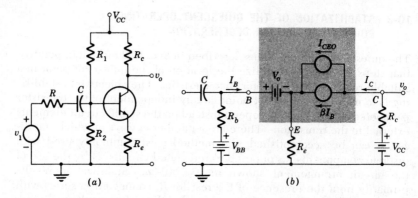

Fig. 10-6 Quiescent-point stabilization with emitter degeneration. (*a*) Circuit; (*b*) model for the circuit.

ence of substantial changes in the transistor parameters. It should also be noted, however, that the circuit stabilizes the collector current equally well against changes caused by the input signal; that is, the stabilization reduces the amplification. Means for overcoming this problem are presented in a subsequent paragraph.

The circuit in Fig. 10-6*b* contains a controlled source βI_B connected in such a way that its input, or controlling, quantity I_B depends on the collector current I_C. But I_C in turn depends on the output of the controlled source βI_B. Thus the input to this controlled source depends on the output from the controlled source. This action, in which the output of a controlled source is transmitted through the circuit in such a way that it affects the input to the same controlled source, is called feedback. Feedback can have a profound effect on the behavior of an amplifier. The resistor R_e in the circuit of Fig. 10-6*b* is responsible for the feedback in that circuit; if R_e is replaced with a short circuit, the feedback is eliminated. This circuit is an example of the use of feedback to improve the performance of a circuit; in this case it is used to stabilize the quiescent operating point against variations in transistor parameters.

The behavior of the circuit can be examined in more detail with the aid of an equation giving the collector current as a function of the circuit parameters. The loop equation for the base circuit is

$$V_{BB} - V_o = R_b I_B + R_e(I_B + I_C) = (R_b + R_e)I_B + R_e I_C \qquad (10\text{-}9)$$

and the collector current is

$$I_C = \beta I_B + I_{CEO} \qquad (10\text{-}10)$$

Solving (10-10) for I_B and substituting the result into (10-9) yields

$$V_{BB} - V_o = (R_b + R_e) \frac{I_C - I_{CEO}}{\beta} + R_e I_C$$

$$= \frac{R_b + (1 + \beta)R_e}{\beta} I_C - \frac{R_b + R_e}{\beta} I_{CEO} \qquad (10\text{-}11)$$

and solving this equation for I_C gives the desired result,

$$I_C = \frac{\beta(V_{BB} - V_o)}{R_b + (1 + \beta)R_e} + \frac{R_b + R_e}{R_b + (1 + \beta)R_e} I_{CEO} \qquad (10\text{-}12)$$

If the circuit is designed so that $\beta R_e \gg R_b$, then the coefficient on I_{CEO} is small, and I_C is nearly independent of I_{CEO}. Furthermore, under this condition the current gain β nearly cancels in the first term on the right of (10-12), and I_C is nearly independent of β. The effect of V_o on I_C is made small by designing the circuit so that V_{BB} is several times as big as V_o. Thus I_C can be made nearly independent of all three transistor parameters by proper design of the circuit in Fig. 10-6.

It should be noted that the results obtained above are independent of the collector-circuit resistor R_c, a fact that is a consequence of representing the collector by two current sources. Thus Eq. (10-12) holds for any value of R_c, including $R_c = 0$, a condition that arises in some of the circuits examined in later chapters.

If I_{CEO} changes to $I_{CEO} + \Delta I_{CEO}$, I_C also changes and is given by

$$I_C + \Delta I_C = \frac{\beta(V_{BB} - V_o)}{R_b + (1 + \beta)R_e} + \frac{R_b + R_e}{R_b + (1 + \beta)R_e} (I_{CEO} + \Delta I_{CEO}) \qquad (10\text{-}13)$$

Subtracting Eq. (10-12) from (10-13) yields

$$\Delta I_C = \frac{R_b + R_e}{R_b + (1 + \beta)R_e} \Delta I_{CEO} \qquad (10\text{-}14)$$

and

$$\frac{\Delta I_C}{\Delta I_{CEO}} = S_e = \frac{R_b + R_e}{R_b + (1 + \beta)R_e} = \frac{1 + R_e/R_b}{1 + (1 + \beta)R_e/R_b} \qquad (10\text{-}15)$$

The quantity S_e, known as the stability factor for the common-emitter amplifier, is a useful measure of the degree of stabilization in the circuit. Equation (10-12) for the collector current can now be written as

$$I_C = \frac{\beta(V_{BB} - V_o)}{R_b + (1 + \beta)R_e} + S_e I_{CEO}$$

The stability factor S_e ranges from a maximum value of unity to a minimum of $1/(1 + \beta)$. In well-designed amplifiers it is 0.1 or less. Thus the second term in the equation for I_C can be neglected with germanium transistors at room temperature and at moderately elevated temperatures, and it can always be neglected with silicon transistors. Hence in well-designed amplifiers

$$I_C = \frac{\beta(V_{BB} - V_o)}{R_b + (1 + \beta)R_e} \approx \frac{V_{BB} - V_o}{R_e} \qquad (10\text{-}16)$$

where it has been assumed that $\beta R_e \gg R_b$ in the approximate expression. The approximate expression is in fact the limiting value approached by I_C as β tends to infinity.

Equation (10-16) is used very often in the analysis and design of BJT amplifiers, and therefore Fig. 10-7 is presented as a mnemonic, or memory aid, for the equation. Figure 10-7a shows conditions existing in the base circuit when the effects of I_{CEO} are negligible. In this case the voltage drop across R_e is $(1 + \beta)I_B R_e$ as shown. This voltage remains unchanged if the collector current βI_B is omitted from the diagram and the emitter resistance is multiplied by $(1 + \beta)$ as shown in Fig. 10-7b; thus Fig. 10-7b is equivalent to Fig. 10-7a insofar as the base current I_B is concerned. The base current calculated from Fig. 10-7b by inspection yields exactly the collector current given by Eq. (10-16) when multiplied by β.

Figure 10-7b also shows how the stabilizing resistor R_e reduces the amplification. Any signal current flowing into the input terminal of the amplifier divides between R_b and $(1 + \beta)R_e$, and only a fraction of it goes into the base of the transistor. If $\beta R_e \gg R_b$, as it must be for good stabilization, then most of the input signal current is diverted through R_b, and very little of it goes into the base of the transistor.

The reduction in amplification introduced by the stabilizing resistor R_e can be eliminated for the time-varying components of current by connecting a large bypass capacitor in parallel with R_e as shown in Fig. 10-8a; this capacitor is chosen to act as a short circuit to the time-

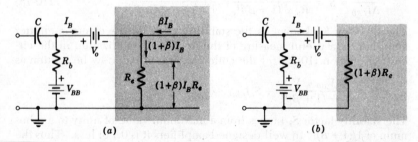

Fig. 10-7   The effect of R_e on the input circuit of a BJT amplifier. (a) Input circuit; (b) equivalent input circuit.

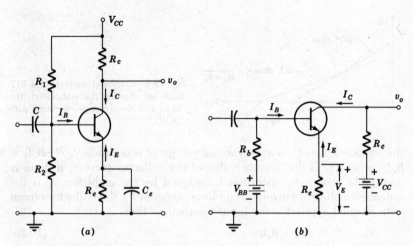

Fig. 10-8 A typical BJT amplifier. (a) Complete circuit; (b) circuit for dc components of current and voltage.

varying components of current, and in effect it removes R_e from the circuit insofar as these components are concerned.

As a result of the presence of the coupling capacitor C and the bypass capacitor C_e in the circuit of Fig. 10-8a, the circuit behaves differently for the dc components and the time-varying, or small-signal, components of current and voltage. For the dc components the circuit reduces to the one shown in Fig. 10-8b; in this representation the base-biasing circuit is replaced by its Thévenin equivalent, V_{BB} and R_b. The loop equation for the collector circuit under quiescent operating conditions is

$$V_{CC} = V_{CE} + R_e(I_C + I_B) + R_c I_C \qquad (10\text{-}17)$$

and since $I_C \gg I_B$, this equation can be written with sufficient accuracy as

$$V_{CC} = V_{CE} + (R_c + R_e)I_C \qquad (10\text{-}18)$$

Thus

$$I_C = \frac{V_{CC}}{R_c + R_e} - \frac{1}{R_c + R_e} V_{CE} = I_o - \frac{1}{R_c + R_e} V_{CE} \qquad (10\text{-}19)$$

This is the equation of the static load line shown in Fig. 10-9. The quiescent collector current is given by Eq. (10-16), and this current locates the quiescent operating point Q on the static load line.

The capacitors C and C_e in Fig. 10-8a are chosen to act as short circuits to the time-varying components of current and voltage. Thus

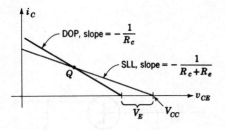

Fig. 10-9 Graphical analysis of a BJT amplifier showing the static load line (SLL) and the dynamic operating path (DOP).

the voltage across C_e is a nonvarying voltage of magnitude $V_E = -R_e I_E \approx R_e I_C$, where I_C is the average value of the collector current; if there is negligible waveform distortion introduced by the amplifier, I_C is the quiescent collector current. The loop equation for the collector circuit with a signal applied at the input terminals is thus

$$V_{CC} - V_E = v_{CE} + R_c i_c \qquad (10\text{-}20)$$

and

$$i_c = \frac{V_{CC} - V_E}{R_c} - \frac{1}{R_c} v_{CE} \qquad (10\text{-}21)$$

This is the equation of the DOP shown in Fig. 10-9; it is a line with a slope $-1/R_c$ that intersects the v_{CE} axis at $V_{CC} - V_E$, and it also passes through the quiescent operating point if the waveform distortion introduced by the amplifier is negligible.

Example 10-1

A typical BJT amplifier is shown in Fig. 10-10a; it uses a silicon pnp transistor with the parameters $\beta = 50$ and $V_o = 0.7$ volt. The

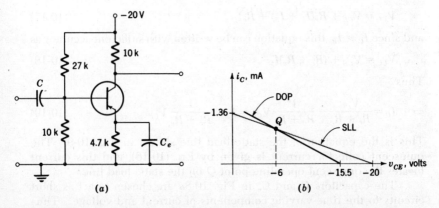

Fig. 10-10 BJT amplifier for Example 10-1. (a) Circuit; (b) graphical construction.

static load line and the DOP are to be constructed, and the stability of the quiescent operating point is to be examined.

Solution The static load line intersects the v_{CE} axis at $V_{CC} = -20$ volts as shown in Fig. 10-10b, and it intersects the i_C axis at

$$I_o = \frac{V_{CC}}{R_c + R_e} = \frac{-20}{10 + 4.7} = -1.36 \text{ mA}$$

The quiescent operating point on the load line can be determined from Eq. (10-16) after V_{BB} and R_b have been determined. Equation (10-7) gives V_{BB} as

$$V_{BB} = -\tfrac{10}{37}(20) = -5.4 \text{ volts}$$

and Eq. (10-8) gives R_b as

$$R_b = \frac{(27)(10)}{37} = 7.3 \text{ kilohms}$$

Then, from Eq. (10-16), with the sign of V_o changed for the *pnp* transistor,

$$I_C = \frac{50(-5.4 + 0.7)}{7.3 + (51)(4.7)} = -0.95 \text{ mA}$$

The corresponding quiescent collector voltage is

$$V_{CE} = V_{CC} - (R_c + R_e)I_C$$

$$= -20 - (10 + 4.7)(-0.95) \approx -6 \text{ volts}$$

The DOP intersects the v_{CE} axis at $V_{CC} - V_E$ as shown in Fig. 10-9, and

$$V_{CC} - V_E = -20 - (4.7)(-0.95) \approx -15.5 \text{ volts}$$

With small distortion the DOP also passes through the quiescent point as shown in Fig. 10-10b.

To examine the stability of the quiescent point, suppose that β increases by a factor of 3 to the value of 150 as a result of changing transistors. Then

$$I_C = \frac{150(-4.7)}{7.3 + (151)(4.7)} = -0.98 \text{ mA}$$

and the corresponding quiescent collector voltage is

$$V_{CE} = -20 - (14.7)(-0.98) = -5.6 \text{ volts}$$

Thus an increase in β by a factor of 3 shifts the quiescent point in Fig. 10-10b by an amount that is barely perceptible. The limiting value of I_C as β increases without limit is

$$I_C = \frac{V_{BB} + V_o}{R_e} = \frac{-5.4 + 0.7}{4.7} = -1.0 \text{ mA}$$

Thus the range of I_C as β changes from 50 to infinity is only from -0.95 to -1.0 mA, a 5 percent change.

10-4   DESIGNING FOR A SPECIFIED QUIESCENT OPERATING POINT WITH A SPECIFIED STABILITY

In designing an amplifier like the one shown in Fig. 10-8, it is known at the outset that the quiescent point will shift as a result of parameter variations, and the designer also knows that he must limit this shift to a suitably small range. Thus one of the first steps in the design is to choose a permissible range of variation for V_{CE} and I_C. This choice is based on a knowledge of how the amplifier will be used, and it is aided by a graphical construction like the one in Fig. 10-9. This choice fixes the maximum ΔI_C to be permitted, and, for example, if the maximum ΔI_{CEO} to be encountered is known, then the required stability factor can be found from Eq. (10-15). The circuit resistances are then chosen to give this stability factor along with the desired quiescent point.

The effect of changes in V_o on the collector current can be estimated from the approximate form of Eq. (10-16),

$$I_C = \frac{V_{BB} - V_o}{R_e} \tag{10-22}$$

When V_o changes to $V_o + \Delta V_o$, the collector current changes from I_C to

$$I_C + \Delta I_C = \frac{V_{BB}}{R_e} - \frac{V_o}{R_e} - \frac{\Delta V_o}{R_e}$$

and

$$\Delta I_C = -\frac{\Delta V_o}{R_e} \tag{10-23}$$

Now let $V_{BB} = nV_o$ so that (10-22) can be written as

$$I_C = \frac{(n-1)V_o}{R_e} \tag{10-24}$$

Using this expression to eliminate R_e in (10-23) yields

$$\Delta I_C = -\frac{\Delta V_o}{n-1}\frac{I_C}{V_o} \tag{10-25}$$

or

$$\frac{\Delta I_C}{I_C} = -\frac{\Delta V_o}{(n-1)V_o} \tag{10-26}$$

Thus the effect of changes in V_o is kept small by making n sufficiently large; n is usually chosen to be 4 or greater.

The collector current is a nonlinear function of β, and hence the effects of changes in β on I_C take a slightly more complicated form than the effects of changes in I_{CEO} and V_o. For one value of β the collector current is given by Eq. (10-16) as

$$I_{C1} = \frac{\beta_1(V_{BB} - V_o)}{R_b + (1 + \beta_1)R_e} \tag{10-27}$$

and for a second value

$$I_{C2} = \frac{\beta_2(V_{BB} - V_o)}{R_b + (1 + \beta_2)R_e} \tag{10-28}$$

The ratio of these currents is

$$\frac{I_{C2}}{I_{C1}} = \frac{\beta_2}{\beta_1} \frac{R_b + (1 + \beta_1)R_e}{R_b + (1 + \beta_2)R_e} \tag{10-29}$$

Subtracting unity from both sides of Eq. (10-29) and rearranging the terms yields

$$\frac{I_{C2}}{I_{C1}} - 1 = \left(\frac{\beta_2}{\beta_1} - 1\right) \frac{R_b + R_e}{R_b + (1 + \beta_2)R_e} \tag{10-30}$$

$$= \left(\frac{\beta_2}{\beta_1} - 1\right) S_{e2} \tag{10-31}$$

where S_{e2} is the value of the stability factor with $\beta = \beta_2$. Defining $\Delta I_C = I_{C2} - I_{C1}$ and $\Delta\beta = \beta_2 - \beta_1$ and substituting these definitions into Eq. (10-31) yields

$$\frac{\Delta I_C}{I_{C1}} = \frac{\Delta\beta}{\beta_1} S_{e2} \tag{10-32}$$

If the range of β to be encountered is known, this relation gives the value of S_{e2} required to limit the range of I_C to the chosen value.

The relations developed above make it possible to design amplifiers like the one in Fig. 10-8 so that the quiescent operating point remains in a chosen range when the parameters of the transistor vary over specified ranges. The details of the design calculations are presented in the illustrative examples that follow.

Example 10-2

An *npn* silicon transistor is used in the amplifier shown in Fig. 10-11. The operating conditions are such that the temperature of the transistor remains essentially constant; however, in mass production, β is expected to range from a low of 30 to a high of 200. The base-to-emitter voltage drop is $V_o = 0.7$ volt. The circuit is to be designed so that the collector voltage is between 4 and 6 volts for all units in mass production, and the collector current is to be 1.5 mA when $V_{CE} = 4$ volts.

Solution The specifications establish three constraints among the four resistors in the circuit; therefore one of the resistors can be chosen to satisfy some other condition. Thus R_e is chosen in this case so that the voltage drop across it is 2 volts (10 percent of V_{CC}) when $I_C = 1.5$ mA. Thus

$$R_e = \frac{2}{1.5} = 1.33 \text{ kilohms}$$

The collector current,

$$I_C = \frac{V_{CC} - V_{CE}}{R_c + R_e}$$

is to be 1.5 mA when $V_{CE} = 4$ volts; thus

$$1.5 = \frac{20 - 4}{R_c + R_e}$$

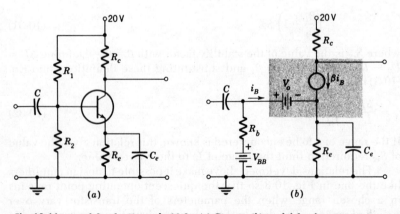

Fig. 10-11 Amplifier for Example 10-2. (*a*) Circuit; (*b*) model for the circuit.

$$R_c + R_e = \frac{16}{1.5} = 10.7 \text{ kilohms}$$

and

$$R_c = 9.4 \text{ kilohms}$$

With $V_{CE} = 6$ volts,

$$I_C = \frac{20 - 6}{10.7} = 1.3 \text{ mA}$$

and thus the permissible range for I_C is

$$\Delta I_C = 1.5 - 1.3 = 0.2 \text{ mA}$$

The low value of I_C is always associated with the low value of β. With $I_{C1} = 1.3$ mA and $\beta_1 = 30$, Eq. (10-32) gives

$$S_{e2} = \frac{\Delta I_C}{I_{C1}} \frac{\beta_1}{\Delta\beta} = \frac{0.2}{1.3} \frac{30}{170} = 0.027$$

From Eq. (10-15) with $\beta = \beta_2$,

$$S_{e2} = 0.027 = \frac{1 + R_e/R_b}{1 + (1 + \beta_2)R_e/R_b}$$

and with $\beta_2 = 200$ this relation yields

$$R_b = \frac{4.4}{0.973} R_e = 6 \text{ kilohms}$$

Now it is necessary to design the voltage divider in the base circuit. With $I_C = 1.5$ mA and $\beta = 200$, the collector current is

$$I_C = 1.5 = \frac{200(V_{BB} - V_o)}{6 + (201)(1.33)}$$

which yields

$$V_{BB} = 2.04 + V_o = 2.74 \text{ volts}$$

As a check on the calculations, I_C is calculated for $\beta = 30$:

$$I_C = \frac{30(2.04)}{6 + (31)(1.33)} = 1.3 \text{ mA}$$

which is the proper value. Now from Eq. (10-8)

$$R_1 = R_b \frac{V_{CC}}{V_{BB}} = 6 \frac{20}{2.74} = 44 \text{ kilohms}$$

and from (10-7)

$$1 + \frac{R_1}{R_2} = \frac{V_{CC}}{V_{BB}} = \frac{20}{2.74}$$

Thus

$R_2 = 7$ kilohms

Summarizing, the resistances in kilohms are $R_e = 1.33$, $R_c = 9.4$, $R_1 = 44$, and $R_2 = 7$. The remainder of the problem, which is not included here, is to convert these values to the nearest standard resistor values and to recheck the collector currents to see if the specifications are still met. Some juggling may be needed at this point, depending on the rigidity of the specifications.

Example 10-3

This example is concerned with the amplifier of Example 10-2 above, but in this case, variations in temperature as well as the production-line spread of β combine to produce the following worst-case conditions:

Minimum I_C: $V_o = 0.7$ volt, $\beta = 30$
Maximum I_C: $V_o = 0.6$ volt, $\beta = 150$

The circuit is to be designed so that $V_{CE} = 6$ volts with the minimum $I_C = 2$ mA and so that V_{CE} is not less than 4 volts with maximum I_C.

Solution   The collector current

$$I_C = \frac{V_{CC} - V_{CE}}{R_c + R_e}$$

is to be 2 mA when V_{CE} is 6 volts; thus

$$2 = \frac{20 - 6}{R_c + R_e}$$

and thus

$R_c + R_e = 7$ kilohms

Then with $V_{CE} = 4$ volts,

$$I_C = \frac{20 - 4}{7} = 2.3 \text{ mA}$$

and the permissible range for I_C is

$$\Delta I_C = 2.3 - 2.0 = 0.3 \text{ mA}$$

In order to keep the effects of the variation in V_o small, a preliminary choice of

$V_{BB} = 5V_o = 3.5$ volts

is made. Then from Eq. (10-22),

$I_C R_e = V_{BB} - V_o = 2.8$ volts

$R_e = \dfrac{2.8}{2} = 1.4$ kilohms

Since the preliminary choice of V_{BB} was arbitrary, the minimum value of I_C was chosen arbitrarily in the calculation of R_e.

Now the effect of changes in V_o can be estimated; from Eq. (10-23)

$\Delta I_C = -\dfrac{\Delta V_o}{R_e} = \dfrac{0.1}{1.4} \approx 0.07$ mA

and this change adds to the change caused by the variation in β. Thus, since the total change allowed in I_C is 0.3 mA, the change due to the variation in β must be limited to about 0.23 mA, and using this value in Eq. (10-32) with $I_{C1} = 2$ mA yields

$S_{e2} = \dfrac{\Delta I_C}{I_{C1}} \dfrac{\beta_1}{\Delta\beta} = \dfrac{0.23}{2} \dfrac{30}{120} = \dfrac{0.23}{8}$

Then with $\beta = \beta_2$, Eq. (10-15) yields

$\dfrac{0.23}{8} = \dfrac{1 + R_e/R_b}{1 + (1 + \beta_2)R_e/R_b}$

and with $\beta_2 = 150$,

$R_b = \dfrac{26.7}{7.77} R_e = 4.8$ kilohms

Now the voltage divider for the base circuit must be designed. For the minimum-current condition,

$I_C = 2 = \dfrac{30(V_{BB} - 0.7)}{4.8 + (31)(1.4)}$

or

$V_{BB} = 3.9$ volts

As a check on the calculations, I_C is determined for the maximum-current conditions:

$$I_C = \frac{150(3.9 - 0.6)}{4.8 + (151)(1.4)} = 2.29 \text{ mA}$$

which is essentially the proper value. Now from Eq. (10-8),

$$R_1 = R_b \frac{V_{CC}}{V_{BB}} = 4.8 \, \frac{20}{3.9} = 24.6 \text{ kilohms}$$

and from Eq. (10-7)

$$1 + \frac{R_1}{R_2} = \frac{V_{CC}}{V_{BB}} = \frac{20}{3.9} = 5.15$$

or

$$R_2 = 6 \text{ kilohms}$$

The collector-circuit resistance is obtained from

$$R_c + R_e = 7 \text{ kilohms}$$

$$R_c = 7 - 1.4 = 5.6 \text{ kilohms}$$

In summary, the resistances in kilohms are $R_e = 1.4$, $R_c = 5.6$, $R_1 = 24.6$, and $R_2 = 6$. The remaining problem, which is not solved here, is to convert these values to the nearest standard values and to recheck the currents to see if the specifications are still met. Some readjustment of the design, or the specifications, or both, will probably be needed.

10-5 STABILIZATION OF THE QUIESCENT OPERATING POINT IN MICROAMPLIFIERS

The special circuit configuration described in Sec. 10-4 for stabilizing the quiescent point is needed because of the production-line spread in the current gain β and because of the temperature dependence of the transistor parameters. Transistors on different wafers processed in the same batch may have quite different current gains, and even if the transistors are on the same wafer, their current gains may be different if they are located in different regions of the wafer. However, if two or more transistors are made side by side on the wafer, they turn out to be almost identical in their electrical properties provided that their junctions have the same areas. Moreover, since the transistors are close together, their temperatures are essentially identical, and their parameters vary almost identically (called tracking) with changes in temperature. These facts are put to good use in microamplifiers where the transistors are close together on a single chip of the original wafer.

Figure 10-12a shows the prototype for a transistor amplifier that is widely used in microamplifiers. Transistor Q_2 is the amplifying transistor in this circuit, and Q_1 provides the bias that establishes a stabilized quiescent point for Q_2. The base of Q_1 is connected to the collector, and the collector junction always has zero voltage across it; thus the volt-ampere characteristic for this two-terminal device is just the characteristic of the emitter junction. The arrangement shown in Fig. 10-12a is therefore called diode biasing. A transistor is used as the biasing element rather than a diode in order to get the best possible matching and tracking with Q_2; it should also be recalled from Chap. 9 that the cost of Q_1 is only 2 or 3 cents. If the temperature of the chip rises, the nearly identical current gains of Q_1 and Q_2 increase. Transistor Q_1 then draws more current, and this action reduces the current flowing into the base of Q_2 to compensate for the increase in the current gain of Q_2. It turns out that, if the current gains are large and matched, the collector current of Q_2 is nearly independent of the current gain. Thus the quiescent point for Q_2 is stabilized against variations in β from one chip to another and from one temperature to another.

The same bias is applied to the emitter junctions of Q_1 and Q_2; hence if the transistors are identical, their base currents will be equal, and

$$I_{B1} = I_{B2} = I_B \tag{10-33}$$

Then, with equal current gains,

$$I_{C1} = I_{C2} = I_C \tag{10-34}$$

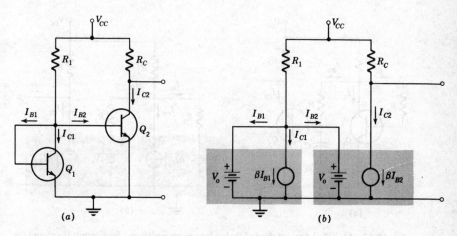

Fig. 10-12 Diode biasing in a microamplifier. (a) Circuit; (b) model for the circuit.

As an aid to further analysis, a model for the circuit is shown in Fig. 10-12b; the transistor model of Fig. 9-9 is used in this representation since both transistors are biased for amplifier operation. A loop equation for this model is

$$V_{CC} = V_o + R_1(I_C + 2I_B) = V_o + R_1\left(1 + \frac{2}{\beta}\right)I_C \qquad (10\text{-}35)$$

and thus

$$I_C = I_{C2} = \frac{V_{CC} - V_o}{R_1(1 + 2/\beta)} \qquad (10\text{-}36)$$

If β is large, this expression reduces to

$$I_C = I_{C2} = \frac{V_{CC} - V_o}{R_1} \qquad (10\text{-}37)$$

The collector current given by (10-37) is still a function of temperature because V_o is temperature dependent. However, V_{CC} is usually 10 times V_o or greater, and thus small changes in V_o have a very small effect on I_{C2}.

The quiescent point for the amplifier in Fig. 10-12 is now established and stabilized. However, there is a problem in applying the input signal to the base of the amplifying transistor Q_2 because Q_1 represents a low-resistance path (a forward-biased diode) from the base of Q_2 to ground; most of the signal current is diverted through Q_1 and little of it goes into the base of Q_2. Figure 10-13a shows a modified form of

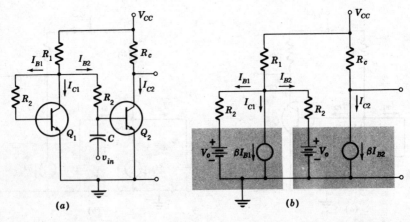

Fig. 10-13 Modified form of diode biasing for microamplifiers. (a) Circuit; (b) model for quiescent operation.

the circuit that eliminates this problem. The resistor R_2 has a relatively high resistance, and it forces most of the input signal current to flow into the base of Q_2. This resistance must be included in series with each base lead in order to preserve the balance of the circuit. Capacitor C is a dc blocking capacitor that prevents the signal source from disturbing the quiescent conditions in the amplifier; it is chosen to act as a short circuit to the input signal.

This circuit operates in essentially the same manner as the circuit in Fig. 10-12a. With identical transistors, the quiescent currents are

$$I_{B1} = I_{B2} = I_B \tag{10-38}$$

and

$$I_{C1} = I_{C2} = I_C \tag{10-39}$$

Now the model in Fig. 10-13b provides a loop equation that is

$$V_{CC} = V_o + R_2 I_B + R_1(I_C + 2I_B)$$

$$= V_o + \left(\frac{R_2}{\beta} + \frac{2R_1}{\beta} + R_1\right) I_C \tag{10-40}$$

Thus

$$I_C = I_{C2} = \frac{V_{CC} - V_o}{R_1 + (2R_1 + R_2)/\beta} \tag{10-41}$$

For large β this expression reduces to

$$I_C = I_{C2} = \frac{V_{CC} - V_o}{R_1} \tag{10-42}$$

which is the same as the current in the circuit of Fig. 10-12 with large β. The quiescent collector voltage for Q_2 is

$$V_{CE2} = V_{CC} - R_c I_{C2}$$

$$= V_{CC} - \frac{R_c}{R_1} (V_{CC} - V_o) \tag{10-43}$$

Thus the collector voltage depends primarily on the supply voltage V_{CC} and the resistance ratio R_c/R_1. This fact is important. The diffused resistors used in microcircuits cannot be held to close tolerances; however, resistance ratios can be controlled quite closely with a tolerance of ±3 percent being typical. Thus V_{CE2} is highly stabilized in the circuit of Fig. 10-13a. If $R_c = R_1/2$ in this circuit,

$$V_{CE2} = V_{CC} - \tfrac{1}{2}(V_{CC} - V_o)$$

$$= \tfrac{1}{2}(V_{CC} + V_o) \tag{10-44}$$

Figure 10-14a shows a form of the diode-biased amplifier that is often used in microamplifiers; the input signal to this circuit consists of equal increments of opposite sign in the two current sources designated I_o in Fig. 10-14a. Under quiescent operating conditions the current sources deliver equal currents, and with identical transistors

$$I_{B1} = I_{B2} = I_B \tag{10-45}$$

$$I_{C1} = I_{C2} = I_C \tag{10-46}$$

and

$$V_{BE1} = V_{BE2} = V_o \tag{10-47}$$

A loop equation for the circuit gives

$$V_{CC} = V_o + R_2(I_B + I_o) + R_1(2I_B + 2I_o + I_C) \tag{10-48}$$

$$= V_o + (R_2 + 2R_1)I_o + \left(\frac{R_2}{\beta} + \frac{2R_1}{\beta} + R_1\right) I_C \tag{10-49}$$

and thus

$$I_C = \frac{V_{CC} - V_o - (R_2 + 2R_1)I_o}{R_1 + (2R_1 + R_2)/\beta} \tag{10-50}$$

For large β this expression reduces to

$$I_C = \frac{V_{CC} - V_o}{R_1} - \left(2 + \frac{R_2}{R_1}\right) I_o \tag{10-51}$$

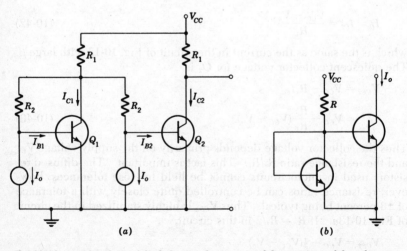

Fig. 10-14 Another form of diode biasing for microamplifiers. (a) Amplifier; (b) current source.

The collector voltage for Q_2 is

$$V_{CE2} = V_{CC} - R_1 I_C$$
$$= V_{CC} - V_{CC} + V_o + (2R_1 + R_2)I_o$$
$$= V_o + (2R_1 + R_2)I_o \qquad (10\text{-}52)$$

The two currents I_o are *in effect* supplied by a circuit like the one shown in Fig. 10-14b. Thus from Eq. (10-37)

$$I_o = \frac{V_{CC} - V_o}{R}$$

and

$$V_{CE2} = V_o + \frac{2R_1 + R_2}{R}(V_{CC} - V_o) \qquad (10\text{-}53)$$

Again, V_{CE2} depends on the ratio of resistances rather than on their absolute values, and therefore it can be controlled closely in production. It is also significant to note that stabilization of the quiescent operating point is achieved in the circuits of Figs. 10-13 and 10-14a without the need for a bypass capacitor. Bypass capacitors of the size generally required in the circuit of Fig. 10-8a cannot be made in integrated-circuit form.

10-6 SMALL-SIGNAL MODELS FOR THE BJT

The basic BJT amplifier is shown in Fig. 10-15. Under normal operating conditions as an amplifier, the transistor currents are given by Eqs. (9-39) and (9-41) as

$$i_B = I_{Bs} \exp \lambda v_{BE} \qquad (10\text{-}54)$$

and

$$i_C = \beta i_B \qquad (10\text{-}55)$$

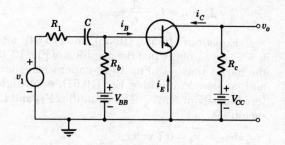

Fig. 10-15 The basic BJT amplifier.

The base voltage and current can also be expressed as

$$v_{BE} = V_{BE} + v_{be} \quad \text{and} \quad i_B = I_B + i_b \tag{10-56}$$

where V_{BE} and I_B are the dc components of voltage and current and v_{be} and i_b are the time-varying components. The current i_b is zero when v_{be} is zero. Thus the base current can be expressed as

$$i_B = I_B + i_b = I_{Bs} \exp \lambda(V_{BE} + v_{be}) \tag{10-57}$$

But with $v_{be} = 0$,

$$i_B = I_B = I_{Bs} \exp \lambda V_{BE} \tag{10-58}$$

Thus in general

$$i_B = I_B + i_b = I_B \exp \lambda v_{be} \tag{10-59}$$

and expanding the exponential in its power series yields

$$i_B = I_B + i_b = I_B[1 + \lambda v_{be} + \tfrac{1}{2}(\lambda v_{be})^2 + \cdots] \tag{10-60}$$

If v_{be} is less than about 5 mV, the square and higher power terms in this series can be neglected as is done in connection with Eq. (8-33), and under this small-signal condition

$$i_B = I_B + i_b = I_B + \lambda I_B v_{be} \tag{10-61}$$

The small-signal component of i_B is thus

$$i_b = \lambda I_B v_{be} \tag{10-62}$$

The quantity

$$r_\pi = \frac{1}{\lambda I_B} \tag{10-63}$$

is the incremental, or small-signal, input resistance of the transistor, and the base current can now be written as

$$i_B = I_B + i_b = I_B + \frac{1}{r_\pi} v_{be} \tag{10-64}$$

Equations (10-55), (10-56), and (10-64), together with the fact that $i_E = -i_C - i_B$, imply that the amplifier of Fig. 10-15 can be represented by the model shown in Fig. 10-16 under small-signal operating conditions, and, as stated in writing Eq. (10-61), small-signal conditions imply v_{be} less than about 5 mV. The quantities V_{BE} and I_B are constant quiescent quantities,

$$V_{BE} \approx V_o = 0.7 \text{ volt} \tag{10-65}$$

and

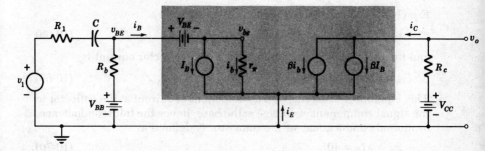

Fig. 10-16 Circuit model for the amplifier in Fig. 10-15 under small-signal operating conditions.

$$I_B = \frac{V_{BB} - V_{BE}}{R_b} \tag{10-66}$$

It is instructive to compare this model with the coarser large-signal model given in Fig. 9-9c.

The model in Fig. 10-16 is a linear circuit, and hence the signal source v_1 can be treated separately from the dc sources by superposition. When the dc sources V_{BE}, I_B, V_{BB}, and V_{CC} are set equal to zero in Fig. 10-16, the small-signal model for the amplifier shown in Fig. 10-17a is obtained. The signal components of voltage and current at all points in the circuit can be calculated from this small-signal model. The small-signal model for the BJT alone is shown in Fig. 10-17b; this model can be used to represent the transistor in any small-signal amplifier circuit. The input resistance of this model is

$$r_\pi = \frac{1}{\lambda I_B} = \frac{\beta}{\lambda I_C} = \frac{\beta}{40 I_C} \tag{10-67}$$

where the value $\lambda = 40$ applies at room temperature. Also, substituting

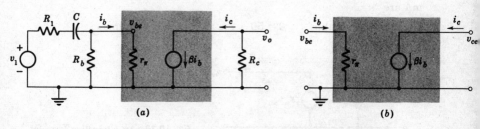

Fig. 10-17 Small-signal models. (a) For the amplifier in Fig. 10-15; (b) for the BJT.

Eq. (10-61) into (10-55) yields

$$i_C = I_C + i_c = \beta I_B + \beta \lambda I_B v_{be} \tag{10-68}$$

and thus the small-signal component of the collector current is

$$i_c = \beta \lambda I_B v_{be} = \lambda I_C v_{be} \tag{10-69}$$

This equation relates the signal component of current at the collector to the signal component of voltage at the base; hence the transconductance, or mutual conductance, of the transistor is defined as

$$g_m = \lambda I_C = 40 I_C \tag{10-70}$$

so that

$$i_c = g_m v_{be} \tag{10-71}$$

This relation implies that the model shown in Fig. 10-18 is equivalent to the one in Fig. 10-17b; both of these forms are used extensively in the analysis and design of small-signal transistor circuits. It follows from Eqs. (10-67) and (10-70) that

$$g_m = \lambda I_C = \frac{\beta}{r_\pi} \tag{10-72}$$

and thus

$$g_m v_{be} = g_m r_\pi i_b = \beta i_b \tag{10-73}$$

Furthermore, from Eqs. (10-54) and (10-55),

$$i_C = \beta I_{Bs} \exp \lambda v_{BE} = I_{Cs} \exp \lambda v_{BE} \tag{10-74}$$

Then

$$\frac{di_C}{dv_{BE}} = \lambda I_{Cs} \exp \lambda v_{BE} = \lambda i_C \tag{10-75}$$

$$= g_m \tag{10-76}$$

The transistor parameters g_m and r_π both depend on the quiescent collector current; typical small-signal transistor parameters for $I_C = 1$ mA are

$$g_m = 40 \text{ mmhos} \qquad \beta = 50 \qquad r_\pi = \frac{\beta}{g_m} = 1.25 \text{ kilohms}$$

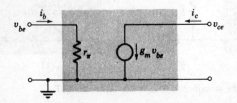

Fig. 10-18 An alternative form for the small-signal BJT model.

The highly nonlinear relation between i_b and v_{be} implied by Eq. (10-59) is a source of waveform distortion in BJT amplifiers. If the Thévenin equivalent of the circuit connected to the base-emitter terminals has a resistance that is much greater than r_π, then the base current is determined primarily by the external resistance and its waveform is essentially undistorted. In this case the waveforms of collector current and output voltage are essentially undistorted because, in a well-made transistor, β is nearly constant even for substantial changes in i_B around its quiescent value. However, if the resistance of the Thévenin equivalent circuit is comparable with or smaller than r_π, the base-to-emitter voltage is an important factor in determining the base current, and waveform distortion will be serious if the input-signal amplitude is excessive, greater than a few millivolts. When the distortion is not too great, it is due almost entirely to the square-law term in Eq. (10-60), and in this case the entire discussion of distortion given in Sec. 6-5 applies also to the bipolar transistor.

Second-order effects in the BJT are discussed in some detail in Sec. 9-3. It is pointed out there that base-width modulation causes the collector current to increase with increasing collector-to-emitter voltage. Under small-signal conditions, the change in collector current is a linear function of the collector voltage, and hence it can be accounted for by a resistance shunting the output of the small-signal model, as shown in Fig. 10-19a. A more detailed study [1] shows that r_o is inversely proportional to the quiescent collector current. Base-width modulation also causes the collector voltage to affect the base current by a very small amount. This effect can be accounted for by a large resistor connected between the output node and the node labeled v in Fig. 10-19a;[1] however, the effect is almost always negligible, and it is customary to omit this complicating element from the model.

The voltage drop caused by the base current flowing through the ohmic resistance of the base of the transistor is also discussed in Sec. 9-3. This voltage drop is accounted for by the resistance r_x in the model

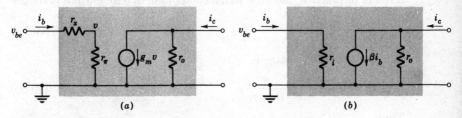

Fig. 10-19 Small-signal BJT models including second-order effects.

of Fig. 10-19a. The remainder of the input voltage, which appears across r_π in the model, is the voltage that appears across the emitter junction; this is the voltage that controls the collector current as indicated by the controlled source $g_m v$ in the model.

The parameters accounting for the second-order effects in the BJT typically lie in the following ranges:

r_o: 20 to 200 kilohms
r_x: 15 to 100 ohms

The parameter r_x is not really constant, and it is hard to measure. However, since it is so small, it does not have an important effect on the behavior of the transistor except at very high frequencies. It is common practice to assume a value of 50 ohms for r_x, or even to neglect it altogether except when very high frequencies are involved.

The small-signal model of Fig. 10-19a is known as the hybrid-π model for the transistor. A detailed discussion of this model and its parameters can be found in Ref. 2 listed at the end of this chapter.

An alternative form of the hybrid-π model is shown in Fig. 10-19b where

$$r_i = r_x + r_\pi \tag{10-77}$$

is the input resistance of the model. Since the voltage v does not appear in this model, the controlled source must be expressed in terms of its dependence on i_b.

Both models shown in Fig. 10-19 apply equally well to pnp transistors without change. However, since the collector current in the pnp transistor is numerically negative, the parameters are given by

$$g_m = \lambda |I_C| \quad \text{and} \quad r_\pi = \frac{\beta}{g_m} \tag{10-78}$$

Figure 10-20 shows the small-signal model of Fig. 10-17a with the second-order effects included and with the input coupling capacitor represented as a short circuit for the signal current. If $R_b \gg r_i$, as it often is in practice, then the signal component of base current is simply

$$i_b = \frac{v_1}{R_1 + r_i} \tag{10-79}$$

Hence the output voltage is

$$v_o = -\beta i_b R = -\beta \frac{R}{R_1 + r_i} v_1 \tag{10-80}$$

Thus

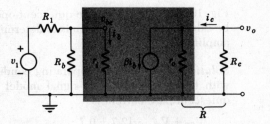

Fig. 10-20 Small-signal model for the amplifier in Fig. 10-15 including second-order effects in the transistor.

$$\frac{v_o}{v_1} = -\beta \frac{R}{R_1 + r_i} = -K_v \qquad (10\text{-}81)$$

and the voltage gain of the amplifier is

$$K_v = \beta \frac{R}{R_1 + r_i} \qquad (10\text{-}82)$$

In practical amplifiers, r_o is often much larger than R_c, and in such cases

$$K_v = \beta \frac{R_c}{R_1 + r_i} \qquad (10\text{-}83)$$

Example 10-4

The *pnp* silicon transistor used in the amplifier of Fig. 10-21*a* is characterized by the following parameters: $\beta = 100$, $r_x = 50$ ohms, and $r_o = 50$ kilohms. The parameter r_i in the small-signal model of

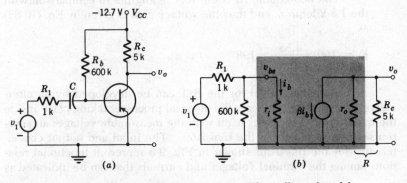

Fig. 10-21 Amplifier for Example 10-4. (*a*) Circuit; (*b*) small-signal model.

Fig. 10-21b depends on the quiescent operating point, and it must be calculated. The problem is to determine the voltage gain of the amplifier.

Solution The quiescent operating conditions can be determined with the aid of the large-signal model of Fig. 9-9c. Thus the quiescent base current is

$$I_B = \frac{V_{CC} + V_o}{R_b} = \frac{-12.7 + 0.7}{600} = -0.02 \text{ mA}$$

where the sign on V_o has been adjusted to correspond to the *pnp* transistor. The quiescent collector current is

$$I_C = \beta I_B = (100)(-0.02) = -2 \text{ mA}$$

and the collector voltage is

$$V_{CE} = V_{CC} - R_c I_C = -12.7 - (5)(-2) = -2.7 \text{ volts}$$

The small-signal parameters under these quiescent conditions are

$$g_m = 40|I_C| = (40)(2) = 80 \text{ mmhos}$$

$$r_\pi = \frac{\beta}{g_m} = \frac{100}{80} = 1.25 \text{ kilohms}$$

$$r_i = r_\pi + r_x = 1.25 + 0.05 = 1.3 \text{ kilohms}$$

and the total resistance shunting the output of the amplifier is

$$R = \frac{(50)(5)}{55} = 4.55 \text{ kilohms}$$

The 600-kilohm R_b is entirely negligible in comparison with the 1.3-kilohm r_i, and thus the voltage gain is given by Eq. (10-82) as

$$K_v = (100) \frac{4.55}{1 + 1.3} = 198$$

A small-signal model for the BJT can be developed by an alternative procedure that ignores the physical processes taking place inside the transistor and concerns itself with the measurable voltages and currents at the terminals of the transistor. The input and output characteristics for the transistor shown in Fig. 9-3 represent functional relations among the terminal voltages and currents that can be indicated as

$$v_{BE} = f_B(i_B, v_{CE}) \quad \text{and} \quad i_C = f_C(i_B, v_{CE}) \tag{10-84}$$

The dependence of v_{BE} on v_{CE} is too small to show in Fig. 9-3, but a small dependence does exist. If the independent variables i_B and v_{CE} are given small increments, the resulting increments in the dependent variables v_{BE} and i_C can be expressed formally as

$$\Delta v_{BE} \approx dv_{BE} = \frac{\partial v_{BE}}{\partial i_B}\, di_B + \frac{\partial v_{BE}}{\partial v_{CE}}\, dv_{CE} \tag{10-85}$$

and

$$\Delta i_C \approx di_C = \frac{\partial i_C}{\partial i_B}\, di_B + \frac{\partial i_C}{\partial v_{CE}}\, dv_{CE} \tag{10-86}$$

These equations merely state that, under small-signal conditions, the dependent variables are linear functions of the independent variables and that the partial derivatives are the constants of proportionality. The numerical information needed to characterize the transistor with these equations is obtained by measuring the partial derivatives experimentally. These measurements must be made at the appropriate quiescent operating point, and they include the second-order as well as the first-order effects. Defining suitable symbols for the partial derivatives and using lowercase subscripts to denote incremental quantities yield

$$v_{be} = h_{ie}i_b + h_{re}v_{ce} \tag{10-87}$$

and

$$i_c = h_{fe}i_b + h_{oe}v_{ce} \tag{10-88}$$

Since the independent variables are a current i_b and a voltage v_{ce}, these equations are called the hybrid equations for the transistor, and the coefficients are called the hybrid parameters, or the h-parameters, of the transistor. The subscripts i, r, f, and o denote input, reverse, forward, and output, respectively, and the subscript e denotes that the parameters apply for operation in the common-emitter connection. Hybrid parameters describing the transistor in the common-base connection use the letter b as the second subscript.

The parameter h_{ie} has the dimensions of resistance, h_{oe} has the dimensions of conductance, and h_{fe} and h_{re} are dimensionless. Thus Eqs. (10-87) and (10-88), along with the relation $i_e = -i_c - i_b$, imply that the transistor can be represented by the small-signal model shown in Fig. 10-22. It follows from this model, and also from Eq. (10-87), that with $v_{ce} = 0$, $v_{be} = h_{ie}i_b$; thus

$$h_{ie} = \frac{v_{be}}{i_b}\bigg|_{v_{ce}=0} = \text{short-circuit input impedance} \tag{10-89}$$

In a similar manner

$$h_{fe} = \frac{i_c}{i_b}\bigg|_{v_{ce}=0} = \text{short-circuit forward current gain} \quad\quad (10\text{-}90)$$
$$\text{(equivalent to } \beta)$$

$$h_{re} = \frac{v_{be}}{v_{ce}}\bigg|_{i_b=0} = \text{open-circuit reverse voltage gain} \quad\quad (10\text{-}91)$$

$$h_{oe} = \frac{i_c}{v_{ce}}\bigg|_{i_b=0} = \text{open-circuit output admittance} \quad\quad (10\text{-}92)$$

The hybrid parameters are easy to measure, and they are often given on BJT data sheets. However, the method by which they are derived gives no insight into the internal behavior of the transistor, and it gives no information about the dependence of the parameters on the quiescent operating point. It is worth noting in passing that BJT data sheets often list a parameter h_{FE}; the capital-letter subscripts indicate that this parameter is the ratio of total collector current i_C to total base current i_B.

The voltage source $h_{re}v_{ce}$ in the model of Fig. 10-22 accounts for the effect of v_{ce} on i_b that results from the base-width modulation discussed in Sec. 9-3; this small effect is neglected in the hybrid-π model of Fig. 10-19a. The validity of neglecting this effect can be examined with the aid of the model in Fig. 10-22 by letting

$$v_{ce} = K_{bc}v_{be} \quad\quad (10\text{-}93)$$

where K_{bc} is the voltage gain from the base terminal to the collector terminal. Then

$$h_{re}v_{ce} = K_{bc}h_{re}v_{be}$$

and

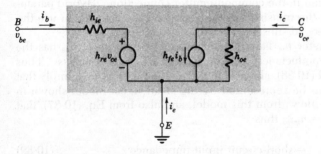

Fig. 10-22 The hybrid small-signal model for the BJT.

$$i_b = \frac{v_{be} - K_{bc}h_{re}v_{be}}{h_{ie}} = \frac{v_{be}}{h_{ie}}(1 - K_{bc}h_{re}) \qquad (10\text{-}94)$$

The parameter h_{re} is almost always less than 5×10^{-4}; thus for

$$K_{bc}h_{re} < 0.1$$

it is necessary that

$$K_{bc} < \frac{0.1}{h_{re}} = 200 \qquad (10\text{-}95)$$

in the worst case. In practical BJT amplifiers the base-to-collector voltage gain is seldom this big, and thus the effect of base-width modulation on i_b is smaller than other uncertainties in the circuit. Hence it is justifiable to neglect the effect, and it is customary to set $h_{re} = 0$ in the hybrid model to obtain the simpler form of the model shown in Fig. 10-23a. The hybrid-π model of Fig. 10-19b is shown in Fig. 10-23b for comparison; it follows from Eqs. (10-87) and (10-88) with $h_{re} = 0$ that these two models are equivalent when

$$h_{fe} = \beta \qquad h_{oe} = \frac{1}{r_o} \qquad h_{ie} = r_i = r_x + r_\pi$$

The BJT is occasionally used in the common-base connection, sometimes as a single transistor but more often in combination with other transistors. The hybrid-π model rearranged for the common-base connection is shown in Fig. 10-24a, and the common-base h-parameter model with $h_{rb} = 0$ is shown in Fig. 10-24b. By using the fact that $r_o \gg r_i$, it can be shown that the parameters in these two models are related by

$$h_{ib} \approx \frac{r_i}{1 + \beta} = \frac{r_x + r_\pi}{1 + \beta} \approx \frac{r_\pi}{\beta} = \frac{1}{g_m} = \frac{1}{\lambda I_C} \qquad (10\text{-}96)$$

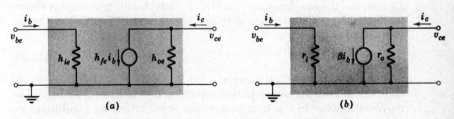

Fig. 10-23 Small-signal models for the BJT. (a) Hybrid model with $h_{re} = 0$; (b) hybrid-π model of Fig. 10-19b.

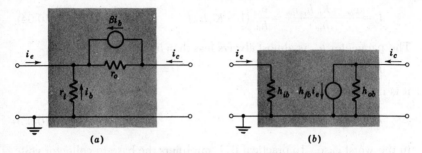

Fig. 10-24 Small-signal common-base models for the BJT. (a) Hybrid-π model; (b) common-base h-parameter model with $h_{rb} = 0$.

$$h_{fb} \approx \frac{-\beta}{1+\beta} \approx -1 \tag{10-97}$$

$$h_{ob} \approx \frac{1}{(1+\beta)r_o} \tag{10-98}$$

Thus under typical operating conditions in the common-base connection, the input resistance is low, a few ohms, and the output resistance is high, usually several megohms. The common-base current gain is less than unity, but the voltage gain may be substantial in a common-base amplifier.

10-7 HIGH-FREQUENCY MODEL FOR THE BJT

The structure of the BJT is represented symbolically in Fig. 10-25a. In this representation the ohmic base resistance r_x is shown external to the transistor, and the voltages across the junctions are designated v'_{BE} and v'_{CB}. The charge stored in the base of the transistor is illustrated in Fig. 10-25b, and the low-frequency hybrid-π model for the transistor is shown in Fig. 10-25c. The voltage drop across the emitter junction can be expressed as

$$v'_{BE} = V'_{BE} + v \tag{10-99}$$

where V'_{BE} is the constant quiescent value of v'_{BE} and v is the small-signal component corresponding to the voltage across r_π in Fig. 10-25c. The currents flowing in the transistor under static operating conditions are given by Eqs. (9-16) and (9-18) as

$$i_B = \frac{q_B}{\tau_B} \quad \text{and} \quad i_C = \frac{q_B}{\tau} \tag{10-100}$$

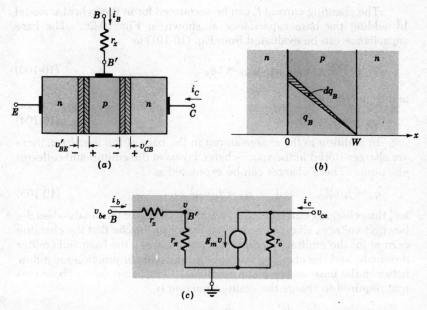

Fig. 10-25 Charge stored in the BJT. (a) The BJT structure; (b) charge stored in the base; (c) static small-signal BJT model.

and Eq. (9-22) gives the charge stored in the base as

$$q_B = q_{Bo}[\exp (\lambda v'_{BE}) - 1] \approx q_{Bo} \exp \lambda v'_{BE} \tag{10-101}$$

Thus in order to change the terminal currents, the charge stored in the base must be changed. The stored charge consists of negative charge associated with electrons injected from the emitter into the base, and space-charge neutrality requires an equal store of positive charge associated with holes that flow into the base at the base terminal. Thus charging currents must flow at the base and emitter terminals to change the stored charge. The base current given by Eq. (10-100) for static conditions feeds the recombination in the base and supplies holes for injection into the emitter. Under dynamic conditions there is an additional component of base current that changes the charge stored in the base; this additional charging current can be expressed as

$$i'_b = \frac{dq_B}{dt} = \frac{dq_B}{dv'_{BE}} \frac{dv'_{BE}}{dt} = C_b \frac{dv}{dt} \tag{10-102}$$

where C_b, the incremental base capacitance, corresponds to the diffusion capacitance defined in Eq. (8-45) for the pn junction.

The charging current i_b' can be accounted for in the hybrid-π model by adding the base capacitance as shown in Fig. 10-26. The base capacitance can be evaluated from Eq. (10-101) as

$$C_b = \frac{dq_B}{dv_{BE}'} = \lambda q_{Bo} \exp \lambda v_{BE}' = \lambda q_B \tag{10-103}$$

and from (10-100)

$$\cdot C_b = \lambda \tau i_C = g_m \tau \tag{10-104}$$

In addition to the charge stored in the base of the transistor, there are charges stored in the space-charge layers at the emitter and collector junctions. These charges can be expressed as

$$q_{je} = f_e(v_{BE}') \qquad \text{and} \qquad q_{jc} = f_c(v_{CB}') \tag{10-105}$$

and thus charging currents must flow at the transistor terminals when the junction voltages change. It is clear from Fig. 10-25a that the charging current for the emitter junction must flow between the base and emitter terminals, and the charging current for the collector junction must flow between the base and collector terminals. The component of base current required to charge the emitter junction is

$$i_b'' = \frac{dq_{je}}{dt} = \frac{dq_{je}}{dv_{BE}'} \frac{dv_{BE}'}{dt} = C_{je} \frac{dv}{dt} \tag{10-106}$$

where C_{je}, the incremental emitter-junction capacitance, corresponds to the incremental capacitance developed in Sec. 5-2 for the pn junction. The total charging current flowing between the base and emitter terminals is

$$i_b' + i_b'' = C_b \frac{dv}{dt} + C_{je} \frac{dv}{dt} = C_\pi \frac{dv}{dt} \tag{10-107}$$

where $C_\pi = C_b + C_{je}$. This current is accounted for by the capacitance C_π shown in the model of Fig. 10-27.

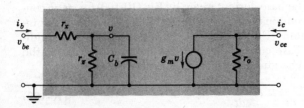

Fig. 10-26 Small-signal BJT model accounting for the charge stored in the base.

The component of collector current required to charge the collector junction is

$$i_c' = -i_b''' = \frac{dq_{jc}}{dt} = \frac{dq_{jc}}{dv_{CB}'} \frac{dv_{CB}'}{dt} = C_\mu \frac{dv_{CB}'}{dt} \tag{10-108}$$

where C_μ is the incremental collector-junction capacitance; this charging current is accounted for by the capacitance C_μ in the model of Fig. 10-27.

The circuit shown in Fig. 10-27 is the complete hybrid-π model for the transistor; it accounts for the small-signal components of voltage and current in the transistor with good accuracy under dynamic and static operating conditions. In real transistors there are additional stray capacitances not mentioned in the foregoing discussion; for example, the package in which the transistor is mounted introduces additional small capacitances. However, the effects of these capacitances are small, and it is usually possible to lump them with the capacitances in the model of Fig. 10-27. A more detailed discussion of these matters can be found in Ref. 2.

Any model that is used to represent a transistor, or a capacitor, or a resistor, is valid only in a limited range of signal frequencies. The range of frequencies in which the hybrid-π transistor model is valid depends on how the transistor is designed and fabricated. As a coarse rule of thumb, the model is valid at signal frequencies up to 100 MHz or more for transistors designed for small-signal, high-frequency operation. At higher frequencies, parasitic elements such as lead inductance, which are not included in the model, may become important.

At low frequencies the capacitors in the model of Fig. 10-27 behave as open circuits and do not affect the performance of the transistor. At high frequencies, however, these capacitors have a relatively low impedance, and they reduce the amplitude of the signal voltage developed at v. This reduction in v causes a reduction in the output of the controlled source $g_m v$ and thus a reduction in the signal component of the collector current. In this way the capacitors reduce the current gain

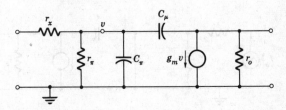

Fig. 10-27 The complete hybrid-π model for the BJT.

of the transistor, and at sufficiently high frequencies the transistor cannot provide any gain at all. It is instructive to examine this matter further by calculating the collector current when the output is short circuited and a sinusoidal current is applied at the base as illustrated in Fig. 10-28. With the output terminals short circuited, C_μ is in parallel with C_π, and if the complex amplitude of the sinusoidal input current is designated I_b, then the complex amplitude of the resulting sinusoidal voltage across C_π is

$$V = \frac{r_\pi}{1 + j\omega(C_\pi + C_\mu)r_\pi} I_b \tag{10-109}$$

The complex amplitude of the short-circuit output current is

$$I_c = g_m V - j\omega C_\mu V \tag{10-110}$$

However, the second term on the right in (10-110) is entirely negligible in comparison with the first term throughout the entire range of frequencies in which the transistor is useful, and thus the output current can be written as

$$I_c = g_m V = \frac{g_m r_\pi}{1 + j\omega(C_\pi + C_\mu)r_\pi} I_b \tag{10-111}$$

$$= \frac{\beta}{1 + j\omega(C_\pi + C_\mu)r_\pi} I_b \tag{10-112}$$

Thus if the amplitude of the input current is held constant as the frequency of the current is increased, the amplitude of the output current decreases and tends to zero at very high frequencies.

At low frequencies the collector current is proportional to the base current at all times, and it is useful to define a current gain β as the ratio of these currents. Equation (10-112) shows, however, that with a high-frequency sinusoidal signal, the collector current is not in phase with the base current, and it follows from this fact that the ratio of collector current to base current is not a constant but varies from instant to instant. Thus for rapidly varying signals in general, it is not possible to define a

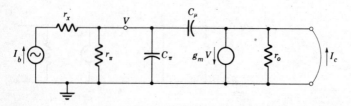

Fig. 10-28   Model for calculating the short-circuit current gain.

current gain, at least not in any simple way. (This statement is true in general for impedance, admittance, transfer functions, and so forth.) However, in the case of a sinusoidal signal the waveform of I_c is the same as the waveform of I_b, and the idea of a current gain can be extended in a useful way by defining it to be the ratio of the complex amplitudes of these currents. Thus the extended definition is

$$h_{fe}(j\omega) = \frac{I_c}{I_b}\bigg|_{V_{ce}=0} = \frac{\beta}{1 + j\omega(C_\pi + C_\mu)r_\pi} \qquad (10\text{-}113)$$

This current gain is a complex number and a function of frequency; it reduces to the real constant β at low frequencies. The magnitude of h_{fe} is the magnitude of the ratio I_c/I_b, and hence it is the amplification of the transistor for sinusoidal signals. The angle of h_{fe} is the phase shift between I_c and I_b for sinusoidal signals. The usefulness of the complex current gain can be extended further to include periodic signals by expanding the periodic signal in its Fourier series and treating each sinusoidal component separately by superposition. Its usefulness can be extended to signals in general with the aid of the Fourier or the Laplace transform.

The quantity $1/(C_\pi + C_\mu)r_\pi$ has the dimensions of frequency, and it has a useful significance as a frequency. Thus it is helpful to define

$$\omega_\beta = \frac{1}{(C_\pi + C_\mu)r_\pi} = 2\pi f_\beta \qquad (10\text{-}114)$$

Substituting this expression into (10-113) yields

$$h_{fe} = \frac{\beta}{1 + j\omega/\omega_\beta} = \frac{\beta}{1 + jf/f_\beta} \qquad (10\text{-}115)$$

The magnitude of h_{fe} is

$$|h_{fe}| = \frac{\beta}{\sqrt{1 + (f/f_\beta)^2}} \qquad (10\text{-}116)$$

But Eq. (10-113) yields

$$|I_c| = |h_{fe}|\,|I_b|$$

Thus if the amplitude of the sinusoidal input signal is known and if the magnitude of h_{fe} is known, the amplitude of the sinusoidal output signal can be calculated. It follows from Eq. (10-116) that, when $f = f_\beta$, $|h_{fe}|$ is reduced to $1/\sqrt{2}$ times its low-frequency value β; therefore f_β serves as a useful measure of the band of frequencies over which the current gain remains reasonably constant. The gain decreases steadily with increasing frequency above f_β, and for this reason f_β is called the β-cutoff frequency.

For frequencies greater than about $4f_\beta$, Eq. (10-116) reduces to

$$|h_{fe}| = \beta \frac{f_\beta}{f} \tag{10-117}$$

An important figure of merit for the transistor is the frequency f_T at which $|h_{fe}|$ is unity, and this frequency, which is much greater than $4f_\beta$, is given by (10-117) as

$$|h_{fe}| = 1 = \beta \frac{f_\beta}{f_T} \tag{10-118}$$

or

$$f_T = \beta f_\beta \tag{10-119}$$

Thus for $f > 4f_\beta$, Eq. (10-117) becomes

$$|h_{fe}| = \frac{f_T}{f} \tag{10-120}$$

From Eq. (10-119),

$$\omega_T = 2\pi f_T = \beta\omega_\beta = \frac{\beta}{(C_\pi + C_\mu)r_\pi}$$

$$= \frac{g_m}{C_\pi + C_\mu} \tag{10-121}$$

All of the parameters in (10-121) depend on the quiescent operating point for the transistor. However, f_T and C_μ are not strongly dependent on the operating point, and many transistor data sheets give information permitting values for f_T and C_μ to be determined. If the quiescent collector current is known, g_m can be determined from Eq. (10-70), and then C_π can be determined from Eq. (10-121). The capacitance to be used for C_μ is usually designated on data sheets as C_{cb} (collector-to-base capacitance), or C_{ob} (output capacitance, common-base connection). Moreover, data sheets usually specify f_T in an indirect way. Although the hybrid-π model is valid for some transistors at signal frequencies up to and beyond f_T, it fails for others at frequencies significantly below f_T. Thus data sheets usually give the value of $|h_{fe}|$ at a stated frequency at which Eq. (10-120) holds, and f_T can be calculated from (10-120). The value of f_T determined in this way is called the extrapolated f_T; it is the correct value to be used in Eq. (10-121) to find the value of C_π for the hybrid-π model.

The reciprocal of Eq. (10-121) is

$$\frac{1}{\omega_T} = \frac{C_\pi + C_\mu}{g_m}$$

and since $C_\pi = C_b + C_{je}$ and $C_b = g_m\tau$, this relation can be written as

$$\frac{1}{\omega_T} = \tau + \frac{C_{je} + C_\mu}{g_m} = \frac{W^2}{2D_{eb}} + \frac{C_{je} + C_\mu}{g_m} \qquad (10\text{-}122)$$

Since C_μ is essentially the collector junction capacitance, this relation shows how f_T depends on the junction capacitances and on the width W of the base region. Furthermore, since $g_m = \lambda I_C$, it shows how f_T varies with the quiescent collector current.

10-8 SUMMARY

This chapter presents several practical BJT amplifier circuits, and it examines the properties of these circuits in some detail. It also develops in some detail small-signal models for the BJT. The more important ideas in this chapter are:

1. *The fact that circuit configurations which stabilize the quiescent operating point must be used in BJT amplifiers*
2. *The fact that the excellent matching between transistors in integrated circuits can be used to good advantage in stabilizing the quiescent point*
3. *The fact that the hybrid-π model is a good representation of the BJT with parameters closely related to the physical processes occurring in the transistor*

PROBLEMS

10-1. *Amplifier design.* The amplifier shown in Fig. 10-29 is to be designed for a quiescent point at $I_C = 2$ mA and $V_{CE} = 4$ volts with $V_{CC} = 20$ volts. The transistor can be represented by the large-signal model shown in Fig. 10-2 with $V_o = 0.7$ volt, $\beta = 50$, and $I_{CEO} = 0$. When the transistor is saturated, $V_{CE} \approx 0$.

(a) Determine the values of R_b and R_c that are required.

(b) If β increases to 80 as a result of a rise in the ambient temperature with R_b and R_c having the values determined in part a, what are the quiescent values of I_C and V_{CE}? *Note:* The model in Fig. 10-2 is valid only for $V_{CE} > 0$.

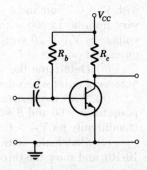

Fig. 10-29 Amplifier for Prob. 10-1.

10-2. *Self-heating.* The parameters in an amplifier like the one shown in Fig. 10-29 are $R_b = 250$ kilohms and $R_c = 4$ kilohms, and the supply voltage is $V_{CC} = 20$ volts. The transistor can be represented by the large-signal model shown in Fig. 10-2 with $V_o = 0.7$ volt, $\beta = 50$, and $I_{CEO} = 0$. The thermal resistance between the transistor and its ambient environment is $\theta_{TA} = 0.5°C/mW$.

If the ambient temperature is $T_A = 25°C$ (77°F), what is the temperature of the transistor under quiescent operating conditions?

10-3. *Amplifier scaling.* The amplifier in Fig. 10-29 has a quiescent point at I_{C1} and V_{CE1}, and the transistor can be represented by the large-signal model of Fig. 10-2 with $I_{CEO} = 0$. If both resistors are multiplied by the factor k, what is the new operating point in terms of I_{C1}, V_{CE1}, and k? Is it possible for the operating point to move out of the active region of the transistor characteristics with this kind of scaling?

10-4. *Maximum permissible dissipation.* The maximum permissible operating temperature for the transistor in Fig. 10-29 is $T_T = 175°C$, and the thermal resistance between the transistor and its ambient environment is $\theta_{TA} = 0.5°C/mW$.

(*a*) What is the maximum permissible device dissipation P_T when the ambient temperature is $T_A = 25°C$ (77°F)?

(*b*) What is the maximum permissible value of P_T when $T_A = 55°C$ (131°F)?

(*c*) The transistor is mounted on a large heat sink with the result that the temperature of the transistor case is held at 25°C. The thermal resistance between the transistor and the case is $\theta_{TC} = 0.15°C/mW$. What is the maximum permissible value of P_T under this condition?

(*d*) For the conditions in part *a*, plot a curve of maximum permissible P_T as a function of T_A for T_A between 25 and 175°C. This curve, called the dissipation-derating curve, is given on many transistor data sheets.

10-5. *Quiescent-point stability.* The transistor in the amplifier of Fig. 10-29 can be represented by the large-signal model of Fig. 10-2 with $V_o = 0.7$ volt and $I_{CEO} = 0$. In mass production the value of β will vary from 60 to 300 with the typical value being 100. The dc supply voltage is $V_{CC} = 20$ volts, and the circuit is to be designed so that the quiescent point is at $I_C = 2$ mA and $V_{CE} = 4$ volts when $\beta = 100$.

(*a*) Determine the values of R_b and R_c that will give the specified quiescent point when $\beta = 100$.

(*b*) With the resistances found in part *a*, determine the quiescent point for $\beta = 60$ and $\beta = 300$. *Note:* The large-signal model of Fig. 10-2 is valid only for $V_{CE} > 0$.

(*c*) Sketch the static load line on the i_C-v_{CE} coordinates (see Fig. 10-10), and show the three quiescent points determined in parts *a* and *b*.

10-6. *Stabilized quiescent point.* The transistor in Prob. 10-5 is to be used in the stabilized circuit of Fig. 10-30. The quiescent point is approximately the same as in the amplifier of Prob. 10-5 when $\beta = 100$.

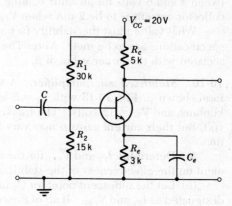

Fig. 10-30 Amplifier for Prob. 10-6.

The performance of this amplifier is to be compared with the performance of the amplifier in Prob. 10-5.

(*a*) Determine the quiescent points for $\beta = 60$, 100, and 300.

(*b*) Sketch the static load line on the i_c-v_{CE} coordinates, and show the three quiescent points found in part *a*.

(*c*) What is the limiting value of I_C as β increases without limit?

10-7. *Amplifier scaling.* The amplifier in Fig. 10-30 has a quiescent point at I_{C1} and V_{CE1}, and the transistor can be represented by the large-signal model shown in Fig. 10-11. If all the resistors are multiplied by the factor k, what is the new quiescent point in terms of I_{C1}, V_{CE1}, and k? Is it possible for the quiescent point to move out of the active region of the transistor characteristics with this kind of scaling? Does the stability factor S_e change with this kind of scaling?

10-8. *Design for a specified quiescent point and stability factor.* A *pnp* silicon transistor is to be used in an amplifier having the form shown in Fig. 10-30 with $V_{CC} = -20$ volts. The transistor can be represented by a large-signal model like the one shown in Fig. 10-11 with the battery V_o reversed (for the *pnp* transistor) and with $\beta = 100$ and $V_o = 0.7$ volt. The quiescent point is to be at $I_C = -1$ mA and $V_{CE} = -4$ volts, and the quiescent voltage drop across R_e is to be 4 volts. In addition, the stability factor is to be $S_e = 0.05$. Determine the required values of R_1, R_2, R_c, and R_e.

10-9. *Determination of the required stability factor.* An *npn* silicon transistor is used in a stabilized amplifier like the one shown in Fig.

10-30 with $V_{CC} = 16$ volts. The transistor parameters are $V_o = 0.7$ volt, $I_{CEO} = 0$, and in mass production the value of β will vary from 60 to 300. However, the quiescent collector voltage V_{CE} must lie in the range between 4 and 6 volts for all units coming off the production line, and the collector current is to be 2 mA when $V_{CE} = 4$ volts.

What value must the stability factor S_e have when $\beta = 300$ if these specifications are to be met? *Note:* The larger value of I_C is always associated with the larger value of β.

10-10. Stabilized microamplifier. A certain microamplifier has the form shown in Fig. 10-13 with $R_1 = 8$ kilohms, $R_c = 4$ kilohms, $R_2 = 2$ kilohms, and $V_{CC} = 12$ volts. The transistors are identical with $V_o = 0.7$ volt, but their current gains β may vary from 50 to 200 in mass production.

(a) Determine I_{C2} and V_{CE2} for the two extreme values of β. Comment on the effectiveness of the stabilization in this circuit.

(b) Let the quiescent point for Q_2 in the circuit described above be designated as I_{C2} and V_{CE2}. If all of the resistors in the circuit are scaled by the factor k, what is the new quiescent point in terms of I_{C2}, V_{CE2}, and k? Do not assume infinite β.

10-11. Microamplifier analysis. A certain microamplifier has the form shown in Fig. 10-14a with $R_1 = 8$ kilohms, $R_2 = 2$ kilohms, and $V_{CC} = 12$ volts. The currents I_o are supplied by two of the current sources shown in Fig. 10-14b with $R = 56$ kilohms.

(a) Determine the quiescent collector current and voltage for Q_2. Large β can be assumed, and $V_o = 0.7$ volt.

(b) If all of the resistors in the amplifier and current sources are multiplied by a factor k, what is the effect on the quiescent point for Q_2?

10-12. Small-signal analysis. An *npn* silicon transistor is used in the amplifier of Fig. 10-31. The circuit parameters and the supply voltage are given in the figure, and the transistor parameters are $\beta = 100$, $V_o = 0.7$ volt, $r_x = 50$ ohms, and $r_o = 50$ kilohms. The amplifier is used at signal frequencies where the internal transistor capacitances have a negligible effect.

(a) Determine the quiescent collector current and voltage.

(b) Determine the input resistance r_i in the small-signal model for the transistor at room temperature.

(c) Give a small-signal model for the amplifier in which C and C_e are treated as short circuits.

(d) Determine the small-signal voltage gain $|v_o/v_1|$ of the amplifier.

10-13. Small-signal models. Four useful transistor circuits are shown in Fig. 10-32. Assuming that the coupling and bypass capacitors be-

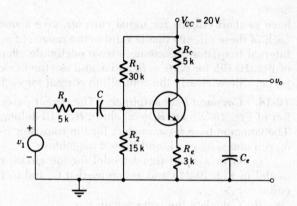

Fig. 10-31 Amplifier for Prob. 10-12.

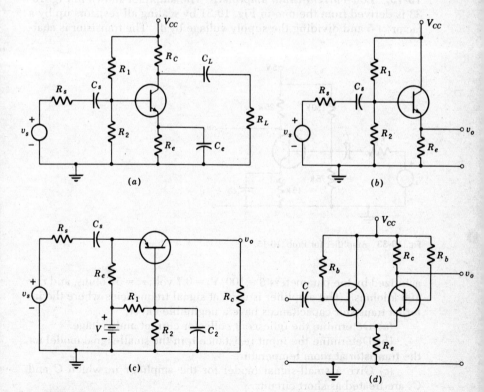

(a)

(b)

(c)

(d)

Fig. 10-32 Amplifiers for Prob. 10-13. (a) Common-emitter amplifier; (b) emitter follower; (c) common-base amplifier; (d) microamplifier.

have as short circuits for signal currents, give a small-signal model for each of these circuits that is valid in the range of frequencies where the internal transistor capacitances have negligible effects. Use the model of Fig. 10-19b for parts a, b, and d, and use the model of Fig. 10-24b for part c. Show clearly the controlling current for each controlled source.

10-14. Common-base amplifier. The circuit parameters in the amplifier of Fig. 10-32c are $R_s = 50$ ohms, $R_e = 1$ kilohm, and $R_c = 1$ kilohm. The common-base h-parameters for the transistor (see Fig. 10-24b) are $h_{ib} = 3$ ohms, $h_{fb} \approx -1$, and $h_{ob} = 2$ megohms.

   (a) Give a small-signal model for the circuit using the transistor model in Fig. 10-24b and assuming that C_s and C_2 behave as short circuits.

   (b) Calculate the voltage gain v_o/v_s.

10-15. Low-current-drain amplifier. The amplifier shown in Fig. 10-33 is derived from the one in Fig. 10-31 by scaling all resistors up by a factor of 5 and dividing the supply voltage by 4. The transistor is char-

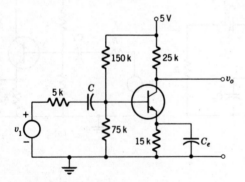

Fig. 10-33 Amplifier for Prob. 10-15.

acterized by the parameters $\beta = 100$, $V_o = 0.7$ volt, $r_x = 50$ ohms, and $r_o = 100$ kilohms. The amplifier is used at signal frequencies where the internal transistor capacitances have a negligible effect.

   (a) Determine the quiescent collector current and voltage.

   (b) Determine the input resistance r_i in the small-signal model for the transistor at room temperature.

   (c) Give a small-signal model for the amplifier in which C and C_e are treated as short circuits.

   (d) Calculate the small-signal voltage gain $|v_o/v_1|$ for the amplifier.

10-16. *Determination of transistor parameters.* The data sheet for a certain *npn* silicon transistor gives the following typical data: $\beta = 150$, $h_{oe} = 1/r_o = 10$ μmhos, $C_{cb} = 2$ pF, $|h_{fe}| = 2$ at a frequency of 100 MHz. These data apply for a quiescent operating point at $I_C = 1$ mA and $V_{CE} = 5$ volts. Assuming $r_x = 50$ ohms, give the room-temperature values for the parameters in the hybrid-π model shown in Fig. 10-27 at this quiescent point.

REFERENCES

1. Gray, P. E., et al.: "Physical Electronics and Circuit Models of Transistors," John Wiley and Sons, Inc., New York, 1964.
2. Searle, C. L., et al.: "Elementary Circuit Properties of Transistors," John Wiley and Sons, Inc., New York, 1964.

11

The Vacuum Tube
as an Amplifier

Historically, the vacuum tube was the first commercially significant electronic device to be developed. It made its appearance during the first decade of this century, and for the next four decades it provided the basis for the development of a very large electronics industry. In comparison with BJTs and FETs, vacuum tubes are large, they dissipate a lot of power, and their reliability is not very great. Their electrical characteristics in most, but not all, respects are inferior to those of transistors. For these reasons vacuum tubes have been displaced by transistors in most electronic systems. In fact, some of today's systems, such as large high-speed digital computers, would not be possible with vacuum tubes as the active elements.

On the other hand, when large signal voltages are encountered, or when large signal power is required at a high signal frequency, the tube retains superiority. Moreover, many electronic systems designed with vacuum tubes before transistors reached their present state of development perform their functions in an entirely satisfactory manner, and they continue to be manufactured with tubes. Such is the case with many of the electronic voltmeters used in academic and industrial laboratories. For the same reasons, in addition to serious cost and circuit-design problems, few television receivers are fully transistorized. However, it is probably correct to say that very little new equipment is being designed with vacuum tubes today, although a substantial amount of tube equipment is still being manufactured.

This chapter gives a brief description of the principal types of vacuum tubes, and it presents their small-signal models. It follows from the models that vacuum tubes as small-signal amplifiers differ from field-effect transistors only in details.

11-1 THE VACUUM DIODE

The vacuum diode is shown pictorially in Fig. 11-1a, and it is represented symbolically in Fig. 11-1b. It consists of two electrodes—a hollow metal rod called the cathode and a metal cylinder called the anode, or plate, surrounding the cathode and concentric with it. The plate and cathode terminals are designated p and k, respectively, in Fig. 11-1b. The diode also contains a cathode heater in the form of a tungsten wire, like the filament in an ordinary incandescent lamp, that is placed inside the hollow cathode rod. These elements are enclosed in an envelope of metal or glass from which as much air has been removed as is economically practical, leaving a very high vacuum in the interelectrode space. When the cathode-heater power is supplied from an ac source through a transformer, it is common practice to provide the transformer with a center-tap connection; this center tap is usually connected to the negative terminal of the plate power supply, as shown in Fig. 11-1b, to fix the potential of the heater relative to the cathode.

When the electrodes are at room temperature, the current I_b is extremely small, no matter how large V_b is made and regardless of the polarity of V_b (assuming, of course, that the electrical insulation outside the tube does not break down). However, if sufficient power is applied to the cathode heater to raise the cathode temperature to about 750°C (for typical small diodes), the behavior of the diode changes. At this temperature electrons escape from the cathode in large numbers in somewhat the same way that water molecules evaporate from the surface of the water in a heated pot. This process of electron emission creates a supply of free electrons in the interelectrode vacuum and thus provides carriers for charge crossing the interelectrode space. The electrons issuing

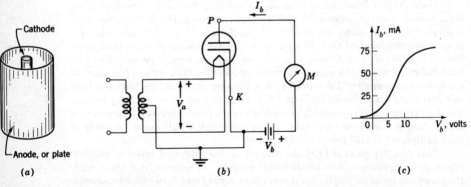

Fig. 11-1 The vacuum diode. (a) Pictorial representation; (b) symbolic representation; (c) volt-ampere characteristic.

from the cathode emerge with some velocity; hence if V_b is zero, electrons arrive at the anode at an appreciable rate, and the meter in Fig. 11-1b indicates an appreciable current. This phenomenon, first observed by Edison in his studies of the incandescent lamp, is known as the Edison effect. It is shown, somewhat exaggerated, in the volt-ampere characteristic of Fig. 11-1c. The Edison effect has been used recently in experimental energy converters for the direct conversion of heat energy to electrical energy, but the efficiencies achieved so far have not been high.

If the polarity of the battery in Fig. 11-1b is reversed, the plate of the diode is held at a negative potential relative to the cathode, and there is an electric field in the interelectrode space that opposes the flow of electrons from cathode to anode. Moreover, since the plate is at a relatively low temperature, it emits electrons at a negligible rate. Hence there is no flow of electrons from plate to cathode, for there is no source of free electrons at the plate. If the reverse voltage applied to the diode is greater than about $\frac{1}{2}$ volt, the diode current is essentially zero.

If the battery is connected in the circuit with the polarity shown in Fig. 11-1b, the plate is held positive relative to the cathode, and the electric field in the interelectrode space accelerates the electrons emitted from the cathode toward the plate. The plate current I_b therefore increases with increasing V_b as shown by the volt-ampere characteristic in Fig. 11-1c. When V_b reaches a sufficiently high value, about 10 volts for the conditions pictured in Fig. 11-1c, electrons are drawn to the plate as fast as they are normally emitted from the cathode, and the diode current increases relatively slowly with further increases in V_b. The slow increase in I_b in this region of the characteristic results from the fact that increasing V_b increases the emission from the cathode somewhat. In this region of the characteristic the diode current is limited by the cathode emission rate. Prolonged operation under these conditions usually results in permanent damage to the cathode; hence vacuum tubes are normally operated on the part of the characteristic curve lying below the knee.

The volt-ampere law for the vacuum diode can be derived from a study of the physical processes occurring inside the diode. For operating currents below the knee of the curve in Fig. 11-1c, and with the plate positive with respect to the cathode, it is

$$I_b = KV_b^{3/2} \tag{11-1}$$

where K, called the perveance of the tube, is a constant depending on the geometry of the electrodes. The derivation of (11-1) assumes, among other things, that the electrons emitted from the cathode emerge with zero initial velocity so that $I_b = 0$ when $V_b = 0$.

11-2 THE VACUUM TRIODE AS AN AMPLIFIER

The basic triode amplifier circuit is shown symbolically in Fig. 11-2.
The cathode and anode structures in triodes designed to be used as
small-signal amplifiers are usually similar to those described in Sec.
11-1 for the vacuum diode. The third electrode, called the grid, is usually a spiral of wire surrounding the cathode in the region between the
cathode and anode. The cathode heater, which plays no active role
in the operation of the tube, is omitted from this and all subsequent diagrams. When the plate is made sufficiently positive with respect to the
cathode, electrons flow from the cathode to the anode through the space
between the grid wires. If v_s is more negative than about $\frac{1}{2}$ volt, all
electrons are repelled from the grid wires, and the grid current is zero.
However, electrons are still able to flow between the grid wires from
cathode to plate to support plate current, and the grid voltage controls
this flow. If v_s is made more negative, the grid repels electrons more
strongly, and the flow of electrons from cathode to plate is reduced. If
v_s is made sufficiently negative, the flow of electrons is completely cut
off and the plate current is zero. A small signal voltage applied at the
input to the amplifier causes variations in the plate current, and an amplified signal is developed at the output terminals.

From an external point of view the triode is much like the n-chan-
nel JFET. As long as the input signal voltage is negative, the input
current is zero. The input voltage controls the amount of current flowing in the output circuit, and in doing so it produces signal amplification. Input current flows in both devices when the input voltage is positive, and hence both devices are usually operated with a negative bias
at the input in amplifier applications.

The convention regarding the current and voltage symbols used in
tube circuits is different from the one used in transistor circuits. In
the earliest days of triode amplifiers, the power for the amplifier was supplied entirely by batteries. The cathode-heating current was sup-

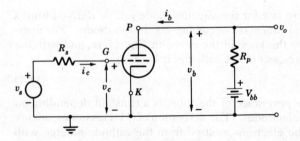

Fig. 11-2 The basic triode amplifier.

plied by the A battery, the plate current by the B battery, and the grid bias was supplied by the C battery. Thus the subscripts b and c are used to designate the total values of voltage and current at the plate and grid, respectively; capital-letter subscripts are not used. The small-signal components of voltage and current at the plate and grid are designated by the subscripts p and g, respectively. As in the case of transistor circuits, lowercase v and i are used for instantaneous values, and capital letters are used for currents and voltages that do not change with time.

Figure 11-3 shows the volt-ampere characteristics for a typical vacuum triode. In small-signal-amplifier applications the triode is usually operated with a negative grid bias so that i_c is always zero. The plate characteristics shown in Fig. 11-3b are the triode counterparts of the FET drain characteristics and the BJT collector characteristics. The plate characteristics can be used for graphical analyses of triode amplifiers, and the procedure is identical to that used with FET and BJT amplifiers.

The circuit diagram of a practical triode amplifier is shown in Fig. 11-4. This circuit is identical in form with the practical JFET amplifier described in Sec. 7-2. Under quiescent conditions with negligible grid current flowing, there is no voltage drop across R_g, and the grid is at ground potential. Resistor R_g is usually chosen to be large, typically 1 megohm, to avoid loading the input-signal source. The cathode current flowing through R_k makes the cathode a few volts positive with respect to ground; thus it makes the grid negative with respect to the cathode, and it provides the grid bias. Bypass capacitor C_k is chosen to

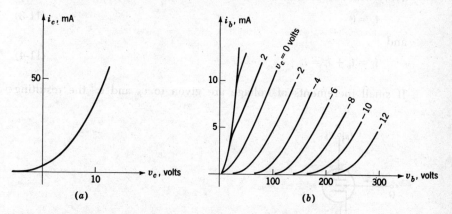

Fig. 11-3 Static triode volt-ampere characteristics. (a) Input characteristic; (b) output, or plate, characteristic.

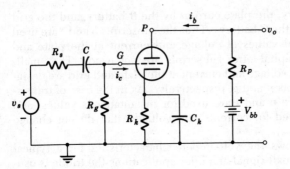

Fig. 11-4 A practical triode amplifier.

act as a short circuit to the signal component of cathode current so that no signal component of voltage appears at the cathode. The signal component of output voltage is developed by the plate current flowing through the plate-circuit resistance R_p.

The small-signal model for the triode can be developed by the method used in Chap. 10 to derive the hybrid model for the BJT. The plate current in the triode of Fig. 11-5 is a function of the plate voltage and the grid voltage as illustrated by the plate characteristic curves in Fig. 11-3b. Thus it can be expressed as

$$i_b = f_b(v_b, v_c) \tag{11-2}$$

In normal operation as a small-signal amplifier, the grid is biased negatively with respect to the cathode, and thus

$$i_c = 0 \tag{11-3}$$

and

$$i_k = i_b + i_c = i_b \tag{11-4}$$

If small increments of voltage are given to v_b and v_c, the resulting

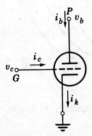

Fig. 11-5 The vacuum triode.

small increment in plate current can be expressed formally as

$$\Delta i_b \approx di_b = \frac{\partial i_b}{\partial v_b}\, dv_b + \frac{\partial i_b}{\partial v_c}\, dv_c \tag{11-5}$$

The partial derivatives in Eq. (11-5) have the dimensions of conductance; thus defining more compact symbols for the derivatives and using the subscripts p and g to designate incremental quantities in (11-5) yields

$$i_p = g_p v_p + g_m v_g = \frac{1}{r_p}\, v_p + g_m v_g \tag{11-6}$$

The quantity r_p is the small-signal plate resistance of the triode, and g_m is the small-signal transconductance.

Equation (11-6), together with the fact that $di_k = di_b = i_p$, implies that the triode can be represented by the model in Fig. 11-6a for small increments in current and voltage. It follows that the amplifier in Fig. 11-4 can be represented by the small-signal model shown in Fig. 11-6b if the coupling and bypass capacitors act as short circuits to the signal components of current and voltage; this model can be used to calculate the signal components of voltage and current in the circuit when the input signal v_s is known.

An alternative form for the small-signal triode model, obtained by making a source conversion in the model of Fig. 11-6a, is shown in Fig. 11-7. The parameter

$$\mu = g_m r_p \tag{11-7}$$

is the small-signal voltage amplification factor for the triode; it is the triode counterpart of the BJT current amplification factor $\beta = g_m r_\pi$.

When the triode is used with signals having high frequencies, greater than a few tens of kilohertz, the charge stored in the tube, both on the electrodes and in the interelectrode space, must be accounted for;

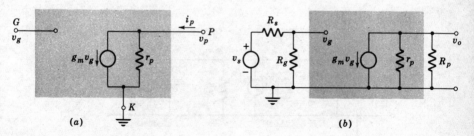

Fig. 11-6 Small-signal models. (a) For the vacuum triode; (b) for the triode amplifier of Fig. 11-4.

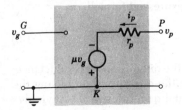

Fig. 11-7 An alternative form for the small-signal triode model.

thus the high-frequency triode model takes the form shown in Fig. 11-8. The capacitances in this model account for the effects of the stored charge, and they are called the interelectrode capacitances of the tube. As in the case of the FET and the BJT, the capacitive coupling between the input and output in the triode destroys the input-output isolation, and it turns out to be a particularly troublesome element.

For triodes designed to be used as small-signal amplifiers, the transconductance g_m ranges from about 1 to 5 mmhos, the plate resistance r_p ranges from about 10 to 100 kilohms, and the interelectrode capacitances are roughly equal at about 5 pF. The triode parameters are thus roughly equal to the corresponding FET parameters.

11-3 THE VACUUM PENTODE AS AN AMPLIFIER

It is stated in Sec. 11-2 that the capacitance coupling the output to the input in amplifiers is a troublesome element. This capacitance is reduced by a substantial factor in MOS transistors by using a dual-gate structure; it is made small, but by no means negligible, in the BJT by making the junction areas as small as possible. It is possible to reduce this capacitance to a negligible value in the vacuum tube. This result is accomplished by placing an electrostatic shield, or screen, in the interelectrode space between the grid and the plate. As indicated in Fig. 11-9, the electrostatic shield takes the form of two additional wire grids,

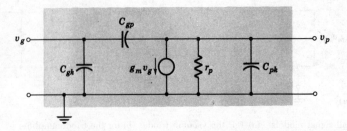

Fig. 11-8 Small-signal triode model accounting for the charge stored in the tube.

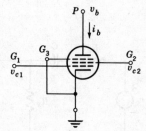

Fig. 11-9 Symbolic representation of the vacuum pentode.

one of which is connected to the cathode with the other being held at a fixed positive potential, typically 100 volts, with respect to the cathode. Electrons flow from cathode to plate through the space between the wires of these grids. Since the tube now has five electrodes, it is called a pentode tube. Grid 1 in Fig. 11-9 is the control grid, grid 2 is called the screen grid, and grid 3 is called the suppressor grid. The combined shielding action of grids 2 and 3 reduces the capacitance between the control grid and the plate to an entirely negligible value.

It is convenient to think of the cathode, grid 1, and grid 2 as forming a triode having a wire grid for its anode. The anode of this triode intercepts about 25 percent of the current leaving the cathode, and the remainder passes through the anode to be collected almost entirely by the plate of the pentode. It also turns out that, because of the shielding provided by grids 2 and 3, the pentode plate current is almost independent of the plate voltage; the plate current is determined primarily by the triode section of the tube. Thus the plate characteristics for a typical pentode have the form shown in Fig. 11-10.

The circuit diagram for a typical pentode amplifier is shown in Fig. 11-11; it is identical with the triode-amplifier circuit shown in Fig. 11-4 except for the addition of the screen-grid-bias circuit consisting of R_2 and C_2. The screen-grid current causes a voltage drop across R_2, and the

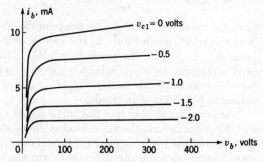

Fig. 11-10 Static plate characteristics for a typical pentode.

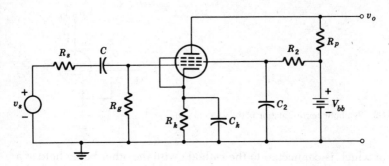

Fig. 11-11   Circuit of a typical pentode amplifier.

value of R_2 is chosen to give the desired value of screen-grid voltage. When a signal voltage is applied at the input to the amplifier, the control grid causes variations in the current leaving the cathode. Since the screen grid intercepts a constant fraction of this current, a signal component of current flows in the screen-grid circuit. Thus, if the bypass capacitor C_2 is omitted, a signal component of voltage is developed at the screen grid. These variations in screen-grid voltage have an effect on the plate current that opposes the effect of the control-grid voltage, and thus they reduce the apparent transconductance of the tube. This undesirable effect is eliminated by C_2, which acts as a short circuit to signal components of current and maintains a constant direct voltage at the screen grid.

With the screen and suppressor grids held at fixed potentials, the plate current can be expressed as a function of v_{c1} and v_p as is done for the triode in Eq. (11-2). Then it follows at once that the increments of voltage and current in the pentode are given by Eq. (11-6),

$$i_p = g_p v_p + g_m v_g = \frac{1}{r_p} v_p + g_m v_g \qquad (11\text{-}8)$$

Thus the small-signal model for the pentode, shown in Fig. 11-12, differs from the triode model only in the values of the model parameters. In particular, $C_{gp} = 0$ in the pentode model. The plate resistance is typically a megohm or more, which is much larger than that for the triode, but the remaining parameters, g_m, C_{gk}, and C_{pk}, have about the same values in pentodes as in triodes.

It is useful to note that the small-signal models for both the triode and the pentode have the same form as the model for the FET. The tube models are different from the FET model only in the values of the parameters, and even there the differences are not major ones. Thus the results of analyzing a small-signal FET circuit can often be applied directly to the vacuum-tube form of the circuit. In fact, high-voltage FETs are

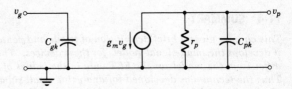

Fig. 11-12 Small-signal model for the pentode.

commercially available that, in many cases, can be substituted for tubes without making any other circuit changes.

Example 11-1

A typical pentode amplifier is shown in Fig. 11-13*a*. The coupling and bypass capacitors act as short circuits for signal components of current, and the tube parameters are $g_m = 3$ mmhos and $r_p = 500$ kilohms. The small-signal voltage gain is to be determined for signals with frequencies at which the interelectrode capacitances have negligible effects.

Solution The small-signal model for the amplifier under the stated conditions is shown in Fig. 11-13*b*. The output voltage is

$$v_o = -g_m R v_g = -(3)(45.5)v_g = -136 v_g$$

To a good approximation, $v_g = v_s$; thus

$$\frac{v_o}{v_s} = -136$$

and the voltage gain is $K_v = 136$.

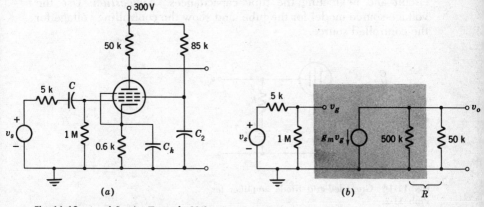

Fig. 11-13 Amplifier for Example 11-1.

11-4   SUMMARY

This chapter gives a brief description of triode and pentode vacuum tubes, and it develops the small-signal models for these devices. The models for both tubes differ from the model for the FET only in the values of the model parameters. Thus the techniques developed for analyzing small-signal FET circuits and the results obtained from analyzing such circuits apply, for the most part, to vacuum-tube circuits as well.

PROBLEMS

11-1. *Triode-amplifier design.* A triode amplifier having the form shown in Fig. 11-4 is to be designed to have a quiescent operating point at $I_b = 2$ mA and $V_b = 100$ volts. At this operating point the tube characteristics show that the grid-to-cathode voltage is −4 volts. The power-supply voltage is $V_{bb} = 250$ volts, and $R_g = 500$ kilohms.

(*a*) Determine the values of R_k and R_p required.

(*b*) Give a small-signal model for the amplifier in which the coupling and bypass capacitors are treated as short circuits. The interelectrode capacitances of the tube can be neglected.

(*c*) The small-signal parameters of the tube at the specified quiescent point are $g_m = 2$ mmhos and $r_p = 10$ kilohms. Determine the small-signal voltage gain for the case in which $R_s \ll R_g$.

11-2. *Grounded-grid-amplifier analysis.* Grounded-grid amplifiers of the general form shown in Fig. 11-14 are often used in TV receivers because they offer certain advantages related to the noise generated by tubes. The tube is characterized by the small-signal parameters μ and r_p.

(*a*) Give a small-signal model for the amplifier treating C as a short circuit and neglecting the tube capacitances. *Suggestion:* Use the voltage-source model for the tube, and show the controlling voltage for the controlled source.

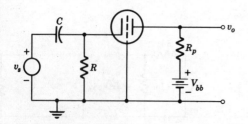

Fig. 11-14 Grounded-grid triode amplifier for Prob. 11-2.

(*b*) Determine the voltage gain of the amplifier in terms of the tube and circuit parameters.

11-3. *Pentode-amplifier design.* A pentode tube is used in an amplifier like the one shown in Fig. 11-11 with $V_{bb} = 300$ volts, $R_s = 10$ kilohms, and $R_g = 1$ megohm. The tube is to be biased so that under quiescent conditions $V_b = 100$ volts, $I_b = 2$ mA, and $V_{c2} = 100$ volts. The tube manual shows that under this condition $V_{c1} = -2$ volts and $I_{c2} = 0.8$ mA.

(*a*) Determine the values of R_k, R_2, and R_p required. Do not overlook the fact that I_{c2} flows through R_k.

(*b*) At the specified quiescent point the tube parameters are $g_m = 2$ mmhos and $g_p = 1/r_p = 0$. Give a small-signal model for the amplifier, treating the coupling and bypass capacitors as short circuits and neglecting the interelectrode capacitances.

(*c*) Determine the small-signal voltage gain of the amplifier.

11-4. *Comparison of amplifiers.* The three amplifiers shown in Fig. 11-15 are biased so that the quiescent plate, drain, and collector currents

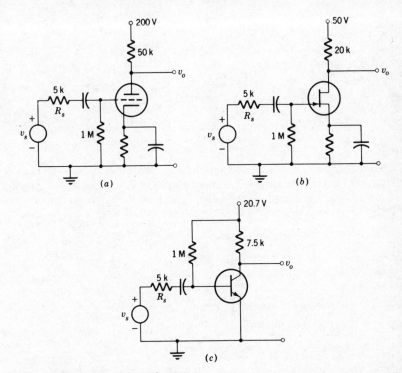

Fig. 11-15 Amplifiers for Prob. 11-4.

are each 2 mA.   The small-signal parameters are: triode, $g_m = 2$ mmhos and $r_p = 10$ kilohms; JFET, $g_m = 2$ mmhos and $r_d = 50$ kilohms; BJT, $g_m = 80$ mmhos, $r_o = 50$ kilohms, $r_x = 50$ ohms, and $\beta = 100$.

(a) The coupling and bypass capacitors act as short circuits for signal currents, and the parasitic device capacitances can be neglected. Determine the small-signal voltage gain for each amplifier.   The results should be compared with each other and with the gain of the pentode amplifier of Example 11-1.

(b) Under what conditions of signal-source resistance R_s might the BJT amplifier be the inferior circuit with respect to voltage gain?

12

Network Theorems and Basic Circuit Configurations

The small-signal models for tube and transistor amplifiers are linear circuits that contain controlled sources in addition to R's, L's, C's, and independent sources. The presence of controlled sources in these circuits can have a surprisingly strong effect on their properties when feedback, as defined in connection with Fig. 10-6b, exists. Furthermore, controlled sources often necessitate minor modifications in the standard procedures of circuit analysis; for example, errors may result in the use of Thévenin's theorem if the theorem is not stated in a way that accounts for the presence of controlled sources. There are also additional network theorems concerning controlled sources; these theorems simplify the analysis of electronic circuits, and they provide the engineer with deeper, clearer insights into the properties of these circuits. One of the objectives of this chapter is to explore in some detail the ways in which controlled sources affect the techniques of circuit analysis.

There are several basic circuit configurations that are used as standard building blocks in electronic systems of many kinds. It is important for the circuit designer to know the properties of these circuits very well and to be able to identify them as components of larger systems. Some of the more important of these circuits are presented in this chapter.

12-1 SOURCE CONVERSIONS AND NODE EQUATIONS

Small-signal models for electronic circuits contain two basically different kinds of sources, independent sources and controlled sources. An independent source is one whose output voltage or current is entirely independent of the circuit in which it is connected; the ·output of such a source is one of the independent variables in the network. A controlled source is one whose output voltage or current depends on a controlling voltage or current which in turn depends on the network in which the source is connected; the output of such a source is one of the dependent variables in the network. Four kinds of controlled sources are encountered in electronic circuits. They are voltage-controlled voltage sources, voltage-controlled current sources, current-controlled current sources, and current-controlled voltage sources. These sources can be transformed from one kind to another just as an independent current source and shunt resistance can be transformed into an equivalent independent voltage source and series resistance. A simple example illustrates some of these transformations.

The small-signal model for the JFET amplifier shown in Fig. 12-1a is given in Fig. 12-1b. The current source $g_m v_{gs}$ in the model can be converted to a voltage source in series with r_d by conventional means with the result shown in Fig. 12-1c; in the process a new symbol

$$\mu = g_m r_d \tag{12-1}$$

has been defined for compactness. It follows from the model that

$$\mu v_{gs} = \mu(v_1 - v_2) = \mu v_1 - \mu v_2 \tag{12-2}$$

Hence the voltage source μv_{gs} in Fig. 12-1c can be replaced by the

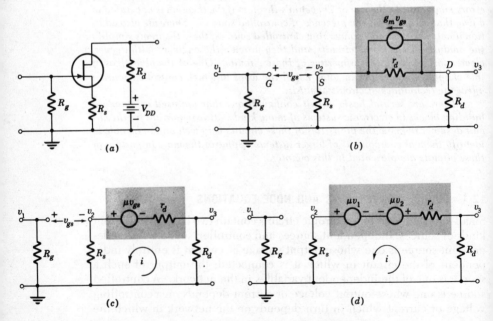

Fig. 12-1 Equivalent representations for a JFET amplifier.

equivalent series connection of two voltage sources shown in Fig. 12-1d.
It is clear from Fig. 12-1d that

$$v_2 = R_s i \quad \text{and} \quad \mu v_2 = \mu R_s i \tag{12-3}$$

Thus the source μv_2 in Fig. 12-1d can be replaced by an equivalent
source having the voltage $\mu R_s i$ as shown in Fig. 12-2a. The voltage
source $\mu R_s i$ can now be converted to a current source in parallel with r_d
by conventional means, and, using the relation given by Eq. (12-1), the
result shown in Fig. 12-2b is obtained. This sequence of source conver-
sions just completed leads from a voltage-controlled current source to a
voltage-controlled voltage source and then to a current-controlled voltage source and finally to a current-controlled current source; thus all
four types of controlled sources are encountered in the sequence.

Two additional source manipulations based on the circuit in Fig.
12-2b are shown in Fig. 12-3; for compactness the symbol $K = g_m R_s$ is
used in Fig. 12-3. In Fig. 12-3a the voltage source μv_1 is replaced by
two identical voltage sources connected in parallel. If the wire joining
them is now cut at the cross mark, the voltages on the left of the current

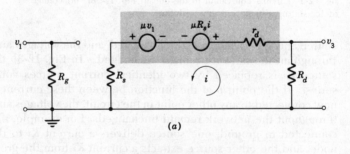

(a)

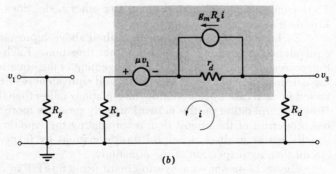

(b)

Fig. 12-2 Additional circuits equivalent to the one in Fig. 12-1b.

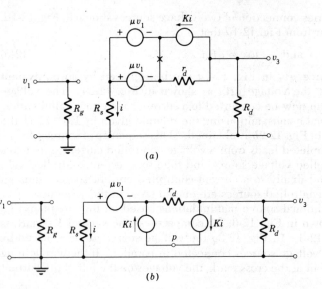

Fig. 12-3 Circuits equivalent to the one in Fig. 12-2b. (a) Voltage
source shifting; (b) current source shifting.

source and the resistor r_d are not changed, and the voltages and currents
throughout the network remain unchanged. In Fig. 12-3b the current
source Ki is replaced by two identical current sources connected in
series. If the point p at the junction between these current sources is
now connected to any other point in the circuit, the voltages and currents
throughout the network remain unchanged. For example, if point p is
connected to ground, one source delivers a current Ki to the ground
node, and the other source extracts a current Ki from the ground node.
Thus connecting p to ground, or to any other node, does not change
Kirchhoff's current law for the node.

Each of the transformations described above represents a mutual
equivalence that can be applied in either direction. Each transformation corresponds to an algebraic rearrangement of the equations relating
the currents and voltages in the circuit, and equivalent results can be obtained in each case by rearranging the equations rather than the network.
However, operating on the network usually provides more insight into
the properties of the circuit than manipulating the equations, and it is
usually easier to discern helpful transformations by inspection of the
circuit than by inspection of the equations.

Figure 12-4a shows a hybrid circuit using a JFET in combination
with a BJT; the JFET provides a high input impedance, and the BJT

provides a low output impedance. In addition, the circuit illustrates
in a simple way how controlled sources can necessitate modifications in
the conventional procedures of circuit analysis. A small-signal model
for the circuit is shown in Fig. 12-4b. The model contains four nodes
designated v_1, v_2, v_3, and ground, and it contains two separate parts. The
separate parts of a network are parts between which there can be no
flow of current; hence the part containing node v_1 in Fig. 12-4b is separate from the rest of the network. Network models for electronic circuits
often contain several separate parts, each of which can be analyzed separately from the rest of the network.

The solution of the left-hand part of the circuit in Fig. 12-4b is
trivial. A systematic procedure for writing the node equations for the
right-hand part is to equate the sum of the currents flowing away from a
node in passive branches to the sum of the currents flowing toward the
node in source branches. Thus for nodes v_2 and v_3 in Fig. 12-4b,

$$g_{d1}v_2 + g_{i2}(v_2 - v_3) = g_{m1}v_{gs} \qquad (12\text{-}4)$$

and

$$(g_{o2} + G_3)v_3 + g_{i2}(v_3 - v_2) = \beta_2 i_b \qquad (12\text{-}5)$$

But these two node equations, which should provide a solution for the
unknown voltages in the circuit, contain four unknowns. The two extra
unknowns, v_{gs} and i_b, are introduced by the controlled sources. Such
extra unknowns can always be eliminated by simple auxiliary equations
that relate them to the proper unknowns, the node-to-datum voltages; in
this case the auxiliary equations are

$$v_{gs} = v_1 - v_2 \qquad \text{and} \qquad i_b = g_{i2}(v_2 - v_3) \qquad (12\text{-}6)$$

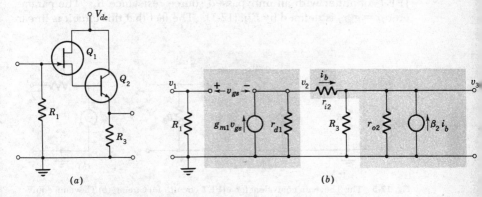

Fig. 12-4 A hybrid JFET-BJT amplifier. (a) Circuit; (b) small-signal model.

Substituting these relations into (12-4) and (12-5) and collecting terms separately on each side yields

$$(g_{d1} + g_{i2})v_2 - g_{i2}v_3 = g_{m1}v_1 - g_{m1}v_2 \qquad (12\text{-}7)$$

and

$$-g_{i2}v_2 + (G_3 + g_{o2} + g_{i2})v_3 = \beta_2 g_{i2}v_2 - \beta_2 g_{i2}v_3 \qquad (12\text{-}8)$$

The terms on the left side of these equations account for the passive branches in the circuit, and the terms on the right side account for the sources; however, the terms on the right side involving v_2 and v_3 are unknowns. Collecting all terms in (12-7) and (12-8) gives

$$(g_{d1} + g_{i2} + g_{m1})v_2 - g_{i2}v_3 = g_{m1}v_1 \qquad (12\text{-}9)$$

and

$$-(1 + \beta_2)g_{i2}v_2 + [G_3 + g_{o2} + (1 + \beta_2)g_{i2}]v_3 = 0 \qquad (12\text{-}10)$$

When the input voltage v_1 is known, these equations can be solved for v_2 and the output voltage v_3.

The discussion given above shows that, when node equations are used in the analysis of a circuit containing controlled sources, it may be necessary to use auxiliary equations to obtain a proper set of unknown variables, the node-to-datum voltages. A similar examination of the loop equations for circuits with controlled sources leads to similar results.

12-2 THÉVENIN'S THEOREM FOR CIRCUITS CONTAINING CONTROLLED SOURCES

The circuit shown in Fig. 12-5a is taken from Fig. 12-1c; it represents a JFET amplifier with an unbypassed source resistance R_s. The parameter $\mu = g_m r_d$ is defined by Eq. (12-1). The fact that this circuit is linear

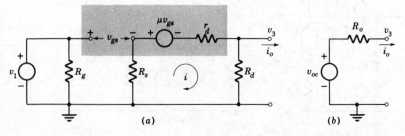

Fig. 12-5 The Thévenin equivalent for a JFET circuit. (a) Circuit; (b) Thévenin equivalent circuit.

ensures that it can be represented by a Thévenin equivalent circuit at any pair of terminals such as the output terminal pair. Special considerations arise, however, in evaluating the internal voltage and resistance for the Thévenin equivalents of circuits containing controlled sources. These considerations can be illustrated by evaluating the parameters in the circuit of Fig. 12-5b as the equivalent of the amplifier in Fig. 12-5a. The open-circuit voltage at the output of the amplifier is

$$v_{oc} = R_d i \tag{12-11}$$

and the loop equation for the amplifier is

$$-\mu v_{gs} = (r_d + R_d + R_s)i \tag{12-12}$$

But v_{gs} depends on i, and

$$\mu v_{gs} = \mu(v_1 + R_s i) \tag{12-13}$$

Substituting this relation into (12-12) and solving for i yields

$$i = \frac{-\mu v_1}{r_d + R_d + (1 + \mu)R_s} \tag{12-14}$$

Thus from Eq. (12-11),

$$v_{oc} = \frac{-\mu R_d v_1}{r_d + R_d + (1 + \mu)R_s} = -K_v v_1 \tag{12-15}$$

This is the voltage of the source in the equivalent circuit in Fig. 12-5b.

A procedure that is commonly used in evaluating the Thévenin equivalent resistance is to determine the resistance between the output terminals with all sources set equal to zero volts or amperes; for the circuit in Fig. 12-5a this resistance is

$$R_o = \frac{R_d(r_d + R_s)}{r_d + R_d + R_s} \tag{12-16}$$

It happens that this is not the correct value of resistance to use in the Thévenin equivalent circuit. Another commonly used procedure for evaluating R_o is to take the ratio of open-circuit voltage to short-circuit current. The current in a short circuit connected across the output terminals of the amplifier is given by Eq. (12-14) with $R_d = 0$; thus

$$i_{sc} = \frac{-\mu v_1}{r_d + (1 + \mu)R_s}$$

and hence

$$R_o = \frac{v_{oc}}{i_{sc}} = \frac{[r_d + (1 + \mu)R_s]R_d}{r_d + R_d + (1 + \mu)R_s} \tag{12-17}$$

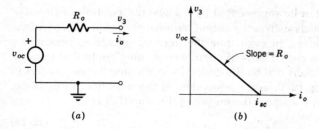

Fig. 12-6 The Thévenin equivalent circuit and its output volt-
ampere characteristic.

This value, which is different from the value given by (12-16), is the correct value for R_o in the Thévenin equivalent circuit.

Now the question that must be answered is: Why is the value of R_o
given by (12-16) incorrect? The Thévenin equivalent circuit and its output volt-ampere characteristic are shown in Fig. 12-6; the slope of this
characteristic is R_o. In circuits containing no controlled sources, all
source voltages and currents are independent of the output current, and
thus the slope of the output characteristic is not affected by these sources.
However, the controlled source μv_{gs} in the amplifier of Fig. 12-5a does
depend on the output current, and the voltage of this source varies as i_o
varies. Thus the slope of the output characteristic depends on the source
μv_{gs}, and setting its voltage equal to zero in arriving at Eq. (12-16)
necessarily leads to the wrong slope and the wrong value of R_o.

These considerations lead to the following modified statement for
Thévenin's theorem: The Thévenin equivalent for a circuit containing
controlled sources consists of a source of voltage equal to the open-circuit voltage appearing at the output terminals and acting in series with
the impedance appearing between the output terminals with all sources
that are independent of the output current adjusted for zero volts or
amperes. The impedance to be used in the Thévenin equivalent for a
circuit containing controlled sources can be evaluated correctly in a
variety of ways; however, the procedure based on the open-circuit voltage and the short-circuit current is often the simplest, especially since
the open-circuit voltage must be determined in any event.

12-3 THE SOURCE FOLLOWER AND THE SUBSTITUTION THEOREM

The JFET circuit shown in Fig. 12-7a has important applications in
practically all phases of electronic circuits. Its important features are
that it has a relatively low output impedance, roughly equal to $1/g_m$, and
in a modified form presented later in this section, it has a relatively high

input impedance, many megohms at low frequencies. A positive increment in voltage applied at the gate causes an increase in drain current,
and this increment in drain current causes a positive increment in the
output voltage at the source of the transistor. However, in order for the
increment in drain current to occur, the increment in the output voltage
must be slightly less than the increment in the input voltage, and the
voltage gain of circuit is slightly less than unity. The output voltage,

$$v_o = v_i - v_{GS} = v_i + v_{SG} \tag{12-18}$$

differs from the input voltage only by the amount of the small gate-to-
source voltage. Thus the variations in voltage at the source of the transistor follow closely the variations in input voltage, and hence the circuit
is known as the source follower.

The loop equation for the output circuit of the source follower is
independent of whether the resistance R_s is on the source side of the
transistor or on the drain side; thus the static load line for the source
follower, shown in Fig. 12-7b, is the same as that for a JFET amplifier.
The voltage transfer characteristic for the circuit is shown in Fig. 12-7c.
When v_i is more negative than V_P, the transistor is cut off and the output

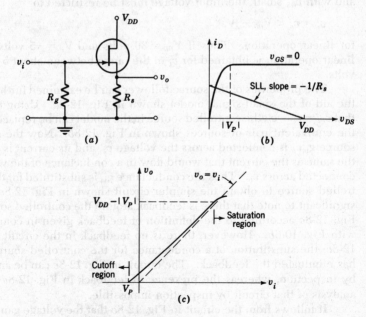

Fig. 12-7 The JFET source follower. (a) Circuit; (b) graphical analysis;
(c) voltage transfer characteristic.

voltage is zero. When v_i is made positive, both v_{GS} and v_{DS} decrease, and when v_{DS} becomes smaller than $|V_P|$, the operating point for the transistor moves into the saturation region of the drain characteristics, as indicated by the load line in Fig. 12-7b. When the operating point enters the saturation region, v_{GS} drops rapidly to zero with further increases in v_i. As is indicated in Fig. 12-7c, the voltage transfer characteristic is quite linear over a substantial range of input voltage between cutoff and saturation.

The operating point does not enter the saturation region provided that

$$v_{DS} = V_{DD} - v_o > |V_P| \tag{12-19}$$

Thus for linear operation the output voltage must be restricted to the range

$$v_0 < V_{DD} - |V_P| \tag{12-20}$$

The input voltage is

$$v_i = v_o + v_{GS} \tag{12-21}$$

and with v_{GS} small, the input voltage must be restricted to

$$v_i \approx v_o < V_{DD} - |V_P| \tag{12-22}$$

for linear operation. Thus if $V_{DD} = 30$ volts and $V_P = -5$ volts, quite linear operation is obtained for v_i in the range between about 5 and 25 volts.

The properties of the source follower can be examined further with the aid of the small-signal model shown in Fig. 12-8a. Using the fact that $v_{gs} = v_i - v_o$, the controlled source in the model can be replaced with the equivalent pair of sources shown in Fig. 12-8b. Now the current source $g_m v_o$ is connected across the voltage v_o, and its current is exactly the same as the current that would flow in a conductance of the value g_m connected across v_o. Thus the conductance g_m is substituted for the controlled source to obtain the simpler circuit shown in Fig. 12-8c. It is significant to note that there is feedback around the controlled source in Fig. 12-8a according to the definition of feedback given in connection with Fig. 10-6b. However, there is no feedback in the circuit of Fig. 12-8c; the substitution of a conductance for the controlled source $g_m v_o$ has eliminated the feedback. The circuit in Fig. 12-8c can be analyzed by inspection, whereas the presence of feedback in Fig. 12-8a makes analysis of that circuit by inspection impossible.

It follows from the circuit in Fig. 12-8c that the voltage gain of the source follower is

$$\frac{v_o}{v_i} = \frac{g_m}{g_m + 1/r_d + 1/R_s} = K_v < 1 \qquad (12\text{-}23)$$

Usually $g_m \gg 1/r_d$, and the expression for the gain can be simplified to

$$K_v = \frac{g_m R_s}{1 + g_m R_s} \qquad (12\text{-}24)$$

The input resistance of the circuit is

$$R_i = R_g \qquad (12\text{-}25)$$

and, neglecting $1/r_d$ in comparison with g_m, the output resistance is

$$R_o = \frac{1}{g_m + 1/R_s} = \frac{R_s}{1 + g_m R_s} \approx \frac{1}{g_m} \qquad (12\text{-}26)$$

The substitution of a conductance for a controlled source in Fig. 12-8 is a specific application of a more general and very useful network theorem known as the substitution theorem. The network N in Fig. 12-9a can be any network having the terminal voltage $v(t)$, an arbitrary function of time, and $Av(t)$ is the current of a voltage-controlled current source. The constant A can be any real number. If the network and source are connected so that the current through the controlled source flows in the direction of the fall in potential across the source, then the

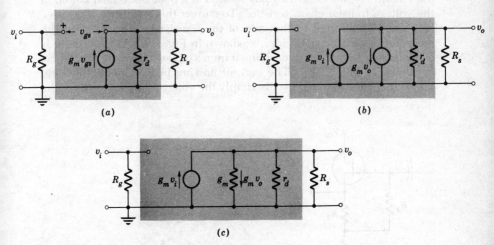

(a)

(b)

(c)

Fig. 12-8 Small-signal models for the source follower. (a) Basic model; (b) an equivalent circuit; (c) a simplified model.

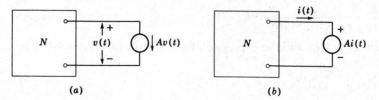

(a)   (b)

Fig. 12-9 Configurations in which the substitution theorem can be used. (a) Voltage-controlled current source; (b) current-controlled voltage source.

currents and voltages in the network remain unchanged when the controlled source is replaced by a conductance of A mhos. The proof of this theorem follows directly from the form of the node equations for the network.

Figure 12-9b illustrates conditions under which the dual form of the substitution theorem can be applied. In this case the current-controlled voltage source can be replaced by a resistance of A ohms without changing the voltages and currents in the circuit. The proof in this case follows from the form of the loop equations for the network.

In most applications for the source follower of Fig. 12-7, it is desirable for the product $g_m R_s$ to be as large as possible so that the gain is close to unity. If R_s is made large, however, the quiescent operating point for the circuit, $v_i = 0$ in Fig. 12-7c, is located at a poor place near cutoff on the voltage transfer characteristic. To correct this condition, a positive bias must be applied to the gate of the transistor. This bias can be derived from a voltage divider, as shown in Fig. 7-9; however, there is considerable advantage in deriving it from a tap on the source resistance, as shown in Fig. 12-10. The current flowing in R_g is negligible, and hence the gate-to-source bias is simply the voltage drop across R_1. The

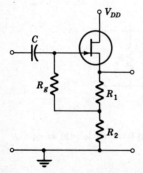

Fig. 12-10. A practical source follower.

dc blocking capacitor C renders the quiescent operating conditions independent of the circuit connected to the input of the source follower.

A positive increment of voltage applied by the input signal to the gate end of R_g increases the current through the transistor, and it thus causes a positive increment in voltage at the junction of R_1 and R_2 that is only slightly smaller than the increment at the gate. Thus the net increment of voltage across R_g is very small, and the increment of current in R_g is much smaller than it would be if R_g were connected between gate and ground. As a consequence the input resistance of the source follower appears to be much greater than R_g. The action by which an increment of voltage applied at the gate end of R_g causes an almost equal increment to appear at the opposite end is called *bootstrap* action in analogy with the man who allegedly lifted himself by his own bootstraps. Bootstrapping circuits are used extensively to obtain high input impedances in electronic circuits.

In designing the circuit of Fig. 12-10, R_1 is chosen to provide the desired gate-to-source bias and thus the desired quiescent drain current. Then R_2 is chosen to provide the desired quiescent value of drain-to-source voltage, assuming that V_{DD} has been specified. The gate resistance R_g is chosen as large as possible without causing an appreciable voltage drop resulting from the leakage current flowing across the gate junctions.

The small-signal model for the source follower of Fig. 12-10 is shown in Fig. 12-11a. Again, $v_{gs} = v_i - v_o$, and following the procedure illustrated in Fig. 12-8, the substitution theorem can be used to obtain the simplified model shown in Fig. 12-11b. The substitution theorem again eliminates the feedback from the circuit and makes analysis by inspection possible. As a result of the bootstrapping action discussed above, the current in R_g is much smaller than the current in R_2. Thus, neglecting the current in R_g gives the node equation at the output terminal as

$$\left(g_m + \frac{1}{r_d} + \frac{1}{R_1 + R_2}\right) v_o = g_m v_i$$

and thus the voltage gain is

$$\frac{v_o}{v_i} = \frac{g_m}{g_m + 1/r_d + 1/(R_1 + R_2)} = K_v < 1 \tag{12-27}$$

Usually $g_m \gg 1/r_d$, and this expression can be simplified to

$$K_v = \frac{g_m(R_1 + R_2)}{1 + g_m(R_1 + R_2)} \tag{12-28}$$

Again neglecting the current in R_g and assuming $g_m \gg 1/r_d$, the resist-

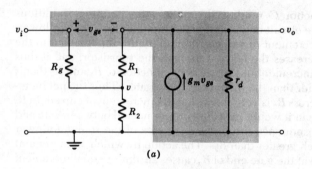

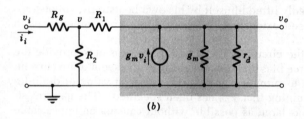

Fig. 12-11 Small-signal models for the source follower of Fig. 12-10. (*a*) Basic model; (*b*) simplified model.

ance seen looking into the output terminals is

$$R_o = \frac{1}{g_m + 1/(R_1 + R_2)} = \frac{R_1 + R_2}{1 + g_m(R_1 + R_2)} \approx \frac{1}{g_m} \qquad (12\text{-}29)$$

The input resistance of the circuit is evaluated by determining the input current i_i as a function of v_i. The input current is

$$i_i = \frac{v_i - v}{R_g} \qquad (12\text{-}30)$$

and, since i_i is much smaller than the current in R_2,

$$v = \frac{R_2}{R_1 + R_2} v_o = \frac{R_2}{R_1 + R_2} K_v v_i = K_v' v_i \qquad (12\text{-}31)$$

Substituting (12-31) into (12-30) yields

$$i_i = \frac{1}{R_g} (1 - K_v') v_i \qquad (12\text{-}32)$$

and

$$\frac{v_i}{i_i} = R_i = \frac{R_g}{1 - K_v'} \qquad (12\text{-}33)$$

The value of K'_v is always less than unity; however, when it is close to unity, the input resistance to the bootstrapped source follower is very high, and the source follower presents a negligible load to the source of input signals.

Example 12-1

The JFET used in a source follower like the one in Fig. 12-10 has the parameters $I_{DSS} = 5$ mA and $V_P = -3$ volts. The power-supply voltage is $V_{DD} = 30$ volts. The circuit is to be designed for a quiescent point at $I_D = 2$ mA and $V_{DS} = 2|V_P|$, and the small-signal performance of the circuit is to be examined.

Solution   The drain current is given by

$$I_D = I_{DSS}\left(1 - \frac{V_{GS}}{V_P}\right)^2$$

Thus

$$2 = 5\left(1 + \frac{V_{GS}}{3}\right)^2$$

which yields

$$V_{GS} = -1.11 \text{ volts}$$

Thus

$$R_1 = -\frac{V_{GS}}{I_D} = \frac{1.11}{2} = 0.55 \text{ kilohm}$$

and

$$R_1 + R_2 = \frac{V_{DD} - V_{DS}}{I_D} = \frac{30 - 6}{2} = 12 \text{ kilohms}$$

or

$$R_2 = 12 - 0.55 \approx 11.5 \text{ kilohms}$$

These values of resistance establish the specified quiescent operating point.

The small-signal transconductance of the transistor is given by Eq. (7-42) as

$$g_m = -\frac{2I_{DSS}}{V_P}\left(1 - \frac{V_{GS}}{V_P}\right) = \frac{(2)(5)}{3}\left(1 - \frac{1.11}{3}\right)$$

$$= 2.1 \approx 2 \text{ mmhos}$$

Assuming that $g_m \gg 1/r_d$, the voltage gain is given by Eq. (12-28) as

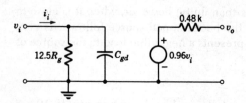

Fig. 12-12 A two-port model for the source follower of Example 12-1.

$$K_v = \frac{(2)(12)}{1 + (2)(12)} = \frac{24}{25} = 0.96$$

From Eq. (12-29) the output resistance is

$R_o = \frac{12}{25} = 0.48$ kilohm

The quantity K_v' defined in Eq. (12-31) has the value

$$K_v' = \frac{11.5}{12} \, 0.96 = 0.92$$

and thus the input resistance is given by Eq. (12-33) as

$$R_i = \frac{R_y}{1 - 0.92} = 12.5R_y$$

Thus if $R_y = 10$ megohms, $R_i = 125$ megohms. However, this statement is somewhat misleading for two reasons. First, if any additional load is connected across the output terminals, K_v' will decrease and hence R_i will decrease. And second, there is also some capacitance, approximately equal to the gate-to-drain capacitance of the transistor, in parallel with the input. Thus a better statement is that the no-load input impedance of this source follower is 125 megohms in parallel with a stated value of capacitance, usually a few picofarads.

The results obtained above imply that the no-load small-signal behavior of the circuit is represented by the two-port model shown in Fig. 12-12.

12-4 IMPEDANCE SCALING AND MICROCIRCUITS

Figure 12-13 shows the circuit diagram of the microamplifier presented in Fig. 10-13. Quiescent conditions in this amplifier are analyzed in Chap. 10, and the quiescent collector current for Q_2 is given by Eq. (10-41) as

$$I_{C2} = I_{C1} = \frac{V_{CC} - V_o}{R_1 + (2R_1 + R_2)/\beta} \tag{12-34}$$

and the collector voltage for Q_2 is

$$V_{CE2} = V_{CC} - R_c I_{C2}$$

$$= V_{CC} - R_c \frac{V_{CC} - V_o}{R_1 + (2R_1 + R_2)/\beta} \tag{12-35}$$

Now if all the resistors in the circuit are multiplied by the same factor k, the currents I_{C1} and I_{C2} are divided by k, but the voltage V_{CE2} remains unchanged. A more detailed examination shows that with this kind of resistance scaling, since the circuit is assumed to be linear, all branch currents in the circuit are divided by k and all branch voltages remain unchanged. The general form of the loop equations for circuits in which all the sources are independent voltage sources is

$$Z_{11}I_1 + Z_{12}I_2 + \cdots + Z_{1L}I_L = V_1$$
$$\cdots\cdots\cdots\cdots\cdots\cdots\cdots\cdots \tag{12-36}$$
$$Z_{L1}I_1 + Z_{L2}I_2 + \cdots + Z_{LL}I_L = V_L$$

If the applied voltages are held constant and if all of the impedances are multiplied by the same factor k, the loop equations are satisfied by dividing all of the loop currents by k. It then follows that, with this kind of impedance scaling, all branch currents are divided by k and all branch voltages remain constant.

If the circuit contains current-controlled current sources and voltage-controlled voltage sources, the theorem holds as stated. If the circuit contains voltage-controlled current sources and current-controlled

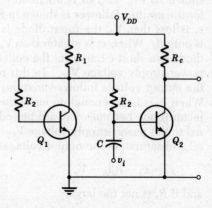

Fig. 12-13 A microamplifier.

voltage sources, the theorem holds if the transconductances and trans-resistances of the sources are scaled with the circuit impedances.

The consequences of this impedance-scaling relation are of great importance to BJT microamplifiers. The diffused silicon resistors used in these circuits cannot be held to close tolerances; variations as great as ±30 percent may be encountered from one amplifier to another. However, the variation is uniform within perhaps ±3 percent for all the resistors in a given amplifier. Thus the variation in resistance values from one microamplifier to another in mass production amounts to a nearly uniform resistance scaling. As a result the voltages across corresponding branches in the amplifiers are nearly equal, although the branch currents may differ considerably. However, the constancy of the branch voltages ensures that the transistors will have suitable quiescent operating points in spite of the substantial variations in the diffused resistances.

A dual form of the impedance-scaling theorem also holds. In this case the currents are held constant while the admittances are multiplied by a factor k and the voltages are divided by the same factor. The proof is the dual of the one given above and is based on the node equations. The special comments made above concerning controlled sources in the circuit also apply in this case.

12-5   THE EMITTER FOLLOWER AND THE REDUCTION THEOREM

Figure 12-14a shows the circuit diagram of an emitter follower. This circuit is the BJT counterpart of the source follower presented in Sec. 12-3, and it has properties that are similar to those of the source follower. When the transistor model of Fig. 9-23 is substituted for the transistor in Fig. 12-14a, the large-signal model for the emitter follower shown in Fig. 12-14b is obtained. The voltage transfer characteristic for the emitter follower is shown in Fig. 12-14c. When the input signal v_s is less than V_o, the input diode is an open circuit, and the transistor is cut off. When v_s is greater than V_o but not greater than V_{CC}, the input diode is a short circuit, but the collector diode is reverse biased by the power-supply voltage V_{CC}. In this region of the transfer characteristic, the output voltage follows variations in the input voltage quite closely. When v_s is large enough to make the output voltage equal V_{CC}, the collector diode becomes forward biased, and it connects the output terminal to the power-supply voltage V_{CC}. In the active region between cutoff and saturation the output voltage is given by

$$v_o = v_s - R_s i_B - V_o \qquad (12\text{-}37)$$

and if R_s is not too large,

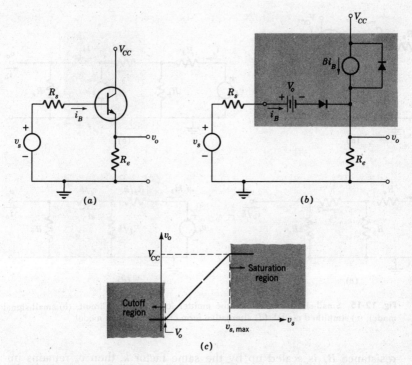

Fig. 12-14 The emitter follower. (a) Circuit; (b) large-signal model; (c) voltage transfer characteristic.

$$v_o \approx v_s - V_o \tag{12-38}$$

For small-signal operation of the emitter follower it is necessary to provide a positive bias voltage at the base of the transistor to locate the operating point at a suitable place on the voltage transfer characteristic of Fig. 12-14c. It is common practice to derive this bias from V_{CC} by means of a voltage divider; V_{BB} and R_b in Fig. 12-15a represent the Thévenin equivalent circuit for the voltage divider.

A small-signal model for the emitter follower of Fig. 12-15a is shown in Fig. 12-15b. The analysis of this circuit is complicated slightly by the fact that it contains feedback; thus it is useful to know that the feedback can be eliminated by a resistance-scaling operation similar to the one presented in Sec. 12-4. If the current in r_i is held constant at the value i_b when the source βi_b is removed from the circuit, then the current flowing toward R_e from the left is scaled down by the factor $k = 1 + \beta$. If at the same time the terminal current i_o is scaled down and the

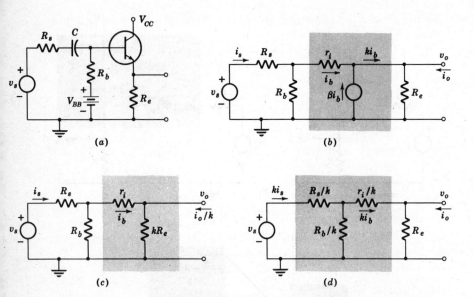

Fig. 12-15 Small-signal models for the emitter follower. (*a*) Circuit; (*b*) small-signal model; (*c*) simplified model; (*d*) alternative form of the simplified model.

resistance R_e is scaled up by the same factor k, then v_o remains unchanged, as shown in Fig. 12-15c, and the voltages remain unchanged throughout the circuit. It is not difficult to show that the node equations for the circuits in Fig. 12-15b and c are identical. Note that there is no feedback in the circuit of Fig. 12-15c; in fact, there is no controlled source in the circuit.

Several properties of the emitter follower can be perceived directly from the model in Fig. 12-15c. First, it is clear that the voltage gain of the circuit is less than unity; however, the current gain can be large. Second, the input resistance depends on the resistance R_e which is connected on the output side of the transistor. However, if $\beta = 100$ and if R_e has the typical value of 5 kilohms, then $kR_e = 500$ kilohms, and the effect of R_e on the input circuit is small.

The output resistance of the emitter follower cannot be seen by inspection of the model in Fig. 12-15c because the output current has been scaled by the factor k. This difficulty can be remedied by rescaling the model as shown in Fig. 12-15d. In making the transition from Fig. 12-15c to 12-15d all resistances have been divided by k and all currents have been multiplied by k; thus the unscaled output current appears in the model of Fig. 12-15d, and the resistance seen at the output terminals of this model is the output resistance of the emitter follower. The output resistance depends on the resistance R_s of the signal source. However, if $\beta = 100$ and if $R_s = 5$ kilohms, then $R_s/k = 50$ ohms, and the output resistance of the circuit will be less than 100 ohms. If $R_s = 0$, the output resistance is

$$R_o = \frac{r_i}{k} = \frac{r_x + r_\pi}{1 + \beta} \approx \frac{r_\pi}{\beta} = \frac{1}{g_m} \qquad (12\text{-}39)$$

The elimination of a controlled source by impedance scaling in Fig. 12-15 is a special case of a more general network theorem known as the reduction theorem, a theorem which in some respects is an extension of the substitution theorem. The substitution theorem permits the elimination of current-controlled voltage sources and voltage-controlled current sources in certain configurations; the reduction theorem permits the elimination of current-controlled current sources and voltage-controlled voltage sources in certain other configurations. The reduction theorem, like the substitution theorem, eliminates feedback and often makes circuit analysis possible by inspection.

The reduction theorem applies to the network configurations shown in Fig. 12-16. The network N_1 in Fig. 12-16a is any linear two-terminal network having the terminal current i_1, an arbitrary function of time, and N_2 is any other linear two-terminal network. The constant A of the current-controlled current source Ai_1 is any real number, positive or negative. The theorem states that all voltages in N_1 and N_2 remain unchanged if the source Ai_1 is removed and if:

1. Each resistance, inductance, and elastance (elastance is $S = 1/C$, the reciprocal of capacitance) in N_1 is divided by $1 + A$ and the current of each current source in N_1 is multiplied by $1 + A$, or
2. Each resistance, inductance, and elastance in N_2 is multiplied by $1 + A$ and the current of each current source in N_2 is divided by $1 + A$.

Proof of the theorem is based on the impedance-scaling theorem of Sec. 12-4. When the source Ai_1 is removed and the parameters of N_1 are scaled as specified in the first part of the theorem, all voltages in N_1 remain unchanged, including the terminal voltage, and all currents in N_1 are multiplied by $1 + A$, including the terminal current. Thus the current and voltage delivered to N_2 are unchanged, and the voltages in N_1 and N_2 are unchanged. After this transformation is completed, all resistances, inductances, and elastances in the composite network N_1-N_2 can be multiplied by $1 + A$ and the currents of all current sources can be divided by $1 + A$ without changing any of the voltages in N_1-N_2. Thus part 2 of the theorem is proved.

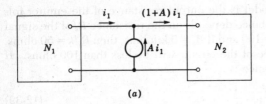

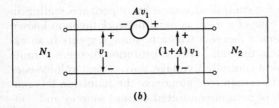

Fig. 12-16 Circuit configurations that can be simplified by the reduction theorem. (a) Current-controlled current source; (b) voltage-controlled voltage source.

When all currents and voltages in the network are sinusoidal and of the same frequency, the constant of the controlled source may be a complex number; the reduction theorem holds in this case also.

The dual form of the reduction theorem is applicable to the network configuration shown in Fig. 12-16b. It states that all currents in the network of Fig. 12-16b remain unchanged if the voltage source Av_1 is replaced with a short circuit and if:

1. Each resistance, inductance, and elastance and the voltage of each voltage source in N_1 is multiplied by $1 + A$, or
2. Each resistance, inductance, and elastance and the voltage of each voltage source in N_2 is divided by $1 + A$.

Proof of the dual form of the theorem is also based on impedance scaling, and it is the dual of the proof outlined above.

In applying the reduction theorem, it is essential that the two networks N_1 and N_2 be properly identified at the outset. The two necessary requirements are: (1) the terminal current (or voltage) of the network identified as N_1 must be the controlling quantity for the controlled source to be eliminated, and (2) no current may enter or leave either network through any terminal other than the two terminals at which the networks and controlled source are joined in parallel (or series). It should be noted that connections joining separate parts of a network carry no current and can therefore be ignored with respect to item 2.

The use of the reduction theorem is illustrated further in the next few paragraphs, where it is used in the analysis of several transistor circuits. These circuits are important for their own sake because they are basic building blocks that are widely used in electronic systems. The first circuit to be examined, the Darlington-connected amplifier, is shown in Fig. 12-17a. This amplifier uses two transistors, and it has a relatively high input resistance; it is used in integrated circuits as well as in circuits made of discrete components. In the interest of simplicity, the biasing resistors are omitted from the diagram in Fig. 12-17a. A small-signal model for the amplifier is shown in Fig. 12-17b; this model assumes that the output resistances r_o of the transistors are large and can be neglected. Figure 12-17c shows the model in a rearranged form with the current source $\beta_1 i_{b1}$ represented by an equivalent pair of sources in accordance with the current-source shifting operation illustrated in Fig. 12-3b.

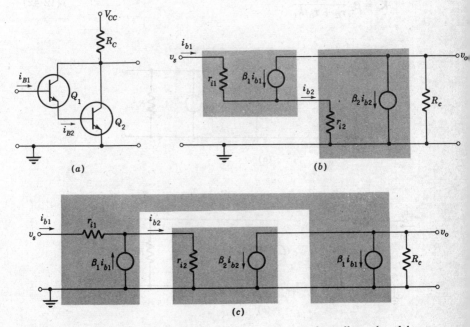

Fig. 12-17 The Darlington-connected BJT amplifier. (a) Circuit; (b) small-signal model; (c) model equivalent to the one in part b.

The circuit in Fig. 12-17c has two separate parts, and the left-hand part is in a form that can be simplified with the reduction theorem. Defining $k_1 = 1 + \beta_1$ and applying the theorem yields the circuit shown in Fig. 12-18a. Using the fact that $i_{b2} = k_1 i_{b1}$, the circuit can be put in the final form shown in Fig. 12-18b. This circuit has no feedback, and it can be analyzed by inspection. The input resistance of the amplifier is

$$R_i = r_{i1} + k_1 r_{i2} = r_{i1} + (1 + \beta_1)r_{i2} \qquad (12\text{-}40)$$

which is much higher than the input resistance of an ordinary common-emitter amplifier because of the second term; thus the Darlington-connected amplifier presents less load to a high-impedance signal source. The output voltage in the circuit of Fig. 12-18b is

$$v_o = - R_c(\beta_1 + k_1 \beta_2)i_{b1} \approx -k_1 \beta_2 R_c i_{b1}$$

$$= -k_1 \beta_2 \frac{R_c}{r_{i1} + k_1 r_{i2}} v_s \qquad (12\text{-}41)$$

Thus the voltage gain is

$$K_v = \beta_2 \frac{R_c}{r_{i2} + r_{i1}/k_1} \qquad (12\text{-}42)$$

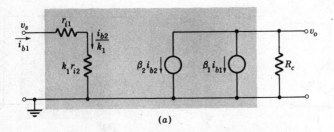

(a)

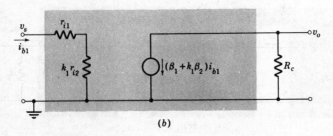

(b)

Fig. 12-18 Simplification of the model for the Darlington-connected amplifier.

This is about the same as the gain of a normal common-emitter amplifier; however, if the signal source has a high internal impedance, the effective gain of the Darlington-connected amplifier may be substantially greater because of its high input resistance.

The circuit shown in Fig. 12-4a can be viewed as a hybrid Darlington-connected emitter follower. The JFET at the input gives it a very high input resistance, and the BJT at the output gives it a very small output resistance. The voltage gain of the circuit is very close to unity. The small-signal model for the circuit is given in Fig. 12-4b, and it is repeated in Fig. 12-19a under the assumption that r_{d1} and r_{o2} are large enough to be neglected. Since $v_{gs} = v_1 - v_2$, the substitution theorem can be used in the manner illustrated in Fig. 12-8 to obtain the simpler model shown in Fig. 12-19b. Now defining $k_2 = 1 + \beta_2$ and applying the reduction theorem leads to the further simplified model of Fig. 12-19c, and a simple source conversion yields the alternative form shown in Fig. 12-19d. The quantity $k_2 g_{m1}$ represents the conductance of the element it stands by.

The input resistance of the circuit is just R_1; it can be made much larger by bootstrapping R_1 in the manner illustrated in Fig. 12-10. The output resistance is effectively

$$R_o \approx \frac{r_{i2}}{k_2} + \frac{1}{k_2 g_{m1}} \tag{12-43}$$

$$\approx \frac{1}{g_{m2}} + \frac{1}{\beta_2 g_{m1}} \tag{12-44}$$

which usually amounts to a few tens of ohms. The resistances in series with R_3 in Fig. 12-19d are usually much smaller than R_3, and hence the voltage gain is close to unity.

It must be noted, however, that the circuit given in Fig. 12-4a is a prototype circuit. Practical circuits will usually require some positive bias voltage at the gate of Q_1.

The circuit of a diode-biased microamplifier is given in Fig. 12-13, and it is repeated in Fig. 12-20a. A small-signal model for the amplifier which is valid at frequencies where the coupling capacitor is a short circuit is shown in Fig. 12-20b. This model assumes that the output resistances r_o of the transistors are large enough to be neglected, and it assumes that the transistors are identical. The model is in a form that can be simplified with the reduction theorem. Thus, defining $k = 1 + \beta$ and applying the theorem yields the reduced model shown in Fig. 12-20c. The output circuit of the amplifier is simple; however, the input circuit is complicated by the elaborate network of bias resistors, most of which make a negligible contribution to the small-signal properties of

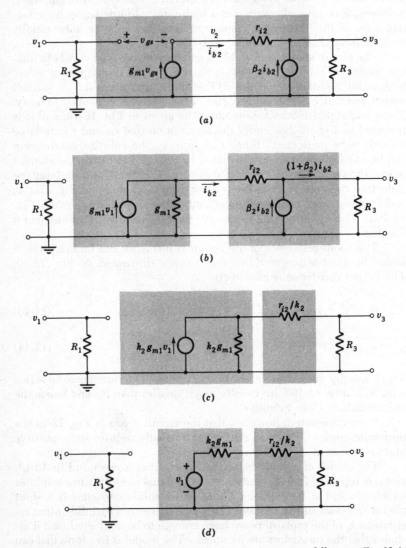

Fig. 12-19 Small-signal models for the hybrid Darlington emitter follower in Fig. 12-4. (a) Model of Fig. 12-4b; (b) model after applying the substitution theorem; (c) model after applying the reduction theorem; (d) alternative form obtained by a source conversion.

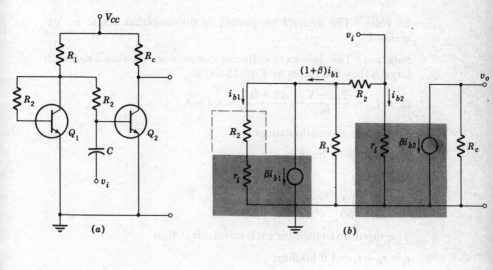

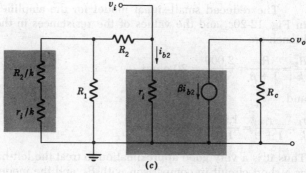

Fig. 12-20 Diode-biased microamplifier. (a) Circuit; (b) small-signal model; (c) reduced model.

the amplifier. The properties of this amplifier are best studied by means of the following numerical example.

Example 12-2

A certain commercial microamplifier has the form shown in Fig. 12-20a. The transistors are identical, and their parameters are $\beta = 100$, $V_o = 0.7$ volt, $r_x = 100$ ohms, and $g_o = 1/r_o \approx 0$. The circuit parameters are $R_c = 5$ kilohms, $R_1 = 8$ kilohms, and $R_2 = 2$ kilohms, and the amplifier is to be used with a power-supply voltage $V_{CC} =$

12 volts. The general properties of the amplifier are to be examined.

Solution The quiescent collector currents are identical, and with large β they are given by Eq. (12-34) as

$$I_{C1} = I_{C2} \approx \frac{V_{CC} - V_o}{R_1} = \frac{12 - 0.7}{8} \approx 1.4 \text{ mA}$$

Then the transconductances are

$$g_m = 40 I_C = 56 \text{ mmhos}$$

and

$$r_\pi = \frac{\beta}{g_m} = \frac{100}{56} \approx 1.8 \text{ kilohms}$$

The input resistance for each transistor is thus

$$r_i = r_x + r_\pi = 1.9 \text{ kilohms}$$

The reduced small-signal model for the amplifier is shown in Fig. 12-20*c*, and the values of the resistances in the left-hand branch are

$$\frac{R_2}{k} = \frac{R_2}{1 + \beta} = \frac{2{,}000}{101} \approx 20 \text{ ohms}$$

and

$$\frac{r_i}{k} = \frac{r_i}{1 + \beta} = \frac{1{,}900}{101} \approx 19 \text{ ohms}$$

Thus it is a very good approximation to treat the left-hand branch as a short circuit in comparison with R_2, and the model takes the form shown in Fig. 12-21.

The input resistance of the amplifier is thus

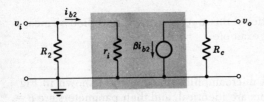

Fig. 12-21 Small-signal model for the amplifier of Example 12-2.

$$R_i = R_2 \| r_i = \frac{(2)(1.9)}{3.9} \approx 1 \text{ kilohm}$$

The output resistance is

$$R_o = R_c = 5 \text{ kilohms}$$

Some amplifiers of this type add an emitter follower at the output to lower the output resistance to a value less than 100 ohms. The output voltage of the amplifier is

$$v_o = -R_c \beta i_{b2} = -R_c \beta \frac{v_i}{r_i}$$

and the voltage gain is thus

$$K_v \doteq \beta \frac{R_c}{r_i} = 100 \frac{5}{1.9} \approx 260$$

12-6 THE DIFFERENCE AMPLIFIER

It is pointed out in Chap. 10 that the design of microamplifiers is greatly aided by the use of circuit configurations that take advantage of the close matching and tracking exhibited by the transistors in microamplifiers. The amplifier of Figs. 12-20 and 10-13 uses this property of the transistors to obtain a well-stabilized quiescent operating point. The difference amplifier shown in Fig. 12-22 is another configuration for accomplishing the same result; this circuit is a basic building block for microcircuits. The fact that the difference amplifier has two input

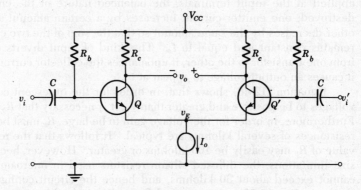

Fig. 12-22 The prototype difference amplifier.

terminals, v_i and v_i' in Fig. 12-22, provides a degree of flexibility that is valuable in many applications.

With a perfectly balanced circuit the current from the current source I_o divides equally between the two transistors; thus under quiescent conditions with $v_i = v_i' = 0$, the transistor currents are

$$I_E = I_E' = \tfrac{1}{2} I_o \approx I_C = I_C' \qquad (12\text{-}45)$$

Treating the base-to-emitter voltage for each transistor as a constant V_o, the voltage at the emitters is

$$V_E = V_{CC} - R_b I_B - V_o$$

$$= V_{CC} - V_o - \frac{R_b}{2\beta} I_o \qquad (12\text{-}46)$$

The collector-to-emitter voltages are then

$$V_{CE} = V_{CE}' = V_{CC} - R_c I_C - V_E \qquad (12\text{-}47)$$

$$= V_o + \left(\frac{R_b}{2\beta} - \frac{R_c}{2} \right) I_o \qquad (12\text{-}48)$$

With equal quiescent collector currents in the balanced circuit, the quiescent output voltage is $v_o = 0$. If two equal signal voltages $v_i = v_i'$ are applied at the input terminals, the balanced nature of the circuit is not disturbed, and the current I_o continues to divide equally between the two transistors. Thus the collector currents remain constant, and the output voltage remains $v_o = 0$. The only change in the circuit is that the voltage v_E at the emitters follows exactly the variations in $v_i = v_i'$. If two voltages $v_i = -v_i'$ that are equal in magnitude but opposite in polarity are applied at the input terminals, the balanced nature of the circuit is destroyed; one emitter current increases by a certain amount and the other decreases by the same amount so that the sum of the two currents remains constant and equal to I_o. This kind of input diverts current from one transistor to the other, it unbalances the collector currents, and it causes an output voltage to appear at v_o.

Equation (12-48) shows that, in order for the quiescent collector voltages to be positive and greater than V_o, it is necessary that $R_b > \beta R_c$. Furthermore, in order for the voltage gain to be large, R_c must be large; resistances of several kilohms are typical. It follows that the required value of R_b may easily be 100 kilohms or greater. However, because of size limitations, the diffused silicon resistors used in microamplifiers cannot exceed about 30 kilohms, and hence the circuit configuration shown in Fig. 12-22 is not practical for realization as a microamplifier.

This problem is eliminated by the slightly modified circuit configuration shown in Fig. 12-23; in this circuit forward bias for the emitter

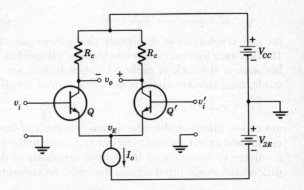

Fig. 12-23 A difference amplifier with positive and negative power-supply voltages.

junctions is obtained by connecting the emitters through the source I_o to a negative voltage rather than by connecting the bases through R_b to a positive voltage. Under quiescent conditions $v_i = v_i' = 0$, and the bases are connected to ground. With a perfectly balanced circuit the quiescent transistor currents are again

$$I_E = I_E' = \tfrac{1}{2}I_o \approx I_C = I_C' \qquad (12\text{-}49)$$

By treating the base-to-emitter voltage drop as a constant V_o, the voltage at the emitters becomes

$$V_E = -V_o \qquad (12\text{-}50)$$

and the quiescent collector-to-emitter voltages are

$$V_{CE} = V_{CE}' = V_{CC} - V_E - R_c I_C$$
$$= V_{CC} + V_o - \tfrac{1}{2}R_c I_o \qquad (12\text{-}51)$$

It is pointed out in Sec. 12-4 that the resistance level may vary considerably from one microamplifier to another. Since V_{CE} is usually the small difference between two much larger quantities as indicated by (12-51), these variations in resistance level can cause intolerable variations in V_{CE}. However, when the current source I_o is approximated by a transistor circuit like the one shown in Fig. 10-14b, the current I_o varies inversely with the resistance level. Thus under this condition V_{CE} is very nearly constant from one amplifier to another and is independent of the transistor parameters and the resistance level. Thus the quiescent collector voltage in the circuit of Fig. 12-23 is highly stabilized.

Input signal voltages of the form

$$v_i = v_i' = v_c \tag{12-52}$$

are called common-mode signals; the voltage gain of the amplifier in Fig. 12-23 is zero for common-mode signals. It must be noted, however, that because of the lack of perfect balance in real amplifiers, the common-mode gain, though small, is not zero. Input signals of the form

$$v_i = -v_i' = v_d \tag{12-53}$$

are called differential-mode signals. Signals of this form unbalance the circuit and divert current from one transistor to the other; thus they cause a voltage v_o to appear at the output terminals of the amplifier. With a differential-mode input signal, the collector currents can be expressed as

$$i_C = I_C + i_c \quad \text{and} \quad i_C' = I_C - i_c \tag{12-54}$$

where I_C is the constant quiescent component of current and i_c is the signal component. The output voltage is then

$$\begin{aligned} v_o &= V_{CC} - R_c i_C' - V_{CC} + R_c i_C \\ &= R_c(i_C - i_C') \\ &= 2R_c i_c \end{aligned} \tag{12-55}$$

It is useful to know that arbitrary input voltages v_i and v_i' can be represented as the superposition of a common-mode pair and a differential-mode pair of voltages; thus arbitrary inputs can be expressed as

$$v_i = v_c + v_d \quad \text{and} \quad v_i' = v_c - v_d \tag{12-56}$$

Adding these equations and then subtracting them yields

$$v_c = \tfrac{1}{2}(v_i + v_i') \quad \text{and} \quad v_d = \tfrac{1}{2}(v_i - v_i') \tag{12-57}$$

When the arbitrary input voltages are known, the common-mode and differential-mode components can be calculated from Eqs. (12-57). The common-mode component makes a negligible contribution to the output voltage.

A small-signal model for the difference amplifier of Fig. 12-23 is shown in Fig. 12-24a; the output resistances r_o of the transistors have been neglected in this model. Redrawing the circuit and using the source-shifting transformation illustrated in Fig. 12-3b leads to the circuit shown in Fig. 12-24b. Kirchhoff's current law applied to the input portion of this circuit gives

$$(1 + \beta)i_b = -(1 + \beta)i_b'$$

and thus

$$i_b = -i_b' \tag{12-58}$$

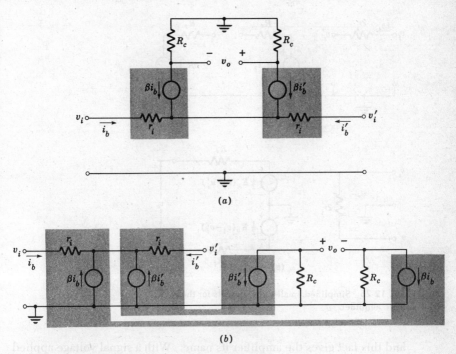

Fig. 12-24 Small-signal models for the difference amplifier.

It follows that the two current sources in the input portion of the circuit cancel exactly, and the simplified circuit shown in Fig. 12-25a is obtained after making source transformations in the output portion of the circuit. The voltage sources are

$$\beta R_c i_b = \beta \frac{R_c}{2r_i}(v_i - v_i') = \tfrac{1}{2}K_v(v_i - v_i') \tag{12-59}$$

and thus the model for the difference amplifier can be put in the final form shown in Fig. 12-25b.

Several different operating modes for the difference amplifier can be examined with the aid of the model in Fig. 12-25b. With a common-mode input signal, $v_i = v_i'$, and $v_o = 0$. With a differential-mode input, $v_i = -v_i'$, and

$$v_o = 2K_v v_i \tag{12-60}$$

With arbitrary input voltages, the output voltage is

$$v_o = K_v(v_i - v_i') \tag{12-61}$$

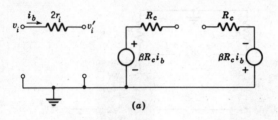

(a)

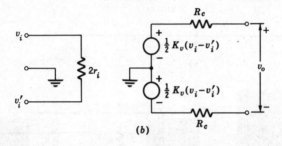

(b)

Fig. 12-25 Simplified small-signal models for the difference amplifier.

and this fact gives the amplifier its name. With a signal voltage applied between one input terminal and ground with the other input terminal grounded, the input signal is said to be single ended; with a single-ended input, the output voltage is

$$v_o = K_v v_i \qquad \text{or} \qquad v_o = -K_v v_i' \qquad (12\text{-}62)$$

Thus the difference amplifier has a (sign) inverting input terminal and a noninverting input terminal. This feature adds a flexibility to the amplifier that is very useful in many applications. The voltage gain of the amplifier is, from Eq. (12-59),

$$K_v = \beta \frac{R_c}{r_i} \qquad (12\text{-}63)$$

The input resistance is

$$R_i = 2r_i \qquad (12\text{-}64)$$

and the output resistance is

$$R_o = 2R_c \qquad (12\text{-}65)$$

The principal output voltage of the difference amplifier appears between the collector terminals of the two transistors, and neither of these terminals is grounded. In most applications, however, it is desirable for one of the output terminals to be grounded. It is possible to take the output voltage of the difference amplifier from one collector terminal and ground, and this is often done. The model in Fig. 12-25b shows that in this case the voltage gain is reduced by a factor of 2, and also in this case, the output voltage is dependent on the supply voltage V_{CC} so that any drift in V_{CC} appears as an unwanted signal at the output terminals. It is therefore of interest to know that the diode-biased amplifier of Fig. 10-14a can be combined with the difference amplifier in such a way that it amplifies the collector-to-collector voltage of the difference amplifier and converts it to a single-ended output voltage with one terminal grounded. This combination is shown in Fig. 12-26. Transistors Q_1 and Q_1' with the current source I_o constitute the difference amplifier, and transistors Q_2 and Q_3 with resistors R_1 and R_2 constitute the diode-biased amplifier. The quiescent operating conditions in the diode-biased amplifier are analyzed in connection with Fig. 10-14. This circuit configuration is used in a number of commercial microamplifiers.

The collector currents in Q_1 and Q_1' can be expressed as

$$i_{C1} = I_{C1} + i_{c1} \quad \text{and} \quad i_{C1}' = I_{C1} - i_{c1} \tag{12-66}$$

and the collector voltage at Q_3 can be expressed as

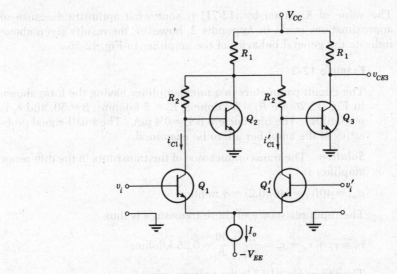

Fig. 12-26 A microamplifier with differential inputs and a single-ended output.

$$v_{CE3} = V_{CE3} + v_o \tag{12-67}$$

where I_{C1} and V_{CE3} are the constant quiescent components of current and voltage and i_{c1} and v_o are the signal components. An approximate analysis of the diode-biased amplifier given in Appendix 2 shows that

$$v_{CE3} \approx \frac{2}{\beta} V_{CC} + V_o + R_1 I_o - 2\beta R_1 i_{c1}$$

and thus the signal component of the output voltage is

$$v_o = -2\beta R_1 i_{c1} \tag{12-68}$$

But $i_{c1} = \beta i_{b1}$, and from Fig. 12-25a

$$i_{c1} = \beta i_{b1} = \beta \frac{v_i - v_i'}{2r_{i1}} \tag{12-69}$$

Hence

$$v_o = -\beta^2 \frac{R_1}{r_{i1}} (v_i - v_i') \tag{12-70}$$

and the voltage gain is

$$K_v = \left| \frac{v_o}{v_i - v_i'} \right| = \beta^2 \frac{R_1}{r_{i1}} \tag{12-71}$$

The value of K_v given by (12-71) is somewhat optimistic because of approximations made in Appendix 2; however, the results given above indicate the general behavior of the amplifier in Fig. 12-26.

Example 12-3

The circuit parameters in a microamplifier having the form shown in Fig. 12-26 are $R_1 = 8$ kilohms, $R_2 = 2$ kilohms, $\beta = 50$, and r_x is negligible. The bias current is $I_o = 0.4$ mA. The small-signal properties of the amplifier are to be examined.

Solution   The transconductance of the transistors in the difference amplifier is

$$g_{m1} = 40 I_{C1} = (40)(0.2) = 8 \text{ mmhos}$$

The input resistance of these transistors is thus

$$r_{i1} = r_x + r_\pi \approx r_\pi = \frac{\beta}{g_{m1}} = \frac{50}{8} = 6.25 \text{ kilohms}$$

Then from Eq. (12-71) the voltage gain is

$$K_v = (2,500) \frac{8}{6.25} = 3,200$$

and from Fig. 12-25 the input resistance of the amplifier is

$R_i = 2r_{i1} = 12.5$ kilohms

Another difference amplifier that is used in commercial microcircuits is shown in Fig. 12-27. Transistors Q_1 and Q_1' form the difference amplifier, and Q_2 is an emitter follower that provides, among other things, a low output impedance. With matched transistors the quiescent collector currents in the difference amplifier are equal, $I_{C1} = I_{C1}'$. Assuming that $v_{BE} = V_o = 0.7$ volt, the voltage at the emitters in the difference amplifier under quiescent conditions is

$$V_E = -V_o = -V_{CC} + 2RI_{C1} \tag{12-72}$$

The quiescent voltage at the collector of Q_1' is

$$V_C = V_{CC} - 2RI_{C1}$$

and from (12-72)

$$V_C = V_o \tag{12-73}$$

The quiescent collector-to-emitter voltage for Q_1' is thus

$$V_{CE1}' = V_C - V_E = 2V_o = 1.4 \text{ volts} \tag{12-74}$$

The output voltage under quiescent conditions is

$$v_o = V_C - V_o = 0$$

Thus zero input voltage gives zero output voltage, a condition that is very useful in many applications. The quiescent collector current in the emitter follower is

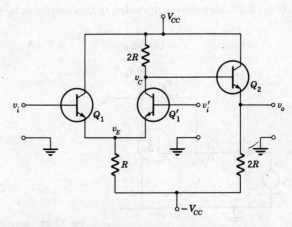

Fig. 12-27 Another difference microamplifier.

$$I_{C2} \approx \frac{V_{CC}}{2R} \approx I_{C1} \tag{12-75}$$

Assuming matched transistors, the quiescent collector voltage V'_{CE1} given by Eq. (12-74) depends on equal power-supply voltages and on the 2:1 ratio of the resistors in the difference amplifier. Since resistance ratios can be held to close tolerances in microamplifiers and since power supplies can be regulated, the value of V'_{CE1} is well stabilized in this circuit.

12-7 SUMMARY

This chapter presents a number of circuit transformations and network theorems related to controlled sources, and it also presents several important amplifier configurations that serve as basic building blocks in electronic systems. The network theorems are important not only because they often simplify the problem of circuit analysis, but also because they aid the engineer in gaining valuable insights into the properties of various circuits. The substitution and reduction theorems are particularly helpful because they remove feedback from the circuit and thereby often make it possible to perceive the important properties of circuits by inspection of the circuit diagram.

The source follower, emitter follower, difference amplifier, and diode-biased amplifier are basic building blocks. It is important for the engineer to understand the properties of these circuits, and it is very helpful if he can identify them when they occur as components in larger electronic systems.

PROBLEMS

12-1. *Nodal analysis.* An enhancement MOST amplifier is shown in Fig. 12-28; gate bias is provided in this amplifier by the resistor R_g. The

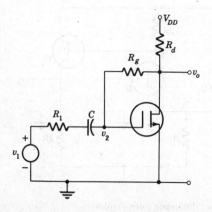

Fig. 12-28 Amplifier for Prob. 12-1.

transistor and circuit parameters are $g_m = 2$ mmhos, $r_d = 50$ kilohms, $R_d = 5$ kilohms, $R_g = 1$ megohm, and $R_1 = 5$ kilohms. The blocking capacitor can be treated as a short circuit for signal components of voltage, and the parasitic capacitances of the transistor can be neglected.

(a) Give a small-signal model for the amplifier.

(b) Write a set of node equations for the circuit, and determine the numerical value of the voltage gain, $K_v = |v_o/v_1|$.

12-2. Nodal analysis. A BJT amplifier is shown in Fig. 12-29. Resistor R_b provides bias for the base, and it also provides a small feedback around the transistor that stabilizes the amplifier somewhat against

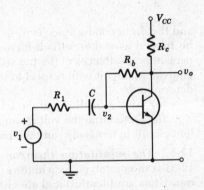

Fig. 12-29 Amplifier for Prob. 12-2.

changes in the transistor parameters. The transistor and circuit parameters are $\beta = 100$, $r_o = 50$ kilohms, $r_i = 1$ kilohm, $R_c = 5$ kilohms, $R_b = 200$ kilohms, and $R_1 = 5$ kilohms. The blocking capacitor can be treated as a short circuit for signal components of voltage, and the parasitic capacitances of the transistor can be neglected.

(a) Give a small-signal model for the amplifier.

(b) Write a set of node equations for the circuit, and give the necessary auxiliary equation relating the unknown base current to the unknown node voltages.

(c) Determine the numerical value of the voltage gain, $K_v = |v_o/v_1|$.

12-3. Input resistance. The output voltage of the amplifier in Fig. 12-28 can be expressed as $v_o = -Kv_2$, where K is the voltage gain from gate to drain.

(a) Determine the current in R_g as a function of v_2, K, and R_g.

(b) Determine the input resistance R_i of the amplifier on the right of the coupling capacitor in terms of R_g and K.

12-4. Thévenin equivalent circuit. The enhancement MOST in the source follower of Fig. 12-30 is characterized by a transconductance g_m,

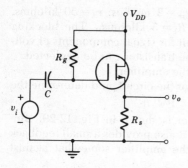

Fig. 12-30 Source follower for Prob. 12-4.

and the drain conductance is $g_d = 1/r_d = 0$. The coupling capacitor can be treated as a short circuit for signal components of voltage, and the parasitic capacitances of the transistor can be neglected. The Thévenin equivalent circuit with respect to the indicated output terminals is to be determined.

(*a*) Give a small-signal model for the circuit.

(*b*) Determine the voltage and resistance for the Thévenin equivalent circuit in terms of v_i and the parameters of the circuit and transistor.

12-5. *The substitution theorem.* The MOST circuit shown in Fig. 12-31 is the prototype for a high-resistance dc electronic voltmeter. The transistors are identical and are characterized by the parameters g_m and r_d; the parasitic capacitances of the transistor can be neglected.

(*a*) What is the value of i_m when $v_i = 0$?

(*b*) Give a small-signal model for the circuit.

(*c*) Apply the substitution theorem to eliminate all controlled sources except one controlled by v_i.

(*d*) Assuming that the microammeter has zero resistance, evaluate the transfer conductance $G_t = i_m/v_i$. If $g_m = 2$ mmhos, what value of v_i is required to produce full-scale deflection with $i_m = 200$ μA?

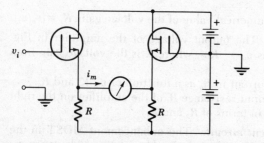

Fig. 12-31 Electronic voltmeter for Prob. 12-5.

12-6. A bootstrap circuit. The voltage source V_{GG} and the 330-kilohm resistor in the source-follower circuit of Fig. 12-32 represent the Thévenin equivalent circuit of a voltage divider that provides gate bias for the JFET. The 3.3-megohm resistor is added to increase the input resistance. The capacitor C acts as a short circuit for signal voltages, and it permits the output voltage v_o to bootstrap the junction between the resistors in the bias circuit.

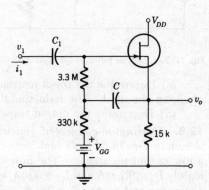

Fig. 12-32 Source follower for Prob. 12-6.

The transistor parameters are $g_m = 2$ mmhos and $g_d = 1/r_d = 0$. Capacitors C and C_1 can be treated as short circuits for signal voltages, and the parasitic capacitances of the transistors can be neglected.

(*a*) Give a small-signal model for the circuit.

(*b*) Calculate the voltage gain $K_v = v_o/v_1$. The tiny current flowing through C can be neglected in this calculation, and the circuit can be simplified with the aid of the substitution theorem.

(*c*) Calculate the input resistance $R_i = v_1/i_1$.

12-7. Large-signal emitter-follower analysis. The voltage $v_{s,\,max}$ in Fig. 12-14c is the value of v_s at which the transistor in the emitter follower just saturates. Determine the value of $v_{s,\,max}$ in terms of the circuit parameters and V_{CC}.

12-8. Emitter-follower design. The emitter follower shown in Fig. 12-33 is to be designed so that the quiescent operating point for the transistor is at $V_{CE} = 5$ volts and $I_C = 1$ mA. In addition, the resistance of R_1 in parallel with R_2 is to be 50 kilohms. The transistor parameters are $\beta = 100$, $V_o = 0.7$ volt, $r_x = 100$ ohms, and $g_o = 1/r_o = 0$.

(*a*) Determine the required values of R_1, R_2, and R_e.

(*b*) Give a small-signal model for the circuit treating the coupling capacitor as a short circuit, and determine the transistor parameter r_i.

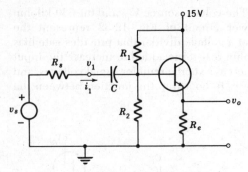

Fig. 12-33 Emitter follower for Prob. 12-8.

(c) Determine the input resistance v_1/i_1 when the output terminals are open circuited. The reduction theorem is helpful.

(d) Determine the output resistance when $R_s = 5$ kilohms.

12-9. Darlington-connected emitter follower. The emitter follower shown in Fig. 12-34 has a high current gain, a high input resistance, and a low output resistance. The quiescent collector currents are approximately $I_{C1} = 0.1$ mA and $I_{C2} = 5$ mA, and under these conditions the transistor parameters are $\beta_1 = 30$, $\beta_2 = 50$, $r_{x1} = r_{x2} = 75$ ohms, and $1/r_{o1} = 1/r_{o2} = 0$. The coupling capacitor can be treated as a short circuit for signal voltages.

(a) Give a small-signal model for the circuit, and determine the value of r_i for each transistor.

(b) Eliminate all controlled sources from the circuit by successive applications of the reduction theorem in such a way that the input current i_1 remains unchanged.

(c) Determine the input resistance $R_i = v_1/i_1$ for the circuit.

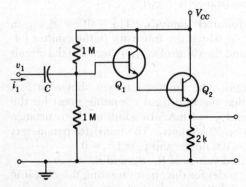

Fig. 12-34 Darlington-connected emitter follower for Prob. 12-9.

12-10. *The reduction theorem.* Figure 12-35 shows the circuit diagram of a grounded-grid triode amplifier. The coupling capacitor can be treated as a short circuit for signal voltages, and the parasitic capacitances of the tube can be neglected.

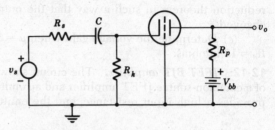

Fig. 12-35 Grounded-grid amplifier for Prob. 12-10.

(*a*) Give a small-signal model for the circuit using the voltage-source representation for the triode given in Fig. 11-7.

(*b*) Use the voltage-source form of the reduction theorem to obtain a simplified circuit in which the output voltage v_o remains unchanged.

(*c*) Calculate the voltage gain in terms of the tube and circuit parameters for the case where $R_s = 0$.

12-11. *The cascode amplifier.* The JFET cascode amplifier shown in Fig. 12-36*a* has a variety of useful applications. Resistors R_1 and R_2 serve only to provide gate bias for Q_2, and the coupling and bypass capacitors can be treated as short circuits for signal voltages. The tran-

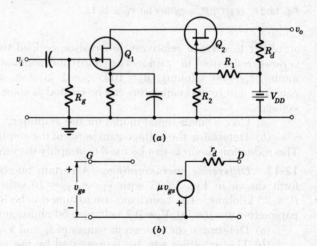

Fig. 12-36 Cascode amplifier for Prob. 12-11.

sistors are identical, and they can be represented by the small-signal model shown in Fig. 12-36b with $\mu = g_m r_d$.

(a) Give the small-signal model for the amplifier.

(b) Simplify the model by applying the voltage-source form of the reduction theorem in such a way that the output voltage remains unchanged.

(c) Determine the voltage gain when $\mu = 50$, $r_d = 50$ kilohms, and $R_d = 15$ kilohms.

12-12. JFET-BJT amplifier. The circuit shown in Fig. 12-37 consists of a common-source JFET amplifier and an emitter follower. The JFET provides a high input resistance, and the emitter follower permits the

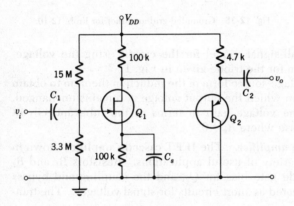

Fig. 12-37 JFET-BJT amplifier for Prob. 12-12.

amplifier to drive a relatively low-resistance load without suffering a serious reduction in gain. The transistor parameters are $g_{m1} = 0.5$ mmho, $r_{d1} = 50$ kilohms, $\beta_2 = 100$, $r_{i2} = 1$ kilohm, and $1/r_{o2} = 0$; the coupling and bypass capacitors can be treated as short circuits for signal voltages.

(a) Give a small-signal model for the amplifier.

(b) Determine the voltage gain $|v_o/v_i|$ of the amplifier. *Suggestion:* The reduction theorem can be used to simplify the circuit considerably.

12-13. *Difference microamplifier.* A certain microamplifier has the form shown in Fig. 12-23 with $V_{CC} = V_{EE} = 10$ volts, $I_o = 40$ μA, and $R_c = 25$ kilohms. The transistors are assumed to be identical, and their parameters are $\beta = 30$, $V_o = 0.7$ volt, $r_x = 50$ ohms, and $1/r_o = 0$.

(a) Determine the quiescent values of I_C and V_{CE}.

(b) The amplifier can be represented by the small-signal model shown in Fig. 12-25b. Determine the values of the parameters in this model.

12-14. *Difference microamplifier.* The small-signal model for the difference amplifier of Fig. 12-27 can be put in the form shown in Fig. 12-38 with the aid of the source-shifting transformation illustrated in Fig. 12-3b. This model can be greatly simplified with the aid of the reduction theorem.

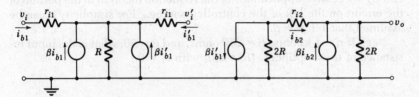

Fig. 12-38 Circuit for Prob. 12-14.

(a) Eliminate the two current sources in the input part of the circuit by successive applications of the reduction theorem to the portion of the circuit on the left of the controlled source. For simplicity it can be assumed that $1 + \beta \approx \beta$.

(b) Eliminate the controlled source βi_{b2} by applying the reduction theorem to the portion of the circuit on the left of the source. Again, $1 + \beta \approx \beta$.

(c) Assuming that $\beta R \gg r_{i1}$ and that $r_{i2}/\beta \ll 2R$, determine the voltage gain $v_o/(v_i - v_i')$ in terms of the circuit parameters.

12-15. *Difference microamplifier.* Difference amplifiers often include small resistors in series with each emitter, as shown in Fig. 12-39,

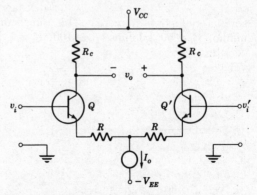

Fig. 12-39 Amplifier for Prob. 12-15.

to add further stabilization to the amplifier and to increase the input resistance. The small-signal properties of such amplifiers are to be studied.

(a) Give a small-signal model for the amplifier that is similar to the one shown in Fig. 12-24b except that it contains the added resistors described above.

(b) Eliminate the two controlled sources in the input part of the circuit by successive applications of the reduction theorem to the portion of the circuit on the left of the controlled source. For simplicity it can be assumed that $1 + \beta \approx \beta$.

(c) If $r_i = 1$ kilohm, $R = 250$ ohms, and $\beta = 50$, what is the input resistance of the amplifier? $(R_i = v_i/i_b$ with $v_i' = 0.)$

13

Multistage Amplifiers and Feedback

The maximum voltage gain obtainable from a single transistor or tube in a practical amplifier circuit usually lies in the range between about 10 and 100. Since there are many applications requiring more amplification than this, it is common practice to connect a number of amplifiers in cascade so that each amplifier amplifies the signal in succession; the total gain is then the product of the gains of the individual amplifiers. Under these conditions the amplification is said to take place in stages, and the overall system is called a multistage amplifier. It is seldom necessary to connect more than three stages in cascade.

Connecting amplifier stages in cascade raises a number of new problems, and a number of circuit configurations have been developed to overcome these problems. Current practice uses feedback extensively to achieve good performance with multistage amplifiers. The objective of this chapter is to present several of the more widely used multistage amplifier configurations and to examine the role of feedback in these circuits.

13-1 MULTISTAGE ENHANCEMENT MOST MICROAMPLIFIERS

Probably the simplest of all multistage amplifiers is the enhancement MOST amplifier; a three-stage example of such an amplifier is shown in Fig. 13-1. When this amplifier is fabricated in microamplifier form, the drain resistors R_d take the form of MOS devices as shown in Fig. 7-11. Under quiescent conditions with identical transistors, all gates and drains are at the same potential V_C; this is the characteristic voltage for the cascade of stages. The blocking capacitors at the input and output ensure that the direct voltages in the amplifier are not disturbed by the signal source and the load connected at the output, and they are chosen to behave as short circuits for signal voltages. The input signal voltage is amplified in turn by Q_1, Q_2, and Q_3, and the overall gain may be as high as several thousand.

MOS transistors in microamplifiers have closely matched charac-

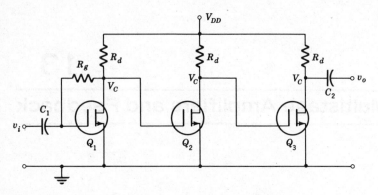

Fig. 13-1 A multistage enhancement MOST microamplifier.

teristics; however, the transistors are not identical. This fact may cause the voltage at the drain of Q_1 to deviate slightly from the characteristic value V_C, and the deviation can be thought of as an unwanted signal at the drain of Q_1. The unwanted signal is amplified by Q_2, and it is amplified again by Q_3. The amplified signal appearing at the drain of Q_3 can easily be large enough to put the operating point for Q_3 at the cut-off point or in the saturation region of the transistor characteristics, and under such conditions Q_3 cannot function properly as an amplifier. This feature is not a special property of MOST amplifiers; it is shared equally by all multistage amplifiers having high gain for direct voltages.

Figure 13-2a shows how feedback can be used to stabilize the quiescent operating points for the transistors in the three-stage MOST amplifier. The resistor R_g providing bias for the gate of Q_1 is connected to the drain of Q_3 instead of to the drain of Q_1, and it introduces overall feedback that acts in the following way. Suppose that there is a positive deviation in the voltage at the drain of Q_1 from the characteristic value V_C. This deviation is amplified and appears as a positive deviation at the drain of Q_3 and also at the gate of Q_1. This action causes Q_1 to draw more current, which in turn tends to restore the voltage at the drain of Q_1 to the proper value. Feedback acting in this manner provides a strong stabilization of the quiescent points throughout the amplifier.

The feedback to the gate of Q_1 causes a reaction in Q_1 that subtracts from the original disturbance appearing at the drain of Q_1; this kind of feedback is called negative feedback. With a different configuration the feedback can cause a reaction in Q_1 that adds to the original disturbance, and this kind of feedback is called positive feedback. Negative feedback usually improves the performance of amplifiers in certain respects, whereas positive feedback usually degrades the performance, even to the extent of turning amplifiers into oscillators in many cases.

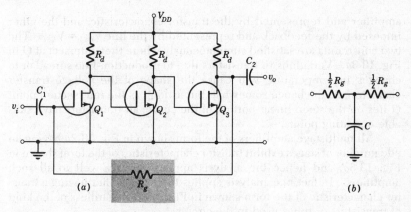

Fig. 13-2 MOST amplifier with stabilized quiescent operating conditions. (a) Circuit; (b) means for removing the feedback for signal voltages.

A useful graphical analysis of the quiescent conditions in the three-stage MOST amplifier is shown in Fig. 13-3a. The curve labeled *amplifier* is the voltage transfer characteristic for the amplifier. In the linear central region where the slope of this curve is large, all of the transistors in the amplifier have a satisfactory operating point. The fact that the slope is negative means that the amplifier introduces a sign reversal; that is, a small negative increment in V_{GS1} produces a positive increment in V_{DS3}, and this fact means that overall feedback around this amplifier is negative feedback. Under quiescent conditions there is no current in resistor R_g in the circuit of Fig. 13-2a, and thus $V_{GS1} = V_{DS3}$; this relation is also plotted in Fig. 13-3a. Now there are two constraints between V_{GS1} and V_{DS3} that must be satisfied simultaneously, one imposed by the

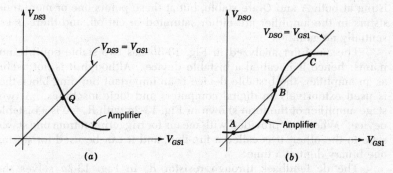

Fig. 13-3 Graphical analysis of the amplifier in Fig. 13-2. (a) Odd number of stages; (b) even number of stages.

amplifier and represented by the transfer characteristic, and the other imposed by the feedback and represented by the line $V_{DS3} = V_{GS1}$. The two constraints are satisfied simultaneously where they intersect at Q in Fig. 13-3a. Variations in transistors due to production-line spread or to changes in temperature will cause the shape of the voltage transfer characteristic to change somewhat, but as long as the intersection point Q lies on the steep linear portion of the curve, all transistors have suitable operating points.

All multistage amplifiers of the form shown in Fig. 13-2a having an odd number of stages exhibit transfer characteristics of the form shown in Fig. 13-3a, and hence the analysis applies equally well to all such amplifiers. In fact, the analysis applies to any amplifier having a transfer characteristic of the form shown in Fig. 13-3a, regardless of the kind of transistors or tubes used in the circuit.

To complete the picture Fig. 13-3b shows a similar graphical analysis for an MOST amplifier having an even number of stages. In this case the slope of the transfer characteristic is positive, the amplifier introduces no net sign reversal, and overall feedback around this amplifier is positive. For the conditions shown in Fig. 13-3b there are three equilibrium points at which the two constraints are satisfied simultaneously; however, the equilibrium at point B is an unstable equilibrium similar to that existing when a chair is balanced on one leg. Any effect that causes the operating point to move away from the equilibrium point at B creates new effects that cause the operating point to move even farther away from B. For example, if a small positive increment of voltage caused by electrical noise appears at V_{GS1}, the increment is amplified and appears as a larger positive increment of voltage at the output. This amplified increment is then fed back to the input where it adds to the initial increment in V_{GS1}. This action continues until the equilibrium point at C is reached. The equilibria existing at both A and C are stable, but at these points one or more transistors in the amplifier are either saturated or cut off, and there is essentially no gain.

The amplifier analyzed in Fig. 13-3b has two stable equilibrium points; hence it is called a bistable device. Although it is not useful as an amplifier, the bistable device is an important building block that is used extensively in digital computers and logic machines. A two-stage amplifier of the form shown in Fig. 13-2a with $R_g = 0$ is a bistable device. When it is provided with means for triggering it from one stable state to the other, it is called a flip-flop, and it can be used for storing one binary digit at a time.

The dc feedback through resistor R_g in Fig. 13-2a solves the quiescent-point problems in that circuit, but it creates a new problem insofar as the small-signal behavior of the circuit is concerned. The small-signal voltage gain from the gate of Q_1 to the drain of Q_3 can be designated $-K_v$, where K_v is a positive number. Thus, assuming that the coupling capacitors are short circuits for signal components of voltage, an input voltage v_i causes a signal component of voltage at the output that is

$$v_o = -K_v v_i \tag{13-1}$$

The signal voltage across R_g is thus $v_i - v_o = (1 + K_v)v_i$, and the current in R_g is

$$i_g = \frac{1 + K_v}{R_g} v_i \tag{13-2}$$

Hence the input resistance of the amplifier is

$$R_i = \frac{v_i}{i_g} = \frac{R_g}{1 + K_v} \tag{13-3}$$

Thus if $R_g = 1$ megohm and $K_v = 1,000$, the input resistance is only 1 kilohm, and it may present an excessive load to the source of signals. The action described here is just the opposite of the bootstrapping action described in Sec. 12-3.

The problem of the low input resistance described above can be eliminated by the simple expedient shown in Fig. 13-2b. Resistor R_g is broken into two parts and the junction between the parts is connected to ground by a capacitor that acts as a short circuit for signal components of voltage. Thus for signal frequencies at which the capacitors act as short circuits, the small-signal model for the amplifier takes the form shown in Fig. 13-4. This model assumes that the drain resistance r_d of the transistor is much larger than R_d; if this is not the case, r_d can be lumped in with R_d. If $R_d \ll \frac{1}{2}R_g$, as normally is the case, the overall voltage gain of the amplifier is

$$\frac{v_o}{v_i} = -K_v = -(g_m R_d)^3 \tag{13-4}$$

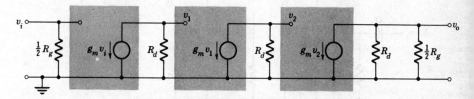

Fig. 13-4 Small-signal model for the amplifier in Fig. 13-2a with the modification shown in Fig. 13-2b.

and the input and output resistances are

$$R_i = \tfrac{1}{2}R_y \quad \text{and} \quad R_o = R_d \qquad (13\text{-}5)$$

The input of this amplifier is isolated from the output in the sense that the input current and voltage are independent of the output voltage and current.

It is appropriate to inquire whether or not there is any limitation on the number of amplifier stages that can be connected in cascade and therefore whether or not there is any limitation on the overall signal gain that can be realized. The noise generated in the first stage of a multistage amplifier is amplified in succession by each stage of the amplifier. As more stages are connected in cascade, the noise voltage at the output of the amplifier increases, and eventually it becomes excessive. For this reason most practical amplifiers do not have gains in excess of about 10^5. In some applications where the input signal is very feeble, as in the case of the signals received from deep space probes, for example, special low-noise circuits and devices are used for the input stage to make greater amplifications possible.

13-2 CAPACITIVELY COUPLED MULTISTAGE AMPLIFIERS

The directly coupled amplifier of Fig. 13-1 has problems related to the quiescent points of the individual transistors; these problems are solved by the use of dc feedback, as illustrated in Fig. 13-2. An alternative way of solving the problems with quiescent points is to provide dc isolation for each stage, as illustrated by the JFET amplifier shown in Fig. 13-5. The blocking, or coupling, capacitors in this circuit prevent the trans-

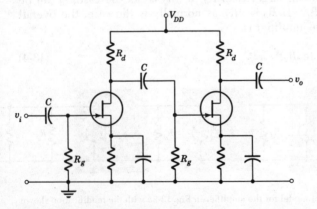

Fig. 13-5 Capacitively coupled JFET stages.

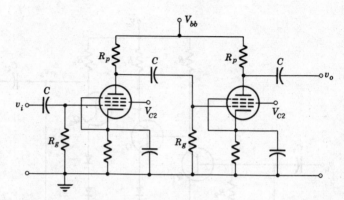

Fig. 13-6 Capacitively coupled pentode stages.

mission of dc signals through the amplifier, and thus a small error in the quiescent point of the first stage is not amplified by succeeding stages. However, the capacitors are chosen large enough to act as short circuits to the desired signals.

In the days before transistors, capacitive coupling was almost universal in vacuum-tube amplifiers. A capacitively coupled multistage pentode amplifier is shown in Fig. 13-6; for simplicity the bias circuits supplying voltage for the screen grids are omitted from this diagram. However, in today's electronic systems using transistors and miniaturized circuits, the size, weight, and cost of coupling capacitors are deemed prohibitive, and their use is kept at a minimum. The use of coupling capacitors in microamplifiers is completely prohibited because it is not feasible to make microcapacitors large enough to serve this function except in a few special cases.

The small-signal models for the amplifiers in Figs. 13-5 and 13-6 have the same form as the model shown in Fig. 13-4 except for the number of stages. Thus the small-signal behavior of these amplifiers is essentially the same as the behavior of the MOST amplifier.

13-3 MULTISTAGE BJT AMPLIFIERS

Figure 13-7 shows the circuit diagram of a three-stage BJT amplifier; this is one of a number of circuit configurations that are used in three-stage amplifiers. The diode in the emitter circuit of Q_2 provides an approximately constant voltage drop $V_o = 0.7$ volt so that the quiescent collector voltage for Q_1 is $V_{CE1} = 2V_o = 1.4$ volt. The two diodes in the emitter circuit of Q_3 perform a similar function with respect to V_{CE2}. The voltage divider R_1-R_1-R_2 provides bias for the base of Q_1, and it also provides

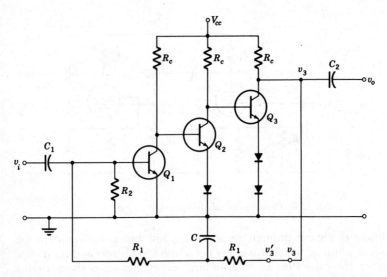

Fig. 13-7   Three-stage BJT amplifier with dc feedback.

overall dc feedback that stabilizes quiescent operating conditions in the amplifier. This amplifier is very similar in many respects to the MOST amplifier shown in Fig. 13-2, and the graphical analysis of its quiescent operating conditions is entirely similar to the one shown in Fig. 13-3a. As in the case of the MOST amplifier, capacitor C acts as a short circuit for signal voltages, and it removes the feedback at signal frequencies.

The statement made above that the graphical analysis of the quiescent operating conditions is similar to the one illustrated in Fig. 13-3a requires a brief clarification. If the small current flowing in R_1 is neglected, then the feedback path can be broken by removing the connection between terminals v_3 and v_3' in Fig. 13-7. Then the voltage transfer characteristic from v_3' to v_3 has the form of the amplifier curve shown in Fig. 13-3a; this is one constraint between v_3 and v_3'. When the connection between these two terminals is restored, $v_3 = v_3'$; this is the second constraint, and it leads to a construction like the one in Fig. 13-3a.

The quiescent properties of the three-stage amplifier can be examined further, and a design procedure can be established with the aid of Fig. 13-8; this figure shows the Thévenin equivalent for the circuit connected to the base of Q_1 under quiescent conditions. It follows from this equivalent circuit that

$$\frac{v_3}{1 + 2R_1/R_2} = V_o + \frac{2R_1 i_{B1}}{1 + 2R_1/R_2} \tag{13-6}$$

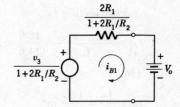

Fig. 13-8 Thévenin equivalent for the circuit connected to the base of Q_1 in Fig. 13-7 under quiescent conditions.

and thus equilibrium is established in the circuit with

$$v_3 = \left(1 + \frac{2R_1}{R_2}\right) V_o + 2R_1 i_{B1} \tag{13-7}$$

The first term on the right side of this equation is very nearly constant except for small variations in V_o; hence if it is made substantially larger than the second term, then v_3 and the quiescent conditions throughout the amplifier are well stabilized. The use of these relations in designing an amplifier is illustrated by the following numerical example.

Example 13-1

A three-stage BJT amplifier having the form shown in Fig. 13-7 with equal collector resistors is to be designed for the following quiescent conditions: $V_{CC} = 20$ volts, $V_{CE3} = 4$ volts, and $I_{C3} = 1$ mA. The transistor parameters are $\beta = 50$ and $V_o = 0.7$ volt.

Solution Assuming that the voltage drops across the diodes in the emitter circuits of Q_2 and Q_3 are all 0.7 volt, the quiescent voltage at the collector of Q_3 is

$$V_3 = V_{CE3} + 2V_o = 4 + 1.4 = 5.4 \text{ volts}$$

Then, neglecting the small current flowing in R_1,

$$R_c = \frac{V_{CC} - V_3}{I_{C3}} = \frac{20 - 5.4}{1} = 14.6 \text{ kilohms}$$

The voltage at the collector of Q_2 with respect to ground is

$$V_{C2} = 3V_o = 2.1 \text{ volts}$$

Thus, neglecting I_{B3},

$$I_{C2} = \frac{V_{CC} - V_{C2}}{R_c} = \frac{17.9}{14.6} = 1.23 \text{ mA}$$

The collector voltage for Q_1 is

$$V_{CE1} = 2V_o = 1.4 \text{ volts}$$

and neglecting I_{B2},

$$I_{C1} = \frac{V_{CC} - V_{CE1}}{R_c} = \frac{18.6}{14.6} = 1.27 \text{ mA}$$

Thus

$$I_{B1} = \frac{I_{C1}}{\beta} = \frac{1.27}{50} \text{ mA}$$

For good stability the first term on the right side of Eq. (13-7) is chosen to be four times the second term; then (13-7) becomes

$$V_3 = 10R_1I_{B1} = 5.4 \text{ volts}$$

and

$$R_1 = 5.4 \frac{50}{12.7} = 21.2 \text{ kilohms}$$

Now, with the choice made above,

$$\left(1 + \frac{2R_1}{R_2}\right) V_o = 8R_1I_{B1} = 4.3 \text{ volts}$$

and

$$R_2 = \frac{2R_1}{5.15} = 8.2 \text{ kilohms}$$

To check the stability of the amplifier, suppose that the current gain of Q_1 is increased from 50 to 100. With feedback stabilization of the quiescent conditions, I_{C1} does not change appreciably, and thus $I_{B1} = 1.27/100$ mA. Then Eq. (13-7) gives V_3 as

$$V_3 = 4.3 + 42.4 \frac{1.27}{100} = 4.84 \text{ volts}$$

which is a slight decrease from the previous value of 5.4 volts. The decrease in V_3 decreases I_{B1} by an amount that almost exactly compensates for the change in β. Changes in the current gains of Q_2 and Q_3 are even more strongly stabilized.

The bias diodes in the amplifier of Fig. 13-7 provide a voltage drop that is almost, but not entirely, constant. Under small-signal conditions the diodes have an incremental resistance given by Eq. (8-28). The effect of the diode resistance on the small-signal behavior of the amplifier can be evaluated with the aid of the small-signal model shown in Fig. 13-9a for the second stage of the amplifier; the resistor r_d in this model

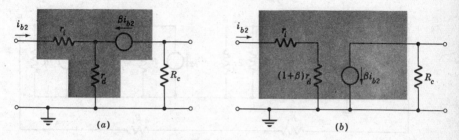

Fig. 13-9 The effect of the diode in the emitter circuit of Q_2 in Fig. 13-7.

represents the diode resistance. With the aid of the source-shifting transformation and the reduction theorem, the model can be transformed into the simpler model shown in Fig. 13-9b. It follows from Eq. (8-28) that

$$r_d = \frac{1}{\lambda I_{C2}}$$

and thus

$$(1 + \beta)r_d \approx \frac{\beta}{\lambda I_{C2}} = r_{\pi 2} \approx r_{i2}$$

Thus the effect of the diode in the emitter circuit of Q_2 is approximately to double the input resistance; this result corresponds to the fact that there are two junctions between the base of Q_2 and ground. Similarly, the two diodes in the emitter circuit of Q_3 increase the apparent input resistance of Q_3 to $3r_{i3}$.

A small-signal model for the three-stage amplifier which is valid for dc and for very low signal frequencies is shown in Fig. 13-10a. The bias diodes are accounted for in this model in accordance with the discussion given above. The voltage v_x represents a small unwanted signal such as a change in base-to-emitter voltage with temperature, for example. Figure 13-10b shows a more compact model for the amplifier. If the feedback is removed from the circuit, the unwanted signal v_x is amplified by the full gain of the amplifier, and it may be large enough at the output to drive the third stage into cutoff or saturation. However, with feedback present, conditions are markedly different. In the usual case $2R_1 \gg R_c > r_i$ and $R_2 \gg r_i$; thus

$$v_i = v_x + \frac{r_i}{r_i + 2R_1} v_3$$

and the output voltage is

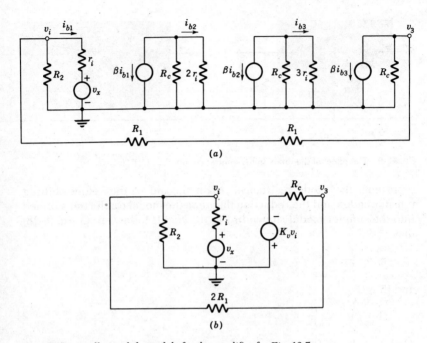

Fig. 13-10 Small-signal dc models for the amplifier for Fig. 13-7.

$$v_3 = -K_v v_i = -K_v \left(v_x + \frac{r_i}{r_i + 2R_1} v_3 \right) \tag{13-8}$$

$$\approx -K_v v_x - \frac{r_i}{2R_1} K_v v_3 \tag{13-9}$$

$$\approx \frac{-K_v}{1 + r_i K_v / 2R_1} v_x \tag{13-10}$$

With $R_1 = \infty$, there is no feedback, and (13-10) reduces to $v_3 = -K_v v_x$. With feedback present the unwanted signal at the output may be reduced substantially; if

$$\frac{r_i K_t}{2R_1} \gg 1$$

$$v_3 \approx -\frac{2R_1}{r_i} v_x \tag{13-11}$$

. Figure 13-11 shows a small-signal model for the three-stage amplifier which is valid at signal frequencies where the coupling and

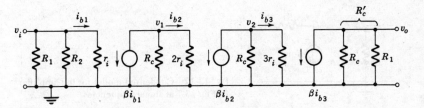

Fig. 13-11 Small-signal model for the amplifier in Fig. 13-7 at signal frequencies.

bypass capacitors act as short circuits. It is convenient to think of BJTs as current amplifiers, and they are represented as such in Fig. 13-11. However, in the building and testing of such amplifiers it turns out that there are many instruments suitable for measuring signal voltages but no instrument that is entirely satisfactory for measuring signal currents. Therefore it is usually more useful to focus attention on the voltage gain than on the current gain. Accordingly, the output voltage of the three-stage amplifier is given by the model in Fig. 13-11 as

$$v_o = -R'_c \beta i_{b3} = -\beta R'_c \left(-\beta \frac{R_c}{R_c + 3r_i} i_{b2} \right)$$

$$= \beta^2 R'_c \frac{R_c}{R_c + 3r_i} \left(-\beta \frac{R_c}{R_c + 2r_i} i_{b1} \right)$$

$$= -\beta^3 R'_c \left(\frac{R_c}{R_c + 3r_i} \right) \left(\frac{R_c}{R_c + 2r_i} \right) \frac{v_i}{r_i}$$

and thus the voltage gain is

$$K_v = \beta^3 \left(\frac{R_c}{R_c + 3r_i} \right) \left(\frac{R_c}{R_c + 2r_i} \right) \frac{R'_c}{r_i} \tag{13-12}$$

The input resistance is

$$R_i = r_i \| R_1 \| R_2 \tag{13-13}$$

and the output resistance is

$$R_o = R'_c \tag{13-14}$$

These relations imply that the three-stage amplifier can be represented by the more compact model shown in Fig. 13-12.

A two-stage BJT amplifier is shown in Fig. 13-13. This amplifier, which can easily provide a voltage gain in excess of 1,000, is one of the more popular building blocks used in electronic systems. In a two-stage amplifier, bias for the base of Q_1 cannot be obtained by feedback from the collector of Q_2; such an arrangement leads to a bistable device, as

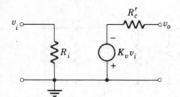

Fig. 13-12 Condensed model for the amplifier
in Fig. 13-7 at signal frequencies.

shown in the discussion of Fig. 13-3b. However, the voltage v_{E2} in Fig.
13-13 differs from the voltage at the collector of Q_1 only by the small
base-to-emitter voltage drop in Q_2, and hence feedback from the emitter
of Q_2 to the base of Q_1 is essentially the same as feedback around one
stage. If the feedback path is broken between terminals v_{E2} and v_{E2}' in
Fig. 13-13, the transfer characteristic from v_{E2}' to v_{E2} has the form shown
in Fig. 13-3a, and the graphical analysis shows that suitable quiescent
operating conditions are obtained when the feedback path is closed.
The bypass capacitor C_e acts as a short circuit for signal voltages, and
thus it removes the feedback for signal voltages.

The analysis of the quiescent conditions in this amplifier is very
similar to the quiescent analysis of the three-stage amplifier. Figure
13-14 shows the Thévenin equivalent for the circuit connected to the
base of Q_1 under quiescent conditions. It follows from this circuit that

$$\frac{v_{E2}}{1 + R_1/R_2} = V_o + \frac{R_1}{1 + R_1/R_2}\, i_{B1} \tag{13-15}$$

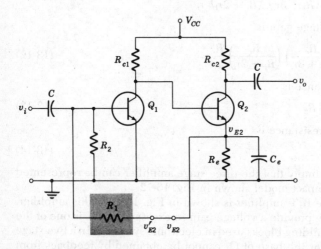

Fig. 13-13 Two-stage BJT amplifier with dc feedback.

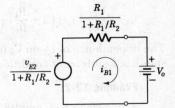

Fig. 13-14 Thévenin equivalent for the circuit connected to the base of Q_1 in Fig. 13-13 under quiescent conditions.

and thus equilibrium is established in the circuit with

$$v_{E2} = \left(1 + \frac{R_1}{R_2}\right) V_o + R_1 i_{B1} \tag{13-16}$$

Again, the first term on the right side of this equation is constant except for small variations in V_o, and if it is made substantially larger than the second term, v_{E2} is well stabilized. With v_{E2} well stabilized, it follows that v_{CE1}, i_{C1}, and i_{C2} are also well stabilized. The collector-to-emitter voltages for the transistors are

$$v_{CE1} = v_{E2} + V_o \tag{13-17}$$

and

$$v_{CE2} = V_{CC} - R_{c2}i_{C2} - v_{E2} \tag{13-18}$$

Since i_{C2} is proportional to v_{E2}, it follows that stabilizing v_{E2} stabilizes both of these voltages.

A more quantitative understanding of the stability of this circuit can be gained by evaluating i_{C1}. Neglecting i_{B2} in comparison with i_{C1},

$$V_{CC} = v_{E2} + V_o + R_{c1}i_{C1} \tag{13-19}$$

$$= \left(1 + \frac{R_1}{R_2}\right) V_o + R_1 i_{B1} + V_o + R_{c1}i_{C1} \tag{13-20}$$

$$= \left(2 + \frac{R_1}{R_2}\right) V_o + \left(\frac{R_1}{\beta_1} + R_{c1}\right) i_{C1} \tag{13-21}$$

and thus

$$
\begin{aligned}
i_{C1} &= \frac{V_{CC} - (2 + R_1/R_2)V_o}{R_{c1} + R_1/\beta_1} \\
&= \frac{V_{CC}}{R_{c1}} \frac{1 - (2 + R_1/R_2)(V_o/V_{CC})}{1 + R_1/\beta_1 R_{c1}}
\end{aligned} \tag{13-22}
$$

This current is independent of β_2, and it can be made nearly independent of β_1 by making $R_1/\beta_1 R_{c1} \ll 1$; when this inequality is satisfied,

$$i_{C1} \approx \frac{V_{CC}}{R_{c1}} \left[1 - \left(2 + \frac{R_1}{R_2} \right) \frac{V_o}{V_{CC}} \right] \tag{13-23}$$

The dependence of i_{C1} on V_o is kept small by making $V_{CC} \gg V_o$. With i_{C1} stabilized, quiescent conditions throughout the amplifier are stabilized.

Example 13-2

A two-stage BJT amplifier having the form shown in Fig. 13-13 is to be designed so that the quiescent operating points for both transistors are at $V_{CE} = 4$ volts and $I_C = 1$ mA when $V_{CC} = 20$ volts. The parameters of the transistors are $\beta = 50$ and $V_o = 0.7$ volt.

Solution Neglecting I_{B2}, the collector resistance needed in the first stage is

$$R_{c1} = \frac{V_{CC} - V_{CE1}}{I_{C1}} = \frac{20 - 4}{1} = 16 \text{ kilohms}$$

The voltage at the emitter of Q_2 is

$$V_{E2} = V_{CE1} - V_o = 4 - 0.7 = 3.3 \text{ volts}$$

and the base current in Q_1 is

$$I_{B1} = \frac{I_{C1}}{\beta} = \frac{1}{50} \text{ mA}$$

To ensure good stability, the first term on the right side of Eq. (13-16) is chosen to be four times the second term; thus

$$\left(1 + \frac{R_1}{R_2} \right) V_o = 4 R_1 I_{B1}$$

and

$$V_{E2} = 3.3 = 5 R_1 I_{B1} = \frac{R_1}{10}$$

Hence

$$R_1 = 33 \text{ kilohms}$$

Now

$$\left(1 + \frac{R_1}{R_2} \right) V_o = 4 R_1 I_{B1} = 4\frac{33}{50} = 2.64 \text{ volts}$$

and

$$R_2 = \frac{R_1}{2.77} = 11.9 \text{ kilohms}$$

Neglecting I_{B2}, the collector current in Q_2 is

$$I_{C2} = \frac{V_{E2}}{R_e} + \frac{V_{E2} - V_o}{R_1} = 1 \text{ mA}$$

$$\frac{3.3}{R_e} + \frac{2.6}{33} = 1$$

and thus

$$R_e = \frac{3.3}{0.92} = 3.6 \text{ kilohms}$$

The collector resistance required in the second stage is

$$R_{c2} = \frac{V_{CC} - V_{CE2} - V_{E2}}{I_{C2}} = \frac{20 - 3.3 - 4}{1}$$

$$= 12.7 \text{ kilohms}$$

These calculations can be checked by using Eq. (13-22) to calculate I_{C1}:

$$I_{C1} = \frac{20}{16} \frac{1 - (2 + 2.77)(0.7/20)}{1 + (33)/(50)(16)}$$

$$= \frac{20}{16} \frac{1 - 0.167}{1 + 0.041} = 1 \text{ mA}$$

It is noteworthy that the two terms depending on V_o and β are small, and hence I_{C1} is essentially independent of these transistor parameters.

Figure 13-15 shows the small-signal model for the two-stage amplifier which is valid at signal frequencies where the coupling and bypass capacitors act as short circuits. The voltage gain, input resistance, and output resistance are easily evaluated from this model when the circuit and transistor parameters are known.

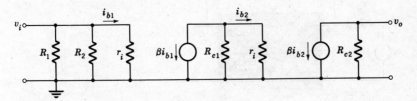

Fig. 13-15 Small-signal model for the two-stage amplifier of Fig. 13-13 at signal frequencies.

13-4   IDEAL OPERATIONAL AMPLIFIERS

The circuit diagram of an operational amplifier is shown in Fig. 13-16; it consists of a high-gain amplifier with feedback introduced by resistors R_s and R_f. Unlike the feedback configurations presented in the preceding sections of this chapter, no bypass capacitor is used to remove the feedback at signal frequencies; the feedback is effective for signal voltages as well as for direct voltages. The preceding sections show that dc feedback can be used to improve the dc performance of electronic amplifiers. This section shows that feedback at signal frequencies also improves the signal performance of amplifiers, although at a certain cost which is examined in the following section.

The operational amplifier was developed originally for the analog computer where it is used for making reasonably accurate calculations and solving differential equations. In this application it is used as a standard building block to add, subtract, integrate, differentiate, generate nonlinear functions, and so forth. It is a unit that performs various mathematical operations; hence it is called an operational amplifier. With the appearance of inexpensive high-gain microamplifiers (less than $10) on the commercial market, the use of the operational amplifier has been rapidly extended to more general signal-processing operations where it has become a basic building block.

The typical operational amplifier is a three-stage amplifier with a differential-mode input and a single-ended output. Its output voltage is zero when its input voltage is zero, and the gain of the amplifier is usually in the range between 10^4 and 10^5. The ideal operational amplifier shown in Fig. 13-16 is an ideal voltage amplifier with infinite input resistance and zero output resistance, and the gain of the amplifier is as large as desired.

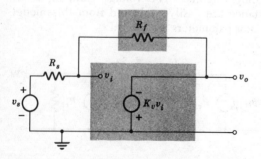

Fig. 13-16 An ideal operational amplifier.

The output voltage from a real amplifier must be limited to a certain range to avoid saturation and cutoff in the last stage; for many transistor amplifiers this range is about ± 10 volts. If the gain of the amplifier is $K_v = 10^6$, then it follows that the input voltage v_i in Fig. 13-16 is restricted to ± 10 μV for the full permissible range of v_o. That is, in comparison with v_s and v_o, $v_i \approx 0$. Moreover, since the input resistance of the ideal amplifier is infinite, the input current to the amplifier is $i_i = 0$. The two statements

$$v_i = 0 \quad \text{and} \quad i_i = 0$$

permit the overall performance of the ideal operational amplifier to be analyzed in a very simple way.

The current flowing through R_s toward node v_i is

$$i_s = \frac{v_s - v_i}{R_s} \approx \frac{v_s}{R_s}$$

where the fact that $v_i \approx 0$ is used. The current flowing through R_f away from node v_i is

$$i_f = \frac{v_i - v_o}{R_f} \approx -\frac{v_o}{R_f}$$

Now since $i_i = 0$, these two currents must be equal, and thus

$$\frac{v_o}{R_f} = -\frac{v_s}{R_s} \tag{13-24}$$

or

$$\frac{v_o}{v_s} = -\frac{R_f}{R_s} \tag{13-25}$$

and the overall voltage gain of the amplifier with feedback is

$$K_{vo} = \frac{R_f}{R_s} \tag{13-26}$$

Thus the overall gain is independent of the transistor parameters and of K_v provided only that K_v is large enough to maintain $v_i \approx 0$. The overall gain depends only on the two resistors, and they can be stable, precision resistors when precision is required. This fact is of great importance in applications where precision amplifiers are required.

Also of importance is the input resistance presented by the amplifier with feedback to the signal source v_s. Since $v_i \approx 0$, this resistance is simply

$$R_{io} = R_s \tag{13-27}$$

Thus if $R_s = 10$ kilohms and $R_f = 1$ megohm, then $R_{io} = 10$ kilohms and $K_{vo} = 100$.

The amplifier of Fig. 13-16 is a sign-inverting amplifier as is indicated by Eq. (13-25). In its practical realization the signal v_i is applied to the inverting input terminal of the amplifier and the noninverting input is grounded. There are many applications in which a noninverting operational amplifier is required. If the signal v_i is shifted to the noninverting input with the inverting input grounded, the feedback around the amplifier is positive feedback, and the conditions discussed in connection with Fig. 13-3b exist. This amplifier is a bistable device that does not act as an operational amplifier. However, the circuit shown in Fig. 13-17 provides the desired noninverting operational amplifier; in this configuration the feedback voltage is applied to the inverting input so that the feedback is negative feedback, and the signal voltage is applied to the noninverting input. In this case the requirement of zero net input voltage to the internal amplifier takes the form

$$v_i - v_i' \approx 0 \tag{13-28}$$

Thus

$$v_i = v_i'$$

and

$$v_s = \frac{R_s}{R_s + R_f}\, v_o$$

The overall voltage gain is then

$$\frac{v_o}{v_s} = 1 + \frac{R_f}{R_s} \tag{13-29}$$

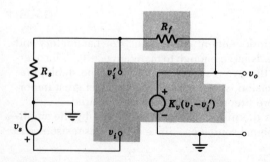

Fig. 13-17   A noninverting operational amplifier.

Again, the gain is independent of the transistor parameters and of K_v as long as K_v is large enough to maintain $v_i - v_i' \approx 0$. The input resistance faced by the signal source in the ideal noninverting operational amplifier is infinite; in a practical circuit the input resistance is very high because the finite input resistance of the amplifier is bootstrapped by the feedback at terminal v_i'.

Figure 13-18 shows the circuit diagram of an integrating operational amplifier; this circuit is extensively used in analog computers and other electronic systems for integrating electrical signals with respect to time. Setting the sum of the currents flowing toward node v_i equal to zero with $v_i = 0$ yields

$$C \frac{dv_o}{dt} + \frac{v_s}{R_s} = 0 \tag{13-30}$$

and thus

$$\frac{dv_o}{dt} = -\frac{v_s}{CR_s}$$

The output voltage is thus

$$v_o = -\frac{1}{CR_s} \int_0^t v_s \, d\tau + v_o(0) \tag{13-31}$$

Again, the performance of the amplifier is independent of the transistor parameters and K_v provided that K_v is large.

In order to obtain an operational amplifier that differentiates the input signal with respect to time, it is necessary merely to interchange R_s and C in the circuit of Fig. 13-18.

Figure 13-19 illustrates a different type of application for the operational amplifier. In this circuit, which is a precision ac voltmeter, a microammeter and bridge rectifier are placed in the feedback path

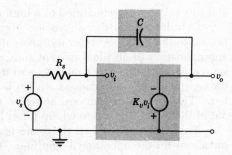

Fig. 13-18 An integrating operational amplifier.

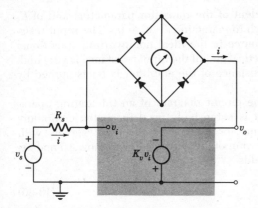

Fig. 13-19 A precision ac electronic-voltmeter rectifier.

around the amplifier; the combined system acts to make the meter current strictly proportional to the input voltage in spite of the nonlinear volt-ampere characteristics of the junction diodes. With $v_i \approx 0$, the current in R_s is

$$i = \frac{v_s}{R_s}$$

and this current is independent of the nonlinear diodes. But this same current must flow through the rectifier circuit, and the amplifier develops whatever value of output voltage v_o is required to drive this current through the nonlinear diodes. As a result of this action and the rectifying action of the diode bridge, the meter current is

$$i_m = \left| \frac{v_s}{R_s} \right| \tag{13-32}$$

and this relation is maintained to a high degree of accuracy.

It follows from the discussion given above that the circuit connected to the bridge-rectifier terminals in Fig. 13-19 acts as a very good approximation to an ideal current source. The current of the current source is determined by v_s and R_s, and the amplifier drives this current through whatever impedance may be placed in the position of the rectifier. This statement assumes, however, that the amplifier is not forced out of the operating region where it has high gain.

The circuits described above are just a few of the many varied applications for the operational amplifier. In its full range of applications it is an extremely useful and versatile device.

13-5 SOME OF THE EFFECTS OF NEGATIVE FEEDBACK

The preceding sections of this chapter have shown that negative feedback can be used to improve the performance of electronic amplifiers to a marked degree. In particular, it has been shown that negative feedback can render the quiescent operating conditions and the small-signal behavior relatively independent of the poorly controlled transistor parameters. However, a price must be paid for these improvements, and the design of a feedback amplifier involves a compromise between the amount of improvement obtained and the price paid. In order to reach the best compromise, it is necessary to have a quantitative formulation of the amount of improvement obtained by the use of feedback and of the price paid for the improvement. Such a quantitative formulation is developed in this section.

One of the first effects of feedback that must be mentioned is the fact that, if a large amount of feedback is introduced around any high-gain amplifier, the amplifier is very likely to become an oscillator. Since oscillations interfere with the proper operation of the circuit as an amplifier, they cannot be permitted. This problem is discussed in some detail in Chap. 16 after additional essential background is developed in the intervening chapters. However, throughout the present chapter there is an implicit assumption that the feedback amplifiers discussed have been designed so that they do not oscillate. The tendency for feedback amplifiers to oscillate increases with the amount of feedback present, and thus the difficulty of designing the amplifier to be stable increases with the amount of feedback. This is an additional reason for seeking a quantitative formulation of the feedback problem that will permit just the required amount of feedback, and no more, to be used.

The small-signal model for a widely used feedback-amplifier configuration is shown in Fig. 13-20a; this is also the configuration of the operational amplifier of Sec. 13-4. Usually the output resistance R_0' of the internal amplifier is much smaller than R_f, and in the interest of algebraic simplicity it is assumed to be zero, as indicated in Fig. 13-20b. The input resistance R_i' is large enough in many cases to be considered as an open circuit, but if it is not, it can be lumped in with v_s' and R_s' by Thévenin's theorem, as shown in Fig. 13-20b. The polarity of the controlled source $K_v'v_i$ ensures that the feedback is negative feedback. Thus a positive voltage at v_s causes a positive voltage at v_i, and this voltage acting through the controlled source causes a negative voltage at v_o. The negative voltage at v_o is fed back to the input where it makes a negative contribution to v_i, thus subtracting from the effect of v_s.

The action described above can be formulated quantitatively in the following way. The output voltage of the circuit in Fig. 13-20b is

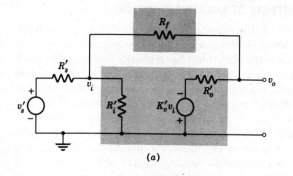

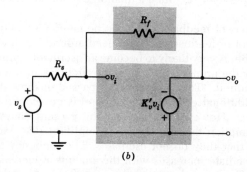

Fig. 13-20 Small-signal models for a feedback amplifier.

$$v_o = -K'_v v_i \qquad (13\text{-}33)$$

and the voltage v_i, obtained by applying superposition to the sources v_s and $K'_v v_i$, is

$$v_i = \frac{R_f}{R_s + R_f} v_s - \frac{R_s}{R_s + R_f} K'_v v_i \qquad (13\text{-}34)$$

More compact symbols for the voltage-divider ratios are defined as

$$\alpha = \frac{R_f}{R_s + R_f} \quad \text{and} \quad \beta = \frac{R_s}{R_s + R_f} \qquad (13\text{-}35)$$

so that

$$v_i = \alpha v_s - \beta K'_v v_i = \alpha v_s + \beta v_o \qquad (13\text{-}36)$$

It follows from (13-35) that $\alpha = 1 - \beta$, and in most cases β is small so that $\alpha \approx 1$. Substituting the right-hand member of Eq. (13-36) into Eq.

(13-33) for v_i yields

$$v_o = -K'_v \alpha v_s - K'_v \beta v_o$$

and thus

$$v_o = \frac{-\alpha K'_v}{1 + \beta K'_v} \, v_s = -K_v v_s \tag{13-37}$$

Hence the net overall voltage gain of the feedback amplifier is

$$K_v = \frac{\alpha K'_v}{1 + \beta K'_v} = \frac{(1 - \beta)K'_v}{1 + \beta K'_v} \tag{13-38}$$

where K'_v = open-loop gain (gain with $R_f = \infty$)
β = feedback ratio
$\beta K'_v = K_L$ = loop gain
$1 + \beta K'_v = F$ = return difference
K_v = closed-loop gain

It is clear from Eq. (13-36) that β is a measure of the voltage trans-mitted from the output of the controlled source $K'_v v_i$ back to its input v_i with $v_s = 0$; thus β is a measure of the amount of feedback in the am-plifier. The quantity K'_v is the voltage gain through the controlled source, and it follows that $\beta K'_v$ is the gain around the complete feedback loop. The significance of the loop gain can be clarified by imagining the feedback loop to be broken at a point where no current flows, as shown in Fig. 13-21. The loop gain is the gain from v'_i to v_i in Fig. 13-21 with $v_s = 0$; it is also a measure of the amount of feedback in the circuit.

Since the loop gain is an important parameter of the amplifier, it is important to note that it is evaluated by determining the gain from v'_i to v_i in Fig. 13-21 *with the input terminals short circuited.* An entirely dif-

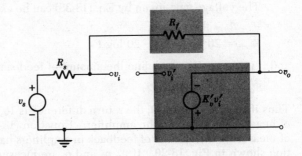

Fig. 13-21 A feedback amplifier with the feedback loop tem-porarily broken.

ferent value is obtained if the loop gain is evaluated with the input terminals open circuited; this gain would be appropriate if the source of signals were a current source. Any internal resistance associated with the signal source v_s must of course be included in R_s.

The denominator in Eq. (13-38), $F = 1 + \beta K_v'$, is the return difference of the feedback loop; it is the difference between v_i' and the voltage that returns to v_i in Fig. 13-21 when 1 volt is applied at v_i' with $v_s = 0$. The return difference appears repeatedly in feedback-amplifier studies, and it is another useful measure of the amount of feedback in the amplifier. The voltage gain given by Eq. (13-38) can be written as

$$K_v = \frac{\alpha K_v'}{F} \approx \frac{K_v'}{F}$$

where $\alpha \approx 1$ has been assumed. Thus the closed-loop gain is the open-loop gain divided by the return difference.

This reduction in the overall gain of the amplifier is a large part of the price paid for the use of negative feedback to improve the performance of the amplifier. When feedback is used solely to stabilize the quiescent operating conditions in the amplifier, it can be removed insofar as signals are concerned with bypass capacitors, as is done in the earlier sections of this chapter. However, when the feedback is used to improve the signal performance of the amplifier, the price of reduced gain must be paid. This price is usually paid cheerfully because it is much easier to get gain from transistors and tubes than it is to get high-quality performance in other respects. Negative feedback provides a means for trading easily gotten gain for other qualities that are not easily built into transistors and tubes. In particular, negative feedback permits the overall performance of the amplifier to be made relatively independent of the poorly controlled, nonlinear parameters of the active devices.

The voltage gain given by Eq. (13-38) can be expressed in decibels as

$$K_{v,\mathrm{dB}} = 20 \log (\alpha K_v') - 20 \log F \tag{13-39}$$

and it is convenient to define the amount of feedback in decibels as

$$F_{\mathrm{dB}} = -20 \log F \tag{13-40}$$

Thus if the loop gain is 9, the return difference is 10, and there is 20 dB of negative feedback in the amplifier. Equation (13-39) suggests a way of measuring the amount of feedback in amplifiers having the configuration shown in Fig. 13-20. If v_s, v_i, and v_o are measured, then $K_v = v_o/v_s$ and $K_v' = v_o/v_i$, and F_{dB} can be calculated from (13-39). This statement assumes that α can be evaluated from the known circuit resistances; often $\alpha \approx 1$.

It is important to note that, if $R_s = 0$, then $\beta = 0$, $F = 1$, and $F_{dB} = 0$; under these conditions there is no feedback in the amplifier. With $R_s = 0$, the input voltage to the controlled source in Fig. 13-20 is determined entirely by the voltage source v_s, and it is independent of the output voltage of the controlled source. Under this condition the feedback resistor R_f serves only to lower the input resistance of the amplifier.

If $K_L = \beta K_v' \gg 1$, Eq. (13-38) reduces to

$$K_v = \frac{\alpha}{\beta} = \frac{R_f}{R_s} \tag{13-41}$$

which is the same as the result obtained in Sec. 13-4 by a simpler reasoning process. In Sec. 13-4 the approximation of a very large loop gain is made at the beginning of the analysis; here the approximation is made at the end. The result shows that, in the case of a very large loop gain, the closed-loop gain is independent of the open-loop gain and of the transistor parameters. (In the case of BJT amplifiers, the input resistance R_i' shown in Fig. 13-20a is usually dependent on the transistor parameters. Thus the closed-loop gain is not independent of the transistor parameters unless R_i' is large enough to make a negligible contribution to R_s. Most BJT amplifiers intended for this kind of service use emitter followers and other special input-circuit configurations to achieve large values for R_i'.) However, since the loop gain is not infinite, this result is only approximate, and it is desirable to have a quantitative formulation of the extent to which K_v is independent of K_v'. Fortunately, such a formulation can be developed without difficulty, and the results are very useful in the design of feedback amplifiers.

The action by which negative feedback renders K_v relatively independent of K_v' is called self-calibration. If K_v' increases with v_s a constant positive voltage, the output of the amplifier in Fig. 13-20 tends to become a larger negative voltage, and it is fed back to the input where it reduces v_i by subtracting from the effect of v_s. This decrease in v_i compensates for the increase in K_v' to an extent that depends on the amount of feedback in the circuit. The degree of self-calibration can be evaluated as follows. When K_v' has a certain initial value K_i', the closed-loop gain has the initial value

$$K_v = K_i = \frac{\alpha K_i'}{1 + \beta K_i'} \tag{13-42}$$

If K_v' then takes on an increment $\Delta K_v'$, the closed-loop gain becomes

$$K_v = K_i + \Delta K_v = \frac{\alpha(K_i' + \Delta K_v')}{1 + \beta(K_i' + \Delta K_v')} \tag{13-43}$$

Subtracting Eq. (13-42) from (13-43) and rearranging the terms in the difference yields

$$\frac{\Delta K_v}{K_i} = \frac{1}{1 + \beta(K_i' + \Delta K_v')}\frac{\Delta K_v'}{K_i'} = \frac{1}{F_i + \Delta F}\frac{\Delta K_v'}{K_i'} \qquad (13\text{-}44)$$

where K_i = initial value of K_v
$\quad\quad K_i'$ = initial value of K_v'
$\quad\quad F_i$ = initial value of F

This equation relates the fractional change in closed-loop gain to the fractional change in open-loop gain; if the amount of feedback in the amplifier is large, F is large, and the fractional change in K_v is much smaller than the fractional change in K_v'. It is useful to note that this equation has the same form as Eq. (10-32) which relates to the stability of the quiescent point in a BJT amplifier. The result given by Eq. (13-44) shows that, in addition to being a measure of the feedback in the amplifier, the return difference is also a measure of the degree of self-calibration provided by the feedback.

The open-loop gain of microamplifiers made in mass production may vary over a considerable range, and often the range is not known. However, in some cases the manufacturer specifies a guaranteed minimum value for the gain. If the minimum value of the open-loop gain is designated K_m', then the minimum value of the closed-loop gain is

$$K_v = K_m = \frac{\alpha K_m'}{1 + \beta K_m'} \qquad (13\text{-}45)$$

As K_v' increases from the minimum value, K_v increases. The maximum possible value for K_v with large K_v' is

$$K_v = K_m + \Delta K_v = \frac{\alpha}{\beta} \qquad (13\text{-}46)$$

Then subtracting (13-45) from (13-46) yields

$$\Delta K_v = \frac{\alpha}{\beta} - \frac{\alpha K_m'}{1 + \beta K_m'} = \frac{K_m}{\beta K_m'} \qquad (13\text{-}47)$$

and the greatest fractional change in the closed-loop gain that can occur for any open-loop gain exceeding K_m' is

$$\frac{\Delta K_v}{K_m} = \frac{1}{\beta K_m'} \qquad (13\text{-}48)$$

This result can be used to design a feedback amplifier so that the closed-loop gain is confined to a specified range of values.

Example 13-3

A certain electronic instrumentation system requires a precision amplifier having a minimum gain $K_m = 99$, and the fractional change in the gain must be $\Delta K_v/K_m < 0.02$. A feedback microamplifier is to be designed for this function.

Solution   From Eq. (13-48),

$$\frac{\Delta K_v}{K_m} = \frac{2}{100} = \frac{1}{\beta K_m'}$$

and thus the minimum loop gain must be $\beta K_m' = 50$.

The minimum closed-loop gain is

$$K_m = 99 = \frac{(1 - \beta)K_m'}{1 + \beta K_m'} = \frac{K_m' - 50}{1 + 50}$$

and thus

$$5,049 = K_m' - 50$$

$$K_m' = 5,100$$

The guaranteed minimum open-loop gain of the microamplifier must have at least this magnitude. Now suppose that the commercial amplifier chosen has $K_m' = 6,000$; then

$$K_m = 99 = \frac{6,000 - \beta K_m'}{1 + \beta K_m'}$$

which yields

$$\beta K_m' = 59$$

This is more than enough loop gain to meet the specifications. Then

$$\beta = \frac{59}{6,000} = \frac{R_s}{R_s + R_f} = \frac{1}{1 + R_f/R_s}$$

and

$$\frac{R_f}{R_s} = \frac{6,000}{59} - 1 = 101$$

The input resistance of the closed-loop amplifier is $R_i \approx R_s$; thus R_s can be chosen for a suitable input resistance, and then R_f can be determined.

Now if K_v' is never less than the guaranteed minimum, K_v lies between 99 and 101 for all values of K_v', and a precision amplifier is obtained. As a check, suppose $K_v' = 15,000$; then

$$K_v = \frac{(1 - 59/6,000)(15,000)}{1 + (59)(15,000)/6,000} = 100$$

Figure 13-22 shows the circuit diagram of a feedback amplifier into which an unwanted disturbance, or corrupting voltage, v_d has been introduced. The disturbance can represent a drift voltage such as a change in base-to-emitter voltage with temperature, or it can represent a noise voltage generated by the random thermal motion of charge carriers, or it can represent a distortion component of voltage generated by nonlinear transistor characteristics in accordance with the discussion of Sec. 6-5. In all cases, however, negative feedback reduces the corrupting voltage that appears at the output. If v_d is a positive voltage, a positive corrupting voltage tends to appear at the output terminals. This voltage is fed back to the input, whereupon it is amplified by K_1' and its sign is reversed so that it appears as an opposing corrupting voltage in series with the original disturbance v_d. In this way the feedback tends to buck out the original disturbance, and the net disturbance appearing in v_1 can be greatly reduced.

These relations can be given a quantitative formulation in the following way. The output voltage of the amplifier is

$$v_o = K_2' v_1 = K_2'(v_d - K_1' v_i) = -K_1' K_2' v_i + K_2' v_d \tag{13-49}$$

and, as before, the input voltage v_i can be expressed as

$$v_i = \alpha v_s + \beta v_o$$

Substituting this relation into Eq. (13-49) yields

$$v_o = -K_1' K_2' \alpha v_s - K_1' K_2' \beta v_o + K_2' v_d \tag{13-50}$$

or

$$v_o = \frac{-\alpha K_1' K_2'}{1 + \beta K_1' K_2'}\, v_s + \frac{K_2'}{1 + \beta K_1' K_2'}\, v_d \tag{13-51}$$

When the loop gain is large, this expression simplifies to

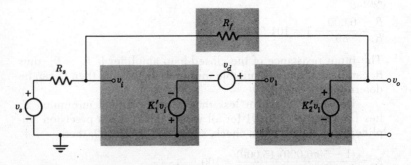

Fig. 13-22 A feedback amplifier with an unwanted corrupting signal.

$$v_o = -\frac{\alpha}{\beta} v_s + \frac{1}{\beta K_1'} v_d \qquad (13\text{-}52)$$

and thus the magnitude of the unwanted signal appearing at the output of the amplifier is reduced by the factor of $1/\beta K_1'$. However, Eq. (13-51) shows that, for any given values of K_1' and K_2', the feedback does not change the *ratio* of wanted to unwanted signal at the output; this ratio can be improved only by increasing the input signal v_s. The value of negative feedback in this respect lies in its ability to reduce the magnitude of the unwanted signal at the output, as for example when v_d represents a drift voltage caused by a change in temperature.

When electronic amplifiers are operated with large signal levels to provide a required large signal voltage at the output, distortion components of voltage may be generated in the last stage, where the signal level is highest. In this case the distortion voltage v_d appears at the output of the amplifier, and the corresponding conditions are described by Eq. (13-51) with $K_2' = 1$,

$$v_o = \frac{-\alpha K_1'}{1 + \beta K_1'} v_s + \frac{1}{1 + \beta K_1'} v_d \qquad (13\text{-}53)$$

Thus the feedback reduces the distortion component of the output voltage, and it reduces the signal component by the same amount. In order to maintain the required large signal voltage at the output, v_s must be increased. When v_s is adjusted to keep the signal component of output voltage constant, the output voltage can be expressed as

$$v_o = v_{os} + \frac{1}{1 + \beta K_1'} v_d \qquad (13\text{-}54)$$

where v_{os} is a constant. Under these conditions the ratio of signal voltage to distortion voltage at the output is improved by the amount of the return difference associated with the feedback. Feedback is widely used in this way to obtain low-distortion, high-power electronic amplifiers; in fact, feedback is almost essential in this area of electronic circuit design.

Feedback also has an effect on the input and output resistances of an amplifier. As an aid in examining this matter, Fig. 13-23 shows a feedback amplifier in which the internal amplifier has input and output resistances R_i' and R_o', respectively. As a result of the feedback network, the input terminals of this amplifier are not isolated from the output terminals, and consequently the input resistance depends on any load resistance that is connected across the output terminals. It is therefore assumed that the effects of any load resistance have been lumped in with K_v' and R_o' by means of Thévenin's theorem.

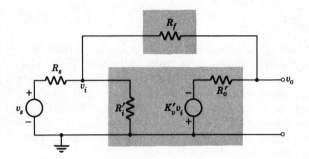

Fig. 13-23 A feedback amplifier.

The clearest insight into the nature of the input resistance of the amplifier in Fig. 13-23 is obtained with the aid of the reduction theorem. Applying the voltage-source form of this theorem leads directly to the reduced circuit shown in Fig. 13-24, and the input resistance can be evaluated by inspection of this simplified circuit.

Network theorems can also be used to gain a clear understanding of the effects of feedback on the output resistance of the amplifier. The output resistance is defined as the resistance seen at the output terminals with $v_s = 0$, and under this condition the voltage v_i is

$$v_i = \beta v_o$$

where β is the voltage-divider ratio associated with R_f, R_s, and R_i'. Then, with $v_s = 0$, the output voltage of the controlled source in Fig. 13-23 is

$$K_v' v_i = \beta K_v' v_o \tag{13-55}$$

and the circuit can be put in the equivalent form shown in Fig. 13-25a. The voltage-source form of the reduction theorem applied to this circuit yields the reduced circuit shown in Fig. 13-25b; the reduced cir-

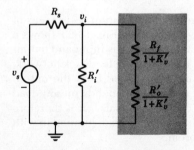

Fig. 13-24 An equivalent circuit for the input resistance of the amplifier in Fig. 13-23.

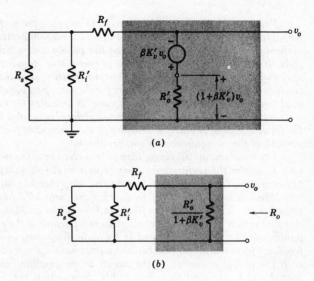

Fig. 13-25 Equivalent circuits for the output resistance of the amplifier in Fig. 13-23.

cuit shows clearly the effect of feedback on the output resistance. Normally R_f is much larger than $R_o'/(1 + \beta K_v')$, and the output resistance is

$$R_o \approx \frac{R_o'}{1 + \beta K_v'} \tag{13-56}$$

If $R_o' \ll R_f$, which it normally is, then $\beta K_v'$ is the loop gain for the amplifier, and $1 + \beta K_v'$ is the return difference. If the loop gain is large, R_o can be quite small, a few ohms or even a fraction of an ohm.

The results obtained above showing the effects of feedback on the input and output resistances of the feedback amplifier apply only to the circuit configuration shown in Fig. 13-23. In other configurations the effects of feedback may be quite the opposite, and in still other configurations feedback may have little or no effect on the input and output resistances.

13-6 SUMMARY

This chapter presents several methods that are used to connect single-stage amplifiers in cascade to form high-gain multistage amplifiers. It is shown that current engineering practice makes extensive use of negative feedback to stabilize the quiescent operating conditions in multistage amplifiers.

Negative feedback can also be used to improve the performance of amplifiers with respect to the signal. It can provide self-calibration that renders the performance relatively independent of the poorly controlled transistor parameters, it can reduce the contribution of corrupting signals to the output voltage, and it can alter the input and output resistances of the amplifier. These consequences of negative feedback are for the most part beneficial, and the price paid for the benefits is a reduction in gain. A quantitative formulation of these features shows that the return difference, which is related to the gain around the feedback loop, is a useful measure of the amount of feedback in the amplifier and of the consequences of this feedback.

The most important single idea in this chapter is the notion of using feedback to render the performance of amplifiers relatively independent of the amplifier parameters provided only that the amplifier has high gain.

When a large amount of feedback exists around a high-gain multistage amplifier, the amplifier is very likely to become an oscillator. An example of this fact which has been experienced by most people is the case of the howling public address system; the howl is an audio-frequency oscillation caused by feedback from the loudspeaker to the microphone. Since oscillations interfere with the proper operation of the circuit as an amplifier, they cannot be permitted. It is assumed throughout this chapter that the feedback amplifiers discussed have been designed so that they do not oscillate. The problem of designing feedback amplifiers so that they do not oscillate is examined in some detail in Chap. 16; this study is based on certain important fundamental ideas developed in Chaps. 14 and 15.

PROBLEMS

13-1. *MOST amplifier design and analysis.* An enhancement MOST amplifier like the one shown in Fig. 13-2a consists of three identical stages with transistors characterized by $i_D = 0.1(v_{GS} - 3)^2$ mA. The power-supply voltage is 30 volts and the circuit is to be designed so that the quiescent drain current for each transistor is 1 mA.

(a) Determine the values of V_{GS} and R_d that are required.

(b) The circuit is provided with a bypass capacitor like the one shown in Fig. 13-2b, and $R_g = 1$ megohm. Give a small-signal model for the amplifier in which the coupling and bypass capacitors are treated as short circuits and the drain resistance r_d of the transistors is treated as an open circuit.

(c) Determine the transconductances of the transistors [see Eq. (7-55)], and determine the overall voltage gain $K_v = |v_o/v_i|$.

13-2. *Input-resistance effects.* The MOST amplifier shown in Fig. 13-26 has a voltage gain $K_v = |v_o/v_i| = 3,000$, and the coupling capacitors can be treated as short circuits for signal voltages. The effect of the input resistance on the performance of the amplifier is to be examined.

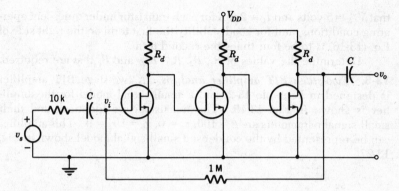

Fig. 13-26   MOST amplifier for Prob. 13-2.

(a) Determine the input resistance of the amplifier.

(b) Determine the voltage gain $K_{vo} = |v_o/v_s|$.

(c) The 1-megohm feedback resistor is divided into two equal parts, and a bypass capacitor is added, as shown in Fig. 13-2b. Assuming that the bypass capacitor acts as a short circuit for signal voltages, determine the voltage gain $K_{vo} = |v_o/v_s|$. Compare this result with the gain obtained in part b.

13-3.   Three-stage BJT amplifier design.   A three-stage BJT amplifier of the form shown in Fig. 13-7 is used with a power-supply voltage $V_{CC} = 15$ volts. The transistors are identical, and they have the parameters $\beta = 50$ and $V_o = 0.7$ volt. The circuit is to be designed to have $V_{CE3} = 3$ volts and $I_{C3} = 2$ mA under quiescent conditions, and for good stability the first term on the right side of Eq. (13-7) is to be four times the second term. The voltage drops across the diodes are assumed to be constant at the value 0.7 volt.

Determine the values of R_c, R_1, and R_2 that are required.

13-4.   BJT amplifier analysis.   A three-stage BJT amplifier is designed in Example 13-1, and a small-signal model for the amplifier is given in Fig. 13-11. The three transistors in the amplifier are identical, and their small-signal parameters are $\beta = 100$, $r_x \approx 0$, and $g_o = 1/r_o \approx 0$. Using the information contained in Example 13-1, determine the small-signal voltage gain $K_v = |v_o/v_i|$. It can be assumed that the collector currents have the values given in Example 13-1.

13-5.   Two-stage BJT amplifier design.   A two-stage BJT amplifier having the form shown in Fig. 13-13 is used with a power-supply voltage $V_{CC} = 25$ volts. The transistors are identical, and they have the parameters $\beta = 50$ and $V_o = 0.7$ volt. The circuit is to be designed so that $V_{CE} = 5$ volts and $I_C = 2$ mA for each transistor under quiescent operating conditions, and for good stability the first term on the right side of Eq. (13-16) is to be four times the second term.

Determine the values of R_1, R_2, R_{c1}, R_{c2}, and R_e that are required.

13-6. *Two-stage BJT amplifier analysis.* A two-stage BJT amplifier is designed in Example 13-2, and a small-signal model for the amplifier is shown in Fig. 13-15. The transistors are identical, and their small-signal parameters are $\beta = 100$, $r_x \approx 0$, $g_o = 1/r_o \approx 0$. This amplifier can be represented by the condensed small-signal model shown in Fig. 13-27.

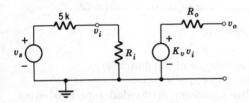

Fig. 13-27 Small-signal model for Prob. 13-6.

(a) Using the information contained in Example 13-2, determine the values of the parameters in the model of Fig. 13-27. It can be assumed that the collector currents have the values given in Example 13-2.

(b) Determine the voltage gain $K_{vo} = v_o/v_s$.

(c) If a 2-kilohm load resistor is connected across the output terminals, what is the value of K_{vo}?

13-7. *The operational amplifier as a current source.* The operational amplifier shown in Fig. 13-28 acts as a nearly ideal current source connected to R_f.

(a) Determine the Norton equivalent for the circuit connected to R_f. Give the parameters of the equivalent circuit in terms of the parameters in Fig. 13-28. *Suggestion:* The reduction theorem and a source conversion provide an easy solution.

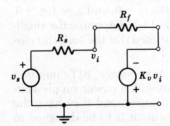

Fig. 13-28 Operational amplifier for Prob. 13-7.

(b) What are the values of the Norton equivalent current and resistance when $R_s = 10$ kilohms, $K_v = 10^4$, and $v_s = 10$ volts?

13-8. Operational-amplifier analysis. The operational amplifier shown in Fig. 13-29 is widely used in analog computers and simulators. The voltage gain K_v is very large so that $v_i \approx 0$. Determine the output voltage v_o as a function of the three input voltages.

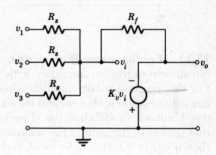

Fig. 13-29 Operational amplifier for Prob. 13-8.

13-9. Operational-amplifier analysis. The operational amplifier shown in Fig. 13-30 has many applications. The voltage gain K_v is very large so that $v_i \approx 0$. Determine the output voltage v_o as a function of the two input voltages.

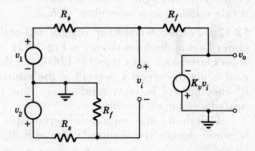

Fig. 13-30 Operational amplifier for Prob. 13-9.

13-10. Feedback-amplifier analysis. The internal amplifier in the circuit of Fig. 13-31 is a microamplifier with a voltage gain $K_v' = 2,000$. The behavior of the amplifier with feedback is to be examined.

(a) Combine R_s, R_i', and voltage source v_s in a Thévenin equivalent circuit.

(b) Calculate the loop gain for the circuit.

(c) What is the overall voltage gain $K_v = |v_o/v_s|$?

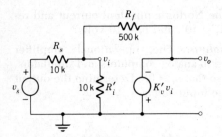

Fig. 13-31 Feedback amplifier for Prob. 13-10.

13-11. Feedback-amplifier design, self-calibration. A certain application requires an amplifier with a minimum voltage gain of 100. For this purpose a microamplifier having a guaranteed minimum voltage gain of 2,500 is chosen, and the excess gain is to be invested in negative feedback to stabilize the closed-loop gain against the production-line spread in the gain of the microamplifier. The circuit configuration shown in Fig. 13-31 is to be used, and R_i' is large enough to be treated as an open circuit.

(a) Determine the ratio R_f/R_s required for a minimum closed-loop gain of 100 when K_r' has the guaranteed minimum value.

(b) What is the closed-loop gain if $K_r' = 5,000$?

(c) What is the range of the closed-loop gain as K_r' goes through the entire range from the guaranteed minimum to very high values? What is this variation as a percentage of K_m?

13-12. Feedback-amplifier design, self-calibration. A feedback amplifier having the form shown in Fig. 13-31 is to be designed for a minimum closed-loop gain $|v_o/v_s| = 1,000$, and the maximum change in this gain is not to exceed 4 percent of the minimum value for any value of K_r' encountered in mass production. The input resistance R_i' can be treated as an open circuit.

Determine the minimum permissible value of open-loop gain K_r', and calculate the required value of R_f/R_s.

13-13. Drift in a feedback amplifier. Figure 13-32 shows the dc small-signal model for a MOST amplifier similar to the one in Fig. 13-2. The voltage gain of each stage is 15, and there is a sign reversal in each stage. The voltage v_d is a small unwanted drift voltage resulting from temperature effects in the first stage.

(a) If the feedback resistor is removed and node v_i is connected to ground, what is the value of v_o in terms of v_d?

(b) With the feedback resistor in place as shown in Fig. 13-32 and the ground connection removed from node v_i, what is the value of v_o in terms of v_d?

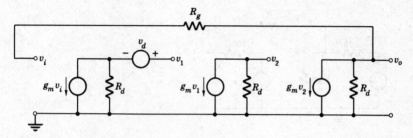

Fig. 13-32 Feedback amplifier for Prob. 13-13.

(c) If $v_d = 0.1$ volt, what do the results obtained above mean in terms of the quiescent operating conditions in the amplifier?

13-14. Feedback-amplifier design, distortion. The voltage v_d in the circuit of Fig. 13-33 represents second-harmonic distortion generated in the last stage of the amplifier. The open-loop gain of the amplifier is

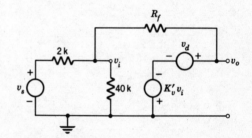

Fig. 13-33 Feedback amplifier for Prob. 13-14.

$K'_v = 3,000$, and the 40-kilohm input resistance can be treated as an open circuit. The feedback resistor is to be adjusted so that the distortion component of the output voltage is $v_d/20$.

(a) Determine the required values of loop gain and feedback resistance R_f.

(b) What is the closed-loop voltage gain?

13-15. Feedback-amplifier study, input and output resistance. The open-loop voltage gain of the amplifier shown in Fig. 13-34 is $K'_v = 30,000$, and the 400-kilohm input resistance can be treated as an open circuit. The feedback resistor is to be adjusted so that the closed-loop gain is $|v_o/v_s| = 100$. It can be anticipated that R_f will be much larger than 150 ohms so that $\beta = R_s/(R_s + R_f)$.

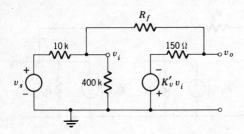

Fig. 13-34 Feedback amplifier for Prob. 13-15.

(a) Determine the value of R_f required.

(b) With this adjustment what is the closed-loop output resistance R_o?

(c) What is the input resistance seen on the right of the 10-kilohm resistor?

(d) What is the input resistance faced by the signal source v_s?

14

Frequency Characteristics of Electronic Amplifiers

The preceding chapters present electronic amplifiers under such operating conditions that the amplifiers can be represented by purely resistive models; the coupling and bypass capacitors are treated as short circuits for signal voltages, and the parasitic capacitances associated with the transistors and tubes are treated as open circuits. All of the amplifiers presented in those chapters have a substantial range of operating conditions under which these approximations are entirely valid. However, when the frequency of the signal voltage is very high, the parasitic capacitances cannot be neglected, and when the frequency of the signal voltage is very low, the coupling and bypass capacitors cannot be neglected. Thus, in general, with low signal frequencies and with high signal frequencies, the voltage gain and the input and output impedances become functions of the signal frequency. An understanding of the basic principles of this frequency dependence is essential for the effective design and utilization of most electronic circuits. A partial exception to this general rule exists in the form of directly coupled amplifiers in which the use of coupling and bypass capacitors is avoided; such amplifiers exhibit frequency dependence only at high frequencies.

The primary objective of this chapter is to develop the concept of the logarithmic gain and phase characteristics as a simple and effective means of characterizing and displaying the frequency dependence of electronic amplifiers. A second objective is to develop useful factual information about the properties of several important circuit configurations; for this reason the basic concepts are presented in terms of specific electronic circuits.

14-1 SOME ELEMENTARY PROPERTIES OF SIGNALS

Before beginning the study of the frequency dependence of specific circuits, it is helpful to examine some of the important properties of typical signals encountered in electronic systems. This study establishes the requirements that amplifiers must meet in order to provide faithful amplification of the signal, and in this way it gives meaning to the results that are obtained subsequently in the examination of specific circuits. This study also provides a means for judging the merit of any particular amplifier after its frequency characteristics have been determined.

An extremely wide variety of signal waveforms is encountered in electronic systems. These range from the very complex video signals in television systems to the relatively simple signals occurring in ac servomechanisms. In addition, there must of necessity be a large uncertainty about signals that carry information, for a signal whose waveform is known completely cannot yield any new information. A serious question thus arises of how these signals can be characterized, or represented, mathematically. The choice of the characterization is governed more by circuit considerations than by signal considerations. Highly developed and highly effective techniques exist for the analysis of linear circuits when the currents and voltages are sinusoidal functions of time or when they are the superposition of a number of sinusoidal components. With only a few exceptions, the analysis of the response of electrical circuits to signals having other mathematical characterizations is a thoroughly intractable problem. Therefore efforts are directed toward the representation of signals as a superposition of sinusoidal components having the appropriate amplitudes, frequencies, and phases. Fortunately, very powerful techniques for this purpose are available; among them are the Fourier series, the Fourier transform, and the Laplace transform. A few simple examples of the representation of complex signals by a superposition of sinusoidal components are described in the following paragraph.

One of the simplest classes of signals is encountered in the ac servomechanism. Under many operating conditions this signal is nearly sinusoidal with a frequency of 60 or 400 Hz; information is carried by this signal in the form of relatively slow variations in the amplitude of the sinusoid. A somewhat more complicated class of signals consists of periodic but nonsinusoidal waveforms. Typical periodic waveforms encountered in electronic systems are square waves, sawtooth waves, and pulse trains. A very complicated periodic signal is the video signal representing a still picture, such as a test pattern, in a television system; this signal has a fundamental frequency of 30 Hz, and it has important sinusoidal components at frequencies up to about 4 MHz. Periodic signals such as these can be represented as the superposition of sinusoidal components by a Fourier series of the form

$$v(t) = V_1 \cos (\omega_s t + \phi_1) + V_2 \cos (2\omega_s t + \phi_2)$$
$$+ V_3 \cos (3\omega_s t + \phi_3) + \cdots \quad (14\text{-}1)$$

Other important signals can be represented either exactly or approximately by a similar superposition of sinusoidal components, although in these cases the sinusoidal components are not harmonically related. For example, it is commonplace to think of sound waves as being composed of sinusoidal components of different frequencies, or pitches. In the case of speech or music, however, the sinusoidal composition of the sound is continually changing. In general, signals that can be approximated by the superposition of a number of sinusoidal components can be expressed as

$$v(t) = V_1 \cos (\omega_1 t + \phi_1) + V_2 \cos (\omega_2 t + \phi_2)$$
$$+ V_3 \cos (\omega_3 t + \phi_3) + \cdots \quad (14\text{-}2)$$

in which the frequencies of the components are not necessarily harmonically related.

The diagram shown in Fig. 14-1a provides a convenient way to picture the sinusoidal composition of a signal; the height of each ordinate represents the amplitude of the sinusoidal component of the signal at the indicated frequency. This diagram represents the frequency spectrum of the signal; it gives the frequency and amplitude of each sinusoidal component of the signal, but it contains no information about the phase angles of the components. When a signal consisting of a sum of sinusoidal components is applied at the input of an amplifier, the output signal can be determined by superposition. The output is calculated separately for each sinusoidal component, and the results are added to obtain the response of the amplifier to the composite signal. In passing through the amplifier, the amplitude of each signal component is changed (amplified), and the phase of each component is shifted. To expedite the calculation of the amplitude and phase of each component at the output, it is helpful to plot curves of gain and phase shift vs. frequency for the amplifier; then the amplitude and phase angle of each sinusoidal component at the output of the amplifier can be

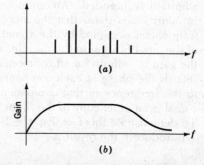

Fig. 14-1 Characterization of signals and amplifiers. (a) Frequency spectrum of a signal; (b) frequency characteristic of an amplifier.

determined directly from the curves. A typical curve of gain vs. frequency is shown in Fig. 14-1b. The gain decreases at high frequencies as a result of parasitic capacitances in the circuit, and it decreases at low frequencies because of the coupling capacitors. Between these extremes there is a band of frequencies in which the gain is independent of frequency; this band of frequencies is called the midband of the amplifier. It is clear that for faithful amplification of the signal the gain of the amplifier should remain constant over the entire band of frequencies occupied by the spectrum of the signal so that all components are amplified by the same amount.

The detailed calculation of the amplitude and phase of each sinusoidal component of the signal at the output of an amplifier is seldom carried out in practice, however. In fact, the exact composition of the input signal is seldom known. In lieu of these detailed calculations it is possible to use certain general properties of signals, together with the gain and phase characteristics, to guide the design of amplifiers and to judge the performance of the circuits after they are built. If the waveform of the signal at the output of the amplifier is to be an exact replica of the waveform at the input, then the relative amplitudes of the sinusoidal components at the output must be the same as those at the input, and the relative phase displacements at the output must be the same as those at the input. (The frequencies of the sinusoidal components can never be altered by a linear amplifier.) Thus an ideal amplifier is one that amplifies all sinusoidal components of the signal by exactly the same amount and produces no phase shifts in the components. The frequency characteristic for an ideal amplifier is thus a horizontal line, and the amplifier is said to have a flat gain characteristic.

The requirements stated above for an ideal amplifier are stringent, and they cannot be realized in practice, although they can be approximated quite well over a restricted band of frequencies. Thus it is fortunate that the requirements can be relaxed somewhat without seriously degrading the performance of the amplifier. The amount by which they can be relaxed depends on the type of service for which the amplifier is intended. An engineering rule of thumb that is applicable in many cases states that the amplification at the edge of the band of frequencies occupied by the signal should not be less than 70 percent of the amplification in the middle of that band. It also turns out that, if the gain is uniform for all components of the signal and if the amplifier retards the phase of each component of the signal in direct proportion to the frequency of that component, then the waveform of the output signal is an exact copy of the input waveform, but it is delayed in time. To demonstrate this fact, Eq. (14-2) is rewritten with a pure time delay T introduced; the result is

$$v(t - T) = V_1 \cos \left[\omega_1(t - T) + \phi_1\right]$$
$$+ V_2 \cos \left[\omega_2(t - T) + \phi_2\right]$$
$$+ V_3 \cos \left[\omega_3(t - T) + \phi_3\right] + \cdots \quad (14\text{-}3)$$

Rearranging this expression and defining $\theta_k = \omega_k T$ yields

$$v(t - T) = V_1 \cos (\omega_1 t - \theta_1 + \phi_1)$$
$$+ V_2 \cos (\omega_2 t - \theta_2 + \phi_2)$$
$$+ V_3 \cos (\omega_3 t - \theta_3 + \phi_3) + \cdots \quad (14\text{-}4)$$

Thus the pure time delay adds a phase lag θ_k to each component which is directly proportional to the frequency of that component. It follows from reversing the above process that if an amplifier introduces a phase lag that increases linearly with frequency $\theta = K\omega$, then the amplifier produces a pure time delay of $T = K$ sec (if ω is in radians per second). If the gain of this amplifier is uniform for all signal frequencies, the signal waveform is not distorted; it is merely delayed in time. In practice it is much easier to approximate a linear phase-shift characteristic of this kind than it is to approximate zero phase shift.

The case in which the signal is an audio signal for the production of sound at the output of the amplifier is one that requires a special comment. Under normal listening conditions the ear is not sensitive to the phase displacement between the signal components. Acoustical reflections and reverberations from surrounding objects cause the phase relations to become random under ordinary conditions. Thus the ear is sensitive to the amplitudes and frequencies of the signal components, but it is not sensitive to the waveform of the signal. For this reason the phase shift is of little interest in the design and testing of audio equipment. In many other electronic systems, however, the waveform of the signal must be preserved faithfully; this is the case, for example, with the video signal in a television system. In such systems the phase shift is important, and a linear phase characteristic is often more important than a flat gain characteristic in preserving the signal waveform.

Thus the analysis and design of linear amplifiers is largely concerned with the response of the amplifier to sinusoidal signals; more specifically, it is concerned with the gain and phase-shift characteristics of amplifiers as a function of frequency. The remainder of this chapter is devoted to developing means for constructing these characteristics with a minimum of computational effort and to presenting techniques for designing amplifiers to have frequency characteristics meeting certain specifications.

14-2 THE PENTODE AMPLIFIER AT HIGH FREQUENCIES

The small-signal model for the pentode amplifier at high frequencies is simple, and the frequency characteristics for this circuit have a very simple form. However, in the process of analyzing this simple circuit it is possible to develop several valuable techniques that are useful in the analysis and design of a wide variety of amplifiers. The circuit of a typical pentode amplifier is shown in Fig. 14-2a. In the usual case the bypass capacitors can be treated as short circuits at frequencies greater than 100 Hz or so, and hence at frequencies above this limit the amplifier can be represented by the small-signal model shown in Fig. 14-2b. This model is valid at frequencies up to several tens of megahertz, depending on the geometry of the tube. The capacitance C_1 accounts for the total parasitic capacitance between the grid and ground, including the wiring capacitance as well as the tube capacitances, and C_2 accounts for the total parasitic capacitance between the plate and ground. The grid-to-plate capacitance of the pentode is negligibly small, and thus the input circuit is isolated from the output circuit.

If the complex amplitude of a sinusoidal input signal is designated V_1, then the complex amplitude of the sinusoidal output signal in the circuit of Fig. 14-2b is

$$V_2 = \frac{-g_m}{g_p + G_2 + j\omega C_2} V_1 \tag{14-5}$$

The comments made in connection with Eq. (10-113) are applicable to all the circuits considered in this chapter, and thus they are repeated

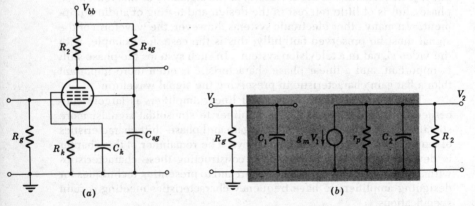

Fig. 14-2 The pentode amplifier at high frequencies. (a) Circuit; (b) high-frequency model.

here. At low frequencies the ratio of output voltage to input voltage is
a constant that does not vary with time; it is the voltage gain of the amplifier. Equation (14-5) shows, however, that with a high-frequency
sinusoidal input signal, the output voltage is not in phase with the input
voltage, and it follows that under these conditions the ratio of output
voltage to input voltage varies from instant to instant. Thus for rapidly
varying signals in general, it is not possible to define a voltage gain, at
least not in any simple way. However, in the case of a sinusoidal input
signal the waveform of V_2 is the same as the waveform of V_1, and the
concept of a voltage gain can be extended in a useful way by defining it
to be the ratio of the complex amplitudes of these voltages. Thus the
extended definition is

$$A_v(\omega) = \frac{V_2}{V_1} = \frac{-g_m}{g_p + G_2 + j\omega C_2} \tag{14-6}$$

This voltage gain is a complex number, and it is a function of frequency.
The magnitude of A_v is the magnitude of the ratio V_2/V_1, and hence it is
the amplification of the circuit for sinusoidal signals. The angle of A_v
is the phase shift between V_1 and V_2. Thus the complex gain $A_v(\omega)$ contains all the information needed to construct the gain and phase-shift
characteristics discussed in Sec. 14-1.

Equation (14-6) can be rewritten in a more convenient form as

$$A_v(\omega) = \frac{-g_m}{g_p + G_2} \frac{1}{1 + j\omega[C_2/(g_p + G_2)]} \tag{14-7}$$

At medium frequencies ω is relatively small, and the second factor in
(14-7) is approximately unity; hence the first factor in (14-7) is the voltage gain at medium frequencies. Furthermore, the quantity $(g_p + G_2)/C_2$
has the dimensions of frequency, and it has an important interpretation
as a frequency. Hence it is convenient to define two new symbols:

$$K_v = \frac{g_m}{g_p + G_2} \qquad \text{and} \qquad \omega_2 = \frac{g_p + G_2}{C_2} \tag{14-8}$$

Substituting these expressions in (14-7) leads to a compact expression
for the voltage gain,

$$A_v(\omega) = -K_v \frac{1}{1 + j\omega/\omega_2} = \frac{-K_v}{\sqrt{1 + (\omega/\omega_2)^2}} \, e^{j\theta} = -|A_v|e^{j\theta} \tag{14-9}$$

Thus the gain and phase shift as functions of frequency are

$$|A_v| = \frac{K_v}{\sqrt{1 + (\omega/\omega_2)^2}} \qquad \text{and} \qquad \theta = -\tan^{-1}\frac{\omega}{\omega_2} \tag{14-10}$$

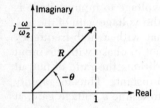

Fig. 14-3 Diagram of the complex number $R = 1 + j\omega/\omega_2$.

Note that the phase shift is defined so that it does not include the constant 180° phase shift associated with the sign reversal introduced by the amplifier.

The dependence of the gain and phase shift on frequency arises entirely from the factor $1 + j\omega/\omega_2$ in the denominator of Eq. (14-9). This factor is a complex number as well as a function of frequency, and the way in which its magnitude and angle vary with frequency is illustrated by the geometric diagram shown in Fig. 14-3. Thus when $\omega = \omega_2$, $\theta = -45°$, $R = \sqrt{2}$, and from (14-10), $|A_v| = K_v/\sqrt{2}$; it follows from these relations that ω_2 is the half-power frequency for the amplifier.

The manner in which A_v varies with frequency can be displayed by calculating A_v for a number of frequencies and plotting the frequency characteristics, $|A_v|$ and θ vs. frequency. It turns out, however, that it is far simpler to plot the gain in decibels,

$$A_{db}(\omega) = 20 \log |A_v(\omega)| \qquad (14-11)$$

and θ as functions of log ω. These logarithmic characteristics possess certain especially simple asymptotic properties that make it possible to construct the complete frequency characteristics without plotting any points in the usual sense. These properties can be developed by substituting (14-10) for $|A_v|$ in (14-11) to obtain

$$A_{dB}(\omega) = 20 \log \frac{K_v}{\sqrt{1 + (\omega/\omega_2)^2}}$$

Defining $K_{dB} = 20 \log K_v$ yields

$$A_{dB}(\omega) = K_{dB} - 20 \log \sqrt{1 + \left(\frac{\omega}{\omega_2}\right)^2} \qquad (14-12)$$

and with $\omega = 2\pi f$, this relation can be written as

$$A_{dB}(f) = K_{dB} - 20 \log \sqrt{1 + \left(\frac{f}{f_2}\right)^2} \qquad (14-13)$$

The asymptotic behavior of A_{dB} can be examined by considering very small and very large values of f. Thus for $f \ll f_2$,

$$A_{dB}(f) \approx K_{dB} - 20 \log 1 = K_{dB} \tag{14-14}$$

and the low-frequency asymptote for the gain characteristic is a constant, K_{dB}, as shown in Fig. 14-4. At high frequencies $f \gg f_2$, and

$$A_{dB}(f) \approx K_{dB} - 20 \log \frac{f}{f_2} \tag{14-15}$$

Thus the high-frequency asymptote is a linear function of $\log (f/f_2)$, and it is a straight line when plotted as a function of that variable or when plotted as a function of f using a logarithmic scale. This asymptote is shown in Fig. 14-4.

The equations of the two asymptotes are (14-14) and (14-15); the intersection of the asymptotes is found by equating these two relations,

$$K_{dB} = K_{dB} - 20 \log \frac{f}{f_2} \tag{14-16}$$

The value of f that satisfies this relation is $f = f_2$, and hence the intersection, or breakpoint, of the asymptotes occurs at the half-power frequency for the circuit. When $f = f_2$, the value of the high-frequency asymptote is given by Eq. (14-15) as $A_{dB} = K_{dB}$, and when $f = 10f_2$, the value of the high-frequency asymptote is $A_{dB} = K_{dB} - 20$; hence the slope of the high-frequency asymptote is -20 dB/decade of frequency, where a decade of frequency is any interval on the frequency scale covering a $10:1$ frequency ratio. Frequencies that stand in the ratio $2:1$ are separated by 1 octave, and it follows directly from Eq. (14-15) that the slope of the high-frequency asymptote is $-6.02 \approx -6$ dB/octave.

The transition of the gain characteristic from one asymptote to the other has the simple form shown in Fig. 14-5a. When $f = f_2$, Eq. (14-13) gives

$$A_{dB}^{*}(f) = K_{dB} - 20 \log \sqrt{2} \approx K_{dB} - 3 \tag{14-17}$$

Thus at the break frequency the gain characteristic is 3 dB below the low-frequency asymptote; hence the break frequency is often referred

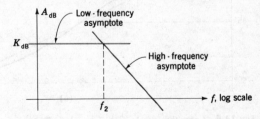

Fig. 14-4 Asymptotic behavior of $A_{dB}(f)$.

to as the 3-dB frequency. When $f = \frac{1}{2}f_2$,

$$A_{dB}(f) = K_{dB} - 20 \log \sqrt{1 + 1/4} = K_{dB} - 10 \log \tfrac{5}{4}$$
$$\approx K_{dB} - 7 + 6 = K_{dB} - 1 \tag{14-18}$$

Thus at 1 octave below the break frequency, the gain characteristic lies 1 dB below the low-frequency asymptote. And finally, when $f = 2f_2$,

$$A_{dB}(f) = K_{dB} - 20 \log \sqrt{1 + 4} \approx K_{dB} - 7 \tag{14-19}$$

Therefore at 1 octave above the break frequency, the gain characteristic lies 7 dB below the low-frequency asymptote. Since the high-frequency asymptote has a slope of -6 dB/octave, the gain characteristic lies 1 dB below it at $f = 2f_2$.

With the aid of these simple relations the gain characteristic can be constructed with negligible effort. For any particular amplifier it is necessary to calculate only the values of K_{dB} and f_2; the asymptotes for

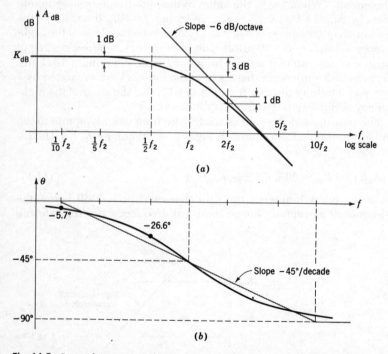

Fig. 14-5 Logarithmic gain and phase characteristics for the pentode amplifier at high frequencies. (a) Gain characteristic; (b) phase characteristic.

the gain characteristic can be constructed with these two numbers, and the characteristic can be constructed by applying the corrections as shown in Fig. 14-5a. The corrections are the same for all gain characteristics of this form. It is also true that for many purposes it is sufficient to construct only the asymptotes for the gain characteristic.

The gain characteristic shows the band of frequencies over which the gain is constant; thus it indicates the ability of the circuit to amplify signals without altering their waveforms. A rule of thumb given in Sec. 14-1 which is useful when the requirements are not too stringent states in effect that the amplifier will give satisfactory performance when all the important sinusoidal components of the signal have frequencies below the half-power frequency for the amplifier. Signals having important components at frequencies above the half-power frequency may be seriously distorted by the amplifier. An amplifier that gives excellent performance with audio signals may be totally inadequate for amplifying video television signals.

The phase-shift characteristic for the pentode amplifier is shown in Fig. 14-5b; it also has simple properties when plotted on a logarithmic frequency scale. The phase angle is given by Eq. (14-10) as

$$\theta = -\tan^{-1} \frac{\omega}{\omega_2} = -\tan^{-1} \frac{f}{f_2} \qquad (14\text{-}20)$$

and it is also represented graphically in Fig. 14-3. The asymptotes for the phase characteristic at low and high frequencies are 0 and $-90°$, respectively, and the phase shift at the half-power frequency f_2 is $-45°$. The phase characteristic is symmetrical about the point at $f = f_2$. As Fig. 14-5b shows, the transition of the phase characteristic from the low-frequency to the high-frequency asymptote can be approximated closely by a straight line beginning 1 decade below the break frequency at the low-frequency asymptote and ending 1 decade above the break frequency at the high-frequency asymptote. The exact phase shifts for $f = f_2/2$ and $f = f_2/10$ are shown in Fig. 14-5b; the phase shifts at $f = 2f_2$ and $f = 10f_2$ are, respectively, the complements of these angles. It is clear that the transition of the phase characteristic from one asymptote to the other occupies a much wider band of frequencies than the transition of the gain characteristic. In certain applications this fact is important and requires that careful attention be paid to the phase characteristic.

The importance of linear phase-shift characteristics is developed in Sec. 14-1. It is to be noted that a linear phase characteristic does not plot as a straight line on the logarithmic frequency scale used in Fig. 14-5. To examine the linearity of the phase characteristic, Eq. (14-20) can be expanded in a power series to obtain

$$\theta(f) = -\frac{f}{f_2} + \frac{1}{3}\left(\frac{f}{f_2}\right)^3 - \frac{1}{5}\left(\frac{f}{f_2}\right)^5 + \cdots \tag{14-21}$$

which converges for $f/f_2 < 1$. Thus it follows that the phase characteristic for this amplifier is a linear function of frequency only at frequencies well below the break frequency f_2.

14-3 CAPACITIVELY COUPLED MICROAMPLIFIERS

There are occasions in which it is desirable to connect microamplifiers
in cascade, either to obtain greater overall gain or to accomplish some
more elaborate signal-processing operation. Some microamplifiers are
designed so that the output of one amplifier can be directly coupled to
the input of a following amplifier. In others, the dc component of voltage at the output does not constitute a suitable input voltage for a following amplifier, and they must be coupled through a dc blocking capacitor.
The small-signal model for two amplifiers coupled through a capacitor
is shown in Fig. 14-6. The low-frequency behavior of this system is
affected by the coupling capacitor.

The voltage V_2 at the input to the second amplifier is

$$V_2 = \frac{R_i}{R_i + R_o + 1/j\omega C} K_{vo}V_1 \tag{14-22}$$

where K_{vo} is the no-load gain of the first amplifier. The complex voltage
gain from V_1 to V_2 can therefore be written as

$$A_v = \frac{V_2}{V_1} = \frac{R_i K_{vo}}{R_i + R_o}\frac{j\omega C(R_i + R_o)}{1 + j\omega C(R_i + R_o)} \tag{14-23}$$

For values of ω large enough to make C behave as a short circuit, corresponding to the midband of the system, the second factor in Eq.
(14-23) is approximately unity, and hence the first factor is the midband
gain. Furthermore, the quantity $1/C(R_i + R_o)$ has the dimensions of

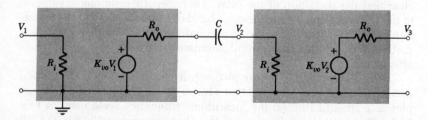

Fig. 14-6　Capacitively coupled microamplifiers.

frequency, and it has a useful interpretation as a frequency. Thus it is convenient to define two new quantities,

$$K_v = \frac{R_i K_{vo}}{R_i + R_o} \quad \text{and} \quad \omega_L = \frac{1}{C(R_i + R_o)} \qquad (14\text{-}24)$$

Then the complex voltage gain becomes

$$A_v = K_v \frac{j\omega/\omega_L}{1 + j\omega/\omega_L} = |A_v| e^{j\theta} \qquad (14\text{-}25)$$

The voltage gain in decibels is

$$A_{\mathrm{dB}} = 20 \log |A_v|$$

Defining $K_{\mathrm{dB}} = 20 \log K_v$ and substituting the appropriate expression for $|A_v|$ yields

$$A_{\mathrm{dB}} = K_{\mathrm{dB}} + 20 \log \frac{\omega}{\omega_L} - 20 \log \sqrt{1 + \left(\frac{\omega}{\omega_L}\right)^2}$$

$$= K_{\mathrm{dB}} + 20 \log \frac{f}{f_L} - 20 \log \sqrt{1 + \left(\frac{f}{f_L}\right)^2} \qquad (14\text{-}26)$$

The first term in Eq. (14-26) is a constant, independent of frequency. The second term has the logarithmic gain characteristic shown as curve A in Fig. 14-7a; it is a straight line having a slope of 6 dB/octave and

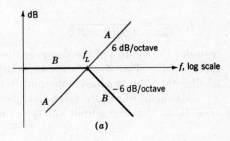

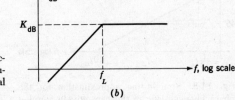

Fig. 14-7 Logarithmic gain characteristic for the capacitively coupled microamplifiers. (a) Individual terms; (b) complete characteristic.

intersecting the zero-gain axis at $f = f_L$. The third term is identical in form with the second term in Eq. (14-13); hence the asymptotes for its gain characteristic have the form shown as curve B in Fig. 14-7a. The complete gain characteristic for A_{dB} is the sum of the gain characteristics for the three terms in Eq. (14-26), and the asymptotic approximation to this characteristic is shown in Fig. 14-7b. The true gain characteristic is obtained by applying the corrections shown in Fig. 14-5a to the asymptotic approximation.

The phase shift introduced by the capacitively coupled amplifiers is obtained from Eq. (14-25) as

$$\theta = 90° - \tan^{-1} \frac{\omega}{\omega_L} = 90° - \tan^{-1} \frac{f}{f_L} \tag{14-27}$$

This phase shift is identical with that given by Eq. (14-20) except for the addition of a constant 90°; hence the straight-line approximation for the phase characteristic has the form shown in Fig. 14-8.

The asymptotic gain characteristic in Fig. 14-7b implies that the gain of the capacitively coupled amplifiers is essentially constant and independent of frequency at frequencies greater than f_L; thus the range of normal operating frequencies for the system lies above f_L. The true gain characteristic is 3 dB below the high-frequency asymptote at $f = f_L$, and hence f_L is the half-power frequency for the system at the low-frequency end of the normal operating band. The value of the half-power frequency,

$$f_L = \frac{\omega_L}{2\pi} = \frac{1}{2\pi C(R_i + R_o)}$$

is under the control of the designer through the choice of the coupling capacitor C. The size of this capacitor must be chosen so that f_L is smaller than the smallest important frequency contained in the signal.

Microamplifiers are often designed to have a small output resistance R_o so that the output voltage is not reduced too much by loads

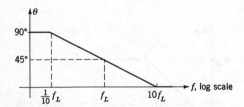

Fig. 14-8 Straight-line approximation for the phase characteristic of the capacitively coupled microamplifiers.

connected across the output terminals. If the input resistance R_i is also small, then a very large value of capacitance C may be required to provide a suitably small value for f_L. This is one of several reasons why a high input resistance is desirable. It is usually easy to have a high input resistance in JFET and MOST amplifiers; however, that is not the case in BJT amplifiers, and many BJT amplifiers have emitter followers at the input for high input resistance along with emitter followers at the output for low output resistance.

14-4 THE JFET AMPLIFIER AT LOW FREQUENCIES

Figure 14-9a shows the circuit diagram of a JFET amplifier using a bypassed resistor in the source circuit to provide gate bias. The performance of this amplifier at low frequencies depends on the bypass capacitor C_s. A small-signal model for the circuit is shown in Fig. 14-9b; this model uses the voltage-source representation for the JFET with $\mu = g_m r_d$. The voltage of the controlled source in the model can also be expressed as

$$\mu V_{gs} = \mu V_1 - \mu V_s \tag{14-28}$$

and thus the controlled source can be replaced by two sources connected in series as shown in Fig. 14-9c. The reduction theorem applied to the circuit in Fig. 14-9c yields the simplified model shown in Fig. 14-9d.

With the aid of the simplified model, the output voltage of the JFET amplifier can be expressed as

$$V_2 = -\frac{R_d}{r_d + R_d + (1+\mu)R_s/(1+j\omega C_s R_s)}\,\mu V_1 \tag{14-29}$$

Factoring $r_d + R_d$ out of the denominator, multiplying the numerator and the denominator by $1 + j\omega C_s R_s$, and dividing through by V_1 yields the complex voltage gain as

$$A_v = \frac{-\mu R_d}{r_d + R_d}\,\frac{1 + j\omega C_s R_s}{1 + (1+\mu)R_s/(r_d + R_d) + j\omega C_s R_s} \tag{14-30}$$

At normal operating frequencies the second factor in Eq. (14-30) is approximately unity, and the first factor is the normal operating voltage gain. Furthermore, the quantity $1/C_s R_s$ has the dimensions of frequency, and it is a break frequency in the gain characteristic. Therefore it is helpful to define three new quantities:

$$K_v = \frac{\mu R_d}{r_d + R_d} \qquad \omega_s = \frac{1}{C_s R_s} \qquad k_s = \frac{(1+\mu)R_s}{r_d + R_d} \tag{14-31}$$

Substituting these relations in (14-30) and factoring $1 + k_s$ out of the denominator yields

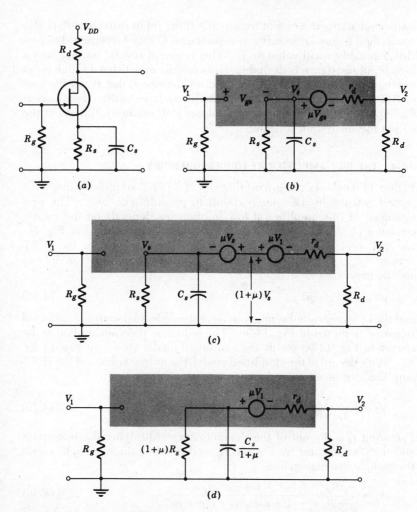

Fig. 14-9 The JFET amplifier at low frequencies. (*a*) Circuit; (*b*) small-signal model; (*c*) rearranged model; (*d*) reduced model.

$$A_v = \frac{-K_v}{1+k_s} \frac{1+j\omega/\omega_s}{1+j\omega/(1+k_s)\omega_s} = -|A_v|e^{j\theta} \tag{14-32}$$

The voltage gain in decibels is

$$A_{dB} = 20 \log |A_v|$$

Defining $K_{dB} = 20 \log K_v$ and substituting the appropriate expression for $|A_v|$ yields

$$A_{dB} = K_{dB} - 20 \log (1 + k_s) + 20 \log \sqrt{1 + \left(\frac{f}{f_s}\right)^2}$$

$$- 20 \log \sqrt{1 + \left[\frac{f}{(1 + k_s)f_s}\right]^2} \quad (14\text{-}33)$$

The first two terms in this expression are constants, independent of frequency. The last term has the same form as the last term in Eq. (14-26); hence the asymptotes for its logarithmic gain characteristic have the form shown as curve A in Fig. 14-10a. The third term in (14-33) has the same form as the last term except for the minus sign; thus its asymptotes are similar to those of the last term except that the high-frequency asymptote goes up, rather than down, with a slope of 6 dB/octave. The asymptotes for this term are shown as curve B in Fig. 14-10a.

The complete gain characteristic is the sum of the characteristics of the four terms in Eq. (14-33); the asymptotic approximation to the complete characteristic is shown in Fig. 14-10b. The true gain characteristic follows the general shape of the asymptotic approximation except for a rounding of the corners at the break frequencies, and it crosses the asymptotic curve at a point midway between the breaks. The true characteristic can be constructed either by constructing the

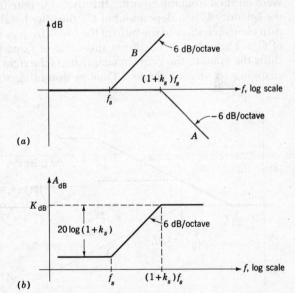

Fig. 14-10 Logarithmic gain characteristic for the JFET amplifier of Fig. 14-9. (a) Individual terms; (b) complete characteristic.

true characteristics in Fig. 14-10*a* and adding them or by adding the appropriate corrections to the asymptotic curve in Fig. 14-10*b*. For most purposes the asymptotic curve is sufficient, and it is not necessary to construct the true characteristic.

The phase shift introduced by the JFET amplifier is obtained from Eq. (14-32) as

$$\theta = \tan^{-1} \frac{f}{f_s} - \tan^{-1} \frac{f}{(1 + k_s)f_s} \qquad (14\text{-}34)$$

Thus the phase characteristic is the difference of two characteristics like the one shown in Fig. 14-5*b*. The construction of a straight-line approximation to the phase characteristic is illustrated in Fig. 14-11, where curves *A* and *B* correspond, respectively, to the second and first terms in (14-34).

The asymptotic gain characteristic in Fig. 14-10*b* implies that the gain is relatively large and nearly constant at frequencies greater than $(1 + k_s)f_s$, and thus the range of normal operating frequencies lies above $(1 + k_s)f_s$. The flat portion of the gain characteristic at frequencies below f_s corresponds to conditions under which the bypass capacitor C_s acts as an open circuit; this is the gain that the amplifier would have if C_s were omitted from the circuit. It is useful to note from Eqs. (14-31) that the quantity k_s is independent of C_s; thus the height of the ramp in the gain characteristic and the ratio of the break frequencies are independent of C_s. The value of C_s affects the value of f_s, and hence changing C_s shifts the ramp in the gain characteristic to the right or to the left without changing its size or shape. Thus in designing such an amplifier the

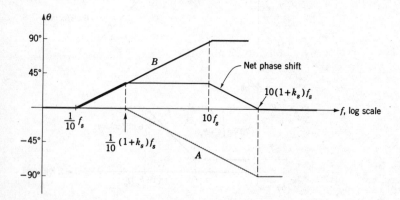

Fig. 14-11 Straight-line approximation to the phase characteristic for the JFET amplifier of Fig. 14-9.

values of R_d and R_s are first chosen to give the desired quiescent point; this act fixes the size of the ramp in the gain characteristic. The value of C_s is then chosen to make the ramp occur in a suitable range of frequencies below the frequency band occupied by the signal.

Example 14-1

A JFET amplifier of the form shown in Fig. 14-9 has the circuit parameters $R_g = 1$ megohm, $R_d = 15$ kilohms, and $R_s = 1.5$ kilohms. The parameters of the transistor are $g_m = 1.5$ mmhos and $r_d = 50$ kilohms. The numerical quantities defining the gain characteristic are to be determined, and the bypass capacitor is to be chosen so that the highest break frequency in the gain characteristic is 100 Hz.

Solution The voltage amplification factor for the transistor is

$$\mu = g_m r_d = (1.5)(50) = 75$$

Thus the midband gain is

$$K_v = \frac{\mu R_d}{r_d + R_d} = \frac{(75)(15)}{50 + 15} = 17.3$$

and its value in decibels is

$$K_{dB} = 20 \log K_v = 20 \log 17.3 = 24.8 \text{ dB}$$

The factor k_s is

$$k_s = \frac{(1 + \mu)R_s}{r_d + R_d} = \frac{(76)(1.5)}{50 + 15} = 1.75$$

and

$$20 \log (1 + k_s) = 20 \log 2.75 = 8.8 \text{ dB}$$

These numbers fix the gain characteristic except for the locations of the break frequencies.

The highest break frequency in the gain characteristic is to be 100 Hz; thus

$$(1 + k_s)f_s = 2.75f_s = 100$$

or

$$f_s = 36.4 \text{ Hz}$$

But also

$$\omega_s = 2\pi f_s = \frac{1}{C_s R_s}$$

and thus

$$C_s = \frac{1}{2\pi f_s R_s} = \frac{1}{(2\pi)(36.4)(1,500)} = 2.9 \ \mu F$$

14-5   TRANSISTOR AMPLIFIERS AT HIGH FREQUENCIES WITH RESISTIVE SOURCE AND LOAD

Figure 14-12a shows the high-frequency small-signal model for an FET amplifier; this model represents MOST amplifiers as well as JFET amplifiers if the small effects of the capacitance shunting the output of the MOS transistor are ignored. Figure 14-12b shows the high-frequency model for a BJT amplifier. In both of these models the output resistance of the transistor is lumped in with R_L, and in the BJT model, the resistance of the circuit supplying bias for the base of the transistor either is large enough to be neglected or is lumped in with R_s and V_s by the use of Thévenin's theorem. The two models in Fig. 14-12 are identical in form, and therefore the results obtained by analyzing the BJT amplifier can be adapted to the FET amplifier by setting $r_x = 0$ and changing the symbols to agree with the FET model. Thus both amplifiers can be analyzed with the labor of analyzing one.

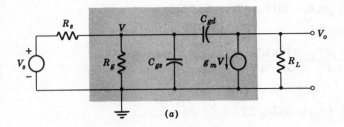

(a)

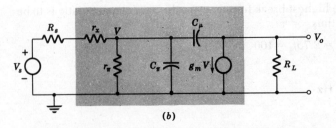

(b)

Fig. 14-12 High-frequency models for transistor amplifiers. (a) FET amplifier; (b) BJT amplifier.

The input terminals are not isolated from the output terminals in either of these amplifiers. The capacitors C_μ and C_{gd} introduce feedback from output to input that complicates the analysis somewhat and that affects the performance of the amplifiers in an important way. In particular, the feedback reduces the input impedance at high frequencies and thereby degrades the high-frequency performance of the amplifiers.

The current flowing through C_μ in Fig. 14-12b is much smaller than $g_m V$ over the entire range of frequencies in which the amplifier is useful. Therefore the output voltage for the BJT amplifier is

$$V_o = -g_m R_L V = -AV \tag{14-35}$$

where A is the voltage gain from V to V_o. Now, in order to get the output voltage as a function of the input voltage V_s, it is necessary to evaluate V as a function of V_s. This task is simplified by using the reduction theorem in the way it is used in Figs. 13-23 and 13-24. The first step is to make a simple source conversion to obtain the equivalent circuit shown in Fig. 14-13a. Now the voltage across the branch containing C_μ and R_L is $(1 + g_m R_L)V$, and the reduction theorem leads at once to the simplified circuit shown in Fig. 14-13b. Note that the output terminals have disappeared in the reduced circuit, and it is not permissible to change the conditions at the output terminals without starting the analysis again from the beginning.

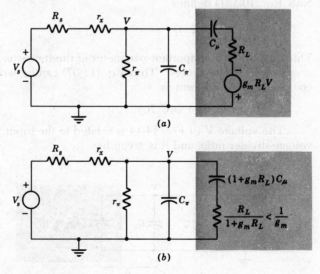

Fig. 14-13 Equivalent circuits for the input impedance of the transistor amplifier.

The resistance in the transformed branch in Fig. 14-13b is

$$\frac{R_L}{1 + g_m R_L} = \frac{1}{1/R_L + g_m} < \frac{1}{g_m}$$

But $1/g_m$ is normally a few tens of ohms or less; thus this resistance is negligible in comparison with the reactance in series with it over the range of frequencies in which the amplifier is useful. This fact leads to the further simplified circuit shown in Fig. 14-14. Thus the feedback through C_μ causes a relatively large capacitance, C_μ multiplied by 1 plus the voltage gain from V to V_o in Fig. 14-12, to appear in parallel with the input to the transistor. This result is entirely analogous to the result obtained in Fig. 13-24. The action by which C_μ is magnified and presented in parallel with the input terminals is often referred to as the Miller effect; in many cases the Miller capacitance is the dominant capacitance in the circuit of Fig. 14-14.

The total capacitance in Fig. 14-14 is

$$C = C_\pi + (1 + g_m R_L)C_\mu \qquad (14\text{-}36)$$

and by regrouping the terms this expression can be put in the form

$$C = (C_\pi + C_\mu)\left(1 + \frac{g_m R_L C_\mu}{C_\pi + C_\mu}\right) \qquad (14\text{-}37)$$

Now Eq. (10-121) defines

$$\omega_T = \frac{g_m}{C_\pi + C_\mu} = 2\pi f_T \qquad (14\text{-}38)$$

This quantity is an important parameter of the transistor, and f_T is often given on BJT data sheets. Thus Eq. (14-37) can be written in a more compact and useful form as

$$C = (C_\pi + C_\mu)(1 + \omega_T C_\mu R_L) \qquad (14\text{-}39)$$

The voltage V in Fig. 14-14 is related to the input voltage V_s by a voltage-divider ratio, and it is given by

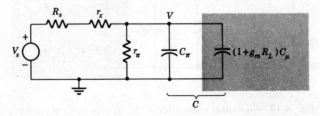

Fig. 14-14 Simplified equivalent circuit for the input impedance of the transistor amplifier.

$$V = \frac{r_\pi/(1 + j\omega C r_\pi)}{R_s + r_x + r_\pi/(1 + j\omega C r_\pi)} \, V_s$$

$$= \frac{r_\pi}{R_s + r_x + r_\pi + j\omega C r_\pi (R_s + r_x)} \, V_s \tag{14-40}$$

The output voltage of the amplifier is now obtained by substituting (14-40) into (14-35); the result is

$$V_o = \frac{-g_m r_\pi R_L}{R_s + r_x + r_\pi + j\omega C r_\pi (R_s + r_x)} \, V_s \tag{14-41}$$

Thus the complex voltage gain can be written as

$$A_v = \frac{V_o}{V_s} = \frac{-g_m r_\pi R_L}{R_s + r_x + r_\pi} \frac{1}{1 + j\omega C R} \tag{14-42}$$

where

$$R = \frac{r_\pi (R_s + r_x)}{R_s + r_x + r_\pi} \tag{14-43}$$

is the net resistance connected in parallel with C. As in the previous circuits studied, it is convenient to define

$$K_v = \frac{g_m r_\pi R_L}{R_s + r_x + r_\pi} \quad \text{and} \quad \omega_H = \frac{1}{CR} \tag{14-44}$$

so that the complex gain can be written in the standard form,

$$A_v = -K_v \frac{1}{1 + j\omega/\omega_H} = -|A_v|e^{j\theta} \tag{14-45}$$

This equation has the same form as Eq. (14-9) describing the pentode at high frequencies; therefore the gain and phase characteristics for the transistor amplifier at high frequencies have the same form as those shown in Fig. 14-5.

The half-power frequency ω_H is an important property of the amplifier, and therefore it is desirable to express its dependence on the circuit parameters in more detail than is given in Eq. (14-44). Thus, using (14-39) and (14-43),

$$\omega_H = \frac{1}{CR} = \frac{R_s + r_x + r_\pi}{(C_\pi + C_\mu)(1 + \omega_T C_\mu R_L)(R_s + r_x)r_\pi} \tag{14-46}$$

But from (14-38)

$$(C_\pi + C_\mu) = \frac{g_m}{\omega_T}$$

and hence (14-46) can be written as

$$\omega_H = \frac{\omega_T}{g_m r_\pi} \frac{R_s + r_x + r_\pi}{(1 + \omega_T C_\mu R_L)(R_s + r_x)} \tag{14-47}$$

or, in a more convenient form for numerical calculations,

$$f_H = \frac{f_T}{g_m r_\pi} \frac{R_s + r_x + r_\pi}{(1 + \omega_T C_\mu R_L)(R_s + r_x)} \tag{14-48}$$

The results obtained above can be summarized and made specific for BJTs and FETs in the following way. For the BJT, $g_m r_\pi = \beta$, and Eqs. (14-44) and (14-48) become

$$K_r = \frac{\beta R_L}{R_s + r_x + r_\pi} \quad \text{and} \quad f_H = \frac{f_T}{\beta} \frac{R_s + r_x + r_\pi}{(1 + \omega_T C_\mu R_L)(R_s + r_x)} \tag{14-49}$$

The factor $f_T/\beta = f_\beta$ is the intrinsic half-power frequency of the transistor, as is indicated by Eq. (10-115). The factor $(1 + \omega_T C_\mu R_L)$ accounts for the Miller effect associated with the feedback through C_μ; it can reduce f_H by a substantial amount if R_L is large.

For the FET, $r_x = 0$, and changing the symbols in Eqs. (14-44) and (14-48) to FET symbols yields

$$K_r = \frac{g_m R_g R_L}{R_s + R_g} \quad \text{and} \quad f_H = \frac{f_T}{g_m R_g R_s} \frac{R_s + R_g}{1 + \omega_T C_{gd} R_L} \tag{14-50}$$

Equation (14-38) in this case becomes

$$\omega_T = \frac{g_m}{C_{gs} + C_{gd}} = \frac{g_m}{C_{iss}} = 2\pi f_T \tag{14-51}$$

The capacitance C_{iss} is given on many FET data sheets. When $R_g \gg R_s$, as it frequently is, Eqs. (14-50) reduce to

$$K_r = g_m R_L \quad \text{and} \quad f_H = \frac{f_T}{g_m R_s} \frac{1}{1 + \omega_T C_{gd} R_L} \tag{14-52}$$

Again, the factor $1 + \omega_T C_{gd} R_L$ accounts for the Miller effect. It appears from Eq. (14-52) that f_H for the FET can be made arbitrarily large by making R_s small enough. This is not the case, however, for, in accordance with the discussion of Fig. 7-23, the transistor model used in the analysis is not valid at arbitrarily high frequencies. With transistors intended for high-frequency applications, the model is valid up to about 100 MHz.

In both of the amplifiers examined above there are two circuit parameters, R_L and R_s, available to permit f_H to be controlled. An examination of Eqs. (14-49) and (14-50) shows, however, that if R_L is chosen to make f_H large, a small gain results, and conversely. A fundamental limitation, crudely analogous to the law of conservation of energy, is involved in this fact; this limitation is examined in the next section.

14-6 GAIN-BANDWIDTH RELATIONS

Small-signal amplifiers are capable of amplifying with little distortion signals whose sinusoidal components have frequencies lying in the band where the gain of the amplifier is essentially uniform and where the phase shift is essentially linear. The width of this band of frequencies is called the bandwidth of the amplifier. For many purposes the bandwidth of an amplifier is considered to be the half-power frequency f_H at the upper edge of the band; in such cases the bandwidth is sometimes designated the 3-dB bandwidth. The bandwidth of an amplifier is a valuable measure of its capabilities. Amplifiers used to amplify audio signals need bandwidths of about 20 kHz; almost any small-signal amplifier will meet this need, and there are no real problems in designing such amplifiers. However, amplifiers used to amplify the video signals in TV systems, for example, require bandwidths of about 4 MHz, and substantial problems are encountered in designing these amplifiers. Some high-performance cathode-ray oscilloscopes have amplifiers with bandwidths of 50 MHz, and microamplifiers with bandwidths of several hundred megahertz are commercially available.

It is desirable for an amplifier to have both high gain and large bandwidth. However, it is pointed out at the end of Sec. 14-5 that transistor amplifiers designed for large bandwidth have relatively small gain, and conversely. This relation is true of amplifiers in general, and thus a compromise between gain and bandwidth must be made. In view of these facts, the gain-bandwidth product of an amplifier becomes an important figure of merit. From Eqs. (14-8) the gain-bandwidth product of the pentode amplifier is

$$K_v \omega_2 = \left(\frac{g_m}{g_p + G_2}\right)\left(\frac{g_p + G_2}{C_2}\right) = \frac{g_m}{C_2} \quad \text{rad/sec} \tag{14-53}$$

This product, in this simple case, is independent of the plate load resistor R_2, although the gain and bandwidth separately are functions of R_2. Thus if R_2 is increased to double the gain, the bandwidth must necessarily be cut in half, and there is no way to avoid this fact with the circuit configuration of Fig. 14-2 and with fixed values of g_m and C_2. The gain-bandwidth trade-off for these conditions is illustrated in Fig. 14-15. It is significant to note that the high-frequency asymptote does not change when R_2 is changed.

To follow up the last remark, the gain of the pentode at high frequencies is given by Eq. (14-9) as

$$A_v = \frac{-K_v}{1 + j\omega/\omega_2}$$

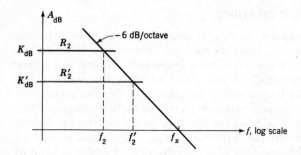

Fig. 14-15 Gain-bandwidth relations for the pentode amplifier.

and the high-frequency asymptote corresponds to

$$|A_v| = \frac{K_v}{\omega/\omega_2} = \frac{K_v\omega_2}{\omega} = \frac{K_vf_2}{f} \tag{14-54}$$

The asymptote depends only on the gain-bandwidth product and frequency, and thus it is independent of R_2. Since the slope of the asymptote is always -6 dB/octave, the asymptote is completely specified by the frequency f_x at which $|A_v| = 1$. This frequency is given by Eq. (14-54) as

$$1 = \frac{K_vf_2}{f_x}$$

or

$$f_x = K_vf_2 = K_v\frac{\omega_2}{2\pi} = \frac{g_m}{2\pi C_2} \qquad \text{Hz} \tag{14-55}$$

Thus f_x, the unity-gain crossover frequency, is equal to the gain-bandwidth product in hertz. The high-frequency asymptote and the frequency f_x work together to impose a fundamental limitation on the pentode amplifier of Fig. 14-2, for the gain characteristic of that amplifier must always lie below the asymptote. Because of this fact, much effort has been spent to achieve the largest possible f_x by designing tubes for the greatest possible g_m/C_2 ratio. Further useful insight into the relationship between the midband gain and the bandwidth is obtained by writing (14-55) as

$$K_v = \frac{f_x}{f_2} \tag{14-56}$$

A similar analysis applied to the transistor amplifier yields similar results. The gain-bandwidth product for the BJT amplifier is obtained from Eqs. (14-49) as

$$K_v f_H = f_x = f_T \frac{R_L}{R_s + r_x} \frac{1}{1 + \omega_T C_\mu R_L} \tag{14-57}$$

and for the FET amplifier Eqs. (14-50) yield

$$K_v f_H = f_x = f_T \frac{R_L}{R_s} \frac{1}{1 + \omega_T C_{gd} R_L} \tag{14-58}$$

In general f_x depends on both R_L and R_s so that changes in circuit design to control the gain or the bandwidth also change f_x. Thus the gain-bandwidth trade-off is not as simple as it is with the pentode amplifier. However, as in the case of the pentode, the gain characteristics for these amplifiers must always lie below the high-frequency asymptote, and thus when large bandwidths are required, large values of f_x are needed. The factor $1 + \omega_T C R_L$ in these expressions accounts for the Miller effect; it can reduce f_x by a large amount if the amplifier is not well designed. Again, Eq. (14-58) implies that f_x for the FET can be made arbitrarily large by making R_s sufficiently small; however, as pointed out in Sec. 14-5, this result is obtained only because the model used in the analysis is not valid at frequencies greater than about 100 MHz.

It is easy to show that the unity-gain crossover frequency f_x is equal to the gain-bandwidth product for all amplifiers in which the cutoff rate, or slope of the high-frequency asymptote, is −6 dB/octave. If the slope of the asymptote has any other value, then f_x is not the gain-bandwidth product. However, f_x is an important figure of merit for all amplifiers, and it is a more fundamental quantity than the gain-bandwidth product. The data sheets for many microamplifiers give a typical value for f_x; these values range from about 1 to 100 MHz. The most effective way of controlling the gain-bandwidth trade-off in microamplifiers is by the use of feedback; this matter is discussed in some detail in Chap. 16. In all cases, however, the crossover frequency f_x imposes a fundamental limit on what can be achieved with any given amplifier.

Example 14-2

The small-signal high-frequency model for a certain BJT amplifier has the form shown in Fig. 14-12b and analyzed in Sec. 14-5. The circuit parameters are $R_s = 5$ kilohms and $R_L = 1$ kilohm, and the transistor parameters are $\beta = 100$, $r_x = 100$ ohms, $f_T = 400$ MHz, and $C_\mu = 3$ pF. The quiescent operating point for the transistor is at $I_C = 5$ mA and $V_{CE} = 5$ volts. The gain and bandwidth of the amplifier are to be determined.

Solution The additional transistor parameters needed are

$g_m = 40 I_C = (40)(5) = 200$ mmhos

and

$$r_\pi = \frac{\beta}{g_m} = \frac{100}{200} = 0.5 \text{ kilohm}$$

The midband gain is then

$$K_v = \frac{\beta R_L}{R_s + r_x + r_\pi} = \frac{(100)(1)}{5 + 0.1 + 0.5} = 17.8$$

and in decibels

$K_{dB} = 20 \log 17.8 = 25$ dB

The bandwidth of the amplifier is

$$f_H = \frac{f_T}{\beta} \frac{R_s + r_x + r_\pi}{(1 + \omega_T C_\mu R_L)(R_s + r_x)}$$

$$= \frac{400}{100} \frac{5.6}{[1 + (2\pi)(4)(10^8)(3)(10^{-12})(10^3)](5 + 0.1)}$$

$$= 0.515 \text{ MHz} = 515 \text{ kHz}$$

The unity-gain frequency for this amplifier is

$f_x = K_v f_H = (17.8)(0.515) = 9.2$ MHz

This frequency is much smaller than f_T, the unity-gain frequency (on a current basis) of the transistor alone, because of an adverse ratio of R_L to R_s and a large Miller effect. Both the gain and the bandwidth can be increased by decreasing R_s, and the bandwidth can be increased further with almost constant f_x by reducing R_L.

14-7 THE EMITTER FOLLOWER AND SOURCE FOLLOWER AT HIGH FREQUENCIES

The gain and phase characteristics of the emitter follower and the source follower are similar to those of the transistor amplifier, except that the gain is less than unity and the bandwidth is correspondingly greater, approaching the transition frequency f_T of the transistor in some cases. With respect to the input impedance, however, matters are quite different. The input impedance is high, of course, but under some conditions it exhibits a negative resistance, and under certain conditions this negative resistance can be a source of self-sustaining oscillations. A completely general analysis of these circuits leads to cumbersome algebraic forms that tend to obscure the properties of the circuit, but by restricting the study to certain operating conditions of practical importance this problem can be removed.

The circuit of a practical source follower is shown in Fig. 12-10. In circuits of this kind a bypass capacitor is sometimes added in parallel with R_1 to enhance the bootstrapping action of the circuit. With this arrangement the small-signal high-frequency model for the circuit takes the form shown in Fig. 14-16a after the substitution theorem is used to simplify the circuit. In a similar way, the small-signal high-frequency model for the emitter follower can be put in the form shown in Fig. 14-16b. A capacitive load C_L is included in these models because these circuits, having a low output impedance, are often used to drive capacitive loads at high frequencies. In both models the output resistance of the transistor is lumped in with the load resistance, and in the emitter follower, the resistance of the circuit supplying bias for the base of the transistor either is large enough to be neglected or is lumped in with R_s and V_s by the use of Thévenin's theorem. Again, the

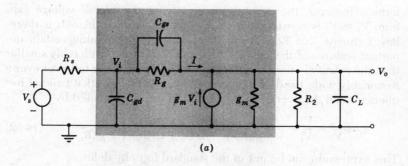

(a)

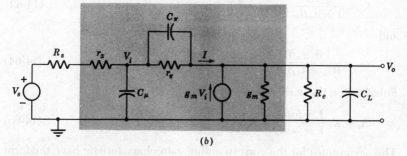

(b)

Fig. 14-16 Small-signal high-frequency models for transistor circuits. (a) FET source follower; (b) BJT emitter follower.

two circuits are identical in form, and thus a single analysis serves for both circuits.

The node equation at the output of the emitter follower in Fig. 14-16b is

$$\left[g_m + \frac{1}{R_e} + \frac{1}{r_\pi} + j\omega(C_\pi + C_L) \right] V_o - \left(\frac{1}{r_\pi} + j\omega C_\pi \right) V_i = g_m V_i \qquad (14\text{-}59)$$

and thus

$$V_o = \frac{g_m + 1/r_\pi + j\omega C_\pi}{g_m + 1/R_e + 1/r_\pi + j\omega(C_\pi + C_L)} V_i \qquad (14\text{-}60)$$

Now the terms g_m and $1/r_\pi$ can be lumped together to form a modified transconductance g'_m, but the effect of $1/r_\pi$ is so small that it will be neglected altogether, and hence (14-60) gives

$$\frac{V_o}{V_i} = \frac{g_m + j\omega C_\pi}{g_m + 1/R_e + j\omega(C_\pi + C_L)} \qquad (14\text{-}61)$$

Thus the voltage gain from V_i to V_o is obtained rather easily in a simple form. However, the general expression for the overall voltage gain from V_s to V_o is considerably more complicated than this. It is therefore fortunate that Eq. (14-61) is adequate for examining certain important features of the circuit. Moreover, if R_s is considerably smaller than the relatively high input impedance of the emitter follower over a reasonably wide band of frequencies, then $V_s \approx V_i$ in that band of frequencies and the overall voltage gain can be approximated by

$$A_r = \frac{V_o}{V_s} \approx \frac{V_o}{V_i} = \frac{g_m R_e}{1 + g_m R_e} \frac{1 + j\omega C_\pi/g_m}{1 + j\omega(C_\pi + C_L)R_e/(1 + g_m R_e)} \qquad (14\text{-}62)$$

This expression can be put in the standard form by defining

$$K_v = \frac{g_m R_e}{1 + g_m R_e} \qquad \omega_\pi = \frac{g_m}{C_\pi} \approx \omega_T \qquad (14\text{-}63)$$

and

$$\omega_H = \frac{1 + g_m R_e}{(C_\pi + C_L)R_e} \approx \frac{g_m}{C_\pi + C_L} \approx \frac{\omega_T}{1 + C_L/C_\pi} \qquad (14\text{-}64)$$

Substituting these relations into (14-62) yields

$$A_v = K_v \frac{1 + j\omega/\omega_\pi}{1 + j\omega/\omega_H} = K_v \frac{1 + jf/f_\pi}{1 + jf/f_H} \qquad (14\text{-}65)$$

The asymptotes for the corresponding gain characteristic have the form shown in Fig. 14-17; the ratio of the two break frequencies is approximately $1 + C_L/C_\pi$.

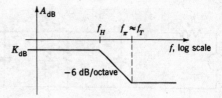

Fig. 14-17 Gain characteristic for the emitter follower and source follower when $V_i \approx V_s$.

These results, obtained by analysis of the emitter follower, apply equally well to the source follower.

The input impedance is to be evaluated next, and in order to keep the algebra manageable, the analysis is restricted to the case in which $C_L \gg C_\pi$ so that $f_H \ll f_\pi$. Hence in the band of frequencies where the circuit is useful,

$$A_v = K_v \frac{1}{1 + j\omega/\omega_H} \tag{14-66}$$

The current I shown in Fig. 14-16b is

$$I = (g_\pi + j\omega C_\pi)(V_i - V_o) \tag{14-67}$$

$$= (g_\pi + j\omega C_\pi)(1 - A_v)V_i \tag{14-68}$$

Thus

$$\frac{I}{V_i} = Y = (g_\pi + j\omega C_\pi)\left(1 - K_v \frac{1}{1 + j\omega/\omega_H}\right)$$

$$= (g_\pi + j\omega C_\pi)\left(\frac{1 - K_v + j\omega/\omega_H}{1 + j\omega/\omega_H}\right) \tag{14-69}$$

Now, since $1 - K_v \ll 1$, the numerator in (14-69) begins to vary with frequency at much lower frequencies than does the denominator, and thus at frequencies well below ω_H,

$$Y = (g_\pi + j\omega C_\pi)\left(1 - K_v + \frac{j\omega}{\omega_H}\right)$$

$$= (g_\pi + j\omega C_\pi)(1 - K_v) + j\omega \frac{g_\pi}{\omega_H} - \omega^2 \frac{C_\pi}{\omega_H}$$

and with $C_L \gg C_\pi$,

$$Y = (g_\pi + j\omega C_\pi)(1 - K_v) + j\omega C_L \frac{g_\pi}{g_m} - \omega^2 \frac{C_\pi C_L}{g_m} \tag{14-70}$$

Thus under the restricted operating conditions stated above, the input impedance of the emitter follower is represented by the circuit shown in Fig. 14-18, and this result also applies to the source follower.

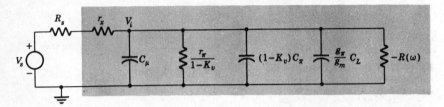

Fig. 14-18 Circuit representing the input impedance of the emitter follower at frequencies below f_H.

The last term in Eq. (14-70) is a resistive term that is negative and a function of frequency. This negative resistance can cause self-sustaining oscillations in the circuit if the source resistance R_s is changed to an inductance; thus the circuit is said to be potentially unstable. Since this condition of an inductive source impedance arises in various electronic systems, unwanted oscillations can be a problem. Frequently it is possible to stop the oscillations by connecting a resistance of a few hundred ohms in series with the base or gate terminal of the transistor. Negative resistance and potential instability are examined in more detail in Chap. 15.

14-8 THE BJT AMPLIFIER AT LOW FREQUENCIES

The circuit diagram for a BJT amplifier made of discrete components is given in Fig. 10-8. A small-signal model for an amplifier of this type that is valid at low frequencies is shown in Fig. 14-19. This model assumes that the output resistance r_o of the transistor is much larger than R_L so that it can be treated as an open circuit, and the reduction theorem has been used to simplify the circuit with $1 + \beta$ approximated by β. It is further assumed that the resistance of the circuit providing bias for the base either is large enough to be treated as an open circuit or is lumped in with R_s and V_s by the use of Thévenin's theorem.

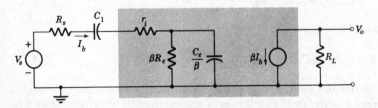

Fig. 14-19 Small-signal model for the BJT amplifier at low frequencies.

 The complex voltage gains examined in the preceding sections all have simple forms; their numerators and denominators are either constants or linear functions of frequency. More complicated circuits, such as the one shown in Fig. 14-19, do not have such simple gain functions. The principal objective of this section is to show how the simple techniques developed in the preceding sections for the construction of gain and phase characteristics can be applied in the analysis of more complicated circuits.

 The output voltage in the circuit of Fig. 14-19 is

$$V_o = -\beta R_L I_b \tag{14-71}$$

and the base current I_b is

$$I_b = \frac{V_s}{R_s + r_i + 1/j\omega C_1 + \beta R_e/(1 + j\omega C_e R_e)} \tag{14-72}$$

$$= \frac{j\omega C_1 V_s}{j\omega C_1 (R_s + r_i) + 1 + j\omega C_1 \beta R_e/(1 + j\omega C_e R_e)} \tag{14-73}$$

For compactness it is convenient to define

$$\omega_e = \frac{1}{C_e R_e} \quad \text{and} \quad \omega_1 = \frac{1}{C_1(R_s + r_i)} \tag{14-74}$$

so that (14-73) can be written as

$$I_b = \frac{j\omega C_1 V_s}{j\omega/\omega_1 + 1 + j\omega C_1 \beta R_e/(1 + j\omega/\omega_e)} \tag{14-75}$$

Multiplying this expression above and below by $\omega_1 \omega_e (1 + j\omega/\omega_e)$ and rearranging the terms yields

$$I_b = \frac{j\omega \omega_1 \omega_e C_1 (1 + j\omega/\omega_e) V_s}{(j\omega)^2 + (\omega_1 + \omega_e + \beta R_e C_1 \omega_1 \omega_e)(j\omega) + \omega_1 \omega_e} \tag{14-76}$$

Substituting (14-76) into (14-71) and noting that $\omega_1 C_1 = 1/(R_s + r_i)$ leads to

$$\frac{V_o}{V_s} = A_v = -\frac{\beta R_L \omega_e}{R_s + r_i} \frac{j\omega(1 + j\omega/\omega_e)}{(j\omega)^2 + (\omega_1 + \omega_e + \beta R_e C_1 \omega_1 \omega_e)(j\omega) + \omega_1 \omega_e}$$

But the voltage gain in the midband where the capacitors in Fig. 14-19 are short circuits is $K_v = \beta R_L/(R_s + r_i)$, and hence

$$A_v = -K_v \omega_e \frac{j\omega(1 + j\omega/\omega_e)}{(j\omega)^2 + (\omega_1 + \omega_e + \beta R_e C_1 \omega_1 \omega_e)(j\omega) + \omega_1 \omega_e} \tag{14-77}$$

The numerator of this expression can be put in a standard form that is convenient at a later point by multiplying and dividing the numerator by ω_1 to obtain

$$A_v = -K_v\omega_1\omega_e \frac{(j\omega/\omega_1)(1 + j\omega/\omega_e)}{(j\omega)^2 + (\omega_1 + \omega_e + \beta R_e C_1 \omega_1 \omega_e)(j\omega) + \omega_1\omega_e} \qquad (14\text{-}78)$$

The two factors in the numerator of Eq. (14-78) are in a familiar form, and their gain and phase characteristics can be constructed by the simple, rapid methods developed in the preceding sections of this chapter. The denominator, on the other hand, is a quadratic rather than a linear function of frequency, and the simple techniques that are applicable to linear functions cannot be used directly. However, any polynomial, such as the denominator of Eq. (14-78), can be factored and expressed as the product of a set of linear factors like those in the numerator of (14-78). Since this operation makes it possible to construct the complete gain and phase characteristics associated with A_v by the simple methods already developed, it is well worth the effort involved.

The quadratic denominator in Eq. (14-78) can be factored with the aid of the quadratic formula. If the coefficients of $j\omega$ are left in literal form, however, the factors are complicated irrational functions of the circuit parameters. Therefore, in the interest of simplicity the procedure is illustrated by giving typical values to the coefficients. The circuit parameters in a certain BJT amplifier of conventional design yield the following numerical results: $K_v = 30$, $\omega_1 = 167$ rps, $\omega_e = 20$ rps, and $\beta R_e C_1 \omega_1 \omega_e = 167$ rps. Substituting these values in (14-78) yields

$$A_v = -10^5 \frac{(j\omega/167)(1 + j\omega/20)}{(j\omega)^2 + 354(j\omega) + 3,340} \qquad (14\text{-}79)$$

If the denominator of (14-79) is factored with the variable expressed as $j\omega$, the roots obtained are in the form $j\omega = -75$, for example, and in other circumstances roots of the form $j\omega = -2 + j\sqrt{3}$ occur. These results are perfectly proper, but they look strange. In order to avoid these unfamiliar forms, and in order to provide a formulation that is consistent with standard practice in other phases of circuit theory, it is desirable to substitute the symbol s for $j\omega$ in (14-79). The resulting expression is

$$A_v(s) = -10^5 \frac{(s/167)(1 + s/20)}{s^2 + 354s + 3,340} \qquad (14\text{-}80)$$

The complex voltage gain for steady-state sinusoidal operation at any frequency ω is obtained from (14-80) by replacing s with $j\omega$.

[The reader who is familiar with the Laplace transform will recognize s as the complex-frequency variable used in transform analyses. The reader who is not familiar with the Laplace transform need not be disturbed by the fact; the transform is not needed for sinusoidal steady-state analysis. The symbol s is used here merely as a substitute variable to facilitate factoring the quadratic polynomial in Eq. (14-80); when the factoring is completed, $j\omega$ is restored as the variable.]

The roots of the equation $s^2 + 354s + 3{,}340 = 0$, which are the values of s that make the denominator of Eq. (14-80) zero, are found from the quadratic formula to be

$$s_1 = -345 \qquad \text{and} \qquad s_2 = -9.7$$

Therefore Eq. (14-80) can be written as

$$A_v(s) = -10^5 \frac{(s/167)(1 + s/20)}{(s + 345)(s + 9.7)}$$

and restoring $j\omega$ as the variable yields

$$A_v(\omega) = -10^5 \frac{(j\omega/167)(1 + j\omega/20)}{(j\omega + 345)(j\omega + 9.7)}$$

Factoring 345 and 9.7 out of the denominator of this expression gives

$$A_v(\omega) = -30 \frac{(j\omega/167)(1 + j\omega/20)}{(1 + j\omega/345)(1 + j\omega/9.7)}$$

and changing all frequencies from radians per second to hertz yields

$$A_v = -30 \frac{(jf/27)(1 + jf/3.2)}{(1 + jf/55)(1 + jf/1.55)} \qquad (14\text{-}81)$$

This expression consists entirely of factors having a familiar form. The gain and phase characteristics for each factor are constructed in the manner developed in Sec. 14-2, and the individual characteristics are added to obtain the complete characteristic for A_v. The asymptotes for the resulting gain characteristic are shown in Fig. 14-20.

The complex voltage gain, input impedance, output impedance, and so forth, for linear circuits under sinusoidal operating conditions can always be put in the form of the ratio of two polynomials in the variable $j\omega$; the proof of this statement follows from the form of the general solutions of the loop and node equations. The coefficients in

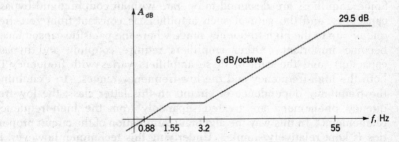

Fig. 14-20 Asymptotic gain characteristic for the BJT amplifier at low frequencies.

these polynomials are combinations of circuit-parameter values, and hence they are real numbers. The fundamental theorem of algebra ensures that these polynomials can be expressed as a product of linear factors; in particular, for any polynomial of degree n,

$$f(s) = a_n s^n + a_{n-1} s^{n-1} + \cdots + a_2 s^2 + a_1 s + a_0$$
$$= a_n(s^n + b_{n-1} s^{n-1} + \cdots + b_2 s^2 + b_1 s + b_0)$$
$$= a_n(s - s_n)(s - s_{n-1}) \cdots (s - s_2)(s - s_1) \tag{14-82}$$

The quantities $s_1, s_2, \ldots, s_n$ are the values of the variable s that make the polynomial zero. There are exactly n of these, although they may not all be distinct, and they are called the zeros of the polynomial.

The zeros of a polynomial having real coefficients may be real, or they may occur in complex-conjugate pairs. The method presented in this chapter for constructing the gain and phase characteristics cannot be applied directly to factors associated with complex zeros. A method for dealing with complex zeros is presented in Chap. 15.

The method presented above for analyzing the BJT amplifier is reasonably simple, and it is quite effective. However, it does not provide much help in solving the real engineering problem, the problem of choosing the coupling and bypass capacitors so as to control the locations of the low-frequency breaks in the gain and phase characteristics. The difficulty arises because the break frequencies are complicated irrational functions of the capacitance values. This matter is not pursued further here because current design practices do not make much use of the circuit configuration of Fig. 14-19. However, a detailed study of the design problem, along with useful engineering results, can be found in Chap. 6 of Ref. 1 listed at the end of this chapter.

14-9 COMBINED LOW-FREQUENCY AND HIGH-FREQUENCY CHARACTERISTICS

Some amplifiers are designed to operate without coupling and bypass capacitors, and the gain of such amplifiers is constant from zero frequency up to the high-frequency range where the parasitic capacitances become important. Other amplifiers require coupling and bypass capacitors, and the gain of these amplifiers varies with frequency in both the high-frequency and the low-frequency ranges. In examining the frequency dependence of circuits in this latter class, the low-frequency phenomena are treated separately from the high-frequency phenomena. In this way the algebraic formulation of the circuit properties is kept relatively simple. Underlying this technique, however, is the tacit assumption that the high-frequency breaks in the gain char-

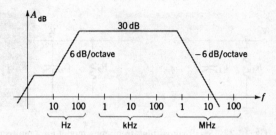

Fig. 14-21 Typical gain characteristic for a single-stage BJT amplifier.

acteristic are separated by a decade of frequency or more from the low-frequency breaks. This condition is satisfied in many important electronic circuits.

The complete frequency characteristics for a circuit can be displayed by plotting the low-frequency and high-frequency characteristics on a single set of coordinates. As an example, the asymptotic approximation for the complete gain characteristic of a BJT amplifier is shown in Fig. 14-21; this characteristic is obtained by adding the high-frequency characteristic to the low-frequency characteristic shown in Fig. 14-20. The numerical values shown in this figure indicate orders of magnitude that may be expected in typical single-stage BJT amplifiers. It should be clear, however, that in any particular amplifier the break frequencies and the gain can be adjusted over fairly wide ranges by adjusting the circuit parameters. In the typical characteristic of Fig. 14-21 the high-frequency breakpoint is more than 3 decades above the highest low-frequency breakpoint; this 3-decade range is the midband for the amplifier.

The complete phase characteristic for the amplifier can be constructed in a similar manner by combining the low-frequency and high-frequency characteristics on a single set of coordinates.

14-10 NONINTERACTING MULTISTAGE AMPLIFIERS

Figure 14-22 shows the small-signal high-frequency model for a two-stage pentode amplifier. The grid-to-plate capacitance of the pentode is negligibly small, and hence the input circuit of each stage is isolated from the output circuit; that is, there is no interaction between the input and output of each stage. As a consequence of this fact, each stage is a separate part of the total circuit, and it can be analyzed separately from the rest of the circuit. Thus the analysis and design of pentode amplifiers is relatively simple.

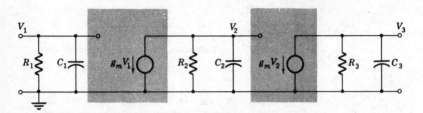

Fig. 14-22 Small-signal high-frequency model for a multistage amplifier with no interaction between stages.

The output voltage of the amplifier is

$$V_3 = -g_m V_2 \frac{R_3}{1 + j\omega C_3 R_3}$$

Letting $\omega_{H2} = 1/C_3 R_3$ gives the voltage gain of the second stage as

$$\frac{V_3}{V_2} = A_{v2} = \frac{-g_m R_3}{1 + j\omega/\omega_{H2}} = -K_{v2} \frac{1}{1 + j\omega/\omega_{H2}} \qquad (14\text{-}83)$$

Similarly, the gain of the first stage is

$$\frac{V_2}{V_1} = A_{v1} = -K_{v1} \frac{1}{1 + j\omega/\omega_{H1}} \qquad (14\text{-}84)$$

The overall voltage gain is thus

$$\frac{V_3}{V_1} = A_v = A_{v1} A_{v2} = K_{v1} K_{v2} \frac{1}{(1 + j\omega/\omega_{H1})(1 + j\omega/\omega_{H2})} \qquad (14\text{-}85)$$

The overall gain characteristic for the two-stage pentode amplifier at high frequencies, constructed from the information contained in Eq. (14-85), is shown in Fig. 14-23a. If the two stages of the amplifier are identical, the gain characteristic takes the form shown in Fig. 14-23b.

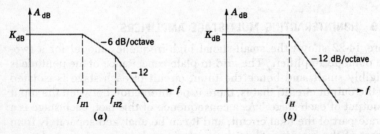

Fig. 14-23 Overall gain characteristics for a two-stage amplifier. (a) Nonidentical stages; (b) identical stages.

When a large bandwidth is needed, it is obtained by using a small load resistance in each stage in accordance with the relations developed in Sec. 14-6. The cost is a reduced gain. If more gain is needed, it can be obtained by adding more stages, and the added stages do not change the behavior of the original stages, for there is isolation between the stages. The added stages do introduce additional breaks in the overall gain characteristic, and each stage contributes -6 dB/octave of slope to the final high-frequency asymptote.

At the break frequency f_H in Fig. 14-23b, the true gain characteristic is 6 dB below the asymptotes, and hence f_H is not the half-power frequency for the two-stage amplifier. The half-power frequency lies somewhere below f_H, and it follows from this fact that connecting identical stages in cascade results in a reduction in the half-power bandwidth. The amount of this reduction for any number of identical stages connected in cascade can be determined as follows. The magnitude of the complex voltage gain for one stage is obtained from Eq. (14-83) as

$$|A_v| = \frac{K_v}{[1 + (\omega/\omega_H)^2]^{1/2}} \tag{14-86}$$

and for n identical stages connected in cascade,

$$|A_v| = \frac{K_v{}^n}{[1 + (\omega/\omega_H)^2]^{n/2}} \tag{14-87}$$

Thus the half-power frequency for n identical stages in cascade is the value of ω that makes the denominator of (14-87) equal $\sqrt{2}$ and the square of the denominator equal 2; thus it is given by

$$\left[1 + \left(\frac{\omega}{\omega_H}\right)^2\right]^n = 2$$

Solving this expression for ω yields

$$\omega = \omega_H \sqrt{2^{1/n} - 1} \qquad \text{or} \qquad f = f_H \sqrt{2^{1/n} - 1} \tag{14-88}$$

Now if the half-power bandwidth of n identical stages in cascade is designated by B_n, then $B_1 = \omega_H$, and in general

$$B_n = B_1 \sqrt{2^{1/n} - 1} \tag{14-89}$$

where B_1 and B_n can be expressed either in hertz or in radians per second. The factor multiplying B_1 is known as the *bandwidth reduction factor* for n identical stages in cascade. For two stages the bandwidth reduction factor is 0.64, for three stages it is 0.51, and for four stages it is 0.43.

14-11 INTERACTING MULTISTAGE AMPLIFIERS

Figure 14-24 shows the small-signal high-frequency model for a two-stage amplifier. This model can represent an FET amplifier, or it can represent a BJT amplifier if the small effects of the base-spreading resistance r_x can be neglected in the second stage. Any resistance shunting the input terminals of the amplifier is lumped in with R_1 and V_s by the use of Thévenin's theorem. The parasitic feedback capacitors C_f associated with the transistors introduce interaction between the input and output of each stage, and as a result there is no isolation between stages. Thus the individual stages cannot be analyzed separately, for each stage affects the performance of the other. In the case of the pentode amplifier examined in Sec. 14-10, the effects of the resistance associated with the source of input signals can be determined by a simple separate calculation, and since these effects are often small, they are omitted from the analysis. In the present case, however, the resistance of signal source is not isolated from the rest of the circuit, and the simplest procedure is to include it in the analysis from the outset.

The small current flowing through C_f in the second stage of the amplifier in Fig. 14-24 is negligible in comparison with $g_m V_2$ throughout the band of frequencies in which the amplifier is useful. Thus the output voltage is

$$V_3 = -g_m R_3 V_2 \qquad\qquad (14\text{-}90)$$

The two node equations for nodes V_1 and V_2 are

$$[G_1 + j\omega(C + C_f)]V_1 - j\omega C_f V_2 - G_1 V_s = 0$$

and $\qquad\qquad\qquad\qquad\qquad\qquad\qquad\qquad\qquad\qquad\qquad$ (14-91)

$$-j\omega C_f V_1 + [G_2 + j\omega(C + 2C_f)]V_2 - j\omega C_f V_3 = -g_m V_1$$

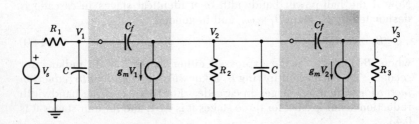

Fig. 14-24 Small-signal high-frequency model for a multistage amplifier with interaction between stages.

Substituting Eq. (14-90) for V_3 in these equations and collecting terms yields

$$[G_1 + j\omega(C + C_f)]V_1 - j\omega C_f V_2 = G_1 V_s$$

and (14-92)

$$(g_m - j\omega C_f)V_1 + [G_2 + j\omega(C + 2C_f + g_m R_3 C_f)]V_2 = 0$$

For compactness it is helpful to define two new symbols,

$$C_1 = C + C_f \quad \text{and} \quad C_2 = C + 2C_f + g_m R_3 C_f \quad (14\text{-}93)$$

Moreover, $g_m \gg \omega C_f$ at all frequencies of interest; thus neglecting ωC_f in comparison with g_m and using Eqs. (14-93) yields for the node equations

$$(G_1 + j\omega C_1)V_1 - j\omega C_f V_2 = G_1 V_s$$

and (14-94)

$$g_m V_1 + (G_2 + j\omega C_2)V_2 = 0$$

When these equations are solved by determinants for V_2 and the result is substituted into Eq. (14-90), the complex voltage gain of the amplifier is obtained as

$$\frac{V_3}{V_s} = A_v = \frac{g_m{}^2 G_1 R_3}{(G_1 + j\omega C_1)(G_2 + j\omega C_2) + j\omega C_f g_m} \quad (14\text{-}95)$$

When $C_f = 0$, the second term in the denominator in Eq. (14-95) vanishes, and the denominator is a quadratic polynomial in factored form. The two factors correspond to two isolated stages in cascade. When $C_f \neq 0$, there is no isolation, and the interaction between stages is accounted for in part by the second term in the denominator in Eq. (14-95) and in part by the last term in the expression for C_2 as given in Eqs. (14-93). In this case the denominator in (14-95) is still a quadratic polynomial, but it is not in factored form. For further compactness, let

$$\omega_1 = \frac{G_1}{C_1} \quad \text{and} \quad \omega_2 = \frac{G_2}{C_2} \quad (14\text{-}96)$$

Note that, when $C_f = 0$, ω_1 and ω_2 reduce to the half-power frequencies of the two isolated stages. Using these definitions, Eq. (14-95) can be put in the following form:

$$A_v = \frac{g_m{}^2 R_3}{C_1 C_2 R_1} \frac{1}{(j\omega)^2 + (\omega_1 + \omega_2 + g_m C_f / C_1 C_2)(j\omega) + \omega_1 \omega_2} \quad (14\text{-}97)$$

Since the midband gain is $K_v = g_m{}^2 R_2 R_3$, A_v can also be written as

$$A_v = K_v \omega_1 \omega_2 \frac{1}{(j\omega)^2 + (\omega_1 + \omega_2 + g_m C_f / C_1 C_2)(j\omega) + \omega_1 \omega_2} \quad (14\text{-}98)$$

Representing the denominator symbolically in factored form gives

$$A_v = K_v \omega_1 \omega_2 \frac{1}{(j\omega + \omega_a)(j\omega + \omega_b)} \tag{14-99}$$

where ω_a and ω_b are given by the quadratic formula as rather complicated functions of the parameters in Eq. (14-98).

The frequencies ω_a and ω_b are the break frequencies in the gain characteristic. If $C_f = 0$, the denominator in Eq. (14-98) is a perfect square, and the break frequencies are simply ω_1 and ω_2. If in addition $R_1 = R_2$, the two break frequencies are equal, and the gain characteristic has the form shown in Fig. 14-23b for two identical isolated stages in cascade. When $C_f \neq 0$, the break frequencies split apart to give a gain characteristic having the form shown in Fig. 14-25. The final high-frequency asymptote corresponds to very large ω in Eqs. (14-97) and (14-98), and under this condition

$$|A_v| = \frac{R_3}{R_1} \frac{g_m^2}{C_1 C_2 \omega^2} = \frac{K_v \omega_1 \omega_2}{\omega^2} \tag{14-100}$$

The slope of the final asymptote is -12 dB/octave, and it crosses the unity gain (0 dB) axis where (14-100) has the value

$$|A_v| = 1 = \frac{R_3}{R_1} \frac{g_m^2}{C_1 C_2 \omega_x^2} = \frac{K_v \omega_1 \omega_2}{\omega_x^2}$$

Thus

$$\omega_x = \frac{g_m}{\sqrt{C_1 C_2 R_1 / R_3}} = \sqrt{K_v \omega_1 \omega_2} \tag{14-101}$$

The break frequencies are often widely separated in amplifiers of this type, and when this is the case, the half-power bandwidth is equal to the lower break frequency. Large bandwidth is obtained by making R_1 and R_2 small so that both ω_1 and ω_2 are large. Making R_3 small decreases C_2 and thereby increases ω_2 for still greater bandwidth. The price paid for additional bandwidth is a reduced midband gain.

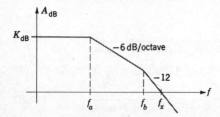

Fig. 14-25 Gain characteristic for the two-stage amplifier of Fig. 14-24.

When more than two interacting stages are connected in cascade, similar conditions exist. However, the analysis in this case is quite complicated, and a complete solution is not at all practical. A rather detailed study of this problem is presented in Ref. 2.

Example 14-3

The circuit diagram for a widely used two-stage BJT amplifier is shown in Fig. 13-13. A small-signal high-frequency model for an amplifier of this type is shown in Fig. 14-26; this model neglects the effects of the base-spreading resistance r_x, and it lumps the output resistance of the transistors with the load resistance for each stage. The transconductance for each transistor is $g_m = 80$ mmhos. The high-frequency performance of this amplifier is to be examined.

Solution First, Thévenin's theorem is used to transform the input circuit to the form shown in Fig. 14-24; the resulting voltage and resistance are

$$V_s = \tfrac{1}{2}V'_s \quad \text{and} \quad R_1 = 0.5 \text{ kilohm}$$

The net resistance shunting the output of the first stage is

$$R_2 = 0.5 \text{ kilohm}$$

The midband gain of the transformed circuit is

$$\frac{V_3}{V_s} = K_v = g_m{}^2 R_2 R_3 = (80)^2 (0.5)^2 = 1,600$$

Because of the voltage-divider action at the input, the midband gain of the actual circuit in Fig. 14-26 is

$$K'_v = \frac{V_3}{V'_s} = \frac{1}{2}\frac{V_3}{V_s} = \frac{1}{2} K_v = 800$$

Thus if K'_v is used in place of K_v in all the relations derived from Fig. 14-24, proper account is taken of the action of the input voltage divider.

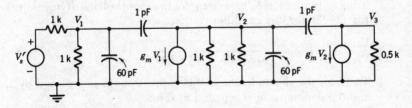

Fig. 14-26 Transistor amplifier for Example 14-3.

The values of the capacitances defined in Eq. (14-93) are

$$C_1 = C + C_f = 60 + 1 = 61 \text{ pF}$$

and

$$C_2 = C + 2C_f + g_m R_3 C_f = 60 + 2 + (80)(0.5)(1) = 102 \text{ pF}$$

Thus the frequencies defined in Eq. (14-96) are

$$\omega_1 = \frac{1}{C_1 R_1} = \frac{1}{(61)(10^{-12})(0.5)(10^3)} = 32.8 \times 10^6 \text{ rps}$$

and

$$\omega_2 = \frac{1}{C_2 R_2} = \frac{1}{(102)(10^{-12})(0.5)(10^3)} = 19.6 \times 10^6 \text{ rps}$$

For purposes of comparison, these values are also calculated for the case in which $C_f = 0$; the results are

$$\omega_1 = \omega_2 = 33.3 \times 10^6 \text{ rps}$$

As is stated in connection with Eq. (14-98), with $C_f = 0$, ω_1 and ω_2 are the break frequencies for the gain characteristic, and with $\omega_1 = \omega_2$, the gain characteristic takes the form shown in Fig. 14-23b. The break frequency in hertz is

$$f_1 = 5.3 \text{ MHz}$$

Also, with $C_f = 0$, the unity-gain frequency is

$$f_x = \frac{1}{2\pi} \sqrt{K_v' \omega_1 \omega_2} = 150 \text{ MHz}$$

With $C_f \neq 0$, the remaining quantity in Eq. (14-98) to be evaluated is

$$\frac{g_m C_f}{C_1 C_2} = \frac{(80)(10^{-3})(1)}{(61)(102)(10^{-12})} = 12.9 \times 10^6 \text{ rps}$$

. Then, expressing all frequencies in megaradians per second and using K_v' in place of K_v to account for the input voltage divider alters Eq. (14-98) for this amplifier so that

$$\frac{V_3}{V_s'} = A_v' = K_v' \omega_1 \omega_2 \frac{1}{(j\omega)^2 + 65.3(j\omega) + 643}$$

To factor the denominator of this expression, $j\omega$ is replaced by s, and the denominator is equated to zero:

$$s^2 + 65.3s + 643 = 0$$

The quadratic formula gives the zeros of this equation as

$s_1 = -12.1$ Mrps and $s_2 = -53.3$ Mrps

Thus the denominator can be written as

$D = (s + 12.1)(s + 53.3)$

Restoring $j\omega$ in place of s and putting this expression into the gain equation as the denominator yields

$$A'_v = K'_v\omega_1\omega_2 \frac{1}{(j\omega + 12.1)(j\omega + 53.3)}$$

It turns out in this case that $\omega_1\omega_2 = (12.1)(53.3)$, and thus if 12.1 and 53.3 are factored out of the denominator, the result is

$$A'_v = K'_v \frac{1}{(1 + j\omega/12.1)(1 + j\omega/53.3)}$$

Converting the frequencies to megahertz and substituting the numerical value for K'_v yields

$$A'_v = 800 \frac{1}{(1 + jf/1.9)(1 + jf/8.5)}$$

Thus the gain characteristic has the form shown in Fig. 14-25 with breaks at 1.9 and 8.5 MHz. These breaks are separated by more than 2 octaves, and the half-power bandwidth is essentially 1.9 MHz.

For a comparison, with $C_f = 0$ there is a double break at 5.3 MHz, and using the bandwidth reduction factor from the end of Sec. 14-10 for two identical stages gives the half-power bandwidth with $C_f = 0$ as

$B_2 = 0.64B_1 = (0.64)(5.3) = 3.4$ MHz

The unity-gain frequency with $C_f \neq 0$ is

$$f_x = \frac{1}{2\pi} \sqrt{K'_v\omega_1\omega_2} = 113 \text{ MHz}$$

as compared with 150 MHz when $C_f = 0$.

14-12 POLE-ZERO PATTERNS

The gain and phase characteristics associated with the complex voltage gain of an amplifier provide a convenient and useful means for displaying the variation of the complex gain with frequency. The pole-zero pattern associated with the complex gain displays the same information in a different and more compact way, and it provides a different kind of insight into the properties of the amplifier. The pole-zero pattern is of greater fundamental importance than the frequency characteristics; however, in principle, given the frequency characteristics, the pole-zero pattern can be constructed, and vice versa.

The concept of the pole-zero pattern can be developed in terms of Eq. (14-32) for the complex voltage gain of a JFET amplifier at low frequencies. This equation can be written as

$$A_v = -K_v \frac{j\omega + \omega_s}{j\omega + (1 + k_s)\omega_s} = -K_v \frac{j\omega + \omega_1}{j\omega + \omega_2} \tag{14-102}$$

For reasons given in Sec. 14-8 it is desirable to replace the symbol $j\omega$ with the symbol s and to define $s_1 = -\omega_1$ and $s_2 = -\omega_2$; then (14-102) can be written as

$$A_v = -K_v \frac{s - s_1}{s - s_2} \tag{14-103}$$

The complex voltage gain for sinusoidal operation at any frequency ω is obtained from (14-103) by letting $s = j\omega$ and inserting the proper values for s_1 and s_2.

All the network functions encountered in the study of the dynamics of linear, lumped-parameter systems, like the voltage gain given by Eq. (14-103), can be expressed as the ratio of two polynomials in the variable s. Functions of this type are called rational functions. The values of s that make such functions zero are called zeros of the function; hence $s = s_1$ is a zero of the function given by (14-103). Similarly, the values of s that make a rational function infinite are called poles of the function, and thus $s = s_2$ is a pole of the function given by (14-103). As is indicated by the form of (14-103), if all the poles and zeros of a rational function are known, the function itself is completely specified except for a constant multiplier. For example, if the zeros of a certain voltage gain are -1 and -3, and if the poles are -2 and -4, then the complex gain is

$$A_v = K \frac{(s + 1)(s + 3)}{(s + 2)(s + 4)} = K \frac{s^2 + 4s + 3}{s^2 + 6s + 8} \tag{14-104}$$

and for steady-state sinusoidal conditions

$$A_v = K \frac{(j\omega + 1)(j\omega + 3)}{(j\omega + 2)(j\omega + 4)}$$

$$= \frac{3K}{8} \frac{(1 + j\omega/1)(1 + j\omega/3)}{(1 + j\omega/2)(1 + j\omega/4)} \tag{14-105}$$

Thus if the poles and zeros of a rational function are known, the logarithmic gain and phase characteristics can be constructed, except for the effect of the constant multiplier. If the constant multiplier is not unity, it adds a constant number of decibels to the gain characteristic; if it is not positive, it adds a constant angle of 180° to the phase shift. (The constant angle is usually omitted from the phase characteristic.) It is also of great significance to note that the poles and zeros of the function correspond to the break frequencies in the gain characteristic expressed in radians per second. A zero corresponds to a break of 6 dB/octave upward in the asymptotic characteristic, and a pole corresponds to a break of 6 dB/octave downward.

The connection between the poles and zeros of the complex gain and the frequency characteristics is displayed in a useful way by the pole-zero diagram. As an example, Fig. 14-27a shows the pole-zero diagram for the voltage gain given by Eq. (14-103) with typical values of s_1 and s_2. The voltage gain for sinusoidal signals is obtained from (14-103) by setting $s = j\omega$. Hence, since s_1 and s_2 are real numbers, $A_v(j\omega)$ contains both real and imaginary numbers, and therefore it is useful to plot the pole-zero diagram on a set of rectangular coordinates known as the plane of complex numbers or, simply, the complex plane. The zero of A_v is indicated by a small circle at the location of s_1, which in this case is on the negative real axis, and the pole is indicated by a small cross at the location of s_2. Specifically, it follows from Eq. (14-102) that $s_1 = -\omega_s$ and $s_2 = -(1 + k_s)\omega_s$.

The important relations between the poles and zeros of $A_v(s)$ and the variable $s = j\omega$ are diagramed in Fig. 14-27b. The complex factors in the numerator and denominator of Eq. (14-103) can be expressed in polar form as

$$s - s_1 = \rho_1 \exp j\theta_1 \quad \text{and} \quad s - s_2 = \rho_2 \exp j\theta_2 \qquad (14\text{-}106)$$

The magnitude and angle of each of these complex numbers are shown geometrically in Fig. 14-27b for the case in which $s = j\omega$. As ω is varied from zero to infinity, the point $s = j\omega$ moves up the imaginary axis from

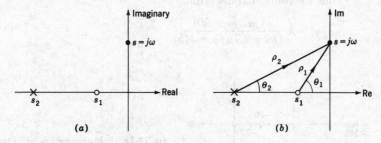

Fig. 14-27 Pole-zero pattern for Eq. (14-103). (a) Diagram; (b) interpretation.

the origin to infinity, and the magnitudes and angles of the vectors vary accordingly.

The way in which the magnitude and phase of A_v vary with frequency can be evaluated qualitatively by inspection of the pole-zero diagram. The exact relations can be formulated by noting that

$$A_v(j\omega) = -|A_v|e^{j\theta} = -K_v \frac{j\omega - s_1}{j\omega - s_2} = -K_v \frac{\rho_1}{\rho_2} \exp j(\theta_1 - \theta_2) \qquad (14\text{-}107)$$

Thus

$$|A_v| = K_v \frac{\rho_1}{\rho_2} \quad \text{and} \quad \theta = \theta_1 - \theta_2 \qquad (14\text{-}108)$$

For any given frequency ω, the corresponding gain and phase angle can be estimated by inspection, or determined more accurately by measurement, of the diagram in Fig. 14-27b. For example, with large values of ω, $\rho_1 \approx \rho_2$, and the gain is approximately K_v. With small values of frequency the gain is approximately $K_v|s_1/s_2|$.

Example 14-4

The pole-zero pattern for the complex voltage gain of a certain amplifier is shown in Fig. 14-28. The voltage gain at high frequencies is 30. The asymptotes for the logarithmic gain characteristic are to be constructed from this information.

Solution The gain characteristic can be constructed by inspection of the pole-zero pattern. However, it is instructive to go through the intermediate steps. The pole-zero pattern indicates that the complex voltage gain is given by

$$A_v = K \frac{(s+0)(s+20)}{(s+9.7)(s+345)}$$

Thus for sinusoidal operation

$$A_v(j\omega) = K \frac{(j\omega)(j\omega + 20)}{(j\omega + 9.7)(j\omega + 345)}$$

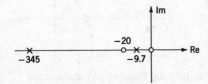

Fig. 14-28 Pole-zero pattern for Example 14-4.

At high frequencies

$$A_v = K = 30$$

and thus the expression for the complex gain can be put in the form

$$A_v = 30 \frac{20}{(9.7)(345)} \frac{j\omega(1 + j\omega/20)}{(1 + j\omega/9.7)(1 + j\omega/345)}$$

$$= 30 \frac{(j\omega/167)(1 + j\omega/20)}{(1 + j\omega/9.7)(1 + j\omega/345)}$$

$$= 30 \frac{(jf/27)(1 + jf/3.2)}{(1 + jf/1.55)(1 + jf/55)}$$

If a minus sign is added to this equation, it becomes identical with Eq. (14-81) for the BJT amplifier at low frequencies, and the corresponding gain characteristic is shown in Fig. 14-20.

14-13 SUMMARY

Generally speaking, the behavior of electronic amplifiers depends on frequency in both the high-frequency and the low-frequency ranges because unavoidable parasitic capacitances become effective at high frequencies and auxiliary circuit elements such as coupling and bypass capacitors become effective at low frequencies. This dependence is displayed by curves called frequency characteristics — plots of gain and phase shift vs. frequency.

The most important result developed in this chapter is the fact that, if a signal is to be amplified without waveform distortion, the amplifier must have uniform gain and a linear phase characteristic over the entire band of frequencies occupied by the signal. Also of primary importance is the technique of using the logarithmic gain and phase characteristics to display the frequency dependence of amplifiers. The value of these characteristics lies in the fact that they have simple asymptotic properties and that complicated characteristics can be constructed by the addition of simple component characteristics.

The pole-zero pattern for the complex gain function provides an alternative way of displaying the frequency dependence of an amplifier. The pole-zero pattern is especially useful when the poles or zeros of the gain function are complex numbers; this fact is developed in considerable detail in Chap. 15.

PROBLEMS

14-1. *Pentode amplifier at high frequencies.* A certain pentode amplifier has the circuit shown in Fig. 14-2. The plate load resistor is $R_2 = 200$ kilohms, and the small-signal tube parameters are $g_m = 1.2$ mmhos and $r_p = 1$ megohm. The total capacitance between plate and ground is 12 pF, a value that includes 7 pF of stray wiring capacitance.

(*a*) Determine the midband gain in decibels and the half-power frequency in hertz for the high-frequency model.

(*b*) Plot the logarithmic gain and phase characteristics for the high-frequency model. Give the true characteristics in addition to the asymptotes. Use semilog coordinate paper, and calibrate the frequency scale in kilohertz.

14-2. *Pentode-amplifier design.* The amplifier of Prob. 14-1 is to be redesigned to have a half-power frequency of 1 MHz.

(*a*) Determine the value of R_2 that is required.

(*b*) What is the midband gain with this value of R_2?

14-3. *Coupling-capacitor calculation.* Two microamplifiers are capacitively coupled as shown in Fig. 14-6. The amplifier parameters are $K_{vo} = 3,000$, $R_o = 100$ ohms, and $R_i = 5$ kilohms, and the coupling capacitor is to be chosen so that the break in the low-frequency gain characteristic occurs at 100 Hz.

(*a*) Find the value of C required.

(*b*) An emitter follower is added at the input of the second amplifier with the result that R_i is increased to 100 kilohms for the second amplifier. What value of C is required in this case?

14-4. *Phonograph-amplifier frequency response.* The source V_p and the series 800-pF capacitor in Fig. 14-29 constitute the model for a cer-

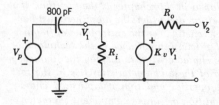

Fig. 14-29 Phonograph amplifier for Prob. 14-4.

tain crystal phonograph pickup. The amplifier is to be designed so that the break in the low-frequency gain characteristic V_2/V_p is below 20 Hz. What is the minimum value that R_i can have?

14-5. *JFET amplifier design.* The JFET used in an amplifier like the one shown in Fig. 14-9 is characterized by $I_{DSS} = 5$ mA and $V_p = -3$ volts. The power-supply voltage is $V_{DD} = 30$ volts.

(*a*) The amplifier is to be designed for a quiescent point at $I_D = 2$ mA and $V_{DS} = 6$ volts. What values of R_d and R_s are required?

(*b*) The small-signal drain resistance is $r_d = 50$ kilohms. Determine the transistor parameters g_m and $\mu = g_m r_d$. Determine the midband gain of the amplifier in decibels.

(c) The highest break frequency in the logarithmic gain characteristic for the low-frequency model is to be 160 Hz. What must be the size of the bypass capacitor?

(d) Sketch and dimension the asymptotes for the low-frequency gain characteristic. Give gains in decibels and frequencies in hertz.

14-6. Frequency characteristics. The complex voltage gain of the low-frequency model for a certain amplifier is

$$A_v = -10\,\frac{1 + jf/50}{1 + jf/100}$$

Plot the logarithmic gain and phase characteristics for the amplifier. Use semilog coordinate paper, and show both the asymptotes and the true characteristics. *Note:* The objective of this problem is to provide practice in applying the corrections to the asymptotes when two break frequencies lie close together.

14-7. JFET input impedance. The JFET amplifier shown in Fig. 14-30 is provided with a 1-megohm potentiometer to serve as a gain control. The effect of the gain control on the high-frequency gain charac-

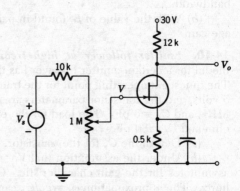

Fig. 14-30 JFET amplifier for Prob. 14-7.

teristic is to be examined. The small-signal transistor parameters are $g_m = 2.5$ mmhos, $r_d = 50$ kilohms, and $C_{gd} = C_{gs} = 5$ pF. The bypass capacitor acts as a short circuit at all frequencies of interest.

(a) The input to the JFET can be represented by an equivalent circuit of the form shown in Fig. 14-14. Determine the values of the parameters in this circuit.

(b) Sketch and dimension the asymptotes for the high-frequency gain characteristic V_o/V_s for the condition that the gain control is set for maximum output voltage. Give gains in decibels and frequencies in kilohertz.

(c) Repeat part b with the gain control set at its midpoint (that is, set so that the potentiometer has 500 kilohms on each side of the slider).

(d) Discuss briefly the significance of the results obtained in parts b and c.

14-8. BJT amplifier at high frequencies. A BJT amplifier is analyzed in Example 14-2. This circuit is modified by the addition of an emitter follower at the input. The bandwidth of the emitter follower is much greater than the bandwidth of the amplifier so that for the purposes of this problem the net effect of the emitter follower is to reduce R_s to 100 ohms.

Calculate K_v, f_H, and f_x for the modified circuit, and compare the results with the values obtained in Example 14-2.

14-9. BJT amplifier design. A certain BJT amplifier is to be designed for a half-power bandwidth f_H of 4 MHz. The high-frequency model for the amplifier has the form shown in Fig. 14-12b with $R_L = 500$ ohms. The transistor parameters are $\beta = 100$, $g_m = 200$ mmhos, $r_x = 100$ ohms, $C_\mu = 1$ pF, and $f_T = 400$ MHz.

(a) Determine the value of R_s that will give the required 4-MHz bandwidth.

(b) With the value of R_s found in part a, what is the midband voltage gain?

14-10. Emitter follower at high frequencies. The high-frequency model for a certain emitter follower has the form shown in Fig. 14-16b. The quiescent operating point for the transistor is at $I_C = 5$ mA and $V_{CE} = 5$ volts, and the transistor parameters are $\beta = 50$, $r_x = 100$ ohms, $f_T = 400$ MHz, and $C_\mu = 3$ pF. The load for the emitter follower is $R_e = 0.5$ kilohm and $C_L = 80$ pF.

(a) Determine C_π for the transistor.

(b) Under the assumption that $V_i \approx V_s$, sketch and dimension the asymptotes for the gain characteristic. Give the frequencies in megahertz. The approximation $\omega_T = g_m/C_\pi$ can be used. *Note:* If the source resistance R_s is as large as 1 kilohm, the assumption that $V_i \approx V_s$ is not valid at frequencies above a few tens of megahertz.

14-11. BJT amplifier at low frequencies. A certain BJT amplifier has a low-frequency model of the form shown in Fig. 14-19. The circuit parameters are $R_s = R_L = 5$ kilohms, $R_e = 2$ kilohms, $C_1 = 2$ μF, and $C_e = 15$ μF. The transistor is biased for a quiescent collector current $I_C = 1$ mA, and its parameters are $\beta = 50$ and $r_x = 50$ ohms.

Sketch the asymptotes for the voltage gain of the low-frequency model. Give the frequency of each break in hertz. It is not required to determine the gain.

14-12. *Decoupling circuit.* The combination of R_2 and C in the amplifier of Fig. 14-31 forms a decoupling network of a type commonly used to reduce the interaction between amplifier stages using a common power supply. The effect of the decoupling network on the low-fre-

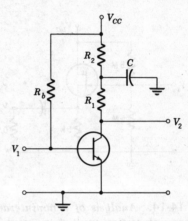

Fig. 14-31 BJT amplifier for Prob. 14-12.

quency gain characteristic is to be examined. The transistor is biased for a quiescent collector current $I_C = 1$ mA, and the transistor parameters are $\beta = 100$, $r_x = 50$ ohms, and $g_o = 1/r_o = 0$.

(*a*) Give a small-signal model for the circuit that is valid at medium and low frequencies, and determine the input resistance r_i for the transistor.

(*b*) Show that the complex voltage gain at low and medium frequencies has the form

$$A_v = \frac{V_2}{V_1} = K \frac{1 + j\omega/\omega_1}{1 + j\omega/\omega_2}$$

Give the values of K, ω_1, and ω_2 in terms of the circuit and transistor parameters.

(*c*) Sketch and dimension the asymptotes for the gain characteristic when $R_b = 100$ kilohms, $R_1 = 5$ kilohms, $R_2 = 2$ kilohms, and $C = 5$ μF. Give frequencies in hertz.

14-13. *Source follower at low frequencies.* Figure 14-32 shows a JFET source follower with a capacitively coupled load. The transistor parameters are $g_{mo} = 5$ mmhos, $I_{DSS} = 5$ mA, and $r_d = 30$ kilohms. The transistor is biased for a quiescent drain current $I_D = 2$ mA.

(*a*) Give a small-signal model for the circuit that is valid at low and medium frequencies, and determine g_m for the stated quiescent drain current.

(b) The voltage gain V_2/V_1 is to have its low-frequency half-power point at 100 Hz. Determine the required value of C. The theorems of Chap. 12 and the analysis in Sec. 14-3 may be helpful.

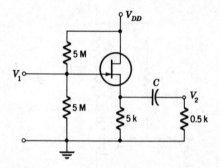

Fig. 14-32 Source follower for Prob. 14-13.

14-14. *Analysis of a noninteracting multistage amplifier.* A certain noninteracting multistage amplifier consists of three identical stages having the form shown in Fig. 14-22. The transconductance for each stage is $g_m = 3$ mmhos, the shunt capacitance is $C = 10$ pF, and the shunt resistance is $R = 15$ kilohms.

(a) Sketch and dimension the asymptotes for the gain characteristic of this amplifier. Give frequencies in megahertz, and give the slope of the high-frequency asymptote.

(b) What is the half-power bandwidth of the amplifier in kilohertz? Do not forget the bandwidth reduction factor.

(c) What is the gain-bandwidth product for one stage of this amplifier in megahertz?

14-15. *Design of a noninteracting multistage amplifier.* A certain amplifier is to consist of a cascade of identical noninteracting stages having the form shown in Fig. 14-22. The overall half-power bandwidth is to be 3.5 MHz, and the overall midband gain is to be at least 1,000. The transconductance in each stage is 5 mmhos, and the shunt capacitance is 10 pF.

(a) Determine the minimum number of stages required, and specify the shunt resistance for each stage. *Suggestion:* Calculate the gain-bandwidth product in MHz [see Eq. (14-55)] for one stage. Then, with the bandwidth reduction factor, the number of stages can be found easily.

(b) What is the overall midband gain when the shunt resistance is adjusted for an overall half-power bandwidth of 3.5 MHz?

14-16. *Analysis of a multistage amplifier with interaction.*   Figure 14-33 shows the circuit of the two-stage BJT amplifier designed in Example 13-2.   The transistors are identical, and their parameters can be

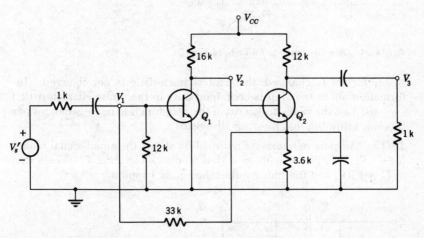

Fig. 14-33   BJT amplifier for Prob. 14-16.

taken as $\beta = 50$, $r_x = 0$, $g_o = 1/r_o = 0$, $f_T = 200$ MHz, and $C_\mu = 2$ pF.   The quiescent collector current in each transistor is 1 mA.

(a) Give a small-signal model for the circuit that is valid at medium and high frequencies where the coupling and bypass capacitors act as short circuits.   The 12- and 33-kilohm bias resistors can be treated as open circuits.

(b) Sketch and dimension the asymptotes for the gain characteristic V_3/V_s' as shown in Fig. 14-25.   Give f_a, f_b, and f_x in megahertz.

14-17. *Frequency characteristics.*   The data sheet for a certain microamplifier gives a gain characteristic having the asymptotes shown in Fig. 14-34, and in addition, the amplifier introduces a sign reversal.

(a) The complex gain function for this amplifier consists of a constant multiplier and a number of linear factors of the form $1 + jf/f_a$.   Determine an expression for the complex gain in which all constants are expressed as numbers.   It is convenient to express frequencies in megahertz.   (Note that, if the sign of the imaginary part in some or all of the

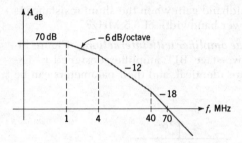

Fig. 14-34 Gain characteristic for Prob. 14-17.

linear factors is changed, the gain characteristic is not changed. Information about the sign is contained only in the phase characteristic.)

(*b*) List the poles and zeros of the gain function in megaradians per second, assuming that they are all negative.

14-18. *All-pass network.* Figure 14-35 shows the small-signal model for an FET amplifier with an added feedback capacitor C. The value of C is 1 μF, and the transconductance g_m is 1 mmho.

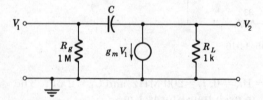

Fig. 14-35 FET amplifier for Prob. 14-18.

(*a*) Sketch and dimension the pole-zero pattern for the complex voltage gain V_2/V_1.

(*b*) Sketch and dimension the asymptotes for the logarithmic gain and phase characteristics.

14-19. *Pole-zero patterns.* The pole-zero patterns for the complex voltage gain of two amplifiers are shown in Fig. 14-36.

(*a*) Which of these amplifiers will not transmit dc signals? Why?

(*b*) Sketch the asymptotes for the gain characteristic corresponding to the pole-zero pattern in Fig. 14-36*b*. The gain at zero frequency is 20 dB, and $\omega_1 = 100$ rps. Give the coordinates of each break in the characteristic, and give the slope of each segment.

(*c*) The two amplifiers corresponding to Fig. 14-36*a* and *b* are connected in cascade. Sketch and dimension the pole-zero pattern for the overall complex gain. The frequency ω_1 has the same value in the two amplifiers.

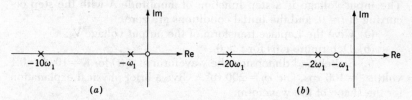

Fig. 14-36 Pole-zero patterns for Prob. 14-19.

14-20. *BJT-amplifier step response (Laplace transform).* The Laplace transform of the output voltage of a BJT amplifier is

$$V_o = A_v V_i = -K_v \frac{1}{1 + s/\omega_H} V_i$$

where V_i is the transform of the input voltage and the initial conditions are zero. The input voltage is a step function of amplitude A with the step occurring at $t = 0$.

(a) Determine $v_o(t)$ for $t > 0$.

(b) Sketch and dimension the waveform of $v_o(t)$ for $K_v = 60$, $\omega_H = 10^6$ rps, and $A = 10$ mV. Use a time scale of about 2 μsec/in.

(c) The amplifier is redesigned for greater bandwidth, and the resulting circuit parameters are $K_v = 12$ and $\omega_H = 5 \times 10^6$ rps. Repeat part b with the same input voltage. Use the same time scale as in part b.

14-21. *Step response of capacitively coupled amplifiers (Laplace transform).* The complex voltage gain of the capacitively coupled amplifiers in Fig. 14-6 is

$$\frac{V_3}{V_1} = A_v = K_v \frac{j\omega/\omega_L}{1 + j\omega/\omega_L}$$

The input voltage is a step function of amplitude A with the step occurring at $t = 0$, and the initial conditions are zero.

(a) Give the Laplace transform of the output voltage V_3.

(b) Determine $v_3(t)$ for $t > 0$.

(c) Sketch and dimension the waveform of $v_3(t)$ for $A = 1/K_v$ and $f_L = \omega_L/2\pi = 80$ Hz. Give a brief physical explanation for the shape of the waveform.

14-22. *Step response (Laplace transform).* The complex voltage gain for a JFET amplifier at low frequencies is given by Eq. (14-32) in the form

$$\frac{V_2}{V_1} = A_v = -K \frac{1 + j\omega/\omega_1}{1 + j\omega/\omega_2}$$

The input voltage is a step function of amplitude A with the step occurring at $t = 0$, and the initial conditions are zero.

(a) Give the Laplace transform of the output voltage V_2.

(b) Determine $v_2(t)$ for $t > 0$.

(c) Sketch and dimension the waveform of $v_2(t)$ for $K = 10, A = 0.1$ volt, $\omega_1 = 100$ rps, and $\omega_2 = 200$ rps. Give a brief physical explanation for the shape of the waveform.

14-23. *Pulse response (Laplace transform).* The complex voltage gain of a certain BJT pulse amplifier is

$$\frac{V_o}{V_i} = A_v = -K_v \frac{1}{1 + jf/f_H}$$

with $K_v = 5$ and $f_H = 8$ MHz. The input voltage is a single rectangular pulse with an amplitude of 0.2 volt and a duration of 0.1 μsec, and the initial conditions are zero.

Sketch and dimension the waveform of the output voltage $v_o(t)$. *Suggestion:* If the time constant of the amplifier is compared with the duration of the pulse, it can be seen that a complete solution of the problem is not needed for making the sketch.

REFERENCES

1. Searle, C. L., et al.: "Elementary Circuit Properties of Transistors," John Wiley and Sons, Inc., New York, 1964.
2. Thornton, R. D., et al.: "Multistage Transistor Circuits," John Wiley and Sons, Inc., New York, 1965.

15

Tuned Amplifiers

T*he amplifiers discussed in Chap. 14 are quite satisfactory for most applications in which it is not required that the signals be transmitted over long distances. Thus they are adequate for servomechanisms, automatic pilots, electronic instruments, audio amplifiers used in public-address and home-entertainment systems, and a wide variety of similar applications. However, when the signals must be transmitted over long distances, as between two cities or between a space vehicle and a ground station, effective use of the transmission medium requires the use of narrowband systems operating at high frequencies. Further aspects of this matter are discussed in Chap. 1, and as is indicated in Fig. 1-2, various systems of this kind operate throughout the frequency spectrum from 10 kHz (radio navigation) to about 10 GHz (radar speed traps), although at the higher end of this range ordinary tubes and transistors are not used. These systems require amplifiers using RLC interstage networks to perform a signal-processing operation in the form of frequency-selective amplification.*

The principal objective of this chapter is to examine the properties of various tuned amplifiers that find application in telecommunication systems. In the course of this study some important new tools for the analysis and design of tuned amplifiers are developed. These tools, which for the most part grow out of the pole-zero patterns presented in Sec. 14-12, provide a very useful insight into the properties of the circuits to which they apply.

15-1 SOME PROPERTIES OF MODULATED SIGNALS

It is pointed out in the introductory paragraphs above and in Chap. 1 that, when signals must be transmitted over long distances, either by wire or by radio waves, efficient utilization of the transmission medium requires the use of high-frequency narrowband signals. To generate these signals a high-frequency carrier wave, usually a sinusoid, is caused to change instant by instant in accordance with the information signal to be transmitted. The process by which the carrier wave is made to

465

change in accordance with the information signal is called modulation. A sinusoid has three parameters that can be modulated; they are the amplitude, the frequency, and the phase, and they give rise to amplitude-modulated, frequency-modulated, and phase-modulated signals. In order to understand the requirements that must be met by tuned amplifiers, it is necessary to examine the nature of these waves and to understand the way in which they are used in telecommunication systems. The properties of AM waves are developed in the paragraphs that follow; although they are not discussed specifically, FM and PM waves have similar properties.

If the information, or modulating, wave is a voltage $v_m(t)$, then one form of amplitude-modulated sinusoidal carrier wave can be expressed as

$$v_s = [V_c + v_m(t)] \cos \omega_c t \tag{15-1}$$

where ω_c is the frequency of the carrier wave and V_c is the amplitude of the unmodulated carrier. A circuit that is capable of generating such a wave is shown in Fig. 2-7, and the waveform of such a signal is shown in Fig. 15-1. Telecommunication systems are usually adjusted so that the magnitude of v_m is always less then V_c; hence the envelope of the modulated wave never drops to zero. It follows from Eq. (15-1) that the envelope of the modulated signal has the same waveform as the modulating signal v_m. Thus if the information in the modulated carrier is to be preserved, the waveform of the envelope must be preserved. The information can be recovered by recovering the waveform of the envelope; the diode demodulator of Sec. 3-4 is often used for this purpose, and in this application it is frequently called an envelope detector.

To examine the properties of the AM wave further, a specific modulating signal, a sinusoid, is chosen. In this case

$$v_m = V_m \cos (\omega_m t + \theta_m) \tag{15-2}$$

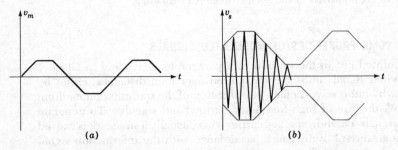

(a)   (b)

Fig. 15-1 Amplitude modulation. (a) Information signal; (b) carrier wave amplitude modulated by the information signal.

and substituting this expression into Eq. (15-1) gives the AM wave as

$$v_s = [V_c + V_m \cos (\omega_m t + \theta_m)] \cos \omega_c t \tag{15-3}$$

$$= V_c[1 + m \cos (\omega_m t + \theta_m)] \cos \omega_c t \tag{15-4}$$

where m, the modulation index, is a number less than unity. Equation (15-4) is a useful representation of the AM wave for some purposes; for example, it shows clearly that the envelope of the wave contains the information in the modulating signal. For other purposes, however, the mathematical form of Eq. (15-4) is not satisfactory; for example, when it is desired to calculate the response of an amplifier to the AM wave, Eq. (15-4) is completely intractable. Except for a few special cases, the response of a linear circuit can be calculated only when the excitation is sinusoidal. The sinusoidal case is capable of considerable extension, however, for, as is suggested in Sec. 14-1, many practical signals can be approximated as the sum of a number of sinusoidal components. For example, if the signal is periodic, it can be expanded in a Fourier series, and the problem can be solved by the superposition of the effects of each sinusoidal component. This process can be applied to Eq. (15-4) if ω_m and ω_c are so related as to yield a periodic wave. However, the special nature of Eq. (15-4) is such that the desired result can be obtained much more directly by performing the indicated multiplication and using the trigonometric identity for the product of two cosines; the result is

$$v_s = V_c \cos \omega_c t + \tfrac{1}{2} m V_c \cos [(\omega_c + \omega_m)t + \theta_m]$$

$$+ \tfrac{1}{2} m V_c \cos [(\omega_c - \omega_m)t - \theta_m] \tag{15-5}$$

Thus the sinusoidally modulated AM wave is seen to be the superposition of three sinusoidal components, a carrier wave at the carrier frequency and two side waves at frequencies on either side of the carrier frequency. The frequency spectrum of a typical sinusoidally modulated AM wave is shown in Fig. 15-2a; it shows the relative amplitudes and frequencies of the carrier and side waves, except that the side waves are usually much closer to the carrier wave than it is possible to show in the diagram. In the more general case there are many pairs of side waves, and these are referred to as the upper and lower sidebands of the AM wave. In most AM waves the carrier frequency is much larger than the highest frequency in the modulating signal. For example, a typical AM radio station broadcasting speech and music with frequencies up to 5 kHz has a carrier frequency of 1 MHz, and the total bandwidth of the signal is 10 kHz centered at 1 MHz. The important feature of such AM signals is that they are narrowband signals; that is, the total bandwidth is small compared to the center frequency.

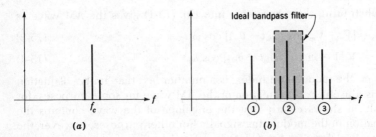

Fig. 15-2 Frequency spectra of modulated waves. (*a*) Sinusoidal carrier modulated by a sinusoidal signal; (*b*) three modulated waves separated in frequency.

The narrowband signal provides the basis for systems transmitting as many as 600 simultaneous telephone conversations on a single pair of conductors in a coaxial cable and for the simultaneous use of space by all radio transmitters. Such systems depend on the possibility of separating the many signals arriving at the receiving end of the system on one pair of wires or on one antenna, and this possibility is provided by frequency-selective amplification, or filtering, at the receiving end. The method of separating signals is illustrated in Fig. 15-2*b*, which shows the frequency spectra of three AM signals. Each signal uses a different carrier frequency, and if the difference between these frequencies is great enough, the spectra of the signals are completely separated and occupy different frequency bands. These signals can be separated by frequency-selective amplification; the idealized gain characteristic for such an amplifier is indicated in Fig. 15-2*b*. Amplifiers of this kind are called bandpass amplifiers. Tuning a radio receiver and switching channels on a TV receiver are operations by which the passband of the amplifier is shifted from one place to another in the frequency spectrum. The study of tuned amplifiers in the sections that follow is concerned with the response of frequency-selective amplifiers to modulated waves and with design techniques for obtaining good performance. Good performance is concerned with the rejection of adjacent channels as well as with faithful amplification of the desired channel.

Frequency-modulated signals have a sinusoidal composition similar to that of AM signals; however, the sideband structure is considerably more complex. FM broadcast signals lie in the frequency range between 88 and 108 MHz, and the total bandwidth of the signal is typically about 200 kHz. Television signals for home entertainment use a combination of AM and FM. The sound is FM, and the picture is AM; however, to conserve bandwidth, one of the AM sidebands is almost completely removed by filtering at the transmitter. Television channels

2 through 13 lie in the frequency band between 54 and 216 MHz, and each channel is 6 MHz wide. The bandwidth of the signal transmitted is less than the channel width, however, perhaps 3 or 4 MHz. Although the discussion that follows is concerned with AM waves, the results apply equally well to FM waves, at least in a qualitative sense.

Equation (15-5) shows that a certain symmetry exists between the sidebands of an AM wave. The side waves have equal amplitudes, and they have phase angles with respect to the carrier that are equal in magnitude but opposite in sign. If these components are to add up to give the original AM wave, this symmetry must be preserved. The significance and importance of the symmetry is illustrated by the diagrams in Fig. 15-3, in which each sinusoidal component of the wave is represented by a rotating phasor. If the carrier phasor is taken as the reference frame and is assumed to be stationary, then the upper side-wave phasor rotates counterclockwise at ω_m rad/sec, and the lower side-wave phasor rotates clockwise at the same speed. In Fig. 15-3a, which represents a normal AM wave, the sum of the two side-wave phasors is at every instant collinear with the carrier phasor, and it gives a sinusoidal variation in the amplitude of the total voltage. It is clear, however, that this result depends on the symmetry between the side waves. To illustrate what may happen when the symmetry is destroyed, Fig. 15-3b shows the case in which the side waves have been shifted 90° in phase with respect to the carrier. In this case the sum of the side-wave phasors is at every instant perpendicular to the carrier phasor, and the amplitude of the total voltage is almost constant as shown in Fig. 15-3c. The signal recovered from this wave by an envelope detector is very small, and it is badly distorted. Transoceanic radio signals are subject to fading and distortion of this nature as a result of the properties of the transmission medium. For the purposes of this book, however, the real point of this discussion is the fact that poorly designed amplifiers can also degrade the signal in this way.

Ideally, then, amplifiers for modulated signals should provide uniform gain over the band of frequencies occupied by the signal, and they

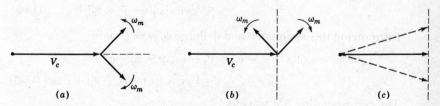

Fig. 15-3 Vector, or phasor, representation of a modulated wave.

should introduce no phase shift in this band. However, to ask for uniform gain in the desired band along with high rejection of signals in adjacent bands is to impose severe requirements on the amplifier. It is therefore useful to note that, if the gain characteristic is symmetrical with respect to the center frequency of the passband, then each of the two side waves is amplified by the same amount, and the amplitude symmetry is preserved. Thus a more reasonable design objective is a symmetrical characteristic with reasonably uniform gain in the band of frequencies occupied by the signal. It also turns out that a linear phase-shift characteristic produces a pure time delay in the envelope waveform without distorting it. This fact can be demonstrated by considering a modulating signal consisting of the sum of a number of sinusoids. In this case

$$v_m = V_1 \cos (\omega_1 t - \theta_1) + V_2 \cos (\omega_2 t - \theta_2)$$
$$+ V_3 \cos (\omega_3 t - \theta_3) + \cdots \quad (15\text{-}6)$$

The corresponding AM wave, obtained by substituting (15-6) into (15-1), is

$$v_s = V_c \cos \omega_c t + \tfrac{1}{2}V_1 \cos [(\omega_c + \omega_1)t - \theta_1]$$
$$+ \tfrac{1}{2}V_1 \cos [(\omega_c - \omega_1)t + \theta_1]$$
$$+ \tfrac{1}{2}V_2 \cos [(\omega_c + \omega_2)t - \theta_2]$$
$$+ \tfrac{1}{2}V_2 \cos [(\omega_c - \omega_2)t + \theta_2] + \cdots \quad (15\text{-}7)$$

The frequency spectrum for this signal is similar to the spectra shown in Fig. 15-2 except that in this case each sideband contains a sinusoidal component corresponding to each sinusoidal component of the modulating signal; each pair of sideband components exhibits the symmetry discussed above.

If the modulating signal is given a pure time delay T, then Eq. (15-6) becomes

$$v_m = V_1 \cos [\omega_1(t - T) - \theta_1] + V_2 \cos [\omega_2(t - T) - \theta_2]$$
$$+ V_3 \cos [\omega_3(t - T) - \theta_3] + \cdots \quad (15\text{-}8)$$

Rearranging this expression and defining $\phi_k = \omega_k T$ yields

$$v_m = V_1 \cos (\omega_1 t - \phi_1 - \theta_1) + V_2 \cos (\omega_2 t - \phi_2 - \theta_2)$$
$$+ V_3 \cos (\omega_3 t - \phi_3 - \theta_3) + \cdots \quad (15\text{-}9)$$

The corresponding AM wave is

$$v_s = V_c \cos \omega_c t + \tfrac{1}{2}V_1 \cos \left[(\omega_c + \omega_1)t - \phi_1 - \theta_1\right]$$
$$+ \tfrac{1}{2}V_1 \cos \left[(\omega_c - \omega_1)t + \phi_1 + \theta_1\right]$$
$$+ \tfrac{1}{2}V_2 \cos \left[(\omega_c + \omega_2)t - \phi_2 - \theta_2\right]$$
$$+ \tfrac{1}{2}V_2 \cos \left[(\omega_c - \omega_2)t + \phi_2 + \theta_2\right] + \cdots \quad (15\text{-}10)$$

where the ϕ_k are directly proportional to the separation of the side frequency from the carrier frequency. Thus if an AM signal is transmitted through an amplifier having a linear phase characteristic with a negative slope, it can be shown by going through this process in reverse that the result is a pure time delay in the envelope. A constant phase shift added to the linear phase characteristic shifts the phases of the carrier and side waves by the same amount, and it does not affect the envelope; this fact follows from the phasor diagrams in Fig. 15-3.

The results obtained in the preceding discussion are useful guides in judging the merits of various tuned amplifiers and in designing these amplifiers for optimum performance. In particular, a good tuned amplifier for use with modulated signals should have a symmetrical gain characteristic with reasonably uniform gain in the band of frequencies occupied by the signal, and it should have a linear phase characteristic in that frequency band.

15-2 TUNED PENTODE AMPLIFIERS

Most tuned amplifiers operate in the frequency range between 100 kHz and 100 MHz, and as a result their analysis and design are complicated somewhat by any feedback capacitance that may exist between the input and output. For this reason the pentode amplifier, which has negligible feedback capacitance, is the simplest tuned amplifier used in electronic systems. The circuit for this amplifier is similar to the one shown in Fig. 14-2 except that the plate load resistor R_2 is replaced by a parallel combination of L and C forming a tuned circuit. A small-signal model for the amplifier is shown in Fig. 15-4a, and the gain and phase characteristics, plotted on a linear frequency scale, are shown in Fig. 15-4b. The amplifier has large gain at frequencies near resonance where the impedance of the parallel-resonant plate circuit is high, and it has small gain at frequencies remote from resonance; thus narrowband filtering is provided. The gain is fairly constant over a narrow band of frequencies near resonance, and the phase characteristic is quite linear in that band. The principal weakness of the circuit is that the gain does not decrease rapidly enough outside the useful passband, and thus for many applications there is inadequate rejection of signals in the adjacent channels.

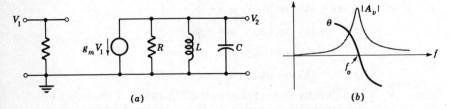

Fig. 15-4 A tuned pentode amplifier. (*a*) Small-signal model; (*b*) frequency character-istics.

It should be noted that the stray and parasitic capacitances, which limit the high-frequency performance of the untuned amplifier, are absorbed into the tuned circuit and merely contribute to the resonance in the tuned amplifier. Thus in principle the tuned amplifier as represented in Fig. 15-4 could operate at arbitrarily high frequencies. In fact, how-ever, as the resonant frequency is increased, the values of L and C be-come very small, and eventually second-order effects not included in the model of Fig. 15-4 become important and limit the performance of the amplifier.

The resonant frequency and the bandwidth of the tuned amplifier depend on the values of the circuit parameters, and they are under the control of the circuit designer. Thus the circuit can be designed to amplify selectively signals in a band of specified width and center fre-quency. The objective of the analysis that follows is to examine the properties of the circuit and to develop relations that are useful in the design of tuned amplifiers.

The output voltage of the amplifier in Fig. 15-4*a* is

$$V_2 = \frac{-g_m V_1}{j\omega C + 1/R + 1/j\omega L}$$

and thus the complex voltage gain can be expressed as

$$\frac{V_2}{V_1} = A_v(j\omega) = -\frac{g_m}{C}\frac{j\omega}{(j\omega)^2 + (1/RC)(j\omega) + 1/LC} \tag{15-11}$$

The analysis that follows shows that $1/LC$ is the square of the resonant frequency in radians per second and $1/RC$ is the half-power bandwidth of the amplifier in radians per second; therefore it is convenient at this point to define two new symbols,

$$\omega_o{}^2 = \frac{1}{LC} \quad \text{and} \quad 2\alpha = \frac{1}{RC} \tag{15-12}$$

so that Eq. (15-11) can be written more compactly as

$$A_v(j\omega) = -\frac{g_m}{C}\frac{j\omega}{(j\omega)^2 + 2\alpha(j\omega) + \omega_o{}^2} \tag{15-13}$$

Now when $\omega = \omega_o$, the voltage gain is

$$A_v(j\omega_o) = -\frac{g_m}{2\alpha C} = -g_m R = -K_o \tag{15-14}$$

and the phase shift is $\theta(j\omega_o) = 0$, apart from the sign reversal introduced by the tube. Thus ω_o is the resonant frequency of the amplifier, and it is the center frequency of the passband of the amplifier. It follows from (15-14) that $g_m/C = 2\alpha K_o$, and hence the general expression for the voltage gain can be written as

$$A_v(j\omega) = -2\alpha K_o \frac{j\omega}{(j\omega)^2 + 2\alpha(j\omega) + \omega_o{}^2}$$

The analysis of the tuned amplifier is aided by expressing the quadratic denominator of the gain function in its factored form, and thus, for reasons set forth in Chap. 14, the variable $j\omega$ is replaced with the symbol s to obtain

$$A_v(s) = -2\alpha K_o \frac{s}{s^2 + 2\alpha s + \omega_o{}^2} \tag{15-15}$$

When s is given the value $j\omega$, Eq. (15-15) is the voltage gain for sinusoidal steady-state operation. When the denominator of (15-15) is expressed in factored form, the gain function becomes

$$A_v(s) = -2\alpha K_o \frac{s}{(s - s_1)(s - s_2)} \tag{15-16}$$

The poles of A_v are obtained from the denominator of (15-15) with the aid of the quadratic formula; they are

$$s_1 = -\alpha + \sqrt{\alpha^2 - \omega_o{}^2} \quad \text{and} \quad s_2 = -\alpha - \sqrt{\alpha^2 - \omega_o{}^2}$$

These poles can be either real or complex, depending on the relative values of the circuit parameters; however, only complex poles can provide a truly narrowband gain characteristic, and hence the remainder of this discussion is concerned only with the case in which the circuit parameters yield complex poles. When the poles are complex, they form a conjugate pair, and it proves convenient to express them as

$$s_1 = -\alpha + j\sqrt{\omega_o{}^2 - \alpha^2} = -\alpha + j\beta \tag{15-17}$$

and

$$s_2 = -\alpha - j\beta \tag{15-18}$$

where

$$\beta^2 = \omega_o{}^2 - \alpha^2 \qquad \alpha = \frac{1}{2RC} \qquad \omega_o{}^2 = \frac{1}{LC} \tag{15-19}$$

When the poles are complex, the pole-zero pattern for A_v has the form shown in Fig. 15-5a. It follows directly from Eqs. (15-19) and the theorem of Pythagoras that ω_o is the hypotenuse of a right triangle having α and β as legs; thus ω_o is the distance of each pole from the origin of the complex plane, as shown in Fig. 15-5a.

The typical AM signal described in Sec. 15-1 has a center frequency that is 100 times the bandwidth of the signal, and an amplifier suitable for use with this signal might have a resonant frequency $\omega_o \approx 100\alpha$. Thus the poles are so close to the imaginary axis in the typical case that it is impossible to draw a meaningful pole-zero diagram to scale. With this fact firmly in mind, consider Fig. 15-5b, which shows the pole-zero diagram for the less typical case in which $\omega_o = 10\alpha$. The frequency dependence of $A_v(j\omega)$ can be determined from the vectors shown in this figure, and of greater importance, the geometry of the diagram shows how the pertinent relations can be simplified. When $\omega_o = 10\alpha$ or more (and certainly when $\omega_o = 100\alpha$), the poles of A_v are much closer to the imaginary axis than they are to the origin of the coordinates. Hence as the variable point $s = j\omega$ moves along the imaginary axis in the vicinity of s_1, the vector $s - s_1$ in Fig. 15-5b is small and experiences large percentage variations, whereas the small percentage variations in s and $s - s_2$ tend to cancel. Sinusoidal operation in the passband of the amplifier corresponds to just these conditions. Thus, for sinusoidal opera-

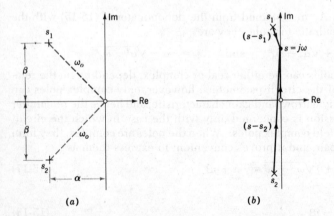

(a)   (b)

Fig. 15-5 Pole-zero pattern for the tuned amplifier. (a) Graphical interpretation of the quantities in Eqs. (15-19); (b) graphical representation of the factors in Eq. (15-16).

tion in the passband,

$$s = j\omega \approx j\omega_o \qquad (15\text{-}20)$$

and

$$s - s_2 = j\omega - s_2 \approx 2j\omega_o \qquad (15\text{-}21)$$

These important equations are the *narrowband approximations;* they are useful in the analysis and design of all narrowband amplifiers and filters. For the typical narrowband amplifier, they are very good approximations. When these approximations are substituted into Eq. (15-16), the simplified result obtained is

$$A_v(s) = -2\alpha K_o \frac{j\omega_o}{(s - s_1)(2j\omega_o)}$$

$$= -\alpha K_o \frac{1}{s - s_1} \qquad (15\text{-}22)$$

With $s = j\omega$ this expression is valid for sinusoidal operation in and near the narrow passband.

Equation (15-22) shows that in narrowband operation the ratio $s/(s - s_2)$ can be treated as a constant, $\frac{1}{2}$. It also shows that the narrowband approximations reduce a quadratic denominator to a first-degree denominator; in general these approximations reduce the degree of the denominator by a factor of 2 or more, a fact of great importance when more complex circuits are considered. Since the voltage gain given by (15-22) depends only on the scale factor αK_o and the variable factor $s - s_1$, its properties can be examined in more detail with the aid of the diagram in Fig. 15-6a; this diagram is an expanded view of Fig. 15-5b in the vicinity of the pole s_1. With the aid of this diagram the gain and phase shift of the amplifier can be determined for various values of $s = j\omega$, and the results can be plotted in the form of the gain and phase character-

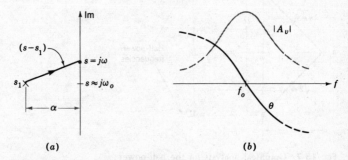

Fig. 15-6 The narrowband approximation. (a) Expanded view of the pole-zero pattern in the vicinity of s_1; (b) frequency characteristics.

istics shown in Fig. 15-6b; the dashed portions of these curves indicate frequency ranges in which the narrowband approximations are not valid.

As in the case of the untuned amplifier, the half-power frequencies of the tuned amplifier provide a convenient measure of the bandwidth. The gain at resonance is given by Eq. (15-14) as $|A_v| = g_m R = K_o$. The half-power frequencies are thus the frequencies at which

$$|A_v| = \frac{K_o}{\sqrt{2}}$$

Hence at the half-power frequencies the length of the vector $s - s_1$ in Fig. 15-6a must be $\sqrt{2}$ times its length at the resonant frequency, and the pertinent relations are shown in Fig. 15-7. There are two half-power frequencies, one above and the other below the resonant frequency, and, as shown in Fig. 15-7, the bandwidth between the half-power frequencies is

$$B = 2\alpha = \frac{1}{RC} \qquad \text{rad/sec} \tag{15-23}$$

It also follows directly from Fig. 15-7 that the phase shift at the half-power frequencies is $\pm 45°$. The half-power bandwidth of this amplifier is independent of the inductance L, and thus changing L shifts the center frequency of the passband without changing the bandwidth. It is also useful to note that the bandwidth in radians per second is the coefficient of the linear term in the denominator of Eq. (15-11).

A useful measure of the relative width of the passband is the ratio

$$\frac{\omega_o}{B} = \frac{\omega_o}{2\alpha} = \frac{\omega_o C}{G} = Q_o = \frac{f_o}{B_{\text{Hz}}} \tag{15-24}$$

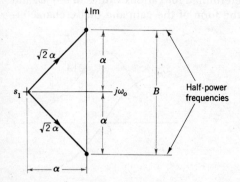

Fig. 15-7 Graphical analysis for the half-power frequencies of the tuned amplifier.

where Q_o is the resonant Q of the tuned circuit and B_{Hz} is the bandwidth in hertz. The resonant Q is thus an important parameter of the tuned circuit.

The pole-zero diagram provides insight into the properties of the amplifier, and it gives a quick overall picture of the frequency characteristics. For detailed calculations of gain and phase shift in the passband, however, it is desirable to use Eq. (15-22), which for this purpose can be put in a more convenient form. For sinusoidal operation in the passband, (15-22) becomes

$$A_v(j\omega) = -\alpha K_o \frac{1}{j\omega - s_1} \tag{15-25}$$

But $s_1 = -\alpha + j\beta \approx -\alpha + j\omega_o$, and thus (15-25) can be written as

$$A_v = -\alpha K_o \frac{1}{j\omega + \alpha - j\omega_o} = -\alpha K_o \frac{1}{\alpha + j(\omega - \omega_o)} \tag{15-26}$$

Letting $\omega - \omega_o = \Delta\omega$ and factoring α out of the denominator produces

$$A_v = \frac{-K_o}{1 + j\,\Delta\omega/\alpha}$$

$$= \frac{-K_o}{1 + j2\,\Delta\omega/B} \tag{15-27}$$

Then, using Eq. (15-24) yields as an alternative form for the complex gain

$$A_v = \frac{-K_o}{1 + j2Q_o\,\Delta\omega/\omega_o} \tag{15-28}$$

$$= \frac{-K_o}{1 + j2Q_o\,\Delta f/f_o} \tag{15-29}$$

Example 15-1

A pentode having a transconductance of 3 mmhos is to be used in a tuned amplifier like the one shown in Fig. 15-4. The center frequency of the passband is to be 10 MHz, and the half-power bandwidth is to be 300 kHz. The total resistance shunting the output of the amplifier is to be 10 kilohms. The problem is to determine the required values of L and C and to determine the gain of the amplifier at resonance.

Solution To obtain the specified bandwidth, the capacitance must be

$$C = \frac{1}{BR} = \frac{1}{(2\pi)(3)(10^5)(10^4)} = 53 \text{ pF}$$

Then to obtain the specified center frequency, the inductance must be

$$L = \frac{1}{\omega_o{}^2 C} = \frac{1}{(2\pi)^2(10^{14})(53)(10^{-12})} = 4.8 \ \mu H$$

The gain at resonance is

$$K_o = g_m R = (3)(10) = 30$$

The shape of the frequency characteristics for the tuned amplifier depends only on the locations of the poles and zeros of the complex voltage gain. Thus the pole-zero diagram provides a compact way of displaying the effects of varying the circuit parameters on the characteristics of tuned amplifiers. For example, if L is increased while R and C are held constant and if the poles are complex, it follows from Eqs. (15-19) that ω_o decreases while α remains constant. Thus the poles must move on a path parallel to the imaginary axis as shown in Fig. 15-8a. If R is decreased while L and C are held constant and if the poles are complex, then α increases while ω_o remains constant. Since ω_o is the distance of the poles from the origin of the complex plane, it follows that the poles must move on a circular path of radius ω_o having its center at the origin as shown in Fig. 15-8b. On the other hand, if R is increased toward infinity, the poles move along the circle toward the imaginary axis. If R is made negative, the poles move along a similar circular path in the right half of the complex plane. If the capacitance C is increased while R and L are held constant and if the poles are complex, then both α and ω_o vary; however, the poles move along the simple circular path shown in Fig. 15-8c. The circular nature of this path is established in

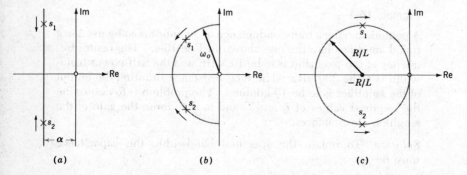

Fig. 15-8 The effect of variations in circuit parameters on the pole-zero pattern. (a) Increasing L; (b) decreasing R; (c) increasing C.

the following manner. If the real and imaginary coordinates of the poles are designated by x and y, respectively, then Eqs. (15-17) to (15-19) yield

$$x = -\alpha = -\frac{1}{2RC} \tag{15-30}$$

and

$$y^2 = \beta^2 = \omega_0{}^2 - \alpha^2 = \frac{1}{LC} - x^2 \tag{15-31}$$

Solving (15-30) for C and substituting the result into (15-31) yields

$$y^2 = -\frac{2R}{L}x - x^2 \tag{15-32}$$

Transferring all terms to the left-hand side and adding $(R/L)^2$ to both sides to complete the square on the left leads to

$$y^2 + \left(x + \frac{R}{L}\right)^2 = \left(\frac{R}{L}\right)^2 \tag{15-33}$$

If R and L are held constant while C is varied, x and y, the coordinates of the poles, must vary in accordance with Eq. (15-33); this equation describes a circle centered at $-R/L$ and having a radius equal to R/L, as shown in Fig. 15-8c. Since increasing C decreases α, the poles must move in the indicated direction with increasing C.

The roots of a quadratic polynomial move on circular paths in the complex plane when the coefficients are varied in a variety of ways. Thus the result expressed by Eq. (15-33) is a special case of a more general theory; the more general case is examined in Sec. 15-3.

In the circuit of Fig. 15-4a the losses associated with the inductor are accounted for by the shunt resistance R. Figure 15-9 shows an alternative configuration in which the losses of the inductor are accounted

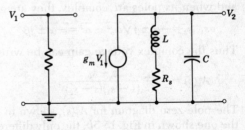

Fig. 15-9 Alternative representation of the losses associated with the tuning inductor.

for by a series resistance R_s. These two representations can be made exactly equivalent at one frequency, and they can be made approximately equivalent over a band of frequencies. The nature of physical inductors is such that at low frequencies the parameters R_s and L in Fig. 15-9 are independent of frequency, whereas R and L in Fig. 15-4a depend rather strongly on frequency; thus the representation in Fig. 15-9 is often preferred in low-frequency systems. However, at high frequencies where skin effect, proximity effect, and winding capacitance become important, the inductance and resistance in both representations are functions of frequency, although they are essentially constant over any limited band of frequencies. This is the condition existing in almost all tuned electronic amplifiers. In this case there is no basic reason to favor either representation, and the wise course of action is to choose the representation that yields the simplest mathematical formulation of the problem. The equivalence of the two representations in narrowband amplifiers is examined in the following paragraphs.

The complex voltage gain of the circuit in Fig. 15-9 is

$$A_v(s) = -\frac{g_m}{C} \frac{s + R_s/L}{s^2 + (R_s/L)s + 1/LC} \tag{15-34}$$

Again it is convenient to let

$$\omega_o{}^2 = \frac{1}{LC} \quad \text{and} \quad 2\alpha = \frac{R_s}{L} \tag{15-35}$$

so that (15-34) can be written more compactly as

$$A_v(s) = -\frac{g_m}{C} \frac{s + 2\alpha}{s^2 + 2\alpha s + \omega_o{}^2} \tag{15-36}$$

This function has a zero at

$$s_0 = -2\alpha = -\frac{R_s}{L} \tag{15-37}$$

and when its poles are complex, they are at

$$s_1, s_2 = -\alpha \pm j\sqrt{\omega_o{}^2 - \alpha^2} = -\alpha \pm j\beta \tag{15-38}$$

Thus the complex voltage gain can be written in factored form as

$$A_v(s) = -\frac{g_m}{C} \frac{s - s_0}{(s - s_1)(s - s_2)} \tag{15-39}$$

The pole-zero diagram for $A_v(s)$, shown in Fig. 15-10, is very similar to the one shown in Fig. 15-5b; the only difference is that the zero is shifted slightly to the left of the origin. In the narrowband, high-Q case, however, the shift is very small compared with ω_o, and its effect in the narrow passband of the amplifier is negligible.

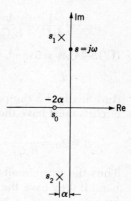

Fig. 15-10 Pole-zero pattern for the complex gain given by Eq. (15-36).

For sinusoidal operation the complex voltage gain is

$$A_v(j\omega) = -\frac{g_m}{C}\frac{j\omega - s_0}{(j\omega - s_1)(j\omega - s_2)} \tag{15-40}$$

It follows from Fig. 15-10 that the narrowband approximations in this case are

$$j\omega - s_0 = j\omega + 2\alpha \approx j\omega_o \tag{15-41}$$

and

$$j\omega - s_2 = j\omega + \alpha + j\beta \approx 2j\omega_o \tag{15-42}$$

Thus for operation in and near the narrow passband

$$A_v(j\omega) = -\frac{g_m}{2C}\frac{1}{j\omega - s_1} \tag{15-43}$$

This equation has the same form as Eq. (15-25) for the parallel RLC amplifier in the passband. Thus the two circuits behave identically in and near the narrow passband if the complex voltage gains have the same poles and the same scale factor. Using the subscript s for the case of the series resistance and the subscript p for the case of the parallel resistance, the requirement of equal scale factors can be expressed as

$$\frac{g_m}{2C_s} = \alpha_p K_{op} = \frac{g_m R_p}{2R_p C_p} = \frac{g_m}{2C_p}$$

and if g_m is the same in both cases, the requirement is

$$C_s = C_p \tag{15-44}$$

In order for the poles to be the same distance from the origin in both cases, the circuits must have the same resonant frequency; thus

$$\omega_0{}^2 = \frac{1}{L_s C_s} = \frac{1}{L_p C_p}$$

If (15-44) is satisfied, this requirement becomes

$$L_s = L_p \tag{15-45}$$

And finally, for the poles to have the same real parts in both cases, the circuits must have the same bandwidths; hence

$$B = 2\alpha = \frac{1}{R_p C_p} = \frac{R_s}{L_s}$$

Thus the two circuits behave identically in and near the narrow passband if they have the same L's and C's and if the resistances are related by

$$R_p C = \frac{L}{R_s} \tag{15-46}$$

Example 15-2

The small-signal model for a tuned amplifier is shown in Fig. 15-11a. The 100-kilohm resistor represents the plate resistance of the tube, and the 100-ohm resistor represents the losses in the coil. The problem is to find two circuits that are equivalent to the model in the narrow passband, one having the form shown in Fig. 15-4, and the other having the form shown in Fig. 15-9.

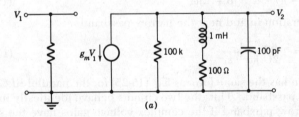

(a)

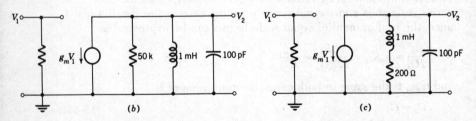

(b)   (c)

Fig. 15-11 Circuits for Example 15-2.

Solution The first step in the solution is to convert the two branches on the right in Fig. 15-11*a* to an equivalent parallel *RLC* circuit. For this purpose Eqs. (15-44) and (15-45) show that the inductance and capacitance remain unchanged, and the equivalent shunt resistance is given by (15-46) as

$$R_p = \frac{L}{CR_s} = \frac{10^{-3}}{(10^{-10})(10^2)} = 100 \text{ kilohms}$$

When this resistance is combined in parallel with the 100-kilohm plate resistance, the circuit shown in Fig. 15-11*b* is obtained. In the conversion of this circuit into the one shown in Fig. 15-11*c* the inductance and capacitance again remain unchanged, and the equivalent series resistance given by Eq. (15-46) is

$$R_s = \frac{L}{CR_p} = \frac{10^{-3}}{(10^{-10})(5)(10^4)} = 200 \text{ ohms}$$

15-3 SOME PROPERTIES OF A PAIR OF COMPLEX-CONJUGATE POLES

Complex poles associated with electric circuits always occur in conjugate pairs, and a number of important circuits are characterized by a single pair of such poles. Furthermore, a number of more complicated circuits are dominated to such an extent by a single pair of complex poles that they can be treated as if these were the only poles. Therefore it is useful to examine the properties of this pole configuration in some detail. The amplifier shown in Fig. 15-12*a* serves as a convenient vehicle for this study. The capacitance *C* represents the input impedance of the amplifier, and the resistance *R* is the output resistance of the signal source V_s. When the inductance *L* is omitted, the high-frequency performance of the amplifier is limited by *R* and *C*. The inductance is added to resonate with *C* just above the half-power frequency of the *RC* amplifier and thereby improve the high-frequency

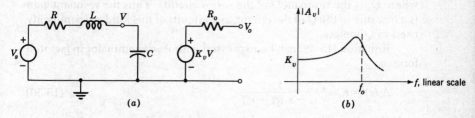

Fig. 15-12 The dynamic behavior associated with a pair of complex poles. (*a*) Amplifier characterized by a pair of complex poles; (*b*) gain characteristic.

performance; the half-power bandwidth can be increased by about 50 percent in this way.

When sufficient inductance is added to the circuit, the poles of the gain function become complex, and if the inductance is large, the poles take positions relatively close to the imaginary axis, as in the case of the tuned amplifier. In this case a resonant peak appears in the gain characteristic as shown in Fig. 15-12b. Such resonant peaks are usually undesirable in amplifiers of this kind (wideband amplifiers), for they result in overemphasis of signal components having frequencies in the vicinity of the peak. In order to understand the dynamic behavior of this amplifier in detail and to be able to design it for optimum perform- ance, it is helpful to know how the poles of the gain function move as the circuit parameters are changed. It is also helpful to know, for any given pole positions, whether or not a peak exists in the gain characteristic, the frequency of the peak if one exists, and the height of the peak. It is therefore significant that all of this information can be related to the pole-zero diagram for the gain function in a relatively simple way, as is shown in the following paragraphs.

The complex voltage gain for the circuit in Fig. 15-12a can be ex- pressed as

$$A_v(s) = \frac{K_v}{LC} \frac{1}{s^2 + (R/L)s + 1/LC} \tag{15-47}$$

As before, it is helpful to define $\omega_o{}^2 = 1/LC$ and $2\alpha = R/L$ so that (15-47) can be written more compactly as

$$A_v(s) = K_v\omega_o{}^2 \frac{1}{s^2 + 2\alpha s + \omega_o{}^2} \tag{15-48}$$

For sinusoidal operation at the resonant frequency, $s = j\omega_o$, the gain is

$$|A_v(j\omega_o)| = K_v \frac{\omega_o}{2\alpha} = K_v Q_o \tag{15-49}$$

where Q_o is the resonant Q of the series circuit. Thus the resonant gain is a measure of the Q of the circuit, and circuits of this kind are commonly used in Q meters.

Equation (15-48) can be expressed with its denominator in factored form as

$$A_v(s) = K_v\omega_o{}^2 \frac{1}{(s - s_1)(s - s_2)} \tag{15-50}$$

and when the poles are complex, they are given by

$$s_1, s_2 = -\alpha \pm j\sqrt{\omega_o{}^2 - \alpha^2} = -\alpha \pm j\beta \tag{15-51}$$

The first task is to determine how the poles of A_v move as the circuit parameters are changed. The special-case solution presented in Sec. 15-2 can be adapted to this case; however, for future use it is desirable to solve a more general problem. Equating the denominator of Eq. (15-48) to zero gives

$$s^2 + 2\alpha s + \omega_0{}^2 = 0 \qquad (15\text{-}52)$$

where α and ω_0 are functions of the circuit parameters. For generality (15-52) is rewritten as

$$s^2 + (k_1 p + k_3)s + k_2 p + k_4 = 0 \qquad (15\text{-}53)$$

where the k's are constants and p is a variable parameter. When the roots of this equation are complex, their real and imaginary coordinates can be designated as x and y, respectively. Then, from (15-51) and (15-53),

$$x = -\alpha = -\tfrac{1}{2}(k_1 p + k_3)$$

and solving this expression for p yields

$$p = \frac{-k_3 - 2x}{k_1} \qquad (15\text{-}54)$$

Similarly, the imaginary coordinate can be expressed as

$$y^2 = \beta^2 = \omega_0{}^2 - \alpha^2 = k_2 p + k_4 - x^2$$

Substituting (15-54) for p into this expression gives

$$y^2 = \frac{-k_2 k_3 - 2k_2 x}{k_1} + k_4 - x^2$$

or

$$y^2 + x^2 + 2\frac{k_2}{k_1} x = -\frac{k_2 k_3}{k_1} + k_4$$

Now adding $(k_2/k_1)^2$ to both sides to complete the square in x on the left yields

$$y^2 + \left(x + \frac{k_2}{k_1}\right)^2 = \left(\frac{k_2}{k_1}\right)^2 - \frac{k_2 k_3}{k_1} + k_4 \qquad (15\text{-}55)$$

or

$$y^2 + \left(x + \frac{k_2}{k_1}\right)^2 = \left(\frac{k_2}{k_1}\right)^2\left(1 - \frac{k_1 k_3}{k_2} + \frac{k_1{}^2 k_4}{k_2{}^2}\right) \qquad (15\text{-}56)$$

As the parameter p is varied, x and y, the real and imaginary coordinates of the roots of (15-53), must vary in accordance with Eq. (15-56). This equation describes a circle with its center at

$$-\frac{k_2}{k_1} = -c \tag{15-57}$$

and having a radius

$$r = c \left(1 - \frac{k_1 k_3}{k_2} + \frac{k_1{}^2 k_4}{k_2{}^2}\right)^{1/2} \tag{15-58}$$

A typical circle of this kind is shown in Fig. 15-13.

If the radius r given by Eq. (15-58) is real, then a circle exists, and the roots of the quadratic are complex for some range of values of the parameter p; if r is imaginary, there is no value of p that produces complex roots, and a circle does not exist.   If $k_2 = 0$, the center of the circle is at the origin of the complex plane, and the radius is $r = \sqrt{k_4}$. If $k_1 = 0$, the center is at infinity, and the circle becomes a straight line parallel to the imaginary axis; it follows from Eq. (15-53) that this line is located a distance $\frac{1}{2}k_3$ to the left of the imaginary axis.

Returning now to the original problem, Eq. (15-47) gives the complex voltage gain of the amplifier in Fig. 15-12a as

$$A_v(s) = \frac{K_v}{LC} \frac{1}{s^2 + (R/L)s + 1/LC}$$

To determine how the complex poles of A_v move as the inductance L is changed, $1/L$ is identified with the parameter p in Eq. (15-53).   The identification is completed by noting that in this case

$$k_1 = R \qquad k_2 = \frac{1}{C} \qquad k_3 = k_4 = 0 \tag{15-59}$$

Substituting these values into Eqs. (15-57) and (15-58) yields

$$c = \frac{k_2}{k_1} = \frac{1}{RC} \tag{15-60}$$

and

$$r = c = \frac{1}{RC} \tag{15-61}$$

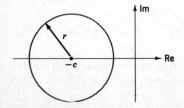

Fig. 15-13   General locus for the complex roots of a quadratic polynomial.

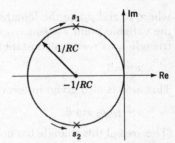

Fig. 15-14 Locus of the complex poles of Eq. (15-47) for increasing L.

Thus a circle exists, and it has the form shown in Fig. 15-14. Increasing L in this circuit reduces $\alpha = R/2L$, and hence the direction in which the poles move with increasing L is as shown in Fig. 15-14.

The next task is to determine the conditions, if any, under which a resonant peak appears in the gain characteristic. The complex voltage gain is given in factored form by Eq. (15-50) as

$$A_v(s) = K_v \omega_o^2 \frac{1}{(s - s_1)(s - s_2)} \tag{15-62}$$

and the corresponding pole-zero pattern is shown in Fig. 15-15a under the assumption that the poles are complex. It follows that under sinusoidal operating conditions the magnitude of the voltage gain can be expressed as

$$A = |A_v(j\omega)| = \frac{K_v \omega_o^2}{\rho_1 \rho_2} \tag{15-63}$$

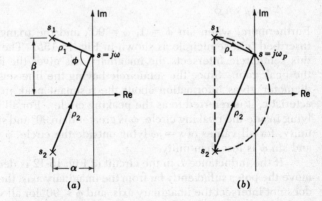

Fig. 15-15 Diagram related to the gain characteristic of a pair of complex poles. (a) Pole-zero diagram; (b) the resonant peaking circle.

where ρ_1 and ρ_2 are the lengths of the vectors shown in Fig. 15-15a. As the variable point $s = j\omega$ moves along the imaginary axis, the area of the triangle s_1s_2s remains constant at the value

$$a = \alpha\beta \tag{15-64}$$

This area is also given in terms of two sides and the included angle by

$$a = \tfrac{1}{2}\rho_1\rho_2 \sin \phi \tag{15-65}$$

(The area of this triangle has no importance in itself; it is merely a device for remembering useful algebraic relations.) Setting the right-hand sides of Eqs. (15-64) and (15-65) equal to each other then yields

$$\frac{1}{\rho_1\rho_2} = \frac{\sin \phi}{2\alpha\beta} \tag{15-66}$$

and substituting this result into (15-63) gives

$$A = \frac{K_v\omega_o^{\,2}}{2\alpha\beta} \sin \phi \tag{15-67}$$

As the operating frequency ω is varied, only the factor $\sin \phi$ in this expression changes.

The gain given by Eq. (15-67) has its maximum value when $\sin \phi$ has its maximum value, and the maximum value of $\sin \phi$ cannot exceed unity. Thus it is convenient to define a peak gain

$$A_p = \frac{K_v\omega_o^{\,2}}{2\alpha\beta} \tag{15-68}$$

so that Eq. (15-67) can be written more compactly as

$$A = A_p \sin \phi \tag{15-69}$$

Furthermore, when $\sin \phi = 1$, $\phi = 90°$, and the triangle s_1s_2s can be inscribed in a semicircle as shown in Fig. 15-15b. The point at which this semicircle intersects the imaginary axis gives the frequency ω_p of the peak gain. Since the semicircle having the line segment s_1s_2 as its diameter gives information about the resonant peak in the gain characteristic, it is referred to as the peaking circle. For all values of $s = j\omega$ lying inside the peaking circle, ϕ is greater than 90° and $\sin \phi$ is less than unity; for all values of $s = j\omega$ lying outside the circle, ϕ is less than 90° and $\sin \phi$ is less than unity.

If the inductance L in the circuit of Fig. 15-12 is decreased so as to move the poles sufficiently far from the imaginary axis, the peaking circle does not intersect the imaginary axis, and $\phi < 90°$ for all values of $s = j\omega$. In this case there is no resonant peak in the gain characteristic, and the gain has its maximum value at $s = 0$. Under these conditions the maximum value of A is less than the value A_p given by Eq. (15-68). A case of special interest arises when L is adjusted so that the peaking circle is tangent to the imaginary axis as shown in Fig. 15-16. This is the maximum value that L can have without a peak appearing in the gain characteristic, and it gives the maximum bandwidth obtainable without a peak. When the circuit is adjusted for this condition, it is said to be maximally flat. It follows directly from Fig. 15-15 that the circuit is maximally flat when $\alpha = \beta$. With the circuit adjusted for maximal flatness, the half-power point occurs when $\sin\phi = 45°$. It is easy to show by geometric means that, under this condition, the triangle $s_1 s_2 s$ is inscribed in a semicircle centered at the origin of the complex plane and passing through s_1 and s_2; hence with the maximally flat adjustment, the half-power bandwidth is ω_o.

The frequency of the resonant peak in the gain characteristic is given by the intersection of the peaking circle with the imaginary axis. Since the radius of the peaking circle is β, it follows that the frequency of the peak is

$$\omega_p{}^2 = \beta^2 - \alpha^2 = \omega_o{}^2 - 2\alpha^2 \tag{15-70}$$

Another important design parameter for these circuits is the peak-to-valley ratio. If the gain for $s = 0$ is designated by A_o and if the corresponding value of ϕ is designated by ϕ_o, then the peak-to-valley ratio is

$$\frac{A_p}{A_o} = \frac{1}{\sin\phi_o} \tag{15-71}$$

The value of ϕ_o is obtained from the relation

$$\tan\frac{\phi_o}{2} = \frac{\beta}{\alpha} \tag{15-72}$$

The concept of the peak-to-valley ratio is meaningful only when a peak appears in the amplitude characteristic, and in that case ϕ_o must be

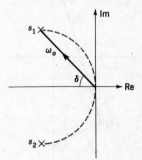

Fig. 15-16 Two poles adjusted for a maximally flat gain characteristic with $\delta = 45°$.

greater than 90°.   A further useful design parameter in the case where a resonant peak occurs is the frequency above the peak at which the gain has the same value as at zero frequency.   When $s = j\omega = 0$, the denominator in Eq. (15-48) has the value ω_o^2.   The high frequency at which the magnitude of the denominator has the same value, and at which $A = A_o$, is found from

$$|-\omega^2 + 2\alpha j\omega + \omega_o^2| = \omega_o^2$$

The result obtained is

$$\omega^2 = 2(\omega_o^2 - 2\alpha^2)$$

and substituting (15-70) into this relation yields the simple result

$$\omega = \sqrt{2}\omega_p \qquad\qquad\qquad\qquad\qquad\qquad\qquad (15\text{-}73)$$

Information about the phase shift can be obtained from the pole-zero diagram with the aid of the construction shown in Fig. 15-17.   If the angles of the two factors in the denominator of Eq. (15-62) are designated by θ_1 and θ_2, then the phase shift of the amplifier is $\theta = -\theta_1 - \theta_2$.   For sinusoidal operation at any frequency ω, the points $s = j\omega$ and $s = -j\omega$ are located on the pole-zero diagram as shown in Fig. 15-17, and the phase shift is then the angle θ shown in that figure.

Figure 15-18 shows the pole-zero pattern for the gain function of the amplifier in Fig. 15-12 for the condition that the inductance L is adjusted to give maximal flatness.   With this adjustment $\delta = 45°$, and, referring to Eq. (15-48) and Fig. 15-18,

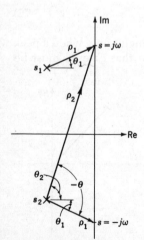

Fig. 15-17   Diagram related to the phase shift of a pair of complex poles. $\theta = -\theta_1 - \theta_2$.

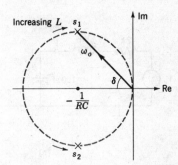

Fig. 15-18 Pole-zero diagram for the amplifier of Fig. 15-12 adjusted for a maximally flat gain characteristic, $\delta = 45°$.

$$\alpha = \frac{R}{2L} = \frac{1}{RC}$$

Thus the value of L required for maximal flatness is

$$L = \tfrac{1}{2}R^2 C \qquad\qquad (15\text{-}74)$$

With $L = 0$, the half-power bandwidth of the amplifier is $B = 1/RC$, and in accordance with the discussion of Fig. 15-16, the half-power bandwidth with L adjusted to give maximal flatness is

$$B' = \omega_o = \frac{1}{\sqrt{LC}}$$

Substituting Eq. (15-74) into this expression for L yields

$$B' = \frac{\sqrt{2}}{RC} = \sqrt{2}B \qquad\qquad (15\text{-}75)$$

Thus adding L to the amplifier and adjusting it for the maximally flat condition increases the half-power bandwidth by $\sqrt{2}$ without affecting the low-frequency gain.

If the amplifier is to be used for amplifying pulses, better results are often obtained by adjusting L to obtain the most linear phase characteristic. It can be shown that this condition results when L is adjusted to make the angle δ in Fig. 15-18 have the value 30°.

Example 15-3

Figure 15-19 shows the small-signal model for an electronic amplifier driven by a source having an inductive output impedance. The performance of this amplifier at high frequencies is to be examined.

Solution The complex voltage gain for this amplifier is given by Eq. (15-47). Hence it follows that the real part of the poles is

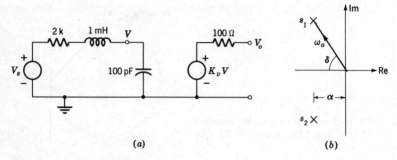

Fig. 15-19 Circuit for Example 15-3.

$$\alpha = \frac{R}{2L} = \frac{2,000}{(2)(10^{-3})} = 10^6$$

and the distance of the poles from the origin of the complex plane is

$$\omega_o = \frac{1}{\sqrt{LC}} = \frac{1}{\sqrt{(10^{-3})(10^{-10})}} = \frac{10^7}{\sqrt{10}} = 3.16(10^6)$$

Also,

$$f_o = \frac{\omega_o}{2\pi} = 0.504(10^6) = 504 \text{ kHz}$$

Since $\omega_o > \sqrt{2}\alpha$, the angle δ in Fig. 15-19b is greater than 45°, and the gain characteristic has a resonant peak. The frequency of this peak is obtained from Eq. (15-70) as

$$\omega_p{}^2 = \omega_o{}^2 - 2\alpha^2 = 10^{13} - 2(10^{12}) = 8(10^{12})$$

or

$$\omega_p = 2.82(10^6) \text{ rps}$$

and

$$f_p = 450 \text{ kHz}$$

The peak-to-valley ratio for the gain characteristic is given by

$$\frac{A_p}{A_o} = \frac{1}{\sin \phi_o} = \frac{1}{\sin 2\delta}$$

Now it follows from Fig. 15-19b that

$$\cos \delta = \frac{\alpha}{\omega_o} = 0.316$$

and thus

$$\delta = 71.6°$$

and

$$\phi_o = 2\delta = 143.2°$$

Hence

$$\sin \phi_o = 0.6$$

and

$$\frac{A_p}{A_o} = 1.67$$

The high frequency at which the gain has the same value as at zero frequency is

$$f = \sqrt{2}f_p = 635 \text{ kHz}$$

The value of source inductance L that would make the gain characteristic maximally flat is

$$L = \tfrac{1}{2}R^2C = \tfrac{1}{2}(4)(10^6)(10^{-10}) = 0.2 \text{ mH}$$

15-4 AMPLIFIERS WITH NONINTERACTING TUNED CIRCUITS AT THE INPUT AND OUTPUT

The principal shortcoming of the amplifier with a single tuned circuit presented in Sec. 15-2 is the fact that its gain does not fall off rapidly enough at frequencies outside the passband; thus for many applications it does not provide sufficient rejection of signals in adjacent channels. Matters can be improved in this respect by adding another tuned circuit at the input to the amplifier as shown in Fig. 15-20. The analysis and design of this amplifier are relatively straightforward because there is no interaction between the tuned circuits at the input and output. The condition of negligible interaction is satisfied by pentode amplifiers

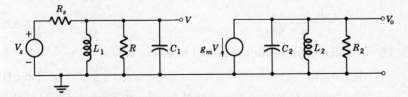

Fig. 15-20 An amplifier with tuned circuits at the input and output.

and by a number of microamplifiers designed for use with tuned circuits; it is also satisfied under many conditions by dual-gate MOS transistors. However, it is not satisfied by ordinary FETs and BJTs except at very low operating frequencies; these amplifiers are examined in Secs. 15-5 and 15-6.

The voltage gain of the amplifier in Fig. 15-20 is to be evaluated in two parts. If the voltage source V_s is converted to a current source and if the parallel combination of R and R_s is designated by R_1, then the complex voltage gain from V_s to V is obtained in a straightforward way as

$$\frac{V}{V_s} = A_{v1} = \frac{1}{R_s C_1} \frac{s}{s^2 + (1/R_1 C_1)s + 1/L_1 C_1} \tag{15-76}$$

$$= \frac{1}{R_s C_1} \frac{s}{s^2 + 2\alpha_1 s + \omega_{o1}{}^2} \tag{15-77}$$

where

$$2\alpha_1 = \frac{1}{R_1 C_1} \quad \text{and} \quad \omega_{o1}{}^2 = \frac{1}{L_1 C_1} \tag{15-78}$$

Equation (15-77) can also be expressed with its denominator in factored form as

$$A_{v1} = \frac{1}{R_s C_1} \frac{s}{(s - s_1)(s - s_2)} \tag{15-79}$$

In a similar way, the complex voltage gain from V to V_o is

$$\frac{V_o}{V} = A_{v2} = \frac{-g_m}{C_2} \frac{s}{s^2 + (1/R_2 C_2)s + 1/L_2 C_2} \tag{15-80}$$

$$= \frac{-g_m}{C_2} \frac{s}{s^2 + 2\alpha_2 s + \omega_{o2}{}^2} \tag{15-81}$$

where

$$2\alpha_2 = \frac{1}{R_2 C_2} \quad \text{and} \quad \omega_{o2}{}^2 = \frac{1}{L_2 C_2} \tag{15-82}$$

With its denominator expressed in factored form, this gain function is

$$A_{v2} = \frac{-g_m}{C_2} \frac{s}{(s - s_3)(s - s_4)} \tag{15-83}$$

The overall complex gain from input to output is now

$$A_v = A_{v1} A_{v2} \tag{15-84}$$

The requirements on the overall frequency characteristics of the amplifier, set forth in Sec. 15-1, are such that the two tuned circuits in the amplifier cannot be adjusted at random. The tuned circuits must be adjusted to give high gain in the passband and low gain outside the passband and to give a symmetrical gain characteristic. One possible adjustment is to design the tuned circuits so that they have identical poles and zeros; this adjustment is called synchronous tuning, and it produces a pole-zero diagram like the one shown in Fig. 15-21. Double symbols are used in this diagram to represent double poles and zeros. To examine the synchronously tuned amplifier in more detail, the narrowband approximations of Eqs. (15-20) and (15-21) are applied separately to A_{v1} and A_{v2}, and the result leads to a simplified expression for the overall complex gain having the form

$$A_v = A_{v1}A_{v2} = \frac{-g_m}{4R_sC_1C_2}\frac{1}{(s-s_1)(s-s_3)} \tag{15-85}$$

But with synchronous tuning $s_1 = s_3$, and (15-85) becomes

$$A_v = \frac{-g_m}{4R_sC_1C_2}\frac{1}{(s-s_1)^2} \tag{15-86}$$

This expression is valid for sinusoidal operation in and near the narrow passband of the amplifier. Corresponding to the narrowband approximations, Fig. 15-22 shows an expanded view of the pole-zero diagram in the vicinity of $s_1 = s_3$. The gain characteristic for the amplifier with two tuned circuits has the same general form as the characteristic for a single tuned circuit. However, the variable factor in the denominator of Eq. (15-86) is the square of the variable factor in the denominator of Eq. (15-22), the equation for the gain of an amplifier with one tuned circuit. Hence the gain of the amplifier with two tuned circuits falls off much more rapidly at frequencies outside the passband, and this amplifier is said to have greater selectivity than the one with a single tuned circuit.

If a given application requires more gain and more selectivity than can be obtained with a single-stage amplifier, stages can be connected in

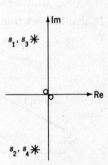

Fig. 15-21 Pole-zero diagram for a synchronously tuned amplifier.

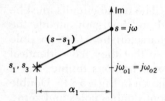

Fig. 15-22 Expanded view of the pole-zero diagram in the vicinity of $s_1 = s_3$.

cascade with tuned circuits at the input, the output, and each interstage point. With no interaction between the input and output in each stage, the tuned circuits can be designed separately; one possible design is to adjust all the tuned circuits for synchronous tuning. With $n - 1$ stages and n tuned circuits connected in cascade, the narrowband approximation for the overall complex gain has the form

$$A_v = \frac{K}{(s - s_1)^n} \tag{15-87}$$

However, when identical tuned stages are connected in cascade, there is a reduction in the half-power bandwidth just as there is when identical untuned stages are cascaded, and the bandwidth reduction factor is identical with the one given in Eq. (14-89). For variety, an alternative proof in terms of the pole-zero diagram is offered here.

It follows from Fig. 15-23 that, when $\omega = \omega_o$, the magnitude of the denominator in Eq. (15-87) is

$$|s - s_1|^n = \alpha_1{}^n \tag{15-88}$$

and by definition, when ω equals the half-power frequency, the magnitude of the denominator is

$$|s - s_1|^n = \sqrt{2}\,\alpha_1{}^n \tag{15-89}$$

Taking the nth root of this expression and squaring the result yields

$$|s - s_1|^2 = (2^{1/n})\alpha_1{}^2 \tag{15-90}$$

It also follows from Fig. 15-23 that

$$|s - s_1|^2 = \alpha_1{}^2 + (\Delta\omega)^2$$

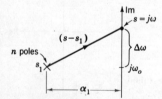

Fig. 15-23 Pole-zero diagram for n narrowband noninteracting tuned circuits in cascade.

where $\Delta\omega$ is the difference between the operating frequency and the center frequency of the amplifier. Substituting (15-90) for the value of $|s - s_1|^2$ at the half-power frequency into this last expression yields

$$(\Delta\omega)^2 = (2^{1/n})\alpha_1{}^2 - \alpha_1{}^2$$

or

$$\Delta\omega = \alpha_1\sqrt{2^{1/n} - 1}$$

at the half-power frequency. There are two half-power frequencies, one above and the other below the center frequency. Thus the half-power bandwidth of a multistage amplifier with n synchronously tuned circuits in cascade is

$$B_n = 2\alpha_1\sqrt{2^{1/n} - 1}$$

But $2\alpha_1 = B_1 =$ bandwidth of one tuned circuit, and thus

$$B_n = B_1\sqrt{2^{1/n} - 1} \qquad (15\text{-}91)$$

As is stated in connection with Eq. (14-89), the bandwidth reduction factor for $n = 2$ is 0.64; for $n = 3$, it is 0.51, and for $n = 4$, it is 0.43.

The synchronous tuning described above is one possible adjustment for amplifiers having tuned circuits at both the input and the output. Synchronous tuning has the advantage of simplicity in that the tuned circuits are identical. However, there is a different design that provides a better approximation to the ideal bandpass gain characteristic at the cost of a somewhat more complex adjustment of the two tuned circuits. This adjustment, called staggered tuning, is illustrated by the pole-zero diagram shown in Fig. 15-24a. With staggered tuning the

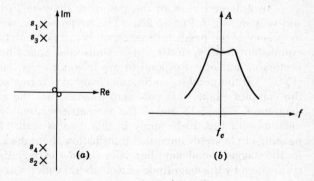

Fig. 15-24 Stagger-tuned amplifier. (a) Pole-zero pattern; (b) gain characteristic.

poles of the two tuned circuits are displaced, or staggered, along a line parallel to the imaginary axis, and the associated gain characteristic may have the form shown in Fig. 15-24b. Poles s_1 and s_2 in Fig. 15-24a are associated with one of the tuned circuits, and poles s_3 and s_4 are associated with the other. The shape of the gain characteristic in the passband depends strongly on the separation between the poles s_1 and s_3, and it is under the control of the circuit designer. As is implied by the root locus shown in Fig. 15-8a, the staggering of the poles along a line parallel to the imaginary axis can be accomplished by adjusting the inductances in the two tuned circuits for slightly different values.

After the narrowband approximations are made, the complex voltage gain for the amplifier with tuned input and tuned output is given by Eq. (15-85) as

$$A_v = \frac{-g_m}{4R_sC_1C_2} \frac{1}{(s - s_1)(s - s_3)} \tag{15-92}$$

and with staggered tuning

$$s_1 = -\alpha + j\beta_1 \quad \text{and} \quad s_3 = -\alpha + j\beta_3 \tag{15-93}$$

It should be noted that the narrowband approximations have reduced a fourth-degree polynomial in the denominator to a second-degree polynomial. These approximations imply that the frequency characteristics of the amplifier in and near the narrow passband are determined by the two poles s_1 and s_3 in Fig. 15-24a; the poles s_2 and s_4 together with the two zeros at the origin contribute a constant $\frac{1}{4}$ in the scale factor for Eq. (15-92). It should also be noted that the form of Eq. (15-92) is exactly the same as the form of Eq. (15-62), and thus many of the results obtained in Sec. 15-3 apply directly to the stagger-tuned amplifier. This matter is examined in some detail in the paragraphs that follow.

An enlarged view of the pole-zero diagram in the vicinity of s_1 and s_3 is shown in Fig. 15-25a. The frequency ω_c, which is the center frequency of the passband, corresponds to a point on the imaginary axis equidistant from s_1 and s_3. For sinusoidal operating conditions, the vectors ρ_1 and ρ_3 correspond to the factors $j\omega - s_1$ and $j\omega - s_3$ in Eq. (15-92). If the point ω_c on the imaginary axis is treated as the origin of the complex plane, this pole-zero diagram has the same form as the diagram in Fig. 15-15, and all the geometric relations developed in connection with Fig. 15-15 apply in this case as well. In particular, the peaking circle yields important information about the gain characteristic for the stagger-tuned amplifier. For sinusoidal operating conditions the gain, given by the magnitude of Eq. (15-92) with $s = j\omega$, can be written as

$$A = |A_v(j\omega)| = \frac{g_m}{4R_sC_1C_2} \frac{1}{\rho_1\rho_3} \tag{15-94}$$

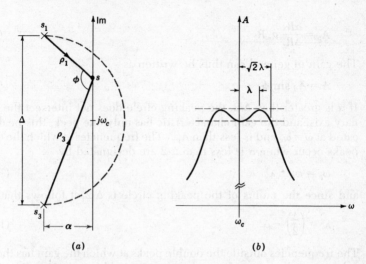

Fig. 15-25 Frequency characteristics for a stagger-tuned amplifier. (a) Geometrical construction; (b) gain characteristic.

Adapting Eq. (15-66) to this case yields

$$\frac{1}{\rho_1 \rho_3} = \frac{\sin \phi}{\alpha \Delta} \tag{15-95}$$

and thus

$$A = \frac{g_m}{4R_s C_1 C_2} \frac{\sin \phi}{\alpha \Delta} \tag{15-96}$$

Now Eq. (15-82) gives $\alpha = \alpha_2 = 1/2R_2C_2$, and hence

$$A = \frac{g_m R_2}{2R_s C_1 \Delta} \sin \phi \tag{15-97}$$

The only factor in this expression that changes when ω changes is $\sin \phi$.

If α is less than $\Delta/2$, the peaking circle intersects the imaginary axis at two points, and the gain characteristic has two peaks, as shown in Fig. 15-25b. The gain at the peaks, obtained from (15-97) with $\sin \phi = 1$, is

$$A_p = \frac{g_m R_2}{2R_s C_1 \Delta}$$

By noting from Eq. (15-78) that $C_1 = 1/2\alpha R_1$, this expression can be put in the alternative form

$$A_p = \frac{\alpha R_1}{\Delta R_s} g_m \dot{R}_2 \tag{15-98}$$

The gain in general can thus be written as

$$A = A_p \sin \phi \tag{15-99}$$

If α is greater than $\Delta/2$, the peaking circle does not intersect the imaginary axis, and the gain characteristic has only one peak; this peak is located at $\omega = \omega_c$ and is less than A_p. The frequencies at which the double peaks occur when α is less than $\Delta/2$ are designated by

$$\omega_p = \omega_c \pm \lambda \tag{15-100}$$

and since the radius of the peaking circle is $\Delta/2$, it follows that

$$\lambda^2 = \left(\frac{\Delta}{2}\right)^2 - \alpha^2 \tag{15-101}$$

The frequencies outside the double peaks at which the gain has the same value as at the center frequency are obtained by adapting Eq. (15-73); they are the triple-point frequencies

$$\omega_t = \omega_c \pm \sqrt{2}\lambda \tag{15-102}$$

The gain at the center frequency is

$$A_c = A_p \sin \phi_c$$

where ϕ_c is the value of ϕ when $\omega = \omega_c$. The peak-to-valley ratio for the double-peaked gain characteristic is thus

$$\frac{A_p}{A_c} = \frac{1}{\sin \phi_c} \tag{15-103}$$

and ϕ_c is related to the pole positions by

$$\tan \frac{\phi_c}{2} = \frac{\Delta}{2\alpha} = r \tag{15-104}$$

The quantity r, which is fixed by the peak-to-valley ratio, is an important design parameter.

When the amplifier is adjusted so that $\alpha = \Delta/2$, the peaking circle is just tangent to the imaginary axis, and the gain characteristic is maximally flat. The amplifier is also said to be flat-staggered, and this condition marks the transition from the overstaggered (double peaks) to the understaggered adjustment. Interest is often centered on the flat-staggered and overstaggered cases; however, it can be shown that the most linear phase characteristic is obtained when $\phi_c = 60°$, corresponding to a slightly understaggered adjustment.

In most of the amplifiers studied up to this point the half-power bandwidth is related in a simple way to the circuit parameters; hence it has been a useful and convenient measure of the performance of the amplifier. In the case of the overstaggered pair of tuned circuits, however, a simpler and in some respects more useful measure of the bandwidth is the triple-point bandwidth,

$$W = 2\sqrt{2}\lambda \tag{15-105}$$

The significance of this bandwidth is illustrated in Fig. 15-25b. It follows from Eq. (15-101) that

$$W^2 = 8\lambda^2 = 8\left[\left(\frac{\Delta}{2}\right)^2 - \alpha^2\right] \tag{15-106}$$

and hence

$$\left(\frac{W}{2\alpha}\right)^2 = 2\left[\left(\frac{\Delta}{2\alpha}\right)^2 - 1\right] = 2(r^2 - 1) \tag{15-107}$$

The design parameter $r = \Delta/2\alpha$ is related directly to the peak-to-valley ratio by Eqs. (15-103) and (15-104).

The relations developed above provide a simple, direct procedure for designing overstaggered amplifiers. The specifications usually fix the center frequency f_c, the bandwidth W, and the peak-to-valley ratio A_p/A_c; these specifications fix the poles of the complex gain, and hence they fix the circuit parameters. In particular, A_p/A_c fixes r; Eq. (15-107) then gives the required value of α, and the required value of Δ is obtained from the relation $\Delta = 2\alpha r$. Thus the pole locations are fixed, and two of the three parameters in each tuned circuit are fixed. The third element in each tuned circuit can be chosen at will or to satisfy some other requirement.

Figure 15-26 shows the pole configuration for a flat-staggered amplifier. In this case $\phi_c = 90°$, $\sin \phi_c = 1$, $r = 1$, and $\Delta = 2\alpha$; thus Eq. (15-98) becomes

$$A_p = A_c = \frac{R_1}{2R_s} g_m R_2 \tag{15-108}$$

This gain is just one-half the center-frequency gain with synchronous tuning. When the amplifier is flat staggered, the triple-point bandwidth is not useful, for its value is zero; however, the half-power bandwidth in this case is related to the pole pattern in a simple way. In accordance with the discussion of Fig. 15-16, the half-power frequencies for the flat-staggered amplifier are those frequencies at which the imaginary axis is intersected by the circle centered on the axis at ω_c and passing through the poles. This circle is shown in Fig. 15-26; since its

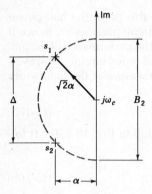

Fig. 15-26 Pole-zero diagram for a flat-staggered amplifier.

radius is $\sqrt{2}\alpha$, the half-power bandwidth is

$$B_2 = 2\sqrt{2}\alpha = \sqrt{2}B_1 \qquad (15\text{-}109)$$

where B_2 is the bandwidth of two flat-staggered tuned circuits and B_1 is the bandwidth of one tuned circuit. For two synchronously tuned circuits, the half-power bandwidth is $0.64B_1$, and thus the flat-staggered bandwidth is 2.2 times the synchronously tuned bandwidth. Since Eq. (15-108) gives the flat-staggered center-frequency gain as one-half the center-frequency gain with synchronous tuning, the gain-band-width product A_cB_2 for flat staggering is 1.1 times its value with synchronous tuning.

The procedure for designing a flat-staggered amplifier for a specified center frequency and half-power bandwidth is straightforward. The bandwidth fixes α through Eq. (15-109), and the required value of Δ for flat staggering is $\Delta = 2\alpha$. This information, together with the specified center frequency, fixes the pole locations, and the pole locations in turn fix two of the three parameters in each tuned circuit.

If a given application requires more gain and more selectivity than can be provided by a single-stage amplifier, stages can be connected in cascade with tuned circuits at the input, the output, and each interstage point. With no interaction between input and output in each stage, the tuned circuits can be designed separately. By a more extensive study it can be shown that, if a maximally flat gain characteristic is desired, the poles of the complex gain must be separated by equal arcs on a semicircle centered on the imaginary axis, as shown in Fig. 15-27a. If the number of poles is n, the size of the arc is

$$a = \frac{180°}{n} \qquad (15\text{-}110)$$

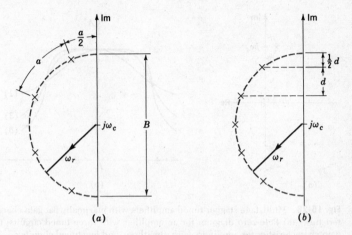

Fig. 15-27 Pole-zero patterns for multistage stagger-tuned amplifiers. (a) Maximally flat gain characteristic; (b) approximation to the most linear phase characteristic.

The center frequency ω_c of the passband corresponds to the center of the circle on which the poles lie, and no matter how many poles there are, the half-power frequencies are those frequencies at which the circle intersects the imaginary axis. The pole locations required to give the most linear phase characteristic have been worked out, but the results cannot be stated in a simple way. However, an excellent approximation to these locations is obtained by separating the poles by equal vertical distances on a semicircle centered on the imaginary axis, as shown in Fig. 15-27b. If the number of poles is n, the vertical separation is

$$d = \frac{2\omega_r}{n} \tag{15-111}$$

The frequency range over which the phase characteristic is linear corresponds approximately to the diameter of the semicircle. Thus with the maximally flat gain adjustment and with the linear phase adjustment, the useful passband for the amplifier corresponds approximately to the diameter of the semicircle on which the poles lie.

The pole-zero diagram for a maximally flat gain characteristic with three tuned circuits is shown in Fig. 15-28a, and Fig. 15-28b shows maximally flat gain characteristics for one, three, and five tuned circuits. The larger the number of poles, the flatter the gain characteristic in the passband and the steeper the characteristic at the edges of the band. It

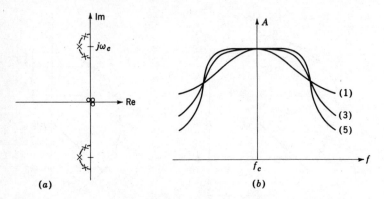

Fig. 15-28 Multistage stagger-tuned amplifiers with maximally flat gain characteristics. (a) Pole-zero diagram for an amplifier with three tuned circuits; (b)
gain characteristics for amplifiers with one, three, and five tuned circuits.

should be mentioned, however, that with a large number of poles, the
transient response of the amplifier is much poorer than when the poles
are adjusted for a linear phase characteristic. The characteristics shown
in Fig. 15-28b should be compared with the ideal filter characteristic
shown in Fig. 15-2.

15-5 FET AMPLIFIER WITH INTERACTING TUNED
CIRCUITS AT THE INPUT AND OUTPUT

Figure 15-29a shows the small-signal model for a tuned FET amplifier,
and Fig. 15-29b shows the same circuit after a source conversion is
made at the input. This model can represent either a JFET amplifier
or a MOST amplifier. The feedback capacitance C_{gd} in this amplifier
introduces interaction between the tuned circuits at the input and output, and this interaction can have a pronounced effect on the behavior
of the amplifier. For example, the interaction can cause the amplifier
to behave as an oscillator generating self-sustaining oscillations. In
order to design such an amplifier, it is necessary to understand the nature of the interaction and to know how its effects can be kept under control.

The current I flowing through the feedback capacitance in Fig.
15-29b is a measure of the interaction, and the nature of the interaction
can be determined by examining this current and the associated input
admittance Y_i indicated in the figure. The current is

$$I = j\omega C_{gd}(V - V_o) \tag{15-112}$$

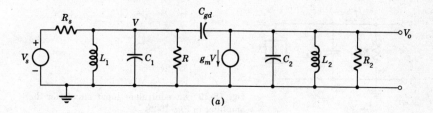

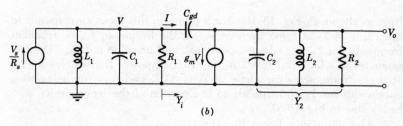

Fig. 15-29 FET tuned amplifier.

This current is usually much smaller than $g_m V$ at all frequencies where the amplifier gives appreciable gain, and thus the output voltage can be expressed as $V_o = -g_m V/Y_2$. Substituting this expression for V_o in (15-112) yields

$$I = \left(j\omega C_{gd} + j\omega C_{gd}\frac{g_m}{Y_2}\right)V \qquad (15\text{-}113)$$

and

$$Y_i = \frac{I}{V} = j\omega C_{gd} + j\omega C_{gd}\frac{g_m}{Y_2} \qquad (15\text{-}114)$$

$$= j\omega C_{gd} + Y \qquad (15\text{-}115)$$

These results imply that Y_i can be represented by the circuit shown in Fig. 15-30.

The admittance Y in Fig. 15-30 depends on the admittance Y_2 of the output tuned circuit, and this admittance varies rapidly with frequency in the vicinity of the resonant frequency. Since Y_2 is a complex number that varies with frequency, it is helpful to plot the value of Y_2 on the complex plane for a wide range of frequencies. It is clear from Fig. 15-29b that, as the frequency of the signal is varied, the real part of Y_2 remains constant at the value $1/R_2$ while the imaginary part varies from large negative values to large positive values. Thus the plot of Y_2 as a function of frequency in the complex plane is a vertical straight

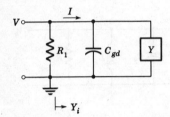

Fig. 15-30 An equivalent input circuit for the amplifier in Fig. 15-29.

line, as shown in Fig. 15-31. Each point on this locus corresponds to the value of Y_2 at some frequency. At the frequency f_o, the resonant frequency for Y_2, the admittance is real, and its value is $1/R_2$. The frequencies at which $|Y_2|$ is $\sqrt{2}/R_2$ are the half-power frequencies for Y_2; at these frequencies the angle of Y_2 is $\pm 45°$, and thus the half-power bandwidth for Y_2 corresponds to the portion of the locus shown as a heavy line in Fig. 15-31.

The admittance Y_2 can be expressed as

$$Y_2 = G_2 + j\omega C_2 + \frac{1}{j\omega L_2} = |Y_2| e^{j\theta} \tag{15-116}$$

and its reciprocal is

$$Z_2 = \frac{1}{Y_2} = \frac{1}{|Y_2|} e^{-j\theta} \tag{15-117}$$

The locus of this impedance as a function of frequency in the complex plane is the inverse of the locus of Y_2, and it is a circle, as shown in Fig. 15-32. The fact that the straight line in Fig. 15-31 inverts into a circular locus can be established by a simple geometric proof. As Eq. (15-117)

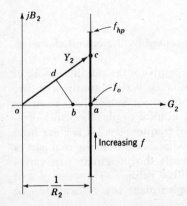

Fig. 15-31 Locus of $Y_2(f)$ in the complex plane.

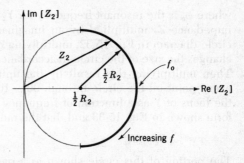

Fig. 15-32   Locus of $Z_2 = 1/Y_2$.

shows, the magnitudes of Z_2 and Y_2 are reciprocals, and the angles are equal but opposite in sign. Consider first the magnitudes, and make the following construction, as shown in Fig. 15-31. Locate the point b such that $ob = 1/oa$. Then choose point c anywhere on the locus of Y_2 other than at point a, and construct od so that $od = 1/oc$. Then it follows from the construction that

$$\frac{ob}{oc} = \frac{od}{oa} \tag{15-118}$$

Thus triangles oac and odb are similar, and the angle odb is a right angle. Thus the triangle odb is inscribed in a semicircle having ob as its diameter, and since point c was chosen arbitrarily on the locus of Y_2, all points on Y_2 invert into points on the circle having ob as its diameter. Now, to satisfy the condition on the angles expressed by Eq. (15-117), it is merely necessary to interchange the upper and lower semicircles. The final circle, shown in Fig. 15-32, has a diameter equal to R_2 and a radius equal to $\frac{1}{2}R_2$; the portion of the circle shown as a heavy line corresponds to the half-power bandwidth of Z_2.

The admittance Y in Fig. 15-30 is given by

$$Y = \frac{j\omega C_{gd} g_m}{Y_2} = j\omega C_{gd} g_m Z_2$$

This expression can be simplified in a very useful way by making a narrowband approximation. Over the entire passband of a narrowband amplifier, the frequency does not vary by more than a few percent from its value at the center frequency, and thus the quantity ωC_{gd} is essentially constant over a range of frequencies in and near the narrow passband. Hence the admittance Y can be approximated by

$$Y = j\omega_o C_{gd} g_m Z_2 \tag{15-119}$$

where ω_o is the resonant frequency for Y_2 and Z_2. Thus Y is simply the impedance Z_2 multiplied by an imaginary constant. In terms of the circle diagram in Fig. 15-32, multiplying Z_2 by the real constant $\omega_o C_{gd} g_m$ changes the size of the circle in accordance with the size of the constant. Then multiplying this result by the unit imaginary number j rotates every point on the circle through 90°. It follows from these facts that the locus of Y as a function of frequency in the complex plane has the form shown in Fig. 15-33 and that the radius of this circle is

$$r = \tfrac{1}{2}\omega_o C_{gd} g_m R_2 \tag{15-120}$$

The portion of this circle shown as a dashed line corresponds to frequencies remote from ω_o where the narrowband approximation used in Eq. (15-119) is not valid.

The results illustrated in Fig. 15-33 show the important fact that the input admittance to the transistor in the tuned amplifier has a negative real part at all frequencies below f_o. This negative conductance can be a source of self-sustaining oscillations, depending on the external circuit connected to the transistor at its input. If the capacitance C_{gd} in Fig. 15-30 is lumped in with the capacitance of the input tuned circuit, then the input circuit, exclusive of L_1 and C_1, has the form shown in Fig. 15-34a. The total admittance Y_A appearing at the input to this circuit is given by the circle diagram shown in Fig. 15-34b. It is clear that the real part of this admittance may or may not be negative in some frequency band, depending on the value of R_1 in relation to the diameter of the circle.

With a sinusoidal current I_A applied to the circuit in Fig. 15-34a, the voltage developed across the input terminals is

$$V = \frac{I_A}{Y_A} = \frac{I_A}{G_A + jB_A} \tag{15-121}$$

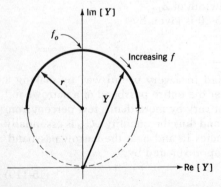

Fig. 15-33 Locus of $Y = j\omega_o C_{gd} g_m Z_2$.

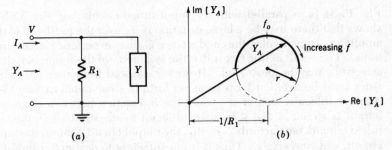

Fig. 15-34 Input admittance of the tuned FET amplifier. (*a*) Equivalent circuit; (*b*) locus of Y_A in the complex plane.

Now suppose that R_1 is adjusted so that G_A is negative at some frequency ω_1. It is possible to connect a suitable susceptance (in this case inductive) across the input terminals to make $B_A = 0$ at this frequency. Then it is possible to connect a suitable positive resistance across the input terminals to make $G_A = 0$ at $\omega = \omega_1$. With these adjustments, the denominator in Eq. (15-121) is zero at $\omega = \omega_1$, and a sinusoidal voltage V can exist across the terminals even when $I_A = 0$. That is, with these adjustments the circuit can support a sinusoidal voltage of frequency ω_1 even when no input signal is applied. Thus when the input conductance is negative at some frequency, the circuit is said to be potentially unstable at that frequency, for the circuit can be made to oscillate just by connecting the appropriate *passive* elements across the input terminals.

The above argument can also be advanced in terms of the Laplace transform. Let Eq. (15-121) be the transform equation giving V in terms of I_A. The adjustments described above make $Y_A(s) = 0$ for sinusoidal operation at the frequency ω_1, and thus $Y_A(s)$ has a pair of complex zeros at $s = \pm j\omega_1$. It then follows from (15-121) that $V(s)$ has a pair of complex poles at $s = \pm j\omega_1$, and thus $v(t)$ has an undamped sinusoidal component with a frequency ω_1.

The condition of potential instability must be avoided in amplifiers of the type under consideration here, for such amplifiers almost always turn out to be oscillators when they are potentially unstable. It follows from Fig. 15-34*b* that potential instability can be avoided by designing the amplifier so that

$$\frac{1}{R_1} > r = \frac{1}{2}\,\omega_o C_{gd} g_m R_2 \qquad (15\text{-}122)$$

where the radius r is given by Eq. (15-120). It turns out, however, that designing to avoid potential instability is not sufficient to guarantee good performance on the part of the amplifier. The admittance Y_A in

Fig. 15-34 is in parallel with the input tuned circuit, and Fig. 15-34b shows that there may be a large variation in Y_A over the passband of the amplifier. In particular, the conductance G_A may experience large variations. The result of this fact is that the symmetry of the frequency characteristics may be destroyed. However, skewed frequency characteristics must be avoided for reasons set forth in some detail in Sec. 15-1. Another difficulty that arises when the interaction between input and output is strong is that it becomes difficult to adjust, or align, the two tuned circuits satisfactorily. Tuning the output circuit detunes the input circuit, and conversely. Thus it is not sufficient to design the amplifier so that the circle diagram in Fig. 15-34b stays out of the negative-conductance region; the input-output interaction must be made small by making the radius of the circle considerably smaller than $1/R_1$ so that Y_A has negligible effect on the input tuned circuit. It is therefore helpful to define an alignability factor

$$k = \frac{r}{1/R_1} = \frac{1}{2}\,\omega_o C_{gd} g_m R_1 R_2 \qquad (15\text{-}123)$$

If potential instability is to be avoided, k must be less than unity. If skewed frequency characteristics and difficult alignment problems are to be avoided, k must be substantially less than unity. The value of k is controlled by the choice of R_1 and R_2. Figure 15-35 illustrates the effect of changing R_2; changing R_2 changes the radius of the circular admittance locus. Changing R_1 has the effect of shifting the circle to the left or to the right, depending on whether R_1 is increased or decreased.

It is important to note that making R_1 and R_2 small to obtain a small alignability factor also results in a small voltage gain. Thus choosing a value for k involves a compromise between small skew in the fre-

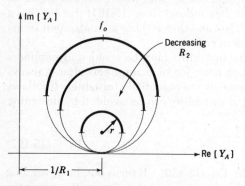

Fig. 15-35 The effect of changing R_2 on the input admittance of the tuned FET amplifier.

quency characteristics and easy alignability on the one hand and high gain on the other. A value of $k = 0.2$ is often chosen as a suitable compromise. With k small, the interaction between input and output can be neglected as an approximation, and the two tuned circuits can be designed as if there were no interaction.

Example 15-4

A synchronously tuned FET amplifier having the form shown in Fig. 15-29 is to be designed for a center frequency $f_o = 1$ MHz and a half-power bandwidth of 25 kHz with an alignability factor $k = 0.2$. The transistor parameters are $C_{gd} = 2$ pF and $g_m = 4$ mmhos. Also, the resistance R in Fig. 15-29a is much larger than R_s so that $R_s = R_1$.

Solution The alignability factor given by Eq. (15-123) is

$$k = 0.2 = \tfrac{1}{2}\omega_o C_{gd} g_m R_1 R_2$$

Thus

$$R_1 R_2 = \frac{0.4}{(2\pi)(10^6)(2)(10^{-12})(4)(10^{-3})}$$

$$= 8(10^6)$$

If R_1 and R_2 are chosen to be equal, then

$$R_1 = R_2 = 2.82 \text{ kilohms}$$

With two synchronously tuned circuits the overall half-power bandwidth is

$$B_2 = 0.64 B_1$$

and thus the bandwidth of each tuned circuit must be

$$B_1 = \frac{B_2}{0.64} = \frac{(2\pi)(25)(10^3)}{0.64} = 2.45(10^5) \text{ rps}$$

But the half-power bandwidth for a tuned circuit is given by Eq. (15-23) as $B = 1/RC$, and thus for the input tuned circuit

$$C_1 = \frac{1}{R_1 B_1} = \frac{1}{(2.82)(10^3)(2.45)(10^5)} = 1.45 \text{ nF}$$

This is the total shunt capacitance required in the input tuned circuit, including the parasitic transistor capacitances. The inductance required is found from

$$L_1 = \frac{1}{\omega_o^2 C_1} = \frac{1}{(2\pi)^2 (10^{12})(1.45)(10^{-9})} = 17.5 \ \mu H$$

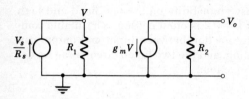

Fig. 15-36   Circuit for Example 15-4.

Since the input and output tuned circuits are identical, $R_2 = R_1$, $C_2 = C_1$, and $L_2 = L_1$.

With both tuned circuits in resonance at 1 MHz, the small-signal model for the amplifier at resonance with $R_1 = R_s$ takes the simple form shown in Fig. 15-36 under the assumption that the input-output interaction is negligibly small. Thus the gain at resonance is

$$A_o = \left| \frac{V_o}{V_s} \right| = g_m R_2 = (4)(2.82) = 11.3$$

In order to obtain a good alignability factor in this amplifier, it is necessary to use rather low values of R_1 and R_2. This fact in turn results in undesirably large values of C and undesirably small values of L in the tuned circuits. Thus in practical amplifiers of this kind it is usually necessary to use an impedance transforming arrangement to transform these values of L and C into more suitable values. This matter is discussed in some detail in Sec. 15-7.

15-6   THE BJT AMPLIFIER WITH TUNED INPUT AND OUTPUT CIRCUITS

A complete discussion of the analysis and design of the tuned BJT amplifier is longer and more involved than is appropriate for this book. However, it is possible to identify the principal problems involved in designing such circuits and to indicate in a general way how they are solved; the reader who is interested in further details can find them in Refs. 1 and 2 listed at the end of the chapter.

For the analysis and design of untuned BJT amplifiers, the hybrid-π model used in Chap. 14 is a convenient characterization of the transistor. Its parameters are independent of frequency, and they change with the quiescent operating point in a simple way. However, for the analysis of narrowband tuned amplifiers, the y-parameter model shown in Fig. 15-37 provides a better characterization. In particular, narrowband approximations applied to this model lead to a relatively simple representation of the BJT in narrowband tuned amplifiers. It should

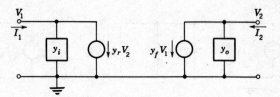

Fig. 15-37 The y-parameter transistor model.

also be noted that the y-parameter model is a general model for two-port devices; it applies equally well to BJTs, FETs, and microamplifiers. The y-parameter model corresponds to the following terminal volt-ampere relations,

$$I_1 = y_iV_1 + y_rV_2$$
$$I_2 = y_fV_1 + y_oV_2$$

(15-124)

The subscripts i, r, f, and o on the parameters indicate input, reverse, forward, and output, respectively. When the parameters characterize a BJT, a second subscript, as in y_{ie} or y_{ib}, may be added to designate the common-emitter or the common-base connection; similarly, when the parameters characterize an FET, the subscript s may be added to designate the common-source connection. The y parameters are complex numbers and functions of frequency; for purposes of amplifier analysis it is convenient to express them in the form

$$y_i = g_i + jb_i \qquad y_r = |y_r| \angle \theta_r$$
$$y_f = |y_f| \angle \theta_f \qquad y_o = g_o + jb_o$$

(15-125)

Data sheets for transistors (BJTs and FETs) that are intended for use in tuned amplifiers frequently give curves of the y parameters as functions of frequency over a wide range of frequencies. These curves show that the parameters are slowly varying functions of frequency that do not change very much over the passband of a narrowband amplifier. Therefore, in the analysis and design of narrowband amplifiers the y parameters can be treated as constants; these are the narrowband approximations for the y-parameter model. Moreover, the susceptances b_i and b_o are usually capacitive, and in narrowband amplifiers they are almost always much smaller than the capacitive susceptances in the tuned circuits connected to the transistor; hence b_i and b_o can usually be neglected. Thus the small-signal model for a narrowband tuned BJT amplifier usually takes the form shown in Fig. 15-38a; the transistor parameters g_i, g_o, y_r, and y_f are treated as constants in and near the nar-

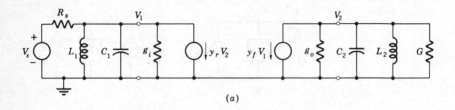

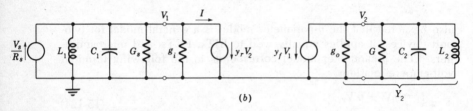

Fig. 15-38 Tuned BJT amplifier.

row passband of the amplifier. After a source conversion at the input, the model takes the form shown in Fig. 15-38b.

The interaction between input and output is accounted for in this model by the parameter y_r, and the current I shown in Fig. 15-38b is a measure of the interaction. This current is

$$I = y_r V_2 = \frac{y_r(-y_f V_1)}{Y_2} = -\frac{y_r y_f}{Y_2} V_1 \tag{15-126}$$

and thus

$$\frac{I}{V_1} = Y = -\frac{y_r y_f}{Y_2} \tag{15-127}$$

With this result it follows from Fig. 15-38b that the input circuit of the amplifier, apart from the tuned circuit $L_1 C_1$, has the form shown in Fig. 15-39a. This circuit is the counterpart of the input circuit for the FET amplifier shown in Fig. 15-34a. Figure 15-39b shows the locus of Y_A as a function of frequency in the complex plane. This locus is the counterpart of the locus shown in Fig. 15-34b, and it is obtained by starting with Eq. (15-127) and following the same procedure that led to Fig. 15-34b. The radius of this circle is

$$r = \frac{|y_f y_r|}{2(g_o + G)} \tag{15-128}$$

and the angle ϕ is

$$\phi = \theta_f + \theta_r + 180° \tag{15-129}$$

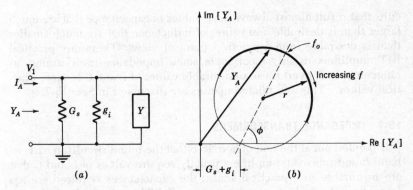

Fig. 15-39 Input admittance of the tuned BJT amplifier. (a) Equivalent circuit; (b) locus of Y_A in the complex plane.

where θ_f and θ_r are defined in Eqs. (15-125). For the FET amplifier the angle ϕ is 90°, as shown in Fig. 15-34b; typical y parameters for the BJT usually give $\phi < 90°$, as shown in Fig. 15-39b.

Figure 15-39b shows that, if $G_s + g_i$ is small enough, the real part of Y_A is negative in a band of frequencies, and the amplifier is potentially unstable. Moreover, if the radius of the circle is large, Y_A experiences a large variation over the passband of the amplifier; as a result the frequency characteristics are skewed, and the alignment problem is difficult. Thus, as in the case of the FET amplifier, it is necessary to design the amplifier so that the radius of the circle is small compared to $G_s + g_i$. Hence it is again useful to define an alignability factor,

$$k = \frac{r}{g_i + G_s} = \frac{|y_f y_r|}{2(g_i + G_s)(g_o + G)} \tag{15-130}$$

where r is given by Eq. (15-128). To avoid skewed frequency characteristics and difficult alignment problems, the circuit must be designed so that k is substantially less than unity. The value of k is determined by the choice of G_s and G, and k is made small by making these conductances large. But making these conductances large reduces the gain of the amplifier, and thus choosing the value of k involves a compromise between small skew in the frequency characteristics and easy alignment on the one hand and high gain on the other. As in the case of the FET amplifier, $k = 0.2$ is often chosen as a good compromise. With the input-output interaction made small by the choice of k, the interaction can be neglected as an approximation, and the input and output tuned circuits can be designed separately as if there were no interaction.

In the design of narrowband tuned BJT amplifiers, the tuned circuits that result almost always have values of capacitance that are much larger than is desirable and values of inductance that are much smaller than is desirable from a practical point of view. Therefore, practical BJT amplifiers usually incorporate some impedance transforming arrangement to convert these undesirable values of L and C to more practical values. This and related matters are discussed in Sec. 15-7.

15-7 IMPEDANCE TRANSFORMERS

It is pointed out at the end of Sec. 15-6 that the tuned circuits in narrow-band tuned transistor amplifiers usually require values of L and C that are unsuitable in a practical sense; the capacitances required are too large, and the inductances are too small. This practical problem is solved by using impedance transformers of various kinds. Figure 15-40a shows a transformer that is widely used in tuned amplifiers for this purpose; it consists of a tapped coil wound on a ferrite core. Since the coil has a high-permeability core, it can be assumed that there is unity coupling among the turns of the coil. Figure 15-40b shows a circuit that is equivalent to the one in Fig. 15-40a with all of the circuit elements referred to the primary side of the transformer. With the voltage ratio for the transformer designated a as shown in Fig. 15-40a, the relations between the two circuits are

$$V_2' = aV_2 \qquad C = a^2 C'$$

$$L = \frac{L'}{a^2} \qquad G = a^2 G' \tag{15-131}$$

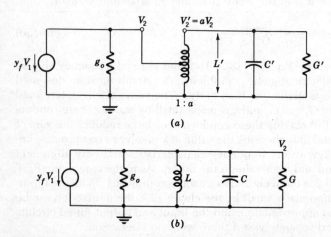

Fig. 15-40 An impedance transformer. (a) Circuit; (b) an equivalent circuit.

Thus with $a > 1$, a practical value of capacitance C' is converted to a much larger value C, and a practical inductance L' is converted to a much smaller value L. It should be noted that the resonant frequency of L and C is the same as that of L' and C'.

The load conductance G reflected into the primary side of the transformer in Fig. 15-40 also depends on the transformer ratio. In amplifiers having interaction between the input and output tuned circuits, this fact can be used in adjusting G to obtain a suitably small interaction by making k small in Eq. (15-130). When the amplifier has no interaction $(y_r = 0)$, the transformer can be used for a different purpose, to obtain maximum output power and output voltage from the amplifier. Figure 15-41 shows the output circuit of an amplifier in which $y_r = 0$. At the resonant frequency the effects of L and C exactly cancel each other, and these two elements can be ignored at resonance. Suppose first that g_o and a are fixed and that G_L is variable. Then V_L is maximum when $G_L = 0$, I_L is maximum when $G_L = \infty$, and the load power P_L is maximum when G_L has some intermediate value between these two extremes. Now suppose that g_o and G_L are fixed and that a is variable. Then

$$P_L = |V_L||I_L| = G_L|V_L|^2 = \frac{|I_L|^2}{G_L} \tag{15-132}$$

and since G_L is fixed, it follows that V_L and I_L have their maximum values when a is adjusted to make P_L a maximum. Thus all three of these quantities are maximum when the load conductance reflected into the primary of the transformer equals the source conductance as shown in Fig. 15-42. Under this condition the load power is

$$P_{L,\text{max}} = \frac{|I'_L|^2}{G'_L} = \frac{|I'_L|^2}{g_o} = \frac{|y_f V_1|^2}{4g_o} \tag{15-133}$$

Any other passive load connected to this amplifier will draw less power than the power given by (15-133); thus it is not possible to draw more power than this from the amplifier, and this power is called the *maximum available power* of the amplifier. Note that the maximum available power is independent of the load conductance.

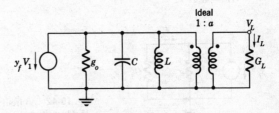

Fig. 15-41　An impedance transformer.

Under the conditions illustrated in Fig. 15-42 the transformer is said to match the load impedance to the output impedance of the amplifier at resonance. This condition of matched impedances is often used in amplifiers having negligible input-output interaction. However, when the interaction is not negligible, it is usually necessary to adjust the circuit for mismatched conditions to keep the effects of the interaction suitably small.

Figure 15-40 shows one kind of transformer that is used in tuned amplifiers. It is possible in addition to identify a class of circuits that are useful in transforming resistances from one value to another. This class is represented by the network N shown in Fig. 15-43a; N is any linear, lossless, passive network with either lumped or distributed parameters. Figure 15-43b shows one member of this class in which N consists simply of a capacitive voltage divider; this transformer is widely used in tuned amplifiers, especially in high-frequency amplifiers. For sinusoidal operation at any fixed frequency, the input impedance to N can be represented by an equivalent parallel combination of resistance and reactance as shown in Fig. 15-43c. Now the fact that N is lossless requires that

$$P_{in} = P_L \tag{15-134}$$

and these powers can be written as

$$\frac{|V_1|^2}{R'} = \frac{|V_2|^2}{R} \tag{15-135}$$

If a voltage ratio is defined as

$$a = \left|\frac{V_1}{V_2}\right| \tag{15-136}$$

then Eq. (15-135) can be written as

$$\frac{|V_1|^2}{R'} = \frac{a^2|V_2|^2}{R'} = \frac{|V_2|^2}{R} \tag{15-137}$$

and thus

$$R' = a^2 R \tag{15-138}$$

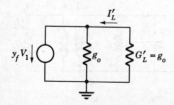

Fig. 15-42 A circuit equivalent to the one in Fig. 15-41 at resonance.

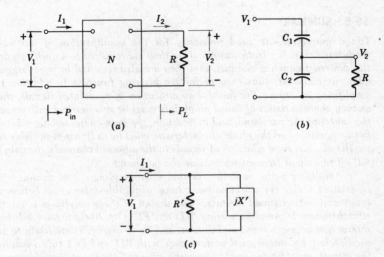

Fig. 15-43 Resistance transformation. (a) General network; (b) a specific example; (c) an equivalent circuit for a and b.

Hence the lossless network in Fig. 15-43a transforms resistances according to the square of its voltage ratio. When this transformer is used in a tuned amplifier, the reactance X' shown in Fig. 15-43c becomes part of a tuned circuit.

For the particular case of the capacitive voltage divider shown in Fig. 15-43b, the voltage ratio is

$$\frac{V_2}{V_1} = \frac{j\omega C_1 R}{1 + j\omega C_1 R + j\omega C_2 R} \tag{15-139}$$

and

$$a = \left|\frac{V_1}{V_2}\right| = \left|\frac{1 + j\omega(C_1 + C_2)R}{j\omega C_1 R}\right| \tag{15-140}$$

If $\omega(C_1 + C_2)R \gg 1$, as is often the case, then the first term in the numerator can be dropped, and

$$a = \frac{C_1 + C_2}{C_1} \tag{15-141}$$

Thus the transformation ratio for the capacitive voltage divider has a very simple form when the inequality is satisfied.

15-8 SUMMARY

Tuned amplifiers are used primarily for the amplification of narrowband modulated signals. In this application they are required to amplify signals in the desired frequency channel, and they are also required to reject signals in adjacent channels; thus they perform a filtering function in addition to an amplifying function. For faithful amplification of modulated signals, the frequency characteristics of tuned amplifiers must be symmetrical with respect to the center of the passband, and in addition, the gain characteristic must be as flat as possible and the phase characteristic must be as linear as possible in the passband. For good rejection of signals in the adjacent channels, the gain must fall off rapidly at frequencies outside the passband.

Amplifiers using pentodes, certain microamplifiers, and dual-gate MOS transistors under certain conditions have negligible interaction between the input and output tuned circuits. The design of these amplifiers is relatively straightforward. Amplifiers using BJTs and FETs are likely to have substantial interaction between input and output, and these amplifiers are likely to act as oscillators. To obtain good performance with BJT and FET tuned amplifiers, the circuits must be designed to keep the effects of the input-output interaction small.

The pole-zero diagram for the complex voltage gain provides a very powerful technique for the analysis and design of tuned amplifiers. It provides useful information about the frequency characteristics of the amplifiers, it shows in a direct way how to make important narrowband approximations, and it often reveals with clarity the effects of varying the parameters of the tuned circuits. The root-locus techniques developed in this chapter are capable of much greater development, and they provide the circuit designer and the research engineer with a very powerful tool.

PROBLEMS

15-1. *Frequency spectrum.* A sinusoidal carrier $v_c = 10 \cos 2\pi(10^6)t$ volts is amplitude modulated with a signal

$$v_m = 3 \cos 2{,}000\pi t + 6 \cos 5{,}000\pi t \qquad \text{volts}$$

Sketch a diagram of the frequency spectrum of the AM wave (see Fig. 15-2). Give the amplitude and the frequency in megahertz for each sinusoidal component of the AM wave.

15-2. *Frequency spectrum.* Two AM radio broadcast stations have carrier frequencies of 1,000 and 1,030 kHz, respectively. If their frequency spectra are not to overlap, what is the highest audio frequency that each can broadcast?

15-3. *Single-tuned-amplifier design.* A single-tuned pentode amplifier like the one shown in Fig. 15-4 is to be designed for a center frequency of 5 MHz and a half-power bandwidth of 200 kHz. The tube parameters are $g_m = 5$ mmhos and $r_p = 1$ megohm, and the total shunt capacitance is to be 50 pF.

(a) Determine the required values of L and R, and find the voltage gain at the center frequency.

(b) Determine the resonant Q of the tuned circuit.

(c) Sketch and dimension the pole-zero diagram for A_v.

(d) If L is multiplied by 4 with R and C remaining constant, what are the new resonant frequency and bandwidth? Give the answers in hertz.

15-4. *Single-tuned amplifier.* The circuit diagram for a narrowband tuned amplifier is shown in Fig. 15-44. The tube parameters are $g_m = 3$ mmhos and $r_p = 500$ kilohms.

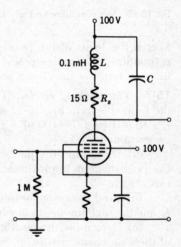

Fig. 15-44 Amplifier for Prob. 15-4.

(a) Determine the value of C required for a resonant frequency $\omega_o = 10^7$ rps (1.6 MHz).

(b) Find the gain at resonance and the half-power bandwidth in hertz. *Suggestion:* Convert the resistance R_s into an equivalent shunt resistance.

15-5. *Root locus.* Figure 15-9 shows the circuit diagram for a single-tuned amplifier.

(a) Sketch the loci of the poles and the zero of A_v as R_s is increased from zero with L and C constant. Cover the range of R_s corresponding to complex poles.

(b) Repeat part *a* for variable C and variable L with $R_s \neq 0$.

15-6. Root locus. The complex voltage gain of the microamplifier circuit shown in Fig. 15-45 is

$$A_v = K_v \omega_o^2 \frac{1}{s^2 + (R/L + 1/R_i C_i)s + (1 + R/R_i)/LC_i}$$

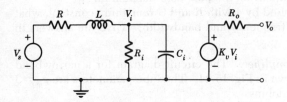

Fig. 15-45 Microamplifier for Prob. 15-6.

Sketch the locus of the poles of A_v as L is varied over the range corresponding to complex poles. Give the important dimensions of this locus in terms of the circuit parameters.

15-7. The peaking circle. The parameter values in a circuit having the form shown in Fig. 15-12 are $K_v = 30$, $R_o = 100$ ohms, $R = 6$ kilohms, $L = 10$ mH, and $C = 100$ pF. The frequency response of the circuit is to be examined.

(a) Determine the poles of the complex voltage gain, and sketch the pole diagram approximately to scale. Show the peaking circle on this sketch.

(b) Find the resonant frequency f_o, the frequency of the peak gain f_p, and the high frequency at which the gain has the same value as at $f = 0$.

(c) Determine the peak gain and the gain at f_o, and make a sketch of the gain characteristic.

15-8. Amplifier design. The amplifier described in Prob. 15-7 has a peak in its gain characteristic that is undesirable in many applications. Hence the circuit is to be modified by changing the inductance L to improve its frequency characteristics.

(a) If the other circuit parameters remain constant at the values given in Prob. 15-7, what value of L is required to make the gain characteristic maximally flat?

(b) With the adjustment of part a, what is the half-power bandwidth? Give the answer in hertz.

(c) What value of L is required to give the most linear phase characteristic ($\delta = 30°$ in Fig. 15-18)?

15-9. Synchronously tuned amplifier design. An amplifier with two tuned circuits like the one shown in Fig. 15-20 is used in a TV receiver.

The tuned circuits are to be synchronously tuned (equal center frequencies and bandwidths) with a center frequency of 5 MHz and an overall half-power bandwidth of 200 kHz. The capacitance in each tuned circuit is to be 50 pF, and the transconductance is $g_m = 10$ mmhos.

(a) Determine the bandwidth required for each of the tuned circuits. Give the answer in hertz.

(b) Determine the values of $R_1, R_2, L_1,$ and L_2 required by the tuned circuits. (R_1 is the parallel combination of R_s and R in Fig. 15-20.)

15-10. *Stagger-tuned amplifier.* An amplifier like the one shown in Fig. 15-20 is stagger tuned with a pole-zero pattern like the ones shown in Figs. 15-24 and 15-25. The pertinent dimensions of the pole pattern are $\omega_c = 10^7$ rps (1.6 MHz), $\alpha = 4.5(10^4)$ rps, and $\Delta = 1.5(10^5)$ rps. The transconductance is $g_m = 5$ mmhos, $R_s = R_2 = 5$ kilohms, and $R \gg R_s$ so that $R_1 = R_s$.

Find the peak gain A_p, the peak-to-valley ratio, and the triple-point bandwidth W. Give the bandwidth in hertz.

15-11. *Flat-staggered amplifier.* Data for an overstaggered amplifier are given in Prob. 15-10. The inductance in each tuned circuit is readjusted to make the overall gain characteristic maximally flat with the same center frequency, 1.6 MHz. All other circuit parameters are left unchanged.

With this adjustment, determine the gain at the center frequency and the half-power bandwidth in hertz.

15-12. *Multistage stagger-tuned amplifier.* The tuned amplifier shown in Fig. 15-46 is used in a TV receiver. It employs two microamplifiers in cascade, and it has a center frequency of 45 MHz. The poles

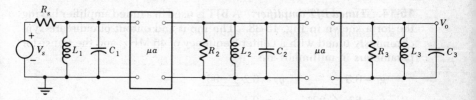

Fig. 15-46 Tuned amplifier for Prob. 15-12.

of the complex voltage gain are adjusted to give a maximally flat overall gain characteristic with a half-power bandwidth of 6 MHz. The pole diagram for the complex gain has the form shown in Figs. 15-27 and 15-28.

(*a*) Sketch a pole diagram like the one in Fig. 15-27 for this amplifier, and designate the poles s_1, s_2, and s_3.

(*b*) Determine the arc *a* and the radius ω_r for this diagram.

(*c*) Each of the poles s_1, s_2, and s_3 is associated with one of the tuned circuits in Fig. 15-46. Determine the half-power bandwidth in megahertz of these three tuned circuits, and compare these bandwidths with the overall bandwidth of the amplifier. (It makes no difference which pole is associated with which tuned circuit.)

15-13. *Tuned dual-gate MOST amplifier.* A dual-gate MOS transistor is used in a tuned amplifier having the form shown in Fig. 15-29. The transistor parameters are $g_m = 1$ mmho and $C_{gd} = 0.01$ pF, and the resistance *R* in Fig. 15-29*a* is much larger than R_s so that $R_1 = R_s$. This amplifier is to be used in an FM radio receiver. The input and output circuits are to be synchronously tuned (equal center frequencies and bandwidths) to give a center frequency of 10 MHz and an overall half-power bandwidth of 250 kHz, and the alignability factor is to be less than 0.2. In addition, the capacitance in each tuned circuit is to be 25 pF.

(*a*) Determine the half-power bandwidth of each tuned circuit.

(*b*) Determine the value of $R_1 = R_2$ required for an alignability factor $k = 0.2$.

(*c*) Assuming that the input-output interaction is negligible, determine the value of $R_1 = R_2$ required for the half-power bandwidth found in part *a*. Is the requirement on *k* satisfied with these resistances?

(*d*) Assuming that the interaction is negligible, determine the value of inductance required in each tuned circuit.

(*e*) Determine the voltage gain at the center frequency when R_1 and R_2 have the values that give the specified bandwidth.

15-14. *Tuned BJT amplifier.* A BJT is used in a tuned amplifier having the form shown in Fig. 15-38. The input and output circuits are synchronously tuned with a center frequency of 45 MHz, and the transistor parameters in millimhos are

$$g_i = 3.5 \qquad y_r = 0.2 \; \underline{/-90^\circ}$$

$$y_f = 83 \; \underline{/-33^\circ} \qquad g_o = 0.1$$

The source resistance is $R_s = 1/G_s = 200$ ohms.

(*a*) Find the value of *G* required to make the alignability factor $k = 0.2$. Give the value of $R = 1/G$.

(*b*) Determine the voltage gain at the center frequency when *G* has the value found in part *a*. *Suggestion:* Use Fig. 15-38*a*. At the center frequency the parallel *LC* branches are open circuits, and with $k = 0.2$ the interaction, and hence the current $y_r V_2$, is negligibly small.

15-15. *Tuned microamplifier.* A certain commercially available micro-amplifier can be represented at 10 MHz by the model shown in the box in Fig. 15-47. The transconductance g_m is 20 mmhos. The input and output circuits are synchronously tuned, the center frequency of the

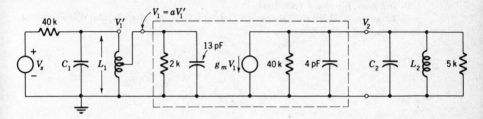

Fig. 15-47 Amplifier for Prob. 15-15.

passband is 10 MHz, and the overall half-power bandwidth is 250 kHz. In addition, the input resistance to the amplifier is matched to the 40-kilohm source resistance to obtain maximum gain, and for this purpose the tapped ferrite-core coil L_1 acts as a lossless transformer.

(*a*) Find the turn ratio *a* that will transform the 2-kilohm input resistance to 40 kilohms.

(*b*) Determine the half-power bandwidth for each tuned circuit.

(*c*) Determine the required values of C_1 and L_1. *Suggestion:* Transfer the 2-kilohm resistor and the 13-pF capacitor to the primary side of the transformer.

(*d*) Determine the voltage gain V_2/V_s at the center frequency.

15-16. *Resistance transformation.* The capacitive voltage divider shown in Fig. 15-48 is used to transform the 200-ohm resistance into a

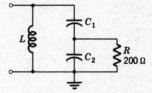

Fig. 15-48 Lossless transformer for Prob. 15-16.

5-kilohm resistance in parallel with the inductor at an operating frequency of 40 MHz. In addition, the capacitors are to be chosen so that $\omega(C_1 + C_2)R = 5$.

Assuming that the inequality associated with Eq. (15-141) is satisfied with this design, determine the required values of C_1 and C_2.

REFERENCES

1. Searle, C. L., et al.: "Elementary Circuit Properties of Transistors," chap. 8, John Wiley and Sons, Inc., New York, 1964.
2. Thornton, R. D., et al.: "Multistage Transistor Circuits," chap. 7, John Wiley and Sons, Inc., New York, 1965.

16

Frequency Characteristics
and the Stability of
Feedback Amplifiers

The use of negative feedback to improve the performance of electronic amplifiers is discussed in some detail in Sec. 13-5. This discussion shows that marked improvements can be obtained through the use of feedback, but it also points out that the existence of feedback raises the possibility that the circuit may generate self-sustaining oscillations. The tuned amplifiers presented in Secs. 15-5 and 15-6 have unintentional feedback as a result of the parasitic capacitance coupling the output to the input in each stage, and again in this case the feedback is a potential source of self-sustaining oscillations. In general, feedback, whether it be intentional or unavoidable, alters the dynamic performance of electronic amplifiers. Thus when feedback is used to obtain a degree of self-calibration or to reduce the effects of nonlinear distortion, the frequency characteristics of the amplifier are also changed. The effect of feedback on the dynamic response can be beneficial; however, unless the feedback is carefully controlled by the circuit designer, it usually degrades the dynamic performance.

The dynamic response of a feedback amplifier, as measured by its complex voltage gain, is directly related to the complex voltage gain of the internal amplifier around which the feedback exists. Thus the complex gain of the internal amplifier, usually called the open-loop gain, is the central quantity in the analysis and design of feedback amplifiers. The objective of this chapter is to study the relationship between the open-loop gain and the overall, or closed-loop, gain, and to develop the basic principles used in the design of feedback amplifiers.

16-1 THE FREQUENCY DEPENDENCE OF THE LOOP GAIN

Figure 16-1 shows the small-signal model for a feedback amplifier. This circuit is identical with the one shown in Fig. 13-20 except that in this case the open-loop voltage gain,

$$\frac{V_o}{V_i} = A_v'$$

is a complex gain and a function of frequency. In the usual case the amplifier is arranged so that A_v' introduces a sign reversal in the band of frequencies where it is independent of frequency. Following the procedure leading to Eq. (13-38), the overall, or closed-loop, complex gain is obtained as

$$\frac{V_o}{V_s} = A_v = \frac{\alpha A_v'}{1 - \beta A_v'} = \frac{(1 - \beta)A_v'}{1 - \beta A_v'} \tag{16-1}$$

where α and β are the voltage-divider ratios defined by Eqs. (13-35). In most cases β is much less than unity, and α is approximately equal to unity. The gain around the feedback loop is $A_L = \beta A_v'$, and thus the closed-loop complex gain can be written as

$$A_v = \frac{\alpha A_v'}{1 - A_L} = \frac{\alpha A_v'}{F} \tag{16-2}$$

where F is the return difference for the feedback loop. The closed-loop gain in decibels is thus

$$|A_v|_{dB} = 20 \log |\alpha A_v'| - 20 \log |F| \tag{16-3}$$

In Chap. 13 the feedback is designated positive or negative according to whether the voltage fed back adds to or subtracts from the input signal voltage. However, with the loop gain a function of frequency as

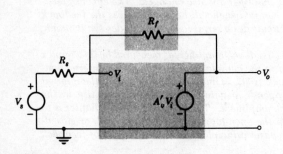

Fig. 16-1 Small-signal model for a feedback amplifier.

in Eq. (16-2), the simple notion of direct addition or subtraction breaks down. In general, for sinusoidal operating conditions the voltage fed back is displaced by some phase angle from the input signal voltage, and this phase angle is a function of frequency. Nevertheless, for sinusoidal operation the notion of positive and negative feedback can be extended in a useful and consistent way by defining the amount of feedback in decibels as

$$F_{dB} = -20 \log |F| \tag{16-4}$$

Note that F_{dB} is defined as the second term on the right-hand side of Eq. (16-3). Now when $|F| > 1$, F_{dB} is negative, and the feedback is said to be negative. Under this condition it follows from (16-2) or (16-3) that the closed-loop gain is less than the open-loop gain. When $|F| < 1$, F_{dB} is positive, and the feedback is said to be positive. Under this condition the feedback increases the closed-loop gain. When $|F| = 1$, $F_{dB} = 0$, and there is no feedback in the circuit.

When feedback is used to improve the performance of an amplifier by providing self-calibration or by reducing the effects of corrupting signals, a large amount of feedback is required if the improvement is to be substantial. But

$$F = 1 - A_L = 1 - \beta A_v' \tag{16-5}$$

and with any physical amplifier, A_L must tend to zero at high frequencies. Thus the amount of feedback varies with frequency, and F_{dB} tends to zero at high frequencies. The design problem therefore involves arranging the circuit so that the feedback remains suitably large over the appropriate band of frequencies. The way in which the return difference, and hence the feedback, in any particular circuit varies with the frequency of a sinusoidal signal can be displayed conveniently by means of a diagram like the one shown in Fig. 16-2. For each different value of frequency the loop gain A_L has a different complex value. A plot in the complex plane of the values of A_L for all frequencies between zero and infinity may take the form of the solid-line contour shown in Fig. 16-2. Each point on this contour corresponds to a particular frequency; the length of the phasor from the origin to the contour is proportional to the magnitude of A_L, and the angle that the phasor makes with the real axis is the phase shift of A_L. The plot in Fig. 16-2 corresponds to a directly coupled amplifier having a sign reversal at zero frequency, and it shows that A_L tends to zero at very high frequencies. The plot of A_L in the complex plane as a function of frequency for sinusoidal signals is called a Nyquist diagram.

The phasor drawn from the Nyquist diagram to the point $1 + j0$ on the positive real axis gives the value of the return difference $F = 1 - A_L$.

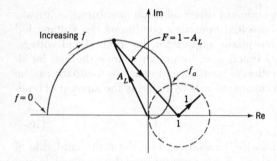

Fig. 16-2 Variation of the loop gain and return difference with frequency.

It follows from this fact that $|F| > 1$ and that the feedback is negative for all frequencies corresponding to portions of the Nyquist plot that lie outside the circle of unit radius centered at the point $1 + j0$. Similarly, the feedback is positive for all frequencies corresponding to points on the Nyquist plot lying inside this unit circle. Positive feedback increases the gain of the amplifier, but it degrades the performance in most other respects. For example, if nonlinearities in the amplifier generate distortion components of voltage having frequencies in the region where the feedback is positive, these distortion voltages are exaggerated by the feedback.

   If the final high-frequency asymptote of the gain characteristic for A_L has a slope of -6 dB/octave, then the maximum change in phase shift introduced by A_L is $90°$, and the Nyquist plot must tend to zero at high frequencies along the positive imaginary axis. In this case the feedback is negative for all frequencies. If the slope of the final asymptote is -12 dB/octave, then the maximum change in phase shift is $180°$, and the Nyquist plot must pass through the unit circle in Fig. 16-2 to approach zero at high frequencies along the positive real axis. If the slope of the final asymptote is -18 dB/octave, then the Nyquist plot must approach zero at high frequencies along the negative imaginary axis. However, it is shown in Sec. 16-3 that, if self-sustaining oscillations are to be avoided, the Nyquist plot must cross the positive real axis on the left of the point $1 + j0$; hence again the Nyquist plot must pass through the region of positive feedback. This last restriction also applies to final slopes greater than -18 dB/octave. Thus it follows that, if the slope of the final asymptote for A_L is greater than -6 dB/octave, the Nyquist diagram must pass through the region of positive feedback.

   Figure 16-3 shows an alternative way of presenting part of the information contained in Fig. 16-2. These curves are based on Eq. (16-3) which gives the closed-loop gain in decibels as

$$|A_v|_{\mathrm{dB}} = 20 \log |\alpha A_v'| - 20 \log |F|$$
$$= 20 \log |\alpha A_v'| + F_{\mathrm{dB}} \qquad (16\text{-}6)$$

With α approximately equal to unity, the first term in this expression is the open-loop gain in decibels; the gain characteristic for this term is shown in Fig. 16-3. At low frequencies the return difference $F = 1 - \beta A_v'$ is constant, and the closed-loop gain is constant, as shown in Fig. 16-3. At high frequencies A_v' tends to zero, and the return difference tends to unity; thus at high frequencies the closed-loop gain tends to

$$|A_v|_{\mathrm{dB}} = 20 \log |\alpha A_v'| = |\alpha A_v'|_{\mathrm{dB}} \qquad (16\text{-}7)$$

That is, at high frequencies the closed-loop gain approaches the open-loop gain, as shown in Fig. 16-3. The transition of the closed-loop gain from its low-frequency to its high-frequency asymptote is simple only when the slope of the final asymptote for A_v' is -6 dB/octave; in this case the transition has the form illustrated in Fig. 14-5. When the slope of the final asymptote is greater than -6 dB/octave, the details of the transition cannot be determined without considerably more work; hence it is shown as a dashed line in Fig. 16-3. However, in accordance with the discussion of Fig. 16-2, when the slope of the final asymptote is greater than -6 dB/octave, there must be a band of frequencies in which the feedback is positive and in which the closed-loop gain characteristic lies above the open-loop characteristic as shown in Fig. 16-3. The frequency f_a, shown in Fig. 16-2 as well as in Fig. 16-3, is the frequency at which the feedback changes from negative to positive.

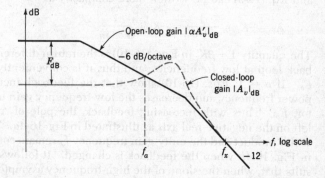

Fig. 16-3 Frequency characteristics for the feedback amplifier of Fig. 16-1.

16-2 THE EFFECT OF FEEDBACK ON THE DYNAMIC RESPONSE

The frequency dependence of the loop gain and some of its consequences are discussed in Sec. 16-1. From this discussion it is possible
to draw some conclusions about the closed-loop dynamic response of the
feedback amplifier. For example, Fig. 16-3 shows the low-frequency
and high-frequency asymptotes for the closed-loop gain characteristic, and it suggests how the transition from one asymptote to the other
might occur. Complete details of the effect of feedback on the dynamic response can be developed in terms of the pole-zero diagram for
the closed-loop complex gain, and a few specific examples are presented
in this section.

For the first example consider the case in which the complex gain
of the internal amplifier in the circuit of Fig. 16-1 is characterized by
a single pole. In this case the open-loop gain can be expressed as

$$A_v' = -K_v'\omega_H' \frac{1}{s + \omega_H'} = -K_v'\omega_H' \frac{1}{s - s_H'} \tag{16-8}$$

where K_v' is the voltage gain at low frequencies and ω_H' is half-power
frequency of the amplifier. [See Eq. (14-45) for an example of a complex gain having this form.] Substituting this expression into Eq.
(16-1) to obtain the closed-loop gain yields

$$A_v = -\alpha K_v'\omega_H' \frac{1}{s + (1 + \beta K_v')\omega_H'} \tag{16-9}$$

Thus the closed-loop half-power frequency and low-frequency gain are

$$\omega_H = (1 + \beta K_v')\omega_H' \quad \text{and} \quad K_v = \frac{\alpha K_v'}{1 + \beta K_v'} \tag{16-10}$$

and Eq. (16-9) can be written more compactly as

$$A_v = -K_v\omega_H \frac{1}{s + \omega_H} = -K_v\omega_H \frac{1}{s - s_H} \tag{16-11}$$

The quantity $1 + \beta K_v'$ in Eqs. (16-10) is the return difference of the feedback loop at low frequencies, and thus it is conveniently designated as
F_o. If $\alpha \approx 1$, Eqs. (16-10) show that the feedback increases the half-
power frequency and decreases the low-frequency gain by the same factor, F_o. Thus with increasing feedback, the pole of A_v moves to the
left on the negative real axis as illustrated in Fig. 16-4a. The gain-bandwidth product and the unity-gain frequency f_x remain constant as shown
in Fig. 16-4b when the feedback is changed. It follows from these results that, when the slope of the high-frequency asymptote is −6 dB/oc-
tave, the transition of the closed-loop gain characteristic from one asymptote to the other has the simple form illustrated in Fig. 14-5.

The pole-zero pattern of Fig. 16-4 also shows the effect of feedback on the transient response of the amplifier. Conventional circuit theory shows that, when A_v has one pole and no zeros, as in Eq. (16-11) and Fig. 16-4a, the unit-step response of the amplifier is

$$v_o(t) = -K_v + c_1 \exp (s_H t) = -K_v + c_1 \exp (-\omega_H t) \qquad (16\text{-}12)$$

where c_1 is a constant depending on the initial conditions existing when the step is applied. If $v_o(0^+) = 0$, then $c_1 = K_v$. The time constant of the exponential term in the step response is $1/\omega_H$, and since ω_H varies inversely with the amount of feedback F_o, it follows that negative feedback reduces the time constant and speeds up the step response of the amplifier.

If the circuit is arranged so that there is no sign reversal in the internal amplifier at low frequencies, then the feedback is positive, and different results are obtained. In this case $\omega_H = (1 - \beta K'_v)\omega'_H$, and the pole of A_v moves toward the origin of the complex plane as β is increased from zero. Thus positive feedback increases the time constant of the amplifier and reduces the speed of response to a step in the input voltage. If β is made large enough, ω_H becomes negative, and the pole of A_v moves into the right half of the complex plane. Under this condition Eq. (16-12) shows that the output voltage contains a growing exponential component. Poles in the right half of the complex plane always give rise to growing transients, and circuits having such transients are said to be unstable. These transients continue to grow in amplitude until the circuit becomes nonlinear because of saturation or cutoff of a transistor in the amplifier, and as a result the circuit usually cannot function properly as an amplifier. Thus instability must normally be avoided in amplifiers.

If the complex gain of the internal amplifier in the circuit of Fig. 16-1 is characterized by two poles and if the amplifier is arranged so that there is a sign reversal at low frequencies, then the open-loop gain can

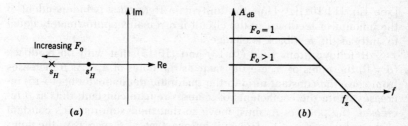

Fig. 16-4 The effect of feedback on the pole-zero pattern of a complex voltage gain having one pole and no zeros. (a) Pole diagram; (b) gain characteristics.

be expressed as

$$A_v' = \frac{-K_v'\omega_1'\omega_2'}{(s + \omega_1')(s + \omega_2')} = \frac{-K_v'\omega_1'\omega_2'}{s^2 + (\omega_1' + \omega_2')s + \omega_1'\omega_2'} \tag{16-13}$$

$$= \frac{-K_v'\omega_1'\omega_2'}{(s - s_1')(s - s_2')} \tag{16-14}$$

[See Eq. (14-99) for an example of a complex gain having this form.] Substituting this expression into Eq. (16-1) to obtain the closed-loop gain yields

$$A_v = -\alpha K_v'\omega_1'\omega_2' \frac{1}{s^2 + (\omega_1' + \omega_2')s + (1 + \beta K_v')\omega_1'\omega_2'} \tag{16-15}$$

Again, $1 + \beta K_v' = F_o$ is the return difference of the feedback loop at low frequencies. The closed-loop gain at low frequencies is

$$K_v = \frac{-\alpha K_v'}{1 + \beta K_v'}$$

and it is also convenient to define

$$\omega_o{}^2 = (1 + \beta K_v')\omega_1'\omega_2'$$

so that (16-15) can be written more compactly as

$$A_v = -K_v\omega_o{}^2 \frac{1}{s^2 + (\omega_1' + \omega_2')s + \omega_o{}^2} \tag{16-16}$$

$$= -K_v\omega_o{}^2 \frac{1}{(s - s_1)(s - s_2)} \tag{16-17}$$

The slope of the final high-frequency asymptote in the gain characteristic for A_v is -12 dB/octave, and it crosses the unity gain (0 dB) axis at the frequency

$$\omega_x = \sqrt{K_v\omega_o{}^2} = \sqrt{\alpha K_v'\omega_1'\omega_2'} \tag{16-18}$$

[see Eq. (14-101)]. Thus the unity-gain frequency is independent of the amount of feedback in the circuit if α remains approximately equal to unity as the feedback is changed.

It follows from Eqs. (16-13) and (16-15) that with no feedback ($\beta = 0$) the poles of A_v are the same as the poles of A_v'. As β increases from zero, the constant term in the quadratic denominator of (16-15) increases while the coefficients of s^2 and s remain constant; thus as β increases, the poles of A_v must move so that their sum remains constant at the value $-(\omega_1' + \omega_2')$. Hence it follows that, as β increases, the poles must move toward each other along the negative real axis as shown in Fig. 16-5 until they meet at $-\frac{1}{2}(\omega_1' + \omega_2')$; they meet when ω_0 in Eq. (16-16)

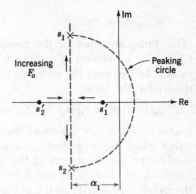

Fig. 16-5 The effect of feedback on a complex gain having two poles and no zeros.

equals $\frac{1}{2}(\omega_1' + \omega_2')$. When the feedback makes ω_0 greater than this value, the poles become complex and move with constant real parts along the vertical line shown in Fig. 16-5. The distance by which the complex poles are separated increases with increasing feedback.

Equation (16-17) has the same form as Eq. (15-62); hence all of the results derived in connection with Eq. (15-62) can also be applied to the two-pole feedback amplifier. In particular, when the poles are complex, important details about the gain characteristic and its dependence on the amount of feedback can be obtained from the peaking circle developed in Sec. 15-3; this peaking circle is shown in Fig. 16-5. If the peaking circle intersects the imaginary axis, the gain characteristic has a maximum at the frequency corresponding to the point of intersection (see Fig. 16-3). The frequency and the height of the peak can be determined easily from the relations developed in Sec. 15-3. The amount of feedback that makes the circle tangent to the imaginary axis is the greatest amount that can be used without a peak appearing in the gain characteristic; with this amount of feedback the gain characteristic is maximally flat. Feedback is often used with amplifiers of this type to increase the half-power bandwidth and at the same time to achieve a maximally flat gain characteristic; the cost of this improved dynamic response is a reduced gain at low frequencies.

The diagram of Fig. 16-5 also shows the effect of the feedback on the transient response of the amplifier. When the poles of A_v are real, conventional circuit theory gives the step response of the circuit as a constant plus two exponential terms, and the time constants of the exponential terms depend on the locations of the poles of A_v. When the feedback is adjusted so that the poles of A_v are complex, the poles can be expressed as

$$s_1, s_2 = -\alpha_1 \pm j\omega_1$$

(The imaginary part of the complex pole, which is written as β in Chap. 15, is written as ω here to avoid confusion with the feedback ratio.) In this case the unit-step response given by conventional circuit theory has the form

$$v_o(t) = -K_v + c_1[\exp(-\alpha_1 t)][\cos(\omega_1 t + \theta_1)] \tag{16-19}$$

where c_1 and θ_1 are constants that depend on the initial conditions existing when the step signal is applied. Thus the transient response is oscillatory; the frequency of the oscillation depends on the amount of feedback, but the rate at which the transient decays is fixed by ω_1' and ω_2', the break frequencies in the open-loop gain characteristic.

If the internal amplifier is arranged so that there is no sign reversal in A_v' at low frequencies, the minus sign disappears from the numerator of Eq. (16-13), and $\omega_o{}^2$ in (16-16) becomes $(1 - \beta K_v')\omega_1'\omega_2'$. In this case $\omega_o{}^2$, which is also the product of the closed-loop poles $s_1 s_2$, decreases with increasing feedback, and the poles of A_v must move apart on the negative real axis. When $\beta K_v' = 1$, $\omega_o = 0$, and one of the poles must lie at the origin of the complex plane. If the feedback is increased further, $\omega_o{}^2$ becomes negative, the pole at the origin moves out on the positive real axis, and the circuit has a growing transient. Thus when A_v' has no sign reversal at low frequencies, the circuit becomes unstable if the low-frequency loop gain is greater than unity. However, when there is a sign reversal in the amplifier, the circuit can never become unstable.

If the complex gain of the internal amplifier in the circuit of Fig. 16-1 is characterized by three identical poles, then the open-loop gain can be expressed as

$$A_v' = -K_v'\omega_1'^3 \frac{1}{(s + \omega_1')^3} = -K_v'\omega_1'^3 \frac{1}{(s - s_1')^3} \tag{16-20}$$

Substituting this expression into Eq. (16-1) to obtain the closed-loop gain yields

$$A_v = -\alpha K_v'\omega_1'^3 \frac{1}{(s - s_1')^3 + \beta K_v'\omega_1'^3} \tag{16-21}$$

$$= -\alpha K_v'\omega_1'^3 \frac{1}{(s - s_1)(s - s_2)(s - s_3)} \tag{16-22}$$

The paths along which the poles move as the amount of feedback is increased can be found with the aid of the diagram shown in Fig. 16-6a. The poles of A_v are the values of s that make the denominator in Eq. (16-21) zero; hence if $s = s_1$ is a pole of A_v, then substituting s_1 for s in (16-21) must yield

$$(s_1 - s_1')^3 + \beta K_v' \omega_1'^3 = 0 \tag{16-23}$$

The complex number $s_1 - s_1'$, shown as a vector in Fig. 16-6a, can be expressed in polar form as $s_1 - s_1' = \rho e^{j\theta}$, where ρ is the length of the vector and θ is the angle that it makes with the real axis. Thus Eq. (16-23) can be rewritten as

$$(s_1 - s_1')^3 = (\rho e^{j\theta})^3 = \rho^3 e^{j3\theta} = -\beta K_v' \omega_1'^3 \tag{16-24}$$

Since the left-hand side of this equation is equal to a negative real number, it follows that 3θ must be an odd multiple of 180°; the values of θ satisfying this requirement are 60, 180, and −60°. Thus as the feedback is increased, the poles move along the paths shown in Fig. 16-6b.

It follows from these results that, when this amplifier has any feedback at all, the closed-loop gain has a pair of complex poles. These complex poles can be expressed as

$$s_1, s_2 = \alpha_1 \pm j\omega_1 \tag{16-25}$$

where α_1 can be either positive or negative depending on the amount of feedback. Thus the unit-step response of this amplifier has the form

$$v_o(t) = -K_v + c_1(\exp s_3 t) + c_2(\exp \alpha_1 t)[\cos (\omega_1 t + \theta_2)] \tag{16-26}$$

where c_1, c_2, and θ_2 are constants depending on the initial conditions existing when the step is applied. It is clear from the diagram in Fig. 16-6b that α_1 is positive when the amount of feedback in the circuit is great enough, and Eq. (16-26) shows that under this condition the transient response of the amplifier contains a growing oscillatory component.

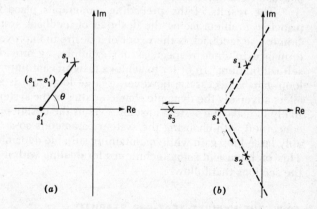

Fig. 16-6 The effect of feedback on a complex gain having three poles and no zeros. (a) Diagram for graphical analysis; (b) the loci of the poles.

That is, when the amount of feedback is great enough, the amplifier acts as an oscillator. When the feedback is adjusted so that the complex poles lie on the imaginary axis, the circuit is said to be on the threshold of instability. The amount of feedback giving the threshold condition can be determined from Fig. 16-6b. Since the paths along which s_1 and s_2 move as the feedback is changed make angles of $60°$ with the real axis, the length of the vector $s_1 - s_1'$ is $\rho = 2|s_1'| = 2\omega_1'$ when the poles are on the imaginary axis. Then it follows from Eq. (16-24) that

$$(2\omega_1')^3 e^{j\pi} = -\beta K_v' \omega_1'^3$$

or

$$\beta K_v' = 8 \qquad\qquad\qquad (16\text{-}27)$$

under the threshold condition. This is not very much feedback.

If the three poles in the open-loop gain function are not identical, the paths that they follow as the amount of feedback is changed are more difficult to construct, for they are not straight lines. However, when the amount of feedback is made large, the poles approach paths similar to the ones shown in Fig. 16-6b as asymptotes. The loci for any number of identical poles can be constructed by the simple method employed above. In particular, if the circuit is always arranged so that there is a sign reversal in the loop transmittance at low frequencies and if the number of identical poles is N, then the loci are straight lines making angles of $n180/N$ degrees with the real axis, where n is an odd integer. The one-pole and two-pole amplifiers treated earlier in this section are particular examples that follow this rule.

The results of the preceding paragraphs place in evidence the principal problem facing the designer of feedback systems. The decision to use feedback is the result of a desire to improve the system performance in some respect, as, for example, by providing a degree of self-calibration. In order to achieve a significant improvement, a large loop gain is necessary; however, a large loop gain is almost certain to affect adversely the dynamic characteristics of the system, and it is quite likely to make the system unstable. Thus the design problem is largely concerned with choosing the system parameters so as to permit a suitably large loop gain while maintaining suitable dynamic characteristics. This problem and some techniques for dealing with it are discussed in the sections that follow.

16-3   THE NYQUIST TEST FOR STABILITY

Circuits composed of controlled sources and positive R's, L's, and C's cannot be unstable if they have no feedback, for such circuits have no sources capable of supplying the energy associated with growing transients. The results of Sec. 16-2 show, however, that when feedback is added to such circuits, instability may develop. In this case the controlled sources enclosed in the feedback loop are capable of supplying the required energy. It can be shown that, if the number of poles of the complex loop gain exceeds the number of zeros by three or more, the circuit is certain to be unstable if the loop gain is made great enough. When the complex loop gain has only two or three poles, the paths along which the closed-loop poles move as the amount of feedback is varied can be determined, and questions related to stability can be answered from the results obtained. The way in which the closed-loop poles move in more complicated circumstances is more difficult to determine, although techniques for constructing the approximate paths of their motion have been developed in considerable detail. However, there are other approaches to the problem that prove to be useful in the analysis and design of more complicated feedback circuits.

Determining whether or not a given circuit is stable involves determining in some manner whether or not the complex closed-loop gain has poles in the right half of the complex plane. The closed-loop gain can be expressed as the ratio of two polynomials in s, and the zeros of the denominator polynomial are the poles of the closed-loop gain. Factoring this polynomial becomes laborious when its degree is greater than three. However, the existence of zeros of the polynomial in the right half of the complex plane can be detected by an examination of its coefficients without actually determining the values of the zeros. For example, it follows directly from the relations between the zeros and the coefficients that, if there is any variation in sign among the coefficients, then there must be at least one zero of the polynomial in the right half of the complex plane. In addition, if the coefficient of any power of s less than the greatest in the polynomial is zero, then there must be at least one zero of the polynomial in the right half plane or on the imaginary axis. However, the fact that all the coefficients are nonzero and of the same sign does not ensure that there are no zeros in the right half plane. The Routh-Hurwitz test is a more detailed examination of the coefficients that discloses the number of right-half-plane zeros of any polynomial. These tests, which are simple to apply, determine whether or not a proposed circuit will be stable. Unfortunately, however, they provide little guidance in the design of feedback amplifiers for stable operation. The Nyquist test for stability, which is developed in the following paragraphs, has a number of features that result in its being especially useful in the design of feedback systems. This test not only shows whether or not the system will be stable but also presents the information in such a way as to aid the designer in arriving at a suitable design. In addition, it permits experimental data to be used in the design of systems that are too complicated for complete analysis.

The complex closed-loop gain for the feedback amplifier of Fig. 16-1 is given by Eq. (16-1) as

$$A_v = \frac{\alpha A_v'}{1 - \beta A_v'} = \frac{\alpha A_v'}{1 - A_L} = \frac{\alpha A_v'}{F} \qquad (16\text{-}28)$$

It is assumed that the quantities α, A_v', and $A_L = \beta A_v'$ are stable quantities and thus that they have no poles in the right half of the complex plane. Since the return difference $F = 1 - A_L$ has the same poles as A_L, it follows that F has no poles in the right half plane. Therefore, if A_v has any poles in the right half plane, they must be right-half-plane zeros of the return difference F, and the stability of the amplifier can be examined by examining the return difference. The Nyquist test for stability is a test for right-half-plane zeros of F.

For concreteness, consider the return difference

$$F = \frac{s - s_1}{s - s_2} = \frac{\rho_1}{\rho_2} \exp j(\theta_1 - \theta_2) \qquad (16\text{-}29)$$

where ρ_1 and ρ_2 are the magnitudes and θ_1 and θ_2 are the angles of the numerator and denominator. The pole-zero pattern for F under one set of conditions is shown in Fig. 16-7a. For any given value of the variable s, a corresponding value of the return difference F can be calculated from Eq. (16-29). If s is given a succession of values corresponding to the movement of the point s around the contour C_1 in Fig. 16-7a, F takes on a succession of values given by Eq. (16-29). A plot of these values of F in the complex plane forms a contour such as C_2 shown in Fig. 16-7b; this contour provides the desired information about the location of the zeros of F. As the variable s makes one complete circuit of the contour C_1, the angles θ_1, θ_2, and $\theta_1 - \theta_2$ go through some variations but finally return to their initial values; there is no net change in these angles. It follows that the angle $\theta_1 - \theta_2$ in Fig. 16-7b experiences no net change as F traces out the contour C_2 and that therefore C_2 does not encircle the origin of the complex plane. This result stems directly from the fact that s_1 and s_2 lie outside the contour C_1. If s_1 lies inside the contour, as shown in Fig. 16-7c, θ_1 changes by 360° in the clockwise direction, and θ_2 experiences no net change as the variable s makes one complete clockwise circuit of C_1. Hence the angle $\theta_1 - \theta_2$ changes by 360°, and the contour C_2 encircles the origin one time in the *clockwise* direction, as shown in Fig. 16-7d. If s_2 is inside the contour C_1, as illustrated in Fig. 16-7e, θ_1 experiences no net change, and θ_2 changes by 360° in the clockwise direction as s makes one complete circuit of C_1 in the clockwise direction. Therefore in this case the angle $\theta_1 - \theta_2$ changes by 360°, and C_2 encircles the origin one time in the *counterclockwise* direction, as shown in Fig. 16-7f. It also follows from this same reasoning that, if both s_1 and s_2 are inside C_1, C_2 does not encircle the origin.

The discussion of the preceding paragraph shows that the number of times that the contour C_2 encircles the origin of the complex plane in the clockwise direction is equal to the excess of zeros of the return difference over poles inside contour C_1. Thus if Z is the number of

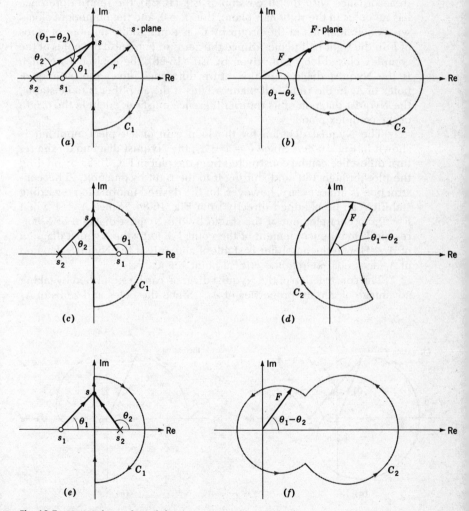

Fig. 16-7 Nyquist's test for stability.

zeros of F inside C_1 and if P is the number of poles inside C_1, then the number of times that C_2 encircles the origin in the clockwise direction is

$$N_{cw} = Z - P \tag{16-30}$$

Now if the radius of the circular portion of C_1 is increased without limit so that C_1 encloses the entire right half of the complex plane, then (16-30) gives the excess of zeros over poles in the entire right half plane. But, in accordance with the discussion of Eq. (16-28), the return difference has no poles in the right half plane; thus $P = 0$, and the number of clockwise encirclements of the origin by C_2 is equal to the number of zeros of F in the right half plane. Since the zeros of F are also the poles of the complex closed-loop gain given by Eq. (16-28), the contour C_2, which is the Nyquist diagram of the return difference, gives the number of poles of A_v in the right half plane. Thus if the amplifier is to be stable, the Nyquist diagram of its return difference must not encircle the origin of the complex plane.

The Nyquist diagram for the loop gain of a typical amplifier is shown in Fig. 16-8a. Since $F = 1 - A_L$, the Nyquist diagram for the return difference can be constructed from the plot in Fig. 16-8a by rotating the plot through 180° and shifting it to the right by one unit. This construction is unnecessary, however, for the desired information regarding stability can be obtained directly from Fig. 16-8a. Since $A_L = 1$ when $F = 0$, an encirclement of the origin by the Nyquist diagram of F corresponds to an encirclement of the point $1 + j0$ by the Nyquist diagram of A_L. Hence the amplifier is stable if and only if the Nyquist diagram of A_L does not encircle the critical point $1 + j0$.

The construction of the Nyquist diagram can be simplified by taking advantage of certain properties of A_L. Since the poles and zeros of A_L

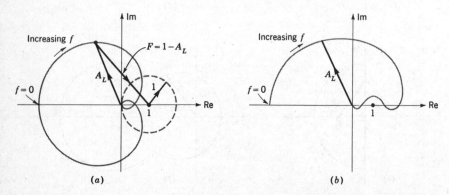

Fig. 16-8 Nyquist diagrams. (a) Stable circuit; (b) conditionally stable circuit.

either are real or occur in conjugate pairs, it follows directly from the pole-zero pattern that $A_L(-j\omega)$ is the conjugate of $A_L(j\omega)$. Hence that portion of the Nyquist plot corresponding to values of s on the negative imaginary axis need not be calculated; it is the mirror image about the real axis of the portion corresponding to values of s on the positive imaginary axis. In addition, because of parasitic elements such as stray capacitance, the loop gain of any physical amplifier must tend to zero as s tends to infinity. Thus in constructing the Nyquist diagram for A_L it is sufficient to consider only values of s on the positive imaginary axis up to a point beyond which $|A_L|$ remains less than unity; since the remainder of the diagram cannot encircle the critical point at $1 + j0$, it can be ignored.

Another important aspect of the Nyquist diagram of the loop gain is the fact that the portion of this diagram corresponding to $s = j\omega$ also corresponds to steady-state sinusoidal operation of the amplifier, and hence it can be measured experimentally by conventional techniques. This fact, coupled with the statements made in the preceding paragraph, means that all of the useful portion of the Nyquist diagram can be determined by experimental measurements. Thus when the circuit is too complicated for a complete analysis, the necessary information can be determined experimentally.

The Nyquist diagram for the loop gain in the circuit of Fig. 16-1 when the internal amplifier is characterized by three identical poles and a sign reversal at low frequencies is shown in Fig. 16-8a. The amplifier is stable under the conditions pictured, and the portion of the curve lying in the second quadrant near the negative real axis corresponds to normal operating frequencies for the amplifier. If the amount of feedback is increased uniformly at all frequencies by increasing the feedback ratio β, then the Nyquist diagram expands without changing its shape. It is clear that if β, and hence $|A_L|$, is increased sufficiently, the critical point is encircled and the amplifier becomes unstable.

If the sign reversal in the internal amplifier at low frequencies is removed by some change in the circuit, then the Nyquist diagram is rotated through 180°, and if $|A_L|$ is greater than unity at zero frequency, the critical point is encircled and the circuit is unstable. In general, if the loop gain does not have a sign reversal at low frequencies, the Nyquist diagram lies primarily in the right half of the complex plane, and the amplifier is unstable if $|A_L|$ is greater than unity at zero frequency.

If the value of β is adjusted so that the Nyquist diagram passes through the critical point, the circuit is at the threshold of instability. This is also the adjustment that places the complex poles in Fig. 16-6b on the imaginary axis.

A special case of interest and importance is illustrated by the Nyquist diagram shown in Fig. 16-8b. For simplicity, only that portion of the diagram corresponding to positive ω is shown; the remainder of the diagram, which is the mirror image about the real axis of the portion shown, can be visualized without difficulty. The critical point is not encircled by this Nyquist diagram, and hence the circuit is stable. However, if the plot is caused to shrink uniformly at all frequencies, a point is reached at which the circuit becomes unstable. If the plot is shrunk still further, another point is reached at which the circuit becomes stable again. Circuits exhibiting such a phenomenon are termed *conditionally stable*.

Conditional stability, which is likely to occur in amplifiers using large amounts of feedback, is important for the following reason. If an excessive signal is applied momentarily to the amplifier, the amplifier tends to saturate, and its effective gain is reduced. If the amplifier is conditionally stable, the momentary reduction in gain may make it unstable. In this case growing oscillations appear, and the amplifier may remain saturated (and thus unstable) as a result of its own oscillations, even after the excessive input signal is removed. When an amplifier is first turned on, the loop gain must grow from zero to its normal operating value as the bias currents in the transistors build up to their normal values. If the Nyquist diagram of the loop gain has the form shown in Fig. 16-8b, the amplifier may have to pass through the unstable condition as the bias currents build up. In such a case oscillations will start, and they may grow to such a magnitude that they saturate the amplifier and prevent the final stable operating conditions from being reached. For these reasons it is usually desirable to avoid conditional stability in feedback amplifiers.

The construction of the Nyquist diagram by computing the loop gain and phase shift for a number of different frequencies is likely to be a tedious procedure. The amount of time and effort required can often be reduced, however, by first constructing the logarithmic gain and phase characteristics for the complex loop gain by the rapid techniques developed in Chap. 14; the loop gain and phase shift can then be read from these curves at various frequencies. But since the gain and phase characteristics contain all the information that the Nyquist diagram contains, there is no need to construct the latter; it is merely necessary to interpret the stability criterion and the feedback relations in terms of the gain and phase characteristics.

The loop gain and phase characteristics for a typical feedback amplifier are shown in Fig. 16-9; it is understood that there is a sign reversal in the loop gain in addition to the phase shift shown by the phase characteristic. It follows directly from the relationship between the char-

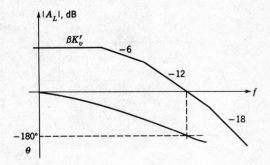

Fig. 16-9 Loop gain and phase characteristics for a feedback amplifier.

acteristics of Fig. 16-9 and the Nyquist plot of A_L that the Nyquist plot does not encircle the critical point if $|A_L|$ drops to 0 dB before the phase shift reaches 180°, for 0 dB corresponds to a numerical ratio of unity. Therefore, except for the case of conditional stability, this is the stability criterion in terms of the loop gain and phase characteristics. It follows that the amplifier having the characteristics shown in Fig. 16-9 is on the threshold of instability.

16-4 THE DESIGN OF OPERATIONAL AMPLIFIERS WITH RESISTIVE FEEDBACK

The small-signal model for an operational amplifier with resistive feedback is shown in Fig. 16-10; this amplifier has the same form as the one shown in Fig. 16-1. The usual practice is to use a commercially available high-gain transistor amplifier as the internal amplifier and to apply the appropriate feedback through an external feedback network. The internal amplifier may be a miniaturized amplifier using discrete components, or it may be a monolithic microamplifier. The open-loop gain and phase characteristics for these amplifiers are determined by experimental measurement, and typical curves are often given by the manufacturer.

The general design problem can be outlined in terms of the loop gain and phase characteristics shown in Fig. 16-9. When feedback is used to improve the performance of the amplifier by providing self-calibration or by reducing distortion, for example, it is first of all necessary to ensure that the loop gain is sufficiently large over the band of frequencies occupied by the signal or by the distortion; this matter is treated in Sec. 13-5. Then it is necessary to control the cutoff character-

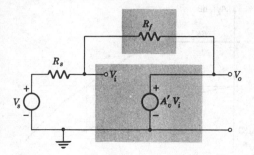

Fig. 16-10 Small-signal model for an operational amplifier.

istic of the feedback loop outside the signal band so that the gain is reduced to 0 dB before the phase shift becomes 180°. The shape of the cutoff characteristic is controlled by RC networks added to the basic amplifier configuration. In shaping the cutoff characteristic some rather severe limitations are encountered, however, because the gain and phase characteristics are interrelated. Any change in the shape of the gain characteristic is usually accompanied by a change in the shape of the phase characteristic, and this latter change may make matters worse rather than better. Certain simple rules of thumb for the guidance of the designer are developed in the following paragraphs; throughout this discussion it is assumed that the loop gain has a sign reversal at low frequencies.

With resistive feedback, the feedback ratio β is independent of frequency, and hence changing β raises or lowers the loop gain characteristic of Fig. 16-9 without changing its shape and without changing the shape of the phase characteristic. Thus the easiest way to ensure the stability of the feedback amplifier is to adjust β so that the loop gain drops to 0 dB before the phase shift reaches 180°. However, this procedure usually results in a very small loop gain, and hence it seldom produces a satisfactory design.

If the complex gain of the internal amplifier in the circuit of Fig. 16-10 is characterized by a single pole, then the slope of the final high-frequency asymptote is −6 dB/octave, and the phase shift approaches a maximum value of 90° at high frequencies. Thus the amplifier can never be unstable, a conclusion that is in agreement with the relations illustrated in Fig. 16-4. However, most feedback amplifiers are characterized by more than one pole.

If the complex gain of the internal amplifier is characterized by two poles, then the slope of the final high-frequency asymptote is −12 dB/octave, and the phase shift approaches a maximum value of 180° at high frequencies. Thus this amplifier can never be unstable, although it may have an undesirable resonant peak in its closed-loop gain characteristic. This conclusion is in agreement with the relations pictured in Fig. 16-5. However, many feedback amplifiers are characterized by more than two poles.

In general, if the final high-frequency asymptote for the gain characteristic has a slope of $-6N$ dB/octave, then the phase shift approaches $90N$ degrees at high frequencies. Thus if N is greater than 2, the phase shift exceeds 180° at high frequencies, and the amplifier is likely to be unstable when the feedback loop is closed. This conclusion is in agreement with the relations shown in Fig. 16-6.

It follows from the foregoing discussion that the stability of the feedback amplifier can be ensured by limiting the cutoff rate for the loop gain to a value somewhat less than -12 dB/octave. Accordingly, it is standard engineering practice in the design of transistor feedback amplifiers to shape the loop gain characteristic so that the cutoff occurs at a rate of -6 dB/octave. This shaping is accomplished by adding RC networks to the basic amplifier configuration, and the process is called compensation. Figure 16-11 shows a typical loop gain characteristic before and after compensation. The -6 dB/octave slope cannot be maintained to arbitrarily high frequencies, for, although it is not obvious, the high-frequency end of this asymptote can be extended only by reducing the low-frequency gain K_v' of the internal amplifier. However, if the -6 dB/octave slope is maintained up to the unity-gain frequency and for some distance beyond, the phase shift at the unity-gain frequency does not exceed 90° by very much, and the feedback amplifier is stable with an adequate margin of safety.

The types of RC networks used to compensate the internal amplifier vary according to the nature of the internal amplifier. Some commercial amplifiers made of discrete components have the compensating

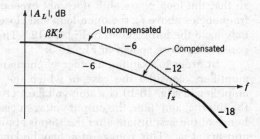

Fig. 16-11   Loop gain characteristics.

networks included as an integral part of the amplifier; others provide terminals to which the user must add the necessary compensation. When external compensation is to be added, the manufacturer often specifies compensating networks that are appropriate for various operating conditions.

If the feedback is to be effective in improving the performance of the amplifier, then the feedback, and thus the loop gain, must be suitably large over the band of frequencies occupied by the signal. The half-power bandwidth f_L of the loop gain is a useful measure of the band of frequencies in which the feedback maintains nearly its full effectiveness. However, it follows from Fig. 16-11 that, if the unity-gain frequency f_x is fixed and if the loop cutoff rate is fixed at -6 dB/octave, then the amount of loop gain that can exist over any specified bandwidth f_L is also fixed. More specifically, with a cutoff rate of -6 dB/octave out to f_x, the loop gain–bandwidth product is

$$\beta K'_v f_L = f_x \qquad (16\text{-}31)$$

where f_x is the frequency at which the loop gain is unity (0 dB). Thus

$$\beta K'_v = \frac{f_x}{f_L} \qquad (16\text{-}32)$$

is the amount of loop gain that will exist in the specified bandwidth f_L. It follows that high-performance amplifiers must have a high unity-gain frequency along with high gain at low frequency. The unity-gain frequency is determined primarily by the internal amplifier, but it is also usually affected by the compensating networks used. The unity-gain frequency of commercial amplifiers intended for this application is often given on the data sheets for the amplifier. After the unity-gain frequency for the feedback loop has been fixed, the designer can only choose the best compromise between low-frequency loop gain and loop bandwidth.

When the cutoff rate for the loop gain is controlled at -6 dB/octave so that the loop phase shift does not exceed 90° appreciably except at frequencies above f_x, the open-loop and closed-loop gain characteristics may have the form shown in Fig. 16-12. These curves should be compared with those shown in Fig. 16-3.

In order to examine another phenomenon that arises in feedback amplifiers, consider the case in which the input voltage $v_s(t)$ for the amplifier in Fig. 16-10 is a step voltage. Owing to the capacitances in the internal amplifier, the output voltage v_o cannot change abruptly, and hence at the first instant after the step is applied, the full step of voltage appears at v_i. This voltage at v_i may be large enough to overdrive the amplifier and cause one or more transistors in the amplifier to saturate

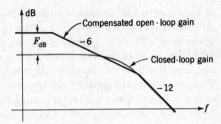

Fig. 16-12 Frequency characteristics for the compensated feedback amplifier.

or cut off. As v_o changes in response to the input, the feedback voltage subtracts from v_i, and when v_o has changed enough, the overdrive condition disappears. While the amplifier is in the overdriven state, the output voltage is changing at the maximum rate permitted by the capacitances in the internal amplifier, and the output is said to be rate limited. The rate limit, also called the slew rate, is sometimes written as

$$\rho = \left(\frac{dv_o}{dt}\right)_{\text{max}} \tag{16-33}$$

The slew rate for commercial amplifiers intended for use with feedback is often given on the data sheet for the amplifier.

Under sinusoidal operating conditions the output voltage is

$$v_o = V_o \cos \omega t$$

and the rate of change of this voltage at any instant is

$$\frac{dv_o}{dt} = -\omega V_o \sin \omega t$$

The maximum value of this rate of change is ωV_o, and when it exceeds the slew rate, the amplifier is overdriven and distortion results. Thus distortion sets in when

$$V_o = \frac{\rho}{\omega} = \frac{\rho}{2\pi f} \tag{16-34}$$

where ρ is the slew rate. Hence at high frequencies the amplitude of the sinusoidal signal must be limited to avoid distortion. (Note that, when the amplifier is overdriven with one or more transistors saturated or cut off, the loop gain is very small or even zero. Thus the feedback cannot reduce this distortion.) The distortion that results when this limit is exceeded causes the output waveform to take on a triangular appearance.

For sinusoidal operating conditions an alternative explanation for the slew-rate distortion can be given. With a sinusoidal signal applied at V_s, a sinusoidal output voltage appears at V_o. At low frequencies the sinusoidal feedback voltage subtracts from V_s, and V_i is a very small sinusoidal voltage. As the frequency of the input sinusoid is increased with the amplitude held constant, the output voltage tends to decrease because the gain of the internal amplifier falls off at high frequencies. The feedback voltage thus decreases, and V_i increases even though the input voltage V_s has a constant amplitude. The voltage V_i continues to increase in this way as the frequency is increased further, and finally V_i becomes large enough to overdrive the amplifier. The overdriven condition at high frequencies is avoided by reducing the amplitude of the input signal V_s.

16-5   A WIDEBAND BJT FEEDBACK AMPLIFIER

When large amounts of feedback are used in a feedback amplifier, there is a strong tendency for the feedback to degrade the dynamic perform-ance of the amplifier; this fact is developed in some detail in Secs. 16-2 and 16-4. However, when the proper amount of carefully controlled feedback is used, it can bring about a substantial improvement in the dynamic response, especially when the complex gain has no more than two poles. This improvement can be expressed as an increase in the bandwidth of the amplifier. Since many high-performance electronic systems require large bandwidths, this fact is important. Cathode-ray oscilloscopes for viewing signals with frequencies up the 20 MHz, for example, must have amplifiers with half-power bandwidths well in excess of 20 MHz. Some high-performance communication systems re-quire bandwidths of several hundred megahertz. Feedback is used ex-tensively in meeting the bandwidth requirements of these systems. The cost of achieving a large bandwidth with a particular transistor is always a reduced gain at low frequencies; feedback provides a con-venient way to exchange gain for bandwidth.

Figure 16-13 shows the circuit of a feedback amplifier that is widely used in wideband systems; since it uses two transistors, it is sometimes called a feedback pair. The basic amplifier for the feedback pair is the two-stage amplifier shown in Fig. 13-13. For simplicity, the biasing network is omitted in Fig. 16-13. Since the basic amplifier has two stages, there is no sign reversal between the input and the output; how-ever, for stable operation as a feedback amplifier, there must be a sign reversal in the feedback loop. This problem is solved by completing the feedback loop through the feedback impedance Z_f from the collector of Q_2 to the emitter of Q_1; with this arrangement, Q_1 does not introduce a sign reversal in the feedback loop, and hence the loop has a sign reversal as a result of the action of Q_2. The open-loop gain of this amplifier is the gain with Z_f replaced by an open circuit, and according to Eq. (14-

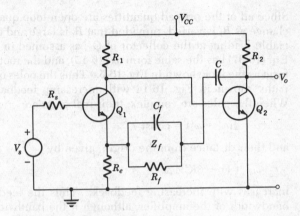

Fig. 16-13 Wideband BJT feedback pair.

99), it is a two-pole function having the form

$$A'_v = \frac{K'_v \omega'_1 \omega'_2}{(s + \omega'_1)(s + \omega'_2)} \qquad (16\text{-}35)$$

The addition of R_e to the basic amplifier as shown in Fig. 16-13 affects the values of K'_v, ω'_1, and ω'_2, but it does not alter the two-pole nature of the open-loop gain. In principle, the feedback through Z_f cannot make this amplifier unstable, although it can produce an undesirable resonant peak when the loop gain is high. In practice, however, parasitic elements not included in the analysis will cause instability when the loop gain is very high. In the usual case the break frequencies ω'_1 and ω'_2 are widely separated as shown in Fig. 14-25 and Example 14-3; thus K'_v, ω'_1, and ω'_2 are easily determined from the measured open-loop gain characteristic.

The complex closed-loop gain is derived in Appendix 3; the result obtained there is

$$A_v = \frac{K'_v \omega'_1 \omega'_2}{s^2 + (\omega'_1 + \omega'_2 + \beta K'_v \omega'_1 \omega'_2/\omega_f)s + (1 + \beta K'_v)\omega'_1 \omega'_2} \qquad (16\text{-}36)$$

where $\beta = R_e/R_f$ is feedback ratio at low frequencies
$\beta K'_v$ = loop gain at low frequencies
$\omega_f = 1/R_f C_f$

Consider first the case in which $C_f = 0$ and thus $\omega_f = \infty$. In this case Eq. (16-36) reduces to

$$A_v = \frac{K'_v \omega'_1 \omega'_2}{s^2 + (\omega'_1 + \omega'_2)s + (1 + \beta K'_v)\omega'_1 \omega'_2} \qquad (16\text{-}37)$$

Since all of the primed quantities are open-loop quantities, they do not change as R_f is varied, provided that R_f is large and does not add appreciable loading at the collector of Q_2, as assumed in Appendix 3. Thus Eq. (16-37) has the same form as (16-15), and the root loci have the same form as the loci shown in Fig. 16-5. Thus the poles of A_v move along the paths shown in Fig. 16-14 with increasing feedback (decreasing R_f). When the poles are complex, their real parts are

$$\alpha = \tfrac{1}{2}(\omega_1' + \omega_2') = \text{const} \tag{16-38}$$

and their distance from the origin, given by

$$\omega_o{}^2 = (1 + \beta K_v')\omega_1'\omega_2' \tag{16-39}$$

increases with increasing feedback. Thus the feedback increases the bandwidth of the amplifier, although if the bandwidth is increased too much, a resonant peak appears in the gain characteristic. When the feedback is adjusted so that the angle δ shown in Fig. 16-14 is 45°, the gain characteristic is maximally flat, as shown in Sec. 15-3, and the half-power bandwidth is $f_o = \omega_o/2\pi$. When the feedback is adjusted so that $\delta = 30°$, the phase characteristic is highly linear at frequencies up to about f_o. Thus with δ between 30 and 45°, f_o is a useful measure of the bandwidth of the amplifier, although it is not the half-power bandwidth except when $\delta = 45°$. When δ is greater than 45°, the gain characteristic has the shape shown in Fig. 16-15, and the peak-to-valley ratio is given by

$$\frac{A_p}{K_v} = \frac{1}{\sin 2\delta} \tag{16-40}$$

The low-frequency closed-loop gain of the amplifier, obtained from

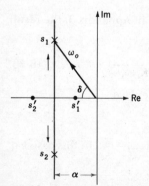

Fig. 16-14 Loci of the poles of A_v with increasing feedback and $C_f = 0$.

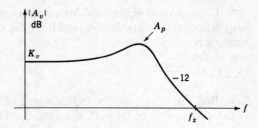

Fig. 16-15 Gain characteristic for the feedback pair with $\delta > 45°$.

(16-37) with $s = 0$, is

$$K_v = \frac{K_v'}{1 + \beta K_v'} \tag{16-41}$$

If the angle δ is too large when the feedback is adjusted to make f_o equal to the desired bandwidth, then the compensating capacitor C_f in Fig. 16-13 is brought into action. As C_f is increased from zero, the only quantity in the general expression for A_v, Eq. (16-36), that changes is ω_f. Hence as C_f is increased, the complex poles of A_v move with

$$\omega_o{}^2 = (1 + \beta K_v')\omega_1'\omega_2' = \text{const} \tag{16-42}$$

That is, the poles move along a circle centered at the origin of the complex plane, as shown in Fig. 16-16. Thus, having adjusted R_f to make f_o equal the desired bandwidth, C_f can be used to reduce the angle δ to a desired smaller value without changing f_o. In this way the amplifier can be adjusted for a desired bandwidth *and* a flat gain characteristic or a linear phase characteristic. Adding the capacitor C_f to the circuit does not alter the low-frequency closed-loop gain of the amplifier, and it remains at the value given by Eq. (16-41).

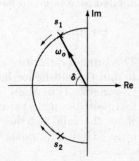

Fig. 16-16 Locus of the complex poles of A_v with increasing C_f.

The equation of the final high-frequency asymptote for the closed-loop gain characteristic is obtained from Eq. (16-36) with $s = j\omega$ and very large; it is

$$|A_v| = \frac{K_v' \omega_1' \omega_2'}{\omega^2} \tag{16-43}$$

Note that this asymptote is independent of R_f and C_f, at least so long as the assumption of large R_f is satisfied. The frequency at which the asymptote crosses the unity-gain (0 dB) axis is obtained from (16-43) by setting $|A_v| = 1$. Hence

$$1 = \frac{K_v' \omega_1' \omega_2'}{\omega_x{}^2}$$

or

$$f_x{}^2 = K_v' f_1' f_2' \tag{16-44}$$

Thus the unity-gain frequency f_x is also independent of R_f and C_f if the assumption about R_f is satisfied. Equation (16-44) can also be written as

$$f_x{}^2 = \frac{K_v'}{1 + \beta K_v'} (1 + \beta K_v') f_1' f_2'$$

Thus it follows from Eqs. (16-41) and (16-42) that

$$f_x{}^2 = K_v f_o{}^2 = K_v' f_1' f_2' \tag{16-45}$$

As stated above, when the angle δ is between 30 and 45°, f_o is a useful measure of the closed-loop bandwidth of the amplifier. Thus under these conditions the bandwidth is obtained from

$$f_o{}^2 = \frac{f_x{}^2}{K_v} = \frac{K_v'}{K_v} f_1' f_2' = (1 + \beta K_v') f_1' f_2' \tag{16-46}$$

It follows that the amount of low-frequency feedback (return difference) needed for any given bandwidth is

$$1 + \beta K_v' = \frac{f_o{}^2}{f_1' f_2'} \tag{16-47}$$

The primed quantities and f_x in these expressions are independent of R_f and C_f, and hence they remain constant as the amount of feedback is varied. Thus these expressions illustrate clearly the compromise that must be made between closed-loop gain and bandwidth, and they show the role that the unity-gain frequency f_x plays in the compromise. High-performance wideband amplifiers require large values for

f_x, and thus f_x is an important figure of merit for the internal amplifier.

When the open-loop quantities K'_v, f'_1, and f'_2 have been measured with a known value of R_e, the feedback amplifier can be designed for specified values of f_o and δ in the following way. The feedback ratio $\beta = R_e/R_f$ is obtained directly from Eq. (16-47), and with R_e known, R_f is found directly from the value of β. Thus the desired value of f_o is obtained.

It follows from the pole-zero diagram of Fig. 16-14 that

$$\alpha = \omega_o \cos \delta \tag{16-48}$$

and since the coefficient on s in the denominator of Eq. (16-36) is 2α, (16-48) can be written as

$$\frac{1}{2}\left(\omega'_1 + \omega'_2 + \frac{\beta K'_v \omega'_1 \omega'_2}{\omega_f}\right) = \omega_o \cos \delta \tag{16-49}$$

All quantities in this expression are known except ω_f; solving for ω_f and dividing through by 2π yields

$$f_f = \frac{\beta K'_v f'_1 f'_2}{2f_o \cos \delta - f'_1 - f'_2} \tag{16-50}$$

and finally,

$$C_f = \frac{1}{\omega_f R_f} \tag{16-51}$$

Since approximations are made in the derivation of Eq. (16-36), this design procedure does not give precisely the desired values of f_o and δ; however, it does give values that are close to the desired ones. When precision is required, it is common practice to trim the values of R_f and C_f experimentally to obtain the desired performance.

The input impedance of the amplifier in Fig. 16-13 is relatively high because the feedback voltage bootstraps the emitter of the input transistor. The feedback also makes the output impedance relatively low, as illustrated in Fig. 13-25. Consequently, if two of these feedback pairs are connected in cascade, the second pair does not load the first pair appreciably, and hence the interaction between the two feedback pairs is small. Thus all of the relations developed above apply individually to the two feedback pairs, and it is relatively easy to adjust the cascade for the desired operation. Figure 16-17 shows the pole-zero diagram for the overall complex gain of two feedback pairs connected in cascade and adjusted, in accordance with Fig. 15-27, for a maximally flat gain characteristic. The poles are separated by equal arcs of 45° on the semicircle. Note that the two feedback pairs can be identical in

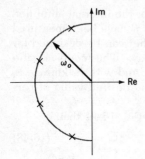

Fig. 16-17 Pole-zero diagram for the overall complex gain of two feedback pairs in cascade with a maximally flat gain characteristic.

every respect except for having different values of feedback capacitance C_f; C_f is used to move the poles along the semicircle to the proper position.

16-6 FEEDBACK OSCILLATORS

The preceding sections of this chapter are concerned with the fact that feedback can cause growing oscillatory transients in electronic amplifiers. These oscillations degrade the performance of feedback amplifiers, and hence they must be avoided in such systems. However, oscillations of this kind are used to advantage in the realization of oscillators and signal generators of the type used in academic and industrial laboratories for experimentation and testing; almost all such oscillators can be viewed as unstable feedback amplifiers. Figure 16-18 shows the small-signal model for a feedback amplifier that can be used either as a frequency-selective amplifier having characteristics similar to those of a tuned amplifier or as an oscillator generating a sinusoidal signal.

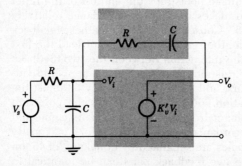

Fig. 16-18 An RC feedback oscillator.

The circuit finds its widest application as an oscillator, and it serves as the prototype for almost every laboratory oscillator operating at frequencies below 1 MHz. In this application the internal amplifier must produce no sign reversal, and it must provide a gain somewhat greater than 3; hence a wideband, two-stage, feedback amplifier of the form shown in Fig. 16-13 is widely used as the internal amplifier in Fig. 16-18. It is assumed in the analysis that follows that the gain of the wideband internal amplifier is independent of frequency in the band of interest.

The output voltage of the amplifier is

$$V_o = K'_v V_i \tag{16-52}$$

and the node equation at V_i is

$$\left(\frac{1}{R} + sC + \frac{sC}{1 + sCR}\right) V_i - \frac{sC}{1 + sCR} V_o = \frac{1}{R} V_s \tag{16-53}$$

Multiplying this expression by R yields

$$\left(1 + sCR + \frac{sCR}{1 + sCR}\right) V_i - \frac{sCR}{1 + sCR} V_o = V_s \tag{16-54}$$

Now defining $\omega_o = 1/CR$ and using Eq. (16-52) to eliminate V_i in (16-54) leads to

$$\frac{V_o}{V_s} = A_v = \omega_o K'_v \frac{s + \omega_o}{s^2 + (3 - K'_v)\omega_o s + \omega_o^2} \tag{16-55}$$

$$= \omega_o K'_v \frac{s - s_0}{(s - s_1)(s - s_2)} \tag{16-56}$$

for the closed-loop complex gain.

Equation (16-55) shows that the poles of the closed-loop complex gain depend on the gain K'_v of the internal amplifier, and it turns out that the poles are complex for a certain range of K'_v. Since $\omega_o = 1/RC$ is independent of K'_v, the complex poles of A_v must move in the complex plane at a constant distance from the origin as K'_v is changed; hence they must move along the circular path shown in Fig. 16-19. Complex poles correspond to values of K'_v in the range between 1 and 5, and the poles lie on the imaginary axis when $K'_v = 3$. When K'_v is made slightly less than 3, the poles lie in the left half of the complex plane, and they are very close to the imaginary axis. Under these conditions the shape of the gain characteristic for values of $s = j\omega$ very close to $j\omega_o$ depends only on the pole s_1, and it is identical to that of a narrowband tuned amplifier. It follows that the bandwidth of this amplifier can be made as small as desired by making K'_v approach the value 3 from below. A practical difficulty prevents the realization of very small bandwidths, however, for

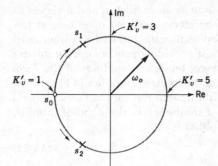

Fig. 16-19 Locus of the complex poles of A_v with increasing K_v'.

if the poles are very close to the imaginary axis, any slight increase in K_v', such as might result from a change in a circuit parameter or a change in line voltage, causes the poles to move into the right half of the complex plane. Under this condition the circuit develops a growing oscillatory transient (if $K_v' < 5$) that interferes with the operation of the circuit as an amplifier.

The growing oscillation that results when K_v' is greater than 3 is profitably employed by using the circuit as an oscillator to generate a sinusoidal signal having a frequency $f_o = \omega_o/2\pi = 1/2\pi RC$. In this application the signal source V_s in Fig. 16-18 is not needed, and it can be replaced by a short circuit. Once started, the oscillatory transient grows until the circuit saturates and becomes nonlinear. The saturation limits further growth of the oscillation, and it may also introduce considerable waveform distortion. The distortion problem can be eliminated by incorporating an automatic gain control in the internal amplifier to maintain K_v' at the value of 3. Automatic gain control is accomplished by controlling the internal negative feedback. A tungsten-filament lamp is commonly used for the resistor R_e shown in Fig. 16-13. When the lamp is cold, its resistance is small, the amount of internal negative feedback is small, and K_v' is large. As oscillations build up in the circuit, the rms voltage across the lamp increases, the filament heats up, and the lamp resistance R_e increases, perhaps by as much as a factor of 10. The increase in R_e increases the negative feedback, and thus it reduces K_v' to the value of 3. If the gain drops below 3, the oscillations begin to die out, the lamp cools, R_e decreases, and K_v' increases. With proper design, this circuit can be made to stabilize K_v' at the critical value of 3 while the signal level in the amplifier is small enough to give good linear operation. The important feature of this automatic-gain-control circuit is the fact that the temperature of the lamp filament cannot change rapidly. The temperature, and hence the resistance R_e, responds to slow variations in the rms signal amplitude over many cycles of oscillation, but R_e does not change throughout one or a few cycles of oscillation. Hence the variations in R_e do not cause waveform distortion. Circuit elements of this kind are sometimes called quasi-linear elements.

Since the radius of the circle on which the poles of A_v move is $\omega_o = 1/RC$, it follows that the frequency of oscillation is ω_o. Thus the frequency of oscillation can be changed by changing R or C, or both.

When the circuit of an oscillator is simple enough, as it is with the oscillator of Fig. 16-18, the loci of the poles of the closed-loop gain can be determined as in Fig. 16-19, and the oscillating conditions can be determined from these loci. However, when the circuit is more complicated, the loci are more difficult to construct, and the oscillating conditions can be determined more easily by an alternative procedure. In order for steady-state oscillations to exist, the circuit must have a pair of poles on the imaginary axis. This fact means that the circuit must be just on the threshold of instability and hence that the Nyquist diagram for the loop gain must pass through the critical point $1 + j0$. Thus the condition for steady-state oscillation can be expressed as

$$A_L(j\omega) = 1 \tag{16-57}$$

Expressed in words, this condition says that, if steady-state oscillations exist in the circuit, the gain around the feedback loop at the oscillation frequency must be unity and the net phase shift must be zero.

The use of Eq. (16-57) in determining the conditions of oscillation can be illustrated by applying it to the circuit of Fig. 16-18. The feedback loop in this circuit consists of the internal amplifier followed by an RC voltage divider; hence with $V_s = 0$, the loop gain is

$$A_L(s) = K'_v \frac{R/(1 + sCR)}{R + 1/sC + R/(1 + sCR)}$$

Letting $\omega_o = 1/CR$ and clearing of fractions yields

$$A_L(s) = \omega_o K'_v \frac{s}{s^2 + 3\omega_o s + \omega_o^2} \tag{16-58}$$

The condition for the threshold of instability is thus

$$A_L(j\omega) = \omega_o K'_v \frac{j\omega}{(j\omega)^2 + 3\omega_o(j\omega) + \omega_o^2} = 1$$

Clearing this expression of fractions and collecting terms yields

$$-\omega^2 + (3 - K'_v)(j\omega\omega_o) + \omega_o^2 = 0 \tag{16-59}$$

Now the real and imaginary parts of the left-hand side of this equation must equal zero separately; thus setting the real part equal to zero yields

$$\omega_o{}^2 - \omega^2 = 0$$

or

$$\omega = \omega_o = \frac{1}{RC} \tag{16-60}$$

This result gives the frequency of oscillation, and it is in agreement with the result obtained from the loci of the poles in Fig. 16-19. Setting the imaginary part of Eq. (16-59) equal to zero yields

$$(3 - K'_v)\omega\omega_o = 0$$

and with $\omega = \omega_o \neq 0$, this condition requires

$$K'_v = 3 \tag{16-61}$$

This result gives the gain that the internal amplifier must have to produce steady-state oscillations, and it is in agreement with the result obtained from Fig. 16-19.

The process of equating the real and imaginary parts of Eq. (16-59) to zero separately greatly simplifies the general problem of determining the conditions for oscillation in feedback oscillators. The separation divides a polynomial of high degree into two polynomials of lower degree, and thus it permits the conditions for oscillation in most oscillator circuits to be determined by relatively simple means. This fact is illustrated by the analysis of the JFET oscillator circuit shown in Fig. 16-20. Oscillators using circuits of this form are called Colpitts oscillators. The JFET introduces a sign reversal in the feedback loop; hence if the circuit oscillates, it must do so at the frequency where the LC network introduces a phase shift of 180°. Thus the oscillation frequency is determined by the LC network.

To facilitate evaluation of the loop gain, Fig. 16-20 shows the feedback loop temporarily broken at a point where no current flows. The loop gain can be determined from two node equations,

$$\left(\frac{1}{r_o} + sC_1 + \frac{1}{sL}\right) V_o - \frac{1}{sL} V'_i = -g_m V_i$$

and

$$-\frac{1}{sL} V_o + \left(\frac{1}{sL} + sC_2\right) V'_i = 0$$

Solving these equations for V'_i yields

$$\frac{V'_i}{V_i} = A_L(s) = \frac{-g_m r_o}{LC_1 C_2 r_o s^3 + LC_2 s^2 + (C_1 + C_2)r_o s + 1}$$

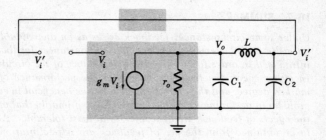

Fig. 16-20   A JFET Colpitts oscillator.

The conditions for oscillation are obtained by setting $s = j\omega$ in this expression and equating the expression to unity; the result obtained is

$$-jLC_1C_2r_o\omega^3 - LC_2\omega^2 + j(C_1 + C_2)r_o\omega + 1 + g_mr_o = 0$$

This is a cubic polynomial in ω; however, when the real and imaginary parts are equated to zero separately, simpler expressions are obtained. Equating the imaginary part to zero gives

$$-LC_1C_2r_o\omega^3 + (C_1 + C_2)r_o\omega = 0$$

or

$$\omega^2 = \frac{C_1 + C_2}{LC_1C_2}$$

This is the square of the oscillation frequency. Similarly, the real part yields

$$-LC_2\omega^2 + 1 + g_mr_o = 0$$

Substituting the expression obtained above for ω^2 gives

$$-LC_2\frac{C_1 + C_2}{LC_1C_2} + 1 + g_mr_o = 0$$

or

$$g_mr_o = \frac{C_2}{C_1}$$

This last expression gives the value of g_m, and hence the gain, that the JFET must supply for steady-state oscillation.

16-7 SUMMARY

Under some circumstances feedback arises as an unavoidable consequence of parasitic circuit elements; under other circumstances feedback is deliberately introduced in order to improve the performance of the circuit in some respect. In either event, the feedback may alter the performance of the circuit to a marked degree, and the alterations may not be beneficial in every respect. The problem in designing a feedback amplifier is primarily that of ensuring that all the effects of feedback are beneficial, or at least tolerable. Among the benefits to be obtained from the use of feedback are self-calibration, the rejection of corrupting signals, the modification of impedance levels, and the modification of dynamic characteristics.

When controlled sources are used in circuits having no feedback, they affect only the constant multiplier of the complex gain function; the poles and zeros of the gain are determined by the passive elements R, L, and C. When feedback is present, however, the controlled sources affect the poles and zeros of the complex gain as well as the constant multiplier. The effect on the pole-zero pattern of introducing feedback may or may not be beneficial. In particular, feedback may cause some of the poles of complex gain to move into the right half of the complex plane. Such circuits exhibit growing transients, a consequence that usually cannot be tolerated except when it is desired to build an oscillator. All except the simplest systems are certain to develop growing transients if the gain around the feedback loop is made great enough.

The design of a feedback amplifier usually consists of two phases: first, the determination of the amount of feedback and the frequency band over which it must exist in order to realize a desired result such as self-calibration, and second, the choice of circuit parameters to meet the above requirement while ensuring a suitable dynamic response for the closed-loop amplifier. The more stringent the requirement imposed by the first phase, the more difficult the solution of the second phase. In the second, and usually more difficult, phase of the problem, two sets of design techniques are particularly valuable. These are the root-locus techniques and the techniques for shaping the gain and phase characteristics of the loop gain. These alternative techniques give different kinds of insight into the problem, and each complements the other. Both of these techniques have been developed and exploited to a far greater degree than it is possible to show in the brief space available for this chapter.

PROBLEMS

16-1. *Dynamic response.* The open-loop gain of a feedback amplifier having the form shown in Fig. 16-21 is given by Eq. (16-13) with $K'_v = 1,000$, $f'_1 = \omega'_1/2\pi = 100$ kHz, and $f'_2 = 1$ MHz.

(a) Sketch and dimension the loci of the poles of the closed-loop gain A_v as the feedback ratio β is increased from zero to unity.

(b) What value of β makes the gain characteristic maximally flat?

(c) With the adjustment of part b, what is the low-frequency value of the return difference F?

(d) Determine the closed-loop half-power bandwidth and low-frequency gain K_v for the adjustment of part b. Compare this half-power bandwidth with the open-loop half-power bandwidth f_1'.

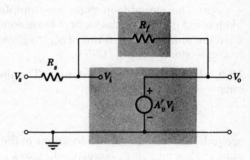

Fig. 16-21 A feedback amplifier.

16-2. Dynamic response. The open-loop gain of a feedback amplifier having the form shown in Fig. 16-21 is given by Eq. (16-20) with $K_v' = 20,000$ and $f_1' = \omega_1'/2\pi = 500$ kHz.

(a) Sketch and dimension the loci of the poles of the closed-loop gain A_v as the feedback ratio β is increased from zero.

(b) What value of β places the circuit on the threshold of instability?

(c) With the adjustment of part b, the circuit can act as an oscillator generating a sinusoidal voltage of constant amplitude. What is the frequency of this oscillation?

16-3. Root locus. The open-loop complex gain of a feedback amplifier having the form shown in Fig. 16-21 is given by

$$A_v' = \frac{-K_v'\omega_1'^4}{(s + \omega_1')^4}$$

with $K_v' = 10,000$ and $f_1' = \omega_1'/2\pi = 100$ kHz.

(a) Following the procedure used in connection with Fig. 16-6, sketch and dimension the loci of the poles of the closed-loop gain as the feedback ratio β is increased from zero.

(b) What value of loop gain $\beta K_v'$ places the amplifier on the threshold of instability?

16-4. Root locus. The open-loop complex gain of a feedback amplifier having the form shown in Fig. 16-21 is given by

$$A_v' = -\frac{1}{4} K_v' \omega_1' \frac{s + 4\omega_1'}{(s + \omega_1')^2}$$

(*a*) Designating the feedback ratio as β, derive an expression for the closed-loop complex gain of the amplifier.

(*b*) The closed-loop poles are complex for a certain range of β. Sketch and dimension the locus of these complex poles as a function of β. Give the dimensions in terms of ω_1'. *Suggestion:* Use the results derived in connection with Eq. (15-53).

16-5. Return-difference contours. The properties of the return difference

$$F(s) = \frac{s - s_1}{s - s_2}$$

are to be studied for various locations of the pole and zero. As the variable s takes on values corresponding to a point moving around the circular contour C_1 in Fig. 16-22*a*, $F(s)$ takes on successive values (complex numbers) corresponding to points on the contour C_2 in Fig. 16-22*b*.

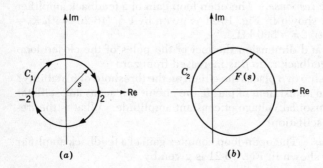

Fig. 16-22 Diagrams for Prob. 16-5.

Because of the simple nature of the function $F(s)$ and the contour C_1, the contour C_2 is always a circle with its center on the real axis. Thus C_2 can be determined by calculating the value of $F(s)$ for $s = 2$ and $s = -2$ and constructing the appropriate circle passing through these two points. The problem to be solved is the construction of C_2 for various locations of the pole and zero of $F(s)$.

(a) Sketch the pole-zero pattern for $F(s)$ when $s_1 = -1$ and $s_2 = -3$. Show the contour C_1 on this sketch.

(b) Repeat part a for the following cases: $s_1 = -3$, $s_2 = 3$; $s_1 = 3$, $s_2 = -1$; and $s_1 = 1$, $s_2 = -1$.

(c) Sketch and dimension the locus of $F(s)$ (the contour C_2) for each of the cases specified above as s moves around the contour C_1.

(d) Indicate on the sketches of part c the direction in which C_2 is traversed when C_1 is traversed in the clockwise direction. *Suggestion:* One way to determine the direction is to evaluate $F(s)$ for $s = j2$.

(e) Comment briefly on the relations between the C_2 contours and the locations of the pole and zero of $F(s)$.

16-6. Pole-zero relations. A certain complex voltage gain $A_r(s)$ has the poles and zeros shown in Fig. 16-23. As the variable s takes on successive values going once completely around the contour C_1 in the indicated direction, what is the net change in the phase angle of $A_r(s)$?

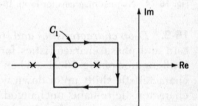

Fig. 16-23 Diagram for Prob. 16-6.

16-7. Poles of the return difference. In the usual case the complex loop gain for a feedback amplifier has the form of a rational function of frequency,

$$A_L(s) = -K \frac{(s - s_1)(s - s_3) \cdots}{(s - s_2)(s - s_4) \cdots}$$

Prove that, if the loop gain has no right-half-plane poles, then the return difference has no right-half-plane poles.

16-8. The Nyquist test. The Nyquist diagrams for four different complex loop gains are shown in Fig. 16-24. These diagrams correspond to the range $0 < \omega < \infty$. In each case the loop gain is known to have no poles in the right half of the complex plane. In each case state whether or not the closed-loop gain is stable. Give the reason for your answer in each case.

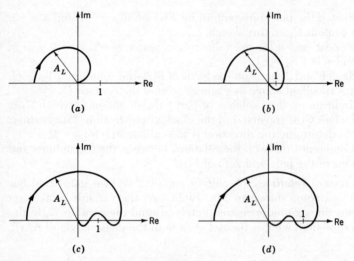

Fig. 16-24   Nyquist diagrams for Prob. 16-8.

16-9.   *Loop characteristics and stability.* Figure 16-25 shows the loop gain and phase characteristics for a microamplifier used in the circuit configuration of Fig. 16-21. When the feedback ratio β is varied, the gain characteristic shifts up or down without changing shape, and the phase characteristic remains unchanged.

(*a*) If β is adjusted so that the closed-loop amplifier is on the threshold of instability, what is the approximate value of the low-frequency loop gain in decibels?

(*b*) If, to provide a margin of safety, β is adjusted so that the loop gain is 0 dB when the loop phase shift is 135°, what is the approximate value of the low-frequency loop gain in decibels?

16-10.   *Loop gain and bandwidth.* The cutoff rate for the compensated loop gain of a certain feedback amplifier is −6 dB/octave (−20 dB/decade) as shown in Fig. 16-11, and the unity-gain frequency f_x is 1 MHz. If the loop gain is required to be 40 dB at low frequencies for good self-calibration, what is the half-power bandwidth of the loop gain?

16-11.   *Rate limiting in a feedback amplifier.* The data sheet for a certain commercial microamplifier gives the slew rate as 10 volts/μsec, and it states that the amplifier can deliver a sinusoidal output voltage with a peak value V_o of 14 volts at low frequencies.

(*a*) What is the maximum frequency at which this output voltage can be delivered without distortion due to rate limiting? Give the answer in kilohertz.

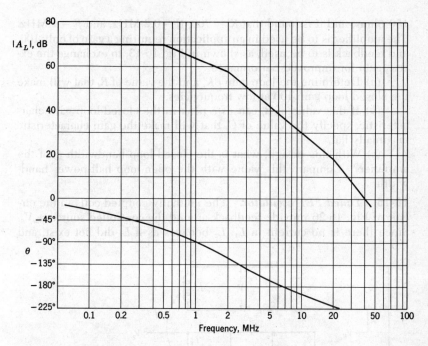

Fig. 16-25 Loop gain and phase characteristics for Prob. 16-9.

(b) What is the maximum value that V_o can have without rate limiting when the signal frequency is 200 kHz?

16-12. Feedback pair. The open-loop gain characteristic for a feedback pair like the one shown in Fig. 16-13 is measured with $R_e = 100$ ohms, and it is found that $K_v' = 500$, $f_1' = 1$ MHz, and $f_2' = 5$ MHz. Feedback is to be used with this amplifier as shown in Fig. 16-13 to obtain a maximally flat closed-loop gain characteristic with a half-power bandwidth of 7 MHz.

(a) Determine the values of R_f and C_f required.

(b) With this design, what is the low-frequency closed-loop gain?

16-13. Feedback pair. The amplifier described in Prob. 16-12 is to be designed for the same closed-loop bandwidth, $f_o = 7$ MHz, but it is to be adjusted for the most linear phase characteristic.

(a) Determine the values of R_f and C_f required.

(b) With this design, what is the low-frequency closed-loop gain?

16-14. Feedback pair. The open-loop gain characteristic for a feedback pair like the one shown in Fig. 16-13 is measured with $R_e = $

150 ohms, and it is found that $K_v' = 700$, $f_1' = 0.2$ MHz, and $f_2' = 1$ MHz. The amplifier is to be used in an application requiring a gain of only 100, and feedback is to be used, as shown in Fig. 16-13, to exchange the excess gain for improved performance.

(a) Determine the loop gain $\beta K_v'$ and the value of R_f that will make the closed-loop gain 100 at low frequencies.

(b) If this value of R_f causes a peak in the closed-loop gain characteristic, specify the value of C_f that will make the gain characteristic maximally flat.

(c) With this design, what is the closed-loop bandwidth f_o of the amplifier? Compare this value with the open-loop half-power bandwidth.

16-15.   *Tuned FET oscillator.* The mutually coupled coils in the circuit of Fig. 16-26 provide feedback around the controlled source $g_m V_g$. Since there is no current in L_1, L_2 behaves as if L_1 did not exist, and

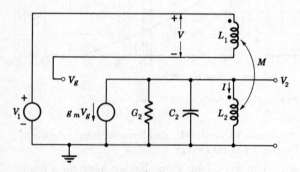

Fig. 16-26   Feedback circuit for Prob. 16-15.

under sinusoidal operating conditions the voltage V is given by $V = j\omega MI$, where I is the current in L_2.

(a) Show that

$$A_v = \frac{V_2}{V_1} = -\frac{g_m}{C_2} \frac{s}{s^2 + As + B}$$

where

$$A = \frac{G_2}{C_2} - \frac{g_m M}{L_2 C_2} \quad \text{and} \quad B = \frac{1}{L_2 C_2}$$

(b) For a certain range of M the poles of A_v are complex. Sketch the locus of the poles (a complete circle) as a function of M in this range. Show the direction of motion of the poles as M increases.

(c) If the circuit is to be used as an oscillator with $V_1 = 0$ to generate a sinusoidal voltage, what condition among the parameters must be satisfied? What is the frequency of oscillation in terms of the circuit parameters?

16-16.   FET Hartley oscillator.   Figure 16-27 shows the small-signal model for an FET in a circuit known as the Hartley oscillator; this oscil-

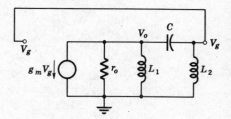

Fig. 16-27   Oscillator for Prob. 16-16.

lator is similar in many respects to the Colpitts oscillator shown in Fig. 16-20.

(a) Find the frequency of oscillation in terms of L_1, L_2, and C when the circuit is adjusted to be on the threshold of oscillation.

(b) Find the value of $g_m r_o$ in terms of the circuit parameters that puts the circuit on the threshold of oscillation. The result should be very similar to the corresponding relation for the Colpitts oscillator.

16-17.   Feedback-amplifier stability.   The open-loop complex gain of a microamplifier similar to the one in Prob. 16-9 can be expressed as

$$A_v' = \frac{-0.5K_v'}{(j\omega + 0.1)(j\omega + 1)(j\omega + 5)}$$

when ω is in megaradians per second. When this circuit is used in the feedback configuration of Fig. 16-21, what value of low-frequency loop gain $\beta K_v'$ puts the circuit on the threshold of instability? *Suggestion:* Use the oscillator condition for the threshold of oscillation, $A_L(j\omega) = 1$.

17
Power Amplifiers

*E*lectronic systems are often called upon to deliver substantial amounts of power to some electrical device that serves as the load for the system. The load may take many different forms; among them are loudspeakers for sound reproduction, servomotors for automatic control systems, and electromechanical transducers for underwater sound systems. The amount of power that a transistor can deliver depends on its maximum current, voltage, and power-dissipation rating and on the nonlinearities in its volt-ampere characteristics. Large, costly transistors with high current, voltage, and power ratings must be used when large output power is required.

The design of power amplifiers is concerned primarily with obtaining a specified output power with the smallest possible transistors while keeping the signal distortion at a suitably low level. It is also concerned with keeping the dc power requirement as small as possible. Apart from the matter of initial cost, the factors of size, weight, and dc power requirement are of further importance in portable and airborne equipment. In the launching of artificial satellites it costs from $1,000 to $10,000 for each pound put into orbit, and since the electronic power supplies account for about half the weight of some communication satellites, the importance of minimizing the power requirements is clear.

In order to obtain large output power with high efficiency, it is usually necessary to employ special circuit configurations. It is the objective of this chapter to examine in some detail the properties of several amplifier configurations that are commonly used to obtain large output power, and to develop techniques for the analysis and design of these circuits.

17-1 CLASS *A* POWER AMPLIFIER

Power amplifiers are classified in three basic categories, designated as class *A*, class *B*, and class *C*, depending on the bias conditions in the circuit. In class *A* operation the bias conditions and the applied signal are such that the operating point for the transistor remains in the linear portion of the collector characteristic and the transistor is never driven

to cutoff. Class B amplifiers use two transistors in a push-pull arrangement, and both transistors are biased just on the edge of cutoff. In
class C operation the transistor is biased in the cutoff region so that a
substantial input signal must be applied to make the transistor draw current; class C operation is normally used only at high frequencies and
with tuned-circuit loads. Class B operation is more efficient than class A,
and class C is more efficient than class B. In addition, there is a hybrid
mode of operation, designated as class AB, in which the operating conditions are intermediate between those of class A and class B; most so-
called class B amplifiers actually operate slightly in the class AB mode.

Figure 17-1a shows the circuit diagram for a class A BJT power
amplifier. Since the bypass capacitor C_e is understood to act as a short
circuit for signal components of current, the circuit can be simplified as
shown in Fig. 17-1b. Since the net effect of the battery V_E in Fig. 17-1b
is to modify the bias applied to the transistor, the circuit can be simplified

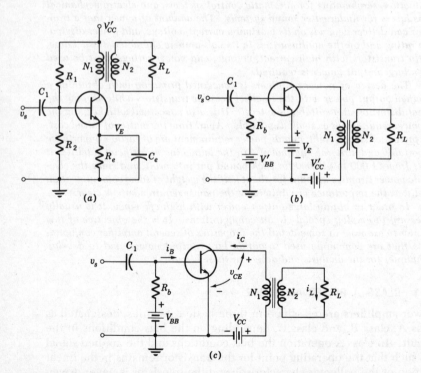

Fig. 17-1 Class A power amplifier. (a) Circuit; (b) an equivalent circuit; (c) a simplified
equivalent circuit.

further, as shown in Fig. 17-1c, by lumping V_E in with V'_{CC} and V'_{BB}. With this simplification, $V_{BB} = V'_{BB} - V_E$ and $V_{CC} = V'_{CC} - V_E$. The function of this circuit is to transmit the input signal v_s to the load R_L and to deliver a specified amount of signal power to the load with small distortion of the signal waveform. But the amount of power that the transistor can deliver to the load with small distortion is limited by the maximum current, voltage, and power-dissipation ratings of the transistor and by the nonlinearities in the transistor characteristics. Transistors with high ratings are costly, and inefficient circuit designs make heavy demands on the dc power supply. Therefore, the design of power amplifiers is concerned with obtaining a specified signal power delivered to the load with the smallest transistor and the most efficient design. In this way size, weight, and initial cost are minimized.

The use of an iron-core transformer to couple the load to the amplifier aids substantially in realizing some of the objectives listed above, although it does not contribute to small weight. With transformer coupling, the dc quiescent collector current does not flow in the load R_L. In most cases the direct current would contribute no useful output, and in many cases, such as with loudspeaker and servomotor loads, the direct current would degrade the performance of the load device. With transformer coupling there is no power dissipated in the load by the quiescent current, and the efficiency of the amplifier is thereby improved. Similarly, there is no voltage drop across the load caused by the quiescent current, and hence a smaller dc supply voltage is adequate. The transformer also serves another important function. It emerges from the analysis that follows that there is an optimum slope for the dynamic operating path on the collector characteristic. This slope depends on the load resistance reflected into the primary of the transformer, and it can be adjusted by the transformer turn ratio when the actual load resistance R_L is not adjustable. This last fact alone is often enough to require the use of transformer coupling.

The circuit of Fig. 17-1c can be simplified further by representing the iron-core transformer with an ideal transformer and a shunt magnetizing inductance L as shown in Fig. 17-2a. In this approximation the transformer winding resistances, the core losses, the leakage inductance, and the winding capacitances are all neglected; thus the approximation is a coarse one, but it is adequate for a first-order analysis. When the load resistance is transferred to the primary side of the ideal transformer, the circuit takes the form shown in Fig. 17-2b; the transferred resistance is

$$R'_L = \left(\frac{N_1}{N_2}\right)^2 R_L \qquad (17\text{-}1)$$

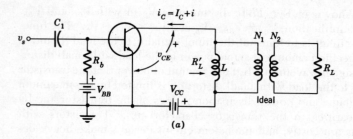

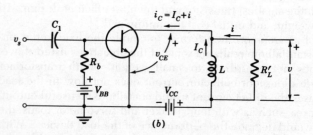

Fig. 17-2 Simplified representation of the iron-core transformer.

The collector current in Fig. 17-2*b* can be expressed as

$$i_C = I_C + i \tag{17-2}$$

where I_C is the average value of collector current and i is the time-varying component having zero average value,

$$i_{av} = 0 \tag{17-3}$$

The inductance L acts as a short circuit to the dc component of collector current, and L is assumed to be so large that it acts as an open circuit to the time-varying component of i_C. Thus the current in L is I_C, and the current in R'_L is i, as shown in Fig. 17-2*b*. It follows that there is no direct voltage drop across R'_L and that the total voltage drop across the transformer primary is just $R'_L i$. Hence the collector-to-emitter voltage can be written as

$$v_{CE} = V_{CC} + v = V_{CC} - R'_L i \tag{17-4}$$

where V_{CC} is the average value of the collector voltage and

$$v = -R'_L i \tag{17-5}$$

is the time-varying component having zero average value,

$$v_{av} = 0 \tag{17-6}$$

Equation (17-4) is the equation of the dynamic operating path for the transistor on the collector characteristics; this path is shown in Fig. 17-3.

It is clear from the construction in Fig. 17-3 that with a signal applied at the input of the amplifier, it is possible for the instantaneous collector voltage to exceed the power-supply voltage V_{CC}. This condition is possible because there is a constant current I_C in the inductor L, as shown in Fig. 17-2b. When the input signal makes i_C less than I_C, the time-varying current i becomes negative, the voltage drop v across R'_L becomes positive, and v adds to V_{CC} to make v_{CE} greater than V_{CC}.

Before continuing with the analysis of the amplifier, it is desirable to make three assumptions about the operating conditions. First, it is assumed that the mode of operation is class A and that the waveform distortion introduced by the amplifier is negligibly small. Second, in the absence of any knowledge to the contrary, it is assumed that the operating point makes excursions of equal magnitude on each side of the quiescent point. And third, it is assumed that in normal operation there will be periods of time during which there is no input signal and during which quiescent conditions prevail in the amplifier.

When the input signal is periodic with period T, the average power drawn from the power supply V_{CC} in Fig. 17-2b is

$$P_{CC} = \frac{1}{T} \int_0^T V_{CC} i_C \, dt$$

Removing the constant V_{CC} from under the integral sign and substituting Eq. (17-2) for i_C yields

$$P_{CC} = V_{CC} \frac{1}{T} \int_0^T (I_C + i) \, dt \tag{17-7}$$

But I_C is constant, and the average value of i is zero; hence

$$P_{CC} = V_{CC} I_C \tag{17-8}$$

Thus in linear class A operation, P_{CC} is constant and independent of the signal amplitude and waveform; it depends only on the quiescent collector current I_C and the power-supply voltage V_{CC}.

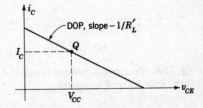

Fig. 17-3 Dynamic operating path for the class A power amplifier.

The average power delivered to the load in Fig. 17-2b when the signal is periodic is

$$P_L = \frac{1}{T}\int_0^T R_L' i^2 \, dt = R_L' \frac{1}{T}\int_0^T i^2 \, dt \qquad (17\text{-}9)$$

The quantity multiplying R_L' is by definition the square of the rms value of i, and hence (17-9) can be written as

$$P_L = R_L'(I_{\text{rms}})^2 \qquad (17\text{-}10)$$

It also follows from Eq. (17-5) that $V_{\text{rms}} = R_L' I_{\text{rms}}$, and thus

$$P_L = \frac{(V_{\text{rms}})^2}{R_L'} \qquad (17\text{-}11)$$

By conservation of energy, the average power dissipated in the transistor, called the collector dissipation, is

$$P_c = P_{CC} - P_L \qquad (17\text{-}12)$$

Under quiescent conditions there are no time-varying currents or voltages in the circuit, and $P_L = 0$. Since P_{CC} is constant, independent of the signal amplitude, it follows from (17-12) that the power dissipated in the transistor has its greatest value under quiescent conditions. Thus

$$P_{c,\text{max}} = P_{CC} = V_{CC}I_C \qquad (17\text{-}13)$$

Figure 17-4a shows the dynamic operating path for the class A power amplifier, and Fig. 17-4b shows the waveform of a periodic signal current. The peak values of signal current and signal voltage are designated

$$I = |i_{\text{max}}| \quad \text{and} \quad V = |v_{\text{max}}| \qquad (17\text{-}14)$$

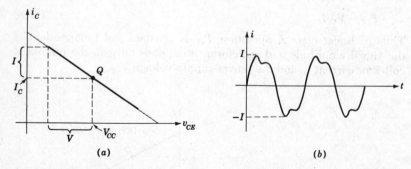

Fig. 17-4 Operating conditions in a class A power amplifier. (a) Dynamic operating path; (b) a periodic signal waveform.

and it follows from Eq. (17-5) that

$$V = R'_L I \tag{17-15}$$

For periodic signals the rms value of current or voltage is directly proportional to the peak value as long as the waveform remains unchanged. Thus for a given waveform a proportionality constant can be defined as

$$K_1 = \frac{(I_{rms})^2}{I^2} = \frac{(V_{rms})^2}{V^2} \tag{17-16}$$

For sine waves $K_1 = \frac{1}{2}$, and for square waves $K_1 = 1$. Now Eq. (17-10) for the average power delivered to the load can be written as

$$P_L = R'_L (I_{rms})^2 = K_1 R'_L I^2 = K_1 \frac{V^2}{R'_L} = K_1 VI \tag{17-17}$$

where V and I are the peak values of voltage and current at the load R'_L.

The collector-circuit efficiency of the amplifier is defined as

$$\eta = \frac{P_L}{P_{CC}} = K_1 \frac{VI}{V_{CC} I_C} \tag{17-18}$$

and hence for high efficiency V and I must be as large as possible. However, the operating point in Fig. 17-4a cannot cross the v_{CE} axis because the transistor cuts off, and it cannot quite reach the i_C axis because the transistor saturates. Thus, since it is assumed that the operating point makes equal excursions on each side of the quiescent point, the greatest possible values of V and I are realized when the quiescent point is at the center of the operating path in Fig. 17-4a and when the input signal is adjusted to drive the operating point to the edge of cutoff at one extreme and to the edge of saturation at the other extreme. Under these maximum-signal conditions, V and I are approximately V_{CC} and I_C, and the collector-circuit efficiency has its maximum value

$$\eta_{max} = K_1 \tag{17-19}$$

Note that this maximum efficiency depends only on the waveform of the signal; it is independent of the output power and of the location of the dynamic operating path on the collector characteristics. For sine waves $\eta_{max} = \frac{1}{2} = 50$ percent, and for square waves $\eta_{max} = 1 = 100$ percent. Also, with this maximum-signal adjustment Eqs. (17-18) and (17-13) give

$$P_L = \eta_{max} P_{CC} = K_1 P_{CC} = K_1 P_{C,max} \tag{17-20}$$

Thus with a given signal waveform, the power delivered to the load under maximum-signal conditions is directly related to the maximum collector dissipation that the transistor must withstand. For sinusoidal signals Eq. (17-20) gives $P_L = \frac{1}{2}P_{C,\text{max}}$, and for square waves it gives $P_L = P_{C,\text{max}}$.

With the collector-circuit efficiency maximized in accordance with the above discussion, the next step is to determine the maximum output power P_L that the amplifier can deliver. Equation (17-20) shows that the output power under maximum-signal conditions is directly proportional to $P_{C,\text{max}}$, the quiescent collector dissipation. Thus to obtain the greatest possible output power, the amplifier must be designed for the greatest possible quiescent dissipation. However, the dynamic operating path must lie within a certain limited region of the collector characteristics, and hence the quiescent dissipation cannot be made arbitrarily large. To understand the limitations on the output power that can be obtained from a class A amplifier, it is necessary to examine the limitations on the region of the collector characteristics in which the dynamic operating path is permitted to lie.

Figure 17-5a shows typical collector characteristics for a transistor, including the collector-breakdown region; collector breakdown is discussed in some detail in connection with Fig. 9-12. As a result of collector breakdown, there is a maximum permissible collector voltage that cannot be exceeded without the occurrence of severe waveform distortion and possible damage to the transistor. Figure 17-5b shows a typical curve of collector current as a function of base current. This curve indicates that the collector current is not proportional to the base current at high current levels, a fact that is discussed in more detail in connection with Fig. 9-11. As a result of this fact, there is a maximum permissible collector current that cannot be exceeded without the occurrence of excessive waveform distortion. In addition, there is a maximum permissible continuous collector dissipation that cannot be ex-

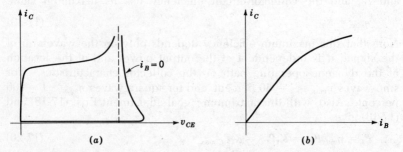

Fig. 17-5 Transistor limitations. (a) Collector breakdown; (b) decrease in current gain at high currents.

ceeded without damage to the transistor; this limitation defines a hyperbola on the collector characteristics as shown in Fig. 9-20. These maximum permissible voltage, current, and power limitations, together with the coordinate axes, form a closed region on the collector characteristics, as shown in Fig. 17-6. In class A operation the dynamic operating path must lie wholly within this region.

Figure 17-6a shows the graphical construction for an amplifier having its operating path entirely within the permitted operating region. To obtain the greatest collector-circuit efficiency under maximum-signal conditions, the quiescent point is located at the center of the operating path, and the input signal drives the operating point from one coordinate axis to the other. Now Eq. (17-20) shows that, to increase the output power from this amplifier under maximum-signal conditions, the quiescent collector dissipation $P_{C.\text{max}}$ must be increased. Consider the case in which the circuit is modified to make the operating path tangent to the hyperbola of maximum permissible collector dissipation, as illustrated in Fig. 17-6b. It is shown in connection with Fig. 10-4b that in this case the center of the operating path is at the point of tangency. Hence under this condition the quiescent collector dissipation has the maximum permissible value, and according to Eq. (17-20) the output power has the greatest possible value. It is not possible to get more output power from the amplifier in class A operation without exceeding the power rating of the transistor.

All operating paths tangent to the same hyperbola yield the same quiescent collector dissipation. Hence, according to Eq. (17-20), all such paths yield the same output power under maximum-signal conditions, provided that they do not intersect the maximum-current or maxi-

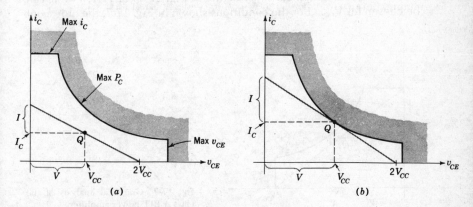

Fig. 17-6 Graphical analysis of the class A BJT power amplifier.

mum-voltage limit. Many such paths are possible on the characteristics of Fig. 17-6b. If the quiescent point is chosen at a small value of V_{CC}, the operating path extends into the region of large collector current where waveform distortion may result, and a relatively large input signal is required. If the quiescent point is chosen at a large value of V_{CC}, the operating path lies in a more linear region of the collector characteristics, and a relatively small input signal is adequate for full output power. After the dynamic operating path has been chosen according to these requirements, all the power relations developed previously apply, and they can be used to work out the details of the circuit.

Figure 17-7 shows an alternative set of conditions that may exist in BJT class A power amplifiers, although the conditions shown in Fig. 17-6b are far more common. Under the conditions of Fig. 17-7 the maximum output power is limited by the maximum voltage and current ratings of the transistor rather than by the maximum collector-dissipation rating. Since the output power is given by Eq. (17-17) as $P_L = K_1VI$, it is easy to show that maximum output power is obtained when the operating path extends from the upper left-hand corner of the permitted region to the lower right-hand corner, for this path gives the greatest possible values for both V and I. With this design the quiescent collector dissipation is somewhat less than the maximum permissible dissipation given by the collector-dissipation hyperbola.

When the power to be delivered to the load with a given signal waveform is specified, Eq. (17-20) can be used to determine $P_{C.max}$. Then a transistor can be chosen having a maximum permissible collector dissipation at least as great as $P_{C.max}$, and its maximum current and voltage ratings can be determined from the data sheets. Then from a consideration of the graphical construction in Fig. 17-6b or 17-7, a value can be chosen for V_{CC}. For the conditions shown in Fig. 17-7, the quiescent

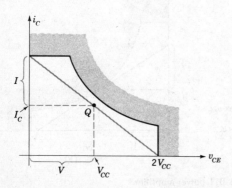

Fig. 17-7 Graphical analysis of the class A BJT power amplifier.

collector current can be determined from maximum-current limit; for the conditions shown in Fig. 17-6b, the quiescent current is

$$I_C = \frac{P_{C,\text{max}}}{V_{CC}} \tag{17-21}$$

where $P_{C,\text{max}}$ is the quiescent collector dissipation. In order for the operating path to have the proper slope, the resistance seen at the primary of the transformer must be

$$R'_L = \frac{V_{CC}}{I_C} = \frac{V_{CC}^2}{P_{C,\text{max}}} \tag{17-22}$$

This value, together with the actual load resistance R_L, fixes the turn ratio for the transformer. Finally, the bias circuit must be designed to give the required quiescent collector current along with suitable stability of the quiescent point. The relations developed in Secs. 10-2 to 10-4 can be used for this purpose. In addition, the thermal relations developed in Sec. 10-2 can be used to determine the temperature rise resulting from a known collector dissipation.

Example 17-1

A silicon power transistor is used in a class A power amplifier having the form shown in Fig. 17-1a. The permitted operating region on the collector characteristics has the form shown in Fig. 17-6; it is bounded by a maximum collector current of 750 mA, a maximum collector voltage of 60 volts, and a hyperbola corresponding to a maximum permissible collector dissipation of 2 watts. The parameters for the large-signal transistor model are $\beta = 50$ and $V_o = 0.8$ volt. The amplifier is to be designed for the maximum output power at maximum efficiency, and for good stabilization of the quiescent point the stability factor S_e is to be 0.05.

Solution If the dynamic operating path extends from the upper left-hand corner of the permitted region to the lower right-hand corner, then the quiescent point is at $I_C = \frac{750}{2} = 375$ mA and $V_{CE} = V_{CC} = \frac{60}{2} = 30$ volts. The corresponding quiescent collector dissipation is 11.25 watts, which exceeds the maximum permissible value, and hence this operating path does not lie wholly within the permitted region. Therefore the operating path must be chosen tangent to the hyperbola of maximum permissible dissipation as shown in Fig. 17-6b, and many different operating paths are possible.

    If the maximum permissible value of 30 volts is chosen for

V_{CC}, then the quiescent collector current must have the value

$$I_C = \frac{P_{C,\max}}{V_{CC}} = \frac{2}{30} = 67 \text{ mA}$$

In order to give the operating path the correct slope, the load resistance seen at the primary of the transformer must be

$$R_L' = \frac{V}{I} = \frac{V_{CC}}{I_C} = \frac{V_{CC}^2}{P_{C,\max}} = \frac{900}{2} = 450 \text{ ohms}$$

With these adjustments and with a sinusoidal signal, the maximum output power obtainable from this amplifier is

$$P_L = K_1 P_{C,\max} = (\tfrac{1}{2})(2) = 1 \text{ watt}$$

The biasing and stabilizing resistors must now be chosen. Equation (10-15) gives the stability factor as

$$S_e = \frac{1 + R_e/R_b}{1 + (1+\beta)R_e/R_b}$$

With $\beta = 50$ and $S_e = 0.05$ as specified, this relation becomes

$$0.05 = \frac{1 + R_e/R_b}{1 + 51R_e/R_b}$$

from which

$$R_b = 1.63 R_e$$

If R_e is chosen to make V_E in Fig. 17-1 be 10 percent of the supply voltage V_{CC}, then

$$R_e = \frac{V_E}{I_C} = \frac{3}{2/30} = 45 \text{ ohms}$$

Then

$$R_b \approx 73 \text{ ohms}$$

The collector current is given by Eq. (10-16) as

$$I_C = \frac{\beta(V_{BB}' - V_o)}{R_b + (1+\beta)R_e} = \frac{50(V_{BB}' - V_o)}{73 + (51)(45)}$$

Thus for $I_C = 67$ mA, $V_{BB}' - V_o = 3.2$ volts and $V_{BB}' = 4.0$ volts. Now with $V_{CC} = 30$ volts and $V_E = 3$ volts, the true power-supply voltage must be $V_{CC}' = 33$ volts. Hence it follows from Eqs. (10-7) and (10-8) that

$$R_1 = R_b \frac{V'_{CC}}{V'_{BB}} = 73\tfrac{33}{4.0} = 600 \text{ ohms}$$

and

$$1 + \frac{R_1}{R_2} = \frac{V'_{CC}}{V'_{BB}} = 8.25$$

from which

$$R_2 = \frac{R_1}{7.25} = \frac{600}{7.25} \approx 82.5 \text{ ohms}$$

In summary, the design calls for $V'_{CC} = 33$ volts, $R'_L = 450$ ohms, $R_e = 45$ ohms, $R_1 = 600$ ohms, and $R_2 = 82.5$ ohms.

It is also instructive to evaluate the power dissipated in the biasing resistors. Under quiescent conditions, neglecting the small base current, this power is

$$P = R_e I_C^2 + \frac{(V'_{CC})^2}{R_1 + R_2} = 0.2 + 1.6 = 1.8 \text{ watts}$$

Thus the power lost in the biasing resistors is greater than the signal power delivered to the load.

17-2 TRANSFORMERLESS PUSH-PULL CLASS B POWER AMPLIFIER

Substantially greater efficiency can be obtained in power amplifiers by operating the transistors in the class B rather than the class A mode. The theoretical maximum efficiency for the class B amplifier with a sinusoidal signal is 78.5 percent, as compared to 50 percent for class A operation. As a result, substantially smaller transistors can deliver a specified power to the load, and substantially less drain is imposed on the power supply. Moreover, in the class B mode the transistors are biased just at the cutoff point, and the power drain is very small when there is no input signal. This fact is important in a number of electronic systems, such as hearing aids, in which the power is supplied by small batteries and yet where it may be necessary to keep the system in operation even when there is no input signal.

The fact that the transistors are biased at cutoff in class B amplifiers introduces a new problem, however. If the amplifier in Fig. 17-1a were operated in the class B mode, the transistor would deliver an output current only during the positive portions of the input signal, and it would clip off all the negative portions of the signal. Thus special circuit configurations must be used for class B operation. Figure 17-8 shows the

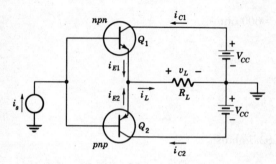

Fig. 17-8 Transformerless class B power amplifier using complementary symmetry.

prototype for a circuit that is widely used as a class B amplifier. It uses two transistors arranged so that one transistor amplifies the positive portions of the signal and the other amplifies the negative portions; hence it is known as a push-pull amplifier. Since one half of the amplifier uses an npn transistor whereas the other half uses a pnp transistor, the circuit is said to have complementary symmetry.

For simplicity it is assumed in Fig. 17-8 that the input signal is supplied by a current source; the more general case is considered in Sec. 17-4. Since Q_1 is an npn transistor, i_{E1} and i_{C1} are either positive or zero, and since Q_2 is a pnp transistor, i_{E2} and i_{C2} are either negative or zero. When the input signal i_s is positive, base current flows into Q_1, and thus Q_1 is turned on. The forward bias developed across the emitter junction of Q_1 is a reverse bias for the emitter junction of Q_2, and thus Q_2 is cut off when i_s is positive. Similarly, when i_s is negative, Q_2 is turned on and Q_1 is cut off. Thus Q_1 amplifies the positive portions of i_s, and Q_2 amplifies the negative portions. If the transistors have closely matched characteristics, the two portions of the signal are amplified equally, and the load current i_L is an amplified version of the input signal current i_s. It follows from these facts that Q_1 and Q_2 operate one at a time as emitter followers delivering the signal to the load resistance R_L.

The relations among the currents in the class B amplifier are illustrated in Fig. 17-9. Figure 17-9a shows a periodic input signal i_s; as in the case of the class A amplifier, in the absence of any information to the contrary it is necessary to assume equal positive and negative peaks in this waveform. The waveforms of the two emitter currents are shown in Fig. 17-9b and c, and the load current $i_L = i_{E1} + i_{E2}$ is shown in Fig. 17-9d.

Further understanding of the class B amplifier can be gained from the dynamic operating paths for the transistors on their collector char-

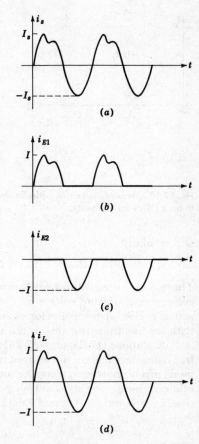

Fig. 17-9 Current waveforms for the amplifier of Fig. 17-8.

acteristics. As a preview of the results to be obtained, Fig. 17-10a shows the operating path for Q_1. The operating path for Q_2 has a similar form. The combined operation of the two transistors is illustrated in Fig. 17-10b where the two sets of characteristics have been brought together and aligned so that the two quiescent points coincide to form the back-to-back characteristics for the amplifier. The equations for these operating paths can be derived from the circuit diagram in Fig. 17-8. With $\beta \gg 1$, the load current is

$$i_L = i_{E1} + i_{E2} \approx i_{C1} + i_{C2} \tag{17-23}$$

Thus

$$v_{CE1} = V_{CC} - R_L i_L = V_{CC} - R_L(i_{C1} + i_{C2}) \tag{17-24}$$

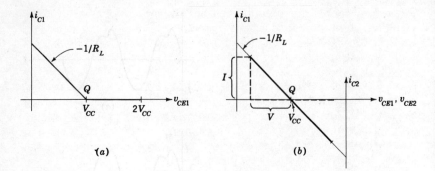

Fig. 17-10 Dynamic operating paths for the class B amplifier. (a) DOP for Q_1; (b) back-to-back DOPs for Q_1 and Q_2.

and similarly,

$$v_{CE2} = -V_{CC} - R_L i_L = -V_{CC} - R_L(i_{C1} + i_{C2}) \qquad (17\text{-}25)$$

These are the equations for the operating paths for Q_1 and Q_2. The collector current and voltage for the pnp transistor Q_2 are negative in the active region of the collector characteristics, and hence the operating path for this transistor lies in the third quadrant of the $v_{CE}\text{-}i_C$ plane.

Equations (17-24) and (17-25) show that the operating path for each transistor depends on the current in the other transistor as well as on its own current; however, since the mode of operation is class B, one or the other of these currents is always zero. When the input signal i_s is positive, $i_{C1} > 0$ and $i_{C2} = 0$, and Eq. (17-24) reduces to

$$v_{CE1} = V_{CC} - R_L i_{C1} = V_{CC} - v_L \qquad (17\text{-}26)$$

This is the normal form for the equation of the operating path of a single transistor, and it gives the portion of the operating path on the left of the quiescent point in Fig. 17-10a. When i_s is negative, $i_{C1} = 0$ and $i_{C2} < 0$, and in this case the two equations

$$i_{C1} = 0 \qquad (17\text{-}27)$$

and

$$v_{CE1} = V_{CC} - R_L i_{C2} = V_{CC} - v_L \qquad (17\text{-}28)$$

give the portion of the operating path on the right of the quiescent point in Fig. 17-10a. The operating path for Q_2 is determined from Eq. (17-25) in a similar manner, and similar results are obtained.

The operating paths for the two transistors can be brought together, as shown in Fig. 17-10b, and aligned so that their quiescent points coincide to form the combined back-to-back characteristics. Then for most purposes the amplifier can be thought of as having a single operating point moving along the operating path shown as a solid line in Fig. 17-10*b*. The operating point may lie on either side of the quiescent point, depending on whether i_s is positive or negative. It is important to note that, as in the class *A* amplifier, the collector break-down voltage of each transistor must be at least $2V_{CC}$. This fact is evident from the operating paths in Fig. 17-10. It is also evident directly from the circuit of Fig. 17-8; when one transistor is at the threshold of satura-tion, the full supply voltage $2V_{CC}$ must appear across the other transistor.

With these results the power relations can be developed for the class *B* amplifier. It follows from Fig. 17-8 that

$$v_L \doteq R_L i_L \approx R_L(i_{C1} + i_{C2})$$

$$v_{CE1} = V_{CC} - v_L \quad \text{and} \quad v_{CE2} = -V_{CC} - v_L$$

A typical waveform for i_L is shown in Fig. 17-9*d*. If the peak values of i_L and v_L are designated I and V, respectively, then it follows that

$$V = R_L I \tag{17-29}$$

The quantities V and I are indicated on the operating path in Fig. 17-10*b*. Using the relations developed in Sec. 17-1, the average power delivered to the load can be expressed as

$$P_L = K_1 R_L I^2 = K_1 \frac{V^2}{R_L} = K_1 VI \tag{17-30}$$

where $K_1 = (I_{rms})^2/I^2 = (V_{rms})^2/V^2$. The total average power drawn from the two power supplies is

$$P_{CC} = \frac{1}{T} \int_0^T V_{CC}|i_L| \, dt \tag{17-31}$$

A typical waveform for $|i_L|$ is shown in Fig. 17-11. Since V_{CC} is constant, Eq. (17-31) can be written as

$$P_{CC} = V_{CC}|i_L|_{av} = V_{CC}I_{av} \tag{17-32}$$

Now for a fixed waveform, I_{av} is proportional to I, and therefore it is con-venient to define a proportionality constant

$$K_2 = \frac{I_{av}}{I} \tag{17-33}$$

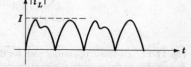

Fig. 17-11 Waveform of $|i_L|$ constructed from Fig. 17-9*d*.

When i_L is a sine wave, $K_2 = 2/\pi$, and when i_L is a square wave, $K_2 = 1$. Equation (17-32) can now be written as

$$P_{CC} = K_2 V_{CC} I \tag{17-34}$$

The collector-circuit efficiency is

$$\eta = \frac{P_L}{P_{CC}} = \frac{K_1 VI}{K_2 V_{CC} I} = \frac{K_1 V}{K_2 V_{CC}} \tag{17-35}$$

Thus for maximum efficiency the input signal must be adjusted so that V has its maximum possible value, $V = V_{CC}$, as illustrated in Fig. 17-12. Under maximum-signal conditions the efficiency is

$$\eta = \frac{K_1}{K_2} \tag{17-36}$$

For sine waves $\eta = \pi/4 = 78.5$ percent, and for square waves $\eta = 1 = 100$ percent. Under maximum-signal conditions Eq. (17-30) gives the output power as

$$P_L = K_1 \frac{V_{CC}^2}{R_L} \tag{17-37}$$

and Eq. (17-29) gives

$$V_{CC} = R_L I \tag{17-38}$$

The maximum power that can be delivered by a class A amplifier is usually limited by the collector dissipation in accordance with the graphical construction in Fig. 17-6b. Since the operating path for the class B amplifier intersects the v_{CE} axis at V_{CC}, as shown in Fig. 17-12,

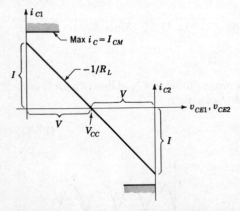

Fig. 17-12 Dynamic operating path for the class B amplifier under maximum-signal conditions.

instead of at $2V_{CC}$, it often turns out that the maximum power that can be delivered by a class *B* amplifier is limited by the current and voltage ratings of the transistors. In this case the maximum output power can be determined easily. Let the maximum permissible collector voltage be designated as V_{CEM}, and let the maximum permissible collector current be designated as I_{CM}. Then the output power is maximum when $V_{CC} = \frac{1}{2}V_{CEM}$ and when the operating path in Fig. 17-12 extends from I_{CM} at the upper left end to $-I_{CM}$ at the lower right end. With this adjustment Eq. (17-30) gives the output power as

$$P_L = K_1 VI = \tfrac{1}{2} K_1 V_{CEM} I_{CM} \tag{17-39}$$

It is not possible to get more output power than this from the amplifier without exceeding the voltage or current limit.

Since the output power from some class *B* amplifiers is limited by collector dissipation, it is necessary to examine that case also. In the class *A* amplifier, the maximum collector dissipation occurs under quiescent operating conditions, and thus it is sufficient to design the amplifier so that the quiescent point lies in a region of permissible collector dissipation. In the class *B* amplifier, however, the collector dissipation is zero under quiescent operating conditions, and hence the maximum dissipation occurs with signal applied. With a periodic signal of constant amplitude there is an average power dissipated over one or more whole cycles of the signal, and this power is not a function of time. There is also an instantaneous power dissipation that varies from instant to instant throughout the cycle of the signal. Since the transistor is such a tiny device, it has a small heat-storage capacity, and hence its temperature can change quite rapidly. As a consequence, the temperature of the transistor can change from instant to instant throughout the cycle, especially when the signal frequency is low. Thus to ensure that excessive temperatures are not generated in low-frequency applications, it is necessary to limit the instantaneous collector dissipation to a safe value.

The instantaneous collector dissipation is simply

$$p_C = v_{CE} i_C \tag{17-40}$$

As the operating point for the amplifier moves along the operating path shown in Fig. 17-13, v_{CE}, i_C, and p_C all change from point to point along the path. It is shown in connection with Fig. 10-4*a* that the maximum value of p_C occurs at the point where the operating path is tangent to a hyperbola of constant collector dissipation, and it is shown in connection with Fig. 10-4*b* that the point of tangency in Fig. 17-13 is located at $v_{CE1} = \frac{1}{2}V_{CC}$ and $i_{C1} = \frac{1}{2}I$. A similar set of conditions exists at the other end of the operating path for transistor Q_2. Thus it follows that under maximum-signal conditions, the maximum instantaneous collector dissipa-

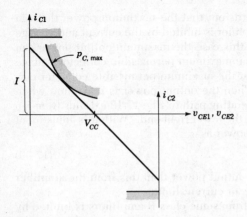

Fig. 17-13 Graphical analysis for the maximum
instantaneous collector dissipation.

tion for each transistor is

$$p_{C,\max} = \frac{1}{4} V_{CC} I = \frac{1}{4} \frac{V_{CC}^2}{R_L} \tag{17-41}$$

Then from Eq. (17-37) the average output power under maximum-signal
conditions is

$$P_L = 4K_1 p_{C,\max} \tag{17-42}$$

Hence the output power is proportional to the dissipation as in the
class A amplifier. If $p_{C,\max}$ is equal to the maximum permissible continuous collector dissipation $P'_{C,\max}$, then certainly the temperature of
the transistor cannot be excessive. In this case the output power can be
expressed as

$$P_L = 4K_1 P'_{C,\max} \tag{17-43}$$

under maximum-signal conditions. This is the maximum power that
can be obtained from the amplifier without having $p_{C,\max}$ exceed $P'_{C,\max}$.
For comparison, the maximum power that can be obtained from the
class A amplifier is

$$P_L = K_1 P'_{C,\max}$$

Thus in class B operation with $p_{C,\max} = P'_{C,\max}$, the output power per
transistor is twice as great as in class A operation. For a sinusoidal
signal, the maximum output power from the class B amplifier is

$$P_L = 2P'_{C,\max} \tag{17-44}$$

In designing the amplifier of Fig. 17-8, the objective is usually to obtain a specified output power P_L with a specified signal waveform (usually sinusoidal) and a specified load resistance R_L. Since there is no transformer available to match the load resistance to the transistors, the transistors must be chosen to match the load resistance. With P_L specified, V_{CC} can be determined from Eq. (17-37), and with R_L specified, the peak current I can be determined from Eq. (17-38). The maximum instantaneous collector dissipation $p_{C.\max}$ can be determined from Eq. (17-42). Then with this information and by remembering that the voltage rating of the transistors must be at least $2V_{CC}$, transistors having suitable voltage, current, and power ratings can be selected.

In some applications it is known that the class *B* amplifier will never be called on to amplify low-frequency signals. In this case the temperature of the transistors cannot change appreciably over a cycle of the signal, and hence there is no need to keep the instantaneous collector dissipation less than the maximum permissible continuous dissipation. Under these conditions safe operation is obtained by designing the amplifier so that the average value of the collector dissipation is less than the maximum permissible continuous dissipation, provided that the current and voltage ratings of the transistor are not exceeded. With this design the amplifier can deliver substantially more output power than the value given by Eq. (17-44). Therefore it is useful to evaluate the average collector dissipation and to relate it to the output power delivered by the amplifier.

Under quiescent conditions the collector dissipation in the class *B* amplifier is zero. As the amplitude of the signal is increased, the average collector dissipation P_C increases, passes through a maximum, and then decreases with further increases in signal amplitude. Thus the collector dissipation must be examined as a function of the signal amplitude to determine its maximum value. It turns out that the graph of P_C as a function of the peak signal voltage V is a parabola that depends on the load resistance R_L; three curves of this function are shown in Fig. 17-14 for different values of R_L. The maximum permissible continuous dissipation $P'_{C.\max}$ is also indicated in Fig. 17-14, and it follows that there is a minimum value that R_L can have without the possibility of excessive collector dissipation at some intermediate signal level. It should be noted that the maximum value that V can have is limited by collector saturation to V_{CC}. For sinusoidal signals, $V = V_{CC}$ corresponds to a point on the parabolas in Fig. 17-14 beyond the maximum but below the point of zero dissipation. For square waves, $V = V_{CC}$ corresponds to the point of zero dissipation. All these facts emerge from the analysis that follows.

The total average dissipation of the two transistors in the class *B*

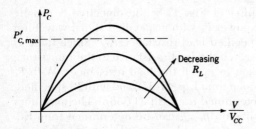

Fig. 17-14 Average collector dissipation in the class
B amplifier as a function of signal amplitude.

amplifier can be written, by conservation of energy, as

$$2P_C = P_{CC} - P_L \tag{17-45}$$

where P_C is the average dissipation in one transistor when the positive and negative halves of the signal waveform are identical. Substituting Eqs. (17-30) and (17-34) into (17-45) yields

$$P_C = \frac{1}{2}\left(K_2 V_{CC}I - K_1 \frac{V^2}{R_L}\right) \tag{17-46}$$

$$= \frac{1}{2}\left(K_2 \frac{V_{CC}V}{R_L} - K_1 \frac{V^2}{R_L}\right)$$

$$= \frac{1}{2}K_1 \frac{V_{CC}{}^2}{R_L}\left[\frac{K_2}{K_1}\left(\frac{V}{V_{CC}}\right) - \left(\frac{V}{V_{CC}}\right)^2\right] \tag{17-47}$$

The output power under maximum-signal conditions is given by Eq.
(17-37); for emphasis it can be written as

$$P_{L,\max} = K_1 \frac{V_{CC}{}^2}{R_L} \tag{17-48}$$

Now substituting this relation into Eq. (17-47) yields

$$P_C = \frac{1}{2} P_{L,\max}\left[\frac{K_2}{K_1}\left(\frac{V}{V_{CC}}\right) - \left(\frac{V}{V_{CC}}\right)^2\right] \tag{17-49}$$

This is the equation for the parabolas in Fig. 17-14. It gives $P_C = 0$
for $V/V_{CC} = 0$ and for $V/V_{CC} = K_2/K_1$, and it has its maximum value when
$V/V_{CC} = K_2/2K_1$. The maximum value of P_C is thus

$$P_{C,\max} = \frac{1}{8} P_{L,\max}\left(\frac{K_2}{K_1}\right)^2 = \frac{K_2{}^2}{8K_1} \frac{V_{CC}{}^2}{R_L} \tag{17-50}$$

For sinusoidal signals,

$$P_{C,\max} = \frac{2}{\pi^2} P_{L,\max} \approx \frac{1}{5} P_{L,\max} \tag{17-51}$$

In the absence of any special knowledge about the signal it must be anticipated that from time to time the signal amplitude will be such as to produce the maximum collector dissipation. If the maximum permissible continuous dissipation for the transistors is $P'_{C,\max}$, then for safe operation it is necessary that

$$P'_{C,\max} > P_{C,\max} = \frac{1}{8} \left(\frac{K_2}{K_1}\right)^2 P_{L,\max} \tag{17-52}$$

For sinusoidal signals this requirement becomes

$$P'_{C,\max} > \tfrac{1}{5} P_{L,\max} \tag{17-53}$$

or

$$P_{L,\max} < 5 P'_{C,\max} \tag{17-54}$$

It must be noted, however, that the results obtained above are derived for periodic signals and that they depend on the waveform of the signal. In some applications the signal is periodic and its waveform is known; this is nearly the case in ac servomechanisms. In most applications, however, the signal is not periodic and very little is known about the details of its waveform; this is the case in electronic sound systems. Nevertheless, the analysis presented above gives a useful insight into the voltage, current, and power relations in the class *B* amplifier.

Example 17-2

A class *B* power amplifier having the form shown in Fig. 17-8 is to be designed to deliver 20 watts of power to a 10-ohm load resistor under maximum-signal conditions with a sinusoidal signal. The problem is to determine the voltage, current, and power ratings required of the transistors.

Solution Equation (17-37) yields

$$V_{CC}^2 = \frac{P_L R_L}{K_1} = (20)(10)(2) = 400$$

or

$$V_{CC} = 20 \text{ volts}$$

Thus the maximum permissible collector-to-emitter voltage must be at least $2V_{CC} = 40$ volts.
     From Eq. (17-38),

$$I = \frac{V_{CC}}{R_L} = \frac{20}{10} = 2 \text{ amp}$$

and hence the maximum permissible collector current must be at least 2 amp.

The maximum instantaneous collector dissipation is given by Eq. (17-42) as

$$p_{C,\max} = \frac{P_L}{4K_1} = \frac{20}{(4)(0.5)} = 10 \text{ watts}$$

Thus safe operation is ensured under all conditions if the maximum permissible continuous collector dissipation is 10 watts or more. Under sinusoidal operating conditions, the greatest value that the average collector dissipation can have is given by Eq. (17-51) as

$$P_{C,\max} = \tfrac{1}{5}P_{L,\max} = \tfrac{20}{5} = 4 \text{ watts}$$

Thus in some applications it would be possible to use transistors with a dissipation rating of only 4 watts.

17-3 PUSH-PULL CLASS B POWER AMPLIFIERS USING TRANSFORMERS

Figure 17-15 shows an alternative circuit configuration that is often used as a class B power amplifier. This circuit does not use complementary symmetry, and hence both transistors can be of the same type, a fact that simplifies matters in some respects. Push-pull class B operation is obtained without complementary symmetry by the use of center-

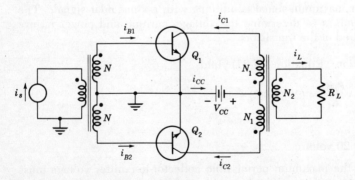

Fig. 17-15 Class B power amplifier using transformers.

tapped transformers at the input and output. These transformers have disadvantages associated with their size, weight, and cost, but they offer a compensating advantage by making impedance transformation possible.

Again, for simplicity, it is assumed that the input signal is supplied by a current source; the more general case is considered in Sec. 17-4. When the input signal i_s is positive, transistor Q_1 is turned on, and a current i_{B1} flows through the upper half of the input-transformer secondary. The forward bias developed across the emitter junction of Q_1 is coupled through the transformer to the emitter junction of Q_2 where it acts as a reverse bias keeping Q_2 cut off. When i_s is negative, Q_2 is turned on, i_{B2} flows in the lower half of the input-transformer secondary, and Q_1 is cut off. Thus, as in the case of the transformerless class *B* amplifier, Q_1 amplifies the positive portions of the signal, and Q_2 amplifies the negative portions. If the two halves of the circuit are identical, the amplified signal is reconstructed by the output transformer.

The waveform of a periodic input signal is shown in Fig. 17-16*a*, and the resulting base-current waveforms are shown in Fig. 17-16*b*. The transistors amplify these base currents and produce the collector-current waveforms shown in Fig. 17-16*c*; the sum of these collector currents is i_{CC}, the current flowing through the power supply V_{CC}. The way in which the output transformer reconstructs the amplified input-signal waveform can be examined with the aid of Fig. 17-17. The total magnetomotive force impressed on the core of the output transformer is, according to the polarity marks on the transformer,

$$\text{mmf} = N_1 i_{C1} - N_1 i_{C2} - N_2 i_L \qquad (17\text{-}55)$$

This expression can be rewritten as

$$N_2 i_L + \text{mmf} = N_1 (i_{C1} - i_{C2}) \qquad (17\text{-}56)$$

Now in a well-designed iron-core transformer, the net magnetomotive force impressed on the core is much smaller than the magnetomotive forces of the individual windings; hence $N_2 i_L \gg \text{mmf}$, and Eq. (17-56) yields

$$i_L = \frac{N_1}{N_2} (i_{C1} - i_{C2}) \qquad (17\text{-}57)$$

The waveform of this current is shown in Fig. 17-16*d*; it is an amplified copy of the input-signal waveform.

It is assumed that all the windings in the iron-core transformer link the same magnetic flux. Thus from Faraday's law the voltages in Fig. 17-17 are related by

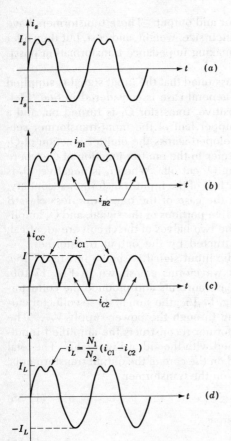

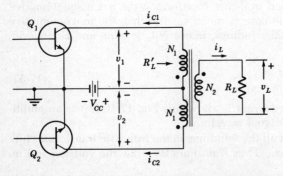

Fig. 17-16 Current waveforms for the amplifier in Fig. 17-15.

Fig. 17-17 Output circuit of the amplifier in Fig. 17-15.

$$\frac{v_1}{N_1} = -\frac{v_2}{N_1} = -\frac{v_L}{N_2}$$

or

$$v_1 = -v_2 = -\frac{N_1}{N_2}\,v_L = -\frac{N_1}{N_2}\,R_L i_L \qquad (17\text{-}58)$$

Now substituting Eq. (17-57) into (17-58) for i_L yields

$$v_1 = -v_2 = -\left(\frac{N_1}{N_2}\right)^2 R_L(i_{C1} - i_{C2}) = -R_L'(i_{C1} - i_{C2}) \qquad (17\text{-}59)$$

where

$$R_L' = \left(\frac{N_1}{N_2}\right)^2 R_L \qquad (17\text{-}60)$$

is the load resistance reflected into one half of the output-transformer primary as indicated in Fig. 17-17. Thus the collector voltage for Q_1 in Fig. 17-17 is

$$v_{CE1} = V_{CC} + v_1 = V_{CC} - R_L'(i_{C1} - i_{C2}) \qquad (17\text{-}61)$$

and for Q_2 it is

$$v_{CE2} = V_{CC} + v_2 = V_{CC} - R_L'(i_{C2} - i_{C1}) \qquad (17\text{-}62)$$

These are the equations for the dynamic operating paths for Q_1 and Q_2; they are equivalent to Eqs. (17-24) and (17-25). The operating path for one transistor is shown in Fig. 17-18a. As in the case of the transformer-less class *B* amplifier, the operating path for each transistor depends on the current in the other transistor as well as on its own current; however, since the mode of operation is class *B*, one or the other of these currents is always zero. When the input signal i_s is positive, $i_{C2} = 0$ and Eq. (17-61) reduces to the normal form for the operating path of a single transistor; this is the portion of the operating path on the left of the quiescent point in Fig. 17-18a. When i_s is negative, $i_{C1} = 0$ and the operating path for Q_1 is the horizontal line on the right of the quiescent point. With i_s negative, the voltage v_{CE1} depends on the nonzero current i_{C2}, as indicated by Eq. (17-61), because i_{C2} flowing in the lower half of the transformer primary induces a voltage in the upper half; the relation $v_1 = -v_2$ holds at all times. Thus, as in the case of the transformer-less class *B* amplifier, under maximum-signal conditions the transistors must be able to withstand a collector voltage of at least $2V_{CC}$.

     The back-to-back characteristics for the amplifier of Fig. 17-17 are shown in Fig. 17-18b; these characteristics are exactly the same in form as the back-to-back characteristics for the transformerless class *B* amplifier. It follows from this fact that all the voltage, current, and power

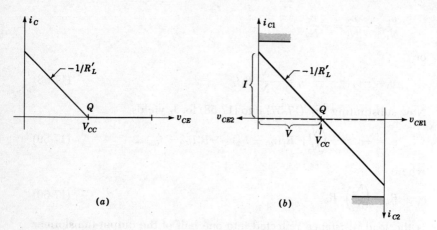

Fig. 17-18 Operating paths for the class B amplifier with transformer output. (a) Operating path for one transistor; (b) operating paths for both transistors on the back-to-back characteristics.

relations derived in Sec. 17-2 apply equally well to the amplifier in Fig. 17-17 when R_L in these relations is replaced by R_L'; hence there is no need to repeat the derivations here. However, in the design of the amplifier of Fig. 17-17 an extra degree of flexibility is available since R_L' can be adjusted by adjusting the turn ratio of the transformer.

Example 17-3

A class B power amplifier having the form shown in Fig. 17-15 uses transistors having the following maximum current, voltage, and dissipation ratings: $I_{CM} = 750$ mA, $V_{CEM} = 60$ volts, and $P'_{C.\text{max}} = 10$ watts. The amplifier is to be designed for maximum output power with maximum efficiency subject to these limitations. A sinusoidal signal can be assumed for the design calculations.

Solution The output power may be limited by the current and voltage ratings of the transistors, or it may be limited by collector dissipation. For a first trial it is assumed that the limit is imposed by the current and voltage ratings. Then Eq. (17-39) gives the maximum-signal output power with a sinusoidal signal as

$$P_L = \tfrac{1}{2} K_1 V_{CEM} I_{CM} = (\tfrac{1}{2})(\tfrac{1}{2})(60)(0.75) = 11.25 \text{ watts}$$

The greatest value that the average collector dissipation can have with a sinusoidal signal is given by Eq. (17-51) as

$$P_{C.\text{max}} = \tfrac{1}{5} P_{L.\text{max}} = \tfrac{11.25}{5} = 2.25 \text{ watts}$$

This value is well below the maximum permissible continuous dissipation of 10 watts. The maximum instantaneous collector dissipation is given by Eq. (17-42) as

$$p_{C.\text{max}} = \frac{P_L}{4K_1} = \frac{(11.25)(2)}{4} = 5.6 \text{ watts}$$

This value is also well below the transistor rating, and thus the assumption that the output power of this amplifier is limited by the maximum current and voltage ratings is a correct assumption.

In order to obtain the maximum output power from the amplifier, the power-supply voltage must be

$$V_{CC} = \tfrac{1}{2}V_{CEM} = 30 \text{ volts}$$

The load resistance that must be seen at the primary of the output transformer is given by Eq. (17-38) as

$$R'_L = \frac{V_{CC}}{I} = \frac{30}{0.75} = 40 \text{ ohms}$$

This quantity fixes the transformer turn ratio when the actual load resistance R_L is specified.

17-4 SOME PRACTICAL ASPECTS OF CLASS *B* TRANSISTOR AMPLIFIERS

When the class *B* amplifier in Fig. 17-15 is driven by a practical signal source rather than by an ideal current source, it can be represented as shown in Fig. 17-19a. In this circuit the two voltage sources v_s and the two resistors R_s represent the equivalent circuits seen when looking into each half of the input-transformer secondary with the other half open circuited. The back-to-back volt-ampere characteristic for the input circuit is shown in Fig. 17-19b. As this characteristic shows, i_{B1} cannot have any appreciable value until v_s is made positive by a few tenths of a volt (see Fig. 9-3), and i_{B2} cannot have any appreciable value until v_s is made negative by a few tenths of a volt. The resulting dead zone near $v_s = 0$ causes waveform distortion called crossover distortion. Typical waveforms of i_{B1} and i_{B2} exhibiting crossover distortion are shown in Fig. 17-19c; these waveforms should be compared with the ideal waveforms shown in Fig. 17-16b. The corresponding waveforms of collector current exhibit the same distortion, and consequently the waveform of the load current is distorted as shown in Fig. 17-19d.

When the input signal for the amplifier is supplied by an ideal current source and an ideal transformer as in Fig. 17-15, the current source forces the waveforms of i_{B1} and i_{B2} to follow the waveform of the input

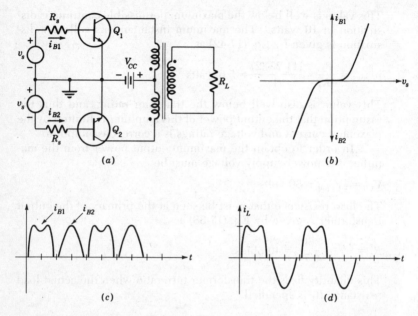

Fig. 17-19 Crossover distortion in class B transistor amplifiers. (*a*) Circuit; (*b*) back-to-back input characteristics; (*c*) base-current waveforms; (*d*) load-current waveform.

current in spite of the nonlinear volt-ampere characteristics of the emitter junctions. Thus to minimize crossover distortion, the input signal source should have a high internal resistance so that it approximates a current source as well as possible.

Crossover distortion can be further reduced by supplying the transistors with a small turn-on bias just sufficient to bring them to the threshold of strong conduction. Figure 17-20*a* shows a widely used circuit in which the turn-on bias is provided by a forward-biased *pn* junction. This diode-biasing scheme is very similar to the one shown in Fig. 10-12, and it has similar properties of stabilizing the quiescent point against variations in temperature. The back-to-back input characteristics for the two transistors with diode bias are shown as colored lines in Fig. 17-20*b*, and it follows from Eq. (17-57) that

$$i_L = \frac{N_1}{N_2}(i_{C1} - i_{C2}) = \beta \frac{N_1}{N_2}(i_{B1} - i_{B2})$$

Thus the load current depends only on the difference $i_{B1} - i_{B2}$, and as is indicated by the black line in Fig. 17-20*b*, this difference is a reason-

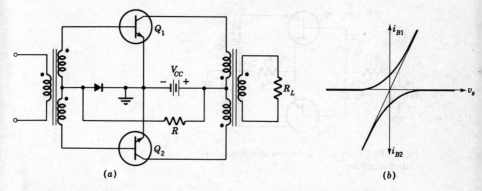

Fig. 17-20 Turn-on bias for the class B amplifier to minimize crossover distortion. (*a*) Circuit; (*b*) back-to-back input characteristics.

ably linear function of the input signal voltage v_s when diode bias is used. Hence the turn-on bias reduces the crossover distortion.

As is pointed out in Sec. 17-3, accurate reconstruction of the signal waveform at the output of the class B amplifier requires that the two halves of the amplifier be well matched. Because of the production-line spread in transistor parameters, it is not feasible to obtain a really good match in mass production. Hence when low distortion is required, it is usually necessary to use substantial amounts of negative feedback to reduce the distortion introduced by the amplifier. The use of feedback for this purpose is examined in some detail in Sec. 13-5.

The maximum output power that can be obtained from a transistor amplifier is often limited by the maximum permissible collector dissipation. It is therefore important that the maximum permissible dissipation for any given transistor type can be increased by the use of a heat sink; the factor by which the dissipation rating is increased may be as great as 5, or even greater. Thus when output powers greater than about 1 watt are required, it is customary to use heat sinks. With the use of heat sinks it is often possible to increase the dissipation rating to the point where the output power is limited by the maximum voltage and current ratings of the transistors. Thermal considerations and heat sinks are discussed briefly in Sec. 10-2.

When the transformerless class B amplifier of Fig. 17-8 is driven by a practical signal source rather than by an ideal current source, it also produces crossover distortion. Again, as in the amplifier of Fig. 17-19a, this distortion can be reduced by providing a turn-on bias for the transistors as shown in Fig. 17-21. The two batteries V_o bias the transistors just at the threshold of strong conduction. In practical circuits

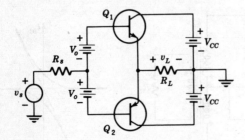

Fig. 17-21 Turn-on bias for the transformerless
class B amplifier.

the bias batteries are replaced by forward-biased diodes or by a suitable
resistive biasing network.

Imperfectly matched transistors in the amplifier of Fig. 17-21 also
cause waveform distortion. Useful insight into the nature of this source
of distortion can be gained by recalling that the two transistors operate
one at a time as emitter followers; thus when the signal voltage v_s is
positive, Q_1 is turned on and Q_2 is turned off. By replacing Q_1 with its
large-signal model, as in Fig. 12-14b, and applying the reduction theorem, the very simple model shown in Fig. 17-22a is obtained. This
model applies when v_s is positive. Similarly, when v_s is negative, the
model shown in Fig. 17-22b applies. These models show clearly the
effect of unbalanced current gains on the output-signal waveform; the
output voltage is simply the input voltage reduced by a resistive voltage
divider. If $\beta_1 \neq \beta_2$, the voltage-divider ratio is not the same for positive
and negative portions of the input signal. However, if both current
gains are large and if R_s is small so that $(1 + \beta)R_L \gg R_s$, then the voltage-
divider ratio is approximately unity for both positive and negative v_s.
Thus the waveform distortion can be made very small by making the
current gains large and by making the source resistance R_s small; then
$v_L = v_s$ to a good approximation, and the output waveform is virtually the
same as the input waveform. A circuit that is used to achieve this result

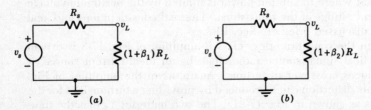

Fig. 17-22 Large-signal models for the amplifier of Fig. 17-21. (a) $v_s > 0$, Q_1
conducts; (b) $v_s < 0$, Q_2 conducts.

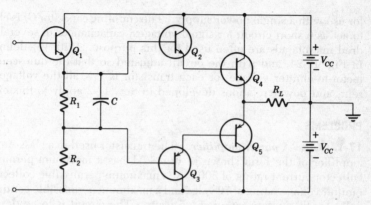

Fig. 17-23 A high-performance transformerless class *B* power amplifier.

is shown in Fig. 17-23. The Darlington-connected pairs Q_2–Q_4 and Q_3–Q_5 have current gains of several thousand, thus satisfying the high-gain requirement. The input signal is obtained from the emitter follower Q_1, and thus the requirement of a small source resistance is met. The voltage drop across R_1 provides the turn-on voltage for the Darlington pairs, and the capacitor C is a bypass capacitor; the bypass capacitor is frequently omitted. This amplifier and several variations of it are widely used in electronic systems.

The transformerless amplifiers discussed above all require two power supplies. In many electronic systems two power supplies are normally available; in many others, however, such as home radio receivers and record players, cost considerations prohibit the use of two power supplies. Figure 17-24 shows the amplifier of Fig. 17-23 modified

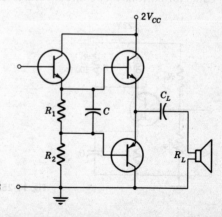

Fig. 17-24 A transformerless class *B* amplifier with a single power supply.

for use with a single power supply. The coupling capacitor C_L is chosen to act as a short circuit for signal voltages; capacitances of several hundred microfarads are often used for this purpose. With V_{CC} defined as in Fig. 17-24, and with the circuit adjusted so that the quiescent collector-to-emitter voltage for each transistor is V_{CC}, all the voltage, current, and power relations developed in Sec. 17-2 apply to this circuit.

PROBLEMS

17-1. *Class A power amplifier.* The transistor used in a class A power amplifier of the form shown in Fig. 17-1 has a maximum permissible collector current rating of 500 mA, a maximum permissible collector-to-emitter voltage rating of 70 volts, and a maximum permissible continuous collector dissipation rating of 3.5 watts. The circuit is to be designed to deliver the greatest possible power in class A operation to a 10-ohm load resistor under sinusoidal operating conditions with maximum efficiency and with the smallest possible input signal i_B.

(*a*) Sketch the permitted operating region on the collector characteristic.

(*b*) Determine the required values of the effective power-supply voltage V_{CC}, the quiescent collector current I_C, and the reflected load resistance R_L'.

(*c*) What is the maximum power that can be delivered to the load with a sinusoidal signal?

17-2. *Class A power amplifier.* The 100-ohm load resistor in the amplifier of Fig. 17-25 provides a suitable dynamic operating path without the need for impedance transformation. The inductor L in parallel with the load keeps the dc component of collector current out of the load; this inductor acts as an open circuit to the time-varying component of collector current. The direct voltage drop across R_e is 2 volts. The am-

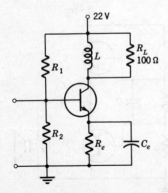

Fig. 17-25 Amplifier for Prob. 17-2.

plifier is to be designed to deliver the greatest possible power to the load in class A operation with a sinusoidal signal.

(a) What is the required value of quiescent collector current?

(b) What are the maximum values of collector current, voltage, and dissipation that the transistor must withstand?

(c) What is the power delivered to the load under maximum-signal conditions with a sinusoidal signal?

17-3. Design of a class A power amplifier. The transistor used in a power amplifier having the form shown in Fig. 17-1 has a maximum permissible collector voltage $v_{CE} = 50$ volts and a maximum permissible collector current $i_C = 250$ mA. The circuit is to be designed for the maximum output power under sinusoidal conditions with the smallest possible input signal i_B. The maximum transistor temperature is not to exceed 175°C when the ambient temperature is 50°C (122°F), and the thermal resistance of the transistor is $\theta = 50$°C/watt.

(a) Determine the maximum permissible collector dissipation when the ambient temperature is 50°C. (See Sec. 10-2.)

(b) Sketch the permitted operating region on the collector characteristics.

(c) Determine the required quiescent collector current I_C, the required quiescent collector voltage V_{CE}, and the required value of R'_L.

(d) What is the maximum output power in class A operation with a sinusoidal signal?

17-4. Design of a class A power amplifier. A transistor amplifier having the form shown in Fig. 17-1 is to be designed to have its quiescent operating point at $I_C = 100$ mA and $V_{CE} = 25$ volts, and the stability factor for the amplifier is to be $S_e = 0.1$. The voltage drop across R_e is to be 2 volts, and the transistor parameters are $\beta = 100$ and $V_o = 0.7$ volt.

Determine the required values of V'_{CC}, R_e, R_1, and R_2.

17-5. Design of a transformerless class B amplifier. Two matched complementary transistors are used in the circuit of Fig. 17-8 to deliver 10 watts to a 15-ohm loudspeaker under sinusoidal operating conditions.

(a) Determine the smallest value of V_{CC} that can be used.

(b) With the power-supply voltage determined in part a, what are the maximum values of i_C and v_{CE} that the transistor must withstand?

(c) What is the maximum instantaneous collector dissipation in each transistor when the amplifier is delivering the specified output power?

17-6. Transformerless class B amplifier. The complementary transistors used in the circuit of Fig. 17-8 have maximum collector voltage ratings of 50 volts and maximum collector current ratings of 3 amp.

When equipped with suitable heat sinks, the transistors have a maximum permissible continuous collector dissipation rating of 30 watts.

(a) What is the maximum output power that the amplifier can deliver with a sinusoidal signal?

(b) Specify the values of V_{CC} and R_L needed to obtain the output power found in part a.

17.-7. Class B amplifier with transformers. A class B amplifier like the one shown in Fig. 17-15 is required to deliver 10 watts to a 4-ohm load with $V_{CC} = 30$ volts under sinusoidal operating conditions. The circuit is to be designed for maximum efficiency $(V = V_{CC})$, and it is to deliver the specified output power with the smallest possible input signal i_B.

(a) Determine the required value of reflected load resistance R_L'.

(b) Determine the maximum instantaneous collector current, voltage, and power dissipation under maximum-signal conditions.

(c) If the sinusoidal signal can have any amplitude between zero and the maximum-signal amplitude, what is the greatest value that the average collector dissipation in each transistor can have?

17-8. Class B amplifier with transformers. The transistors used in a class B amplifier like the one shown in Fig. 17-15 have maximum collector voltage ratings of 100 volts and maximum collector current ratings of 1.0 amp. In addition, the instantaneous collector power dissipation is not to exceed 10 watts. The amplifier is to be designed to deliver maximum power to a 10-ohm load under sinusoidal operating conditions with maximum efficiency and the smallest possible input signal i_B.

(a) Determine the values of V_{CC} and R_L' required.

(b) What is the maximum-signal output power with a sinusoidal signal?

17-9. Distortion in a transformerless class B amplifier. Two complementary transistors are used in the amplifier of Fig. 17-21 with $V_{CC} = 20$ volts. The load is a 20-ohm loudspeaker, the resistance of the signal source is $R_s = 100$ ohms, and the batteries V_o bias the transistors just at the threshold of strong conduction. The current gains of Q_1 and Q_2 are $\beta_1 = 50$ and $\beta_2 = 100$, respectively.

(a) If $v_s = 15 \cos \omega t$, determine the positive and negative peak values of the voltage across the load.

(b) The two transistors are replaced by Darlington-connected pairs, and as a result the effective current gains are $\beta_1 = 2,000$ and $\beta_2 = 4,000$. Repeat the calculation of part a.

Derivation
of the Volt-ampere Law
of the FET

The conventional treatment of the junction FET involves a rather long mathematical analysis that is by no means trivial. The classical result obtained for the pinched-off drain current is given by Ref. 1 listed at the end of this appendix as

$$i_D = I_{DSS}\left[1 - 3\left(\frac{v_{GS}}{V_P}\right) + 2\left(\frac{v_{GS}}{V_P}\right)^{3/2}\right] \quad (A1\text{-}1)$$

This complicated expression is then approximated by the simpler square-law relation given in Eq. (6-11),

$$i_D = I_{DSS}\left(1 - \frac{v_{GS}}{V_P}\right)^2 \quad (A1\text{-}2)$$

These two equations are compared in Chap. 1 of Ref. 1, and the comparison shows that the two equations give results that are in surprisingly close agreement. The simpler form is almost universally used in engineering calculations.

The simplified derivation presented here is based on one given by Middlebrook in Ref. 2; it makes the key approximation at the *beginning* of the study and then proceeds by a relatively simple analysis to Eq. (A1-2).

Figure A1-1a illustrates conditions existing inside an n-channel FET when a negative bias is applied to the gate with $v_{DS} = 0$, and Fig. A1-1b shows the conditions when a positive voltage is applied to the drain in addition to the negative gate bias. The shaded regions in these diagrams represent the depletion layers at the gate junctions. The

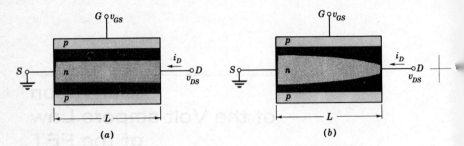

Fig. A1-1 Conditions inside a junction FET. (a) $v_{GS} < 0$, $v_{DS} = 0$; (b) $v_{GS} < 0$, $v_{DS} > 0$.

mobile carriers in the channel are free electrons, and the magnitude of the total charge associated with these carriers is designated as Q. If the drain is biased positively with respect to the source, then the free electrons move from source to drain, and if the average transit time for carriers traversing the channel is designated as τ, then the channel current is

$$i_D = \frac{Q}{\tau} \tag{A1-3}$$

By the definition of the pinchoff voltage, the channel ceases to exist and Q is just reduced to zero when $v_{GS} = V_P$. When v_{GS} is less negative than V_P, Q is greater than zero, and the magnitude of Q depends on the gate voltage.

Consider first the case, illustrated in Fig. A1-1a, in which $v_{DS} = 0$. There is space-charge neutrality in the channel, and the concentration of carriers in the channel is equal to the concentration of donor atoms. Thus the charge Q depends on the thickness of the channel, and the thickness of the channel depends in turn on the thickness of the depletion layers at the gate junctions. The thickness of the depletion layers is a nonlinear function of the voltage across the junction as indicated by Eq. (5-15). However, at this point the simplifying approximation is made that Q is a linear function of the junction voltage; the principal justification for this linear approximation is the fact that it yields results that are in reasonably good agreement with the more exact analysis and with the measured behavior of real transistors. Thus, since $Q = 0$ when $v_{GS} = V_P$, the charge in the channel is expressed as

$$Q = C(v_{GS} - V_P) \tag{A1-4}$$

where C is the constant gate capacitance resulting from the linear charge-voltage approximation. In normal operation, both v_{GS} and V_P are negative numbers.

Conditions existing in the transistor when v_{DS} is positive, but not positive enough to produce pinchoff, are illustrated in Fig. A1-1b. In this case the voltage across the junction varies from point to point along the channel. At the source end it is

$$v_j = v_{GS} \tag{A1-5}$$

and at the drain end it is

$$v_j = v_{GS} - v_{DS} \tag{A1-6}$$

The average value of v_j giving Q under the linear approximation is a value lying between these two extremes, and it can be expressed as

$$v_j = v_{GS} - \frac{1}{n} v_{DS} \tag{A1-7}$$

where n is a number greater than unity. Thus when v_{DS} is not zero, Eq. (A1-4) becomes

$$Q = C \left(v_{GS} - \frac{1}{n} v_{DS} - V_P \right)$$

and substituting this value of Q into Eq. (A1-3) yields

$$i_D = \frac{C}{\tau} \left(v_{GS} - V_P - \frac{1}{n} v_{DS} \right) \tag{A1-8}$$

The average transit time in Eq. (A1-8) is a function of v_{DS}, and this relationship must be evaluated to complete the volt-ampere law for the FET. The net flow of carriers along the channel is due to their drift velocity, and this velocity is

$$v_d = -\mu \mathscr{E} \tag{A1-9}$$

where μ is the mobility of the carriers and $\mathscr{E}$ is the electric field strength established in the channel by v_{DS}. This velocity is directed along the electric field lines. The distance traveled by a carrier in the time dt is

$$ds = v_d \, dt = -\mu \mathscr{E} \, dt \tag{A1-10}$$

in the direction of v_d. As the carrier travels from the source to the drain, the time elapsed is the transit time; thus

$$\int_S^D ds = \int_0^\tau -\mu \mathscr{E} \, dt \tag{A1-11}$$

where the integral is evaluated along an electric field line. Since the length of the channel is much greater than its thickness, the field lines are nearly parallel to the axis of the channel, and the length of every path along a field line is approximately the length L of the axis. Thus

Eq. (A1-11) is approximately

$$L = \int_0^\tau -\mu \mathscr{E} \, dt \tag{A1-12}$$

In order to evaluate the integral on the right, it is necessary to know how the field experienced by the carrier varies with time while the carrier is in transit. This problem is circumvented by multiplying both sides of (A1-12) by ds and integrating again over the same path; the result is

$$\int_S^D L \, ds = \int_S^D \int_0^\tau -\mu \mathscr{E} \, dt \, ds \tag{A1-13}$$

Evaluating the integral on the left and interchanging the order of integration on the right yields

$$L^2 = \mu \int_0^\tau \int_S^D -\mathscr{E} \, ds \, dt \tag{A1-14}$$

Now the inner integral on the right is the definition of v_{DS}; hence

$$L^2 = \mu v_{DS} \int_0^\tau dt = \mu v_{DS} \tau \tag{A1-15}$$

and

$$\tau = \frac{L^2}{\mu v_{DS}} \tag{A1-16}$$

Substituting the value of τ given by Eq. (A1-16) into Eq. (A1-8) yields

$$i_D = \frac{\mu v_{DS} C}{L^2} \left(v_{GS} - V_P - \frac{1}{n} \, v_{DS} \right) \tag{A1-17}$$

$$= \frac{\mu C}{nL^2} \left[n(v_{GS} - V_P)v_{DS} - v_{DS}{}^2 \right] \tag{A1-18}$$

At the pinchoff point, $v_{DS} = v_{GS} - V_P$, and Eq. (A1-18) reduces to

$$i_D = \frac{\mu C}{nL^2} (n - 1)(v_{GS} - V_P)^2 \tag{A1-19}$$

For $n = 2$, (A1-18) becomes

$$i_D = \frac{\mu C}{2L^2} \left[2(v_{GS} - V_P)v_{DS} - v_{DS}{}^2 \right] \tag{A1-20}$$

This is the relation given by Eq. (6-8) for the drain current when the channel is not pinched off. With this value of n, (A1-19) becomes

$$i_D = \frac{\mu C}{2L^2}(v_{GS} - V_P)^2 \tag{A1-21}$$

$$= I_{DSS}\left(1 - \frac{v_{GS}}{V_P}\right)^2 \tag{A1-22}$$

This expression is identical with Eqs. (A1-2) and (6-11).

The derivation presented above is formulated in terms of the junction FET. Similar reasoning can be applied to both depletion and enhancement MOS transistors with similar results. In fact, the more exact analysis of the enhancement MOST leads to the results given by Eq. (A1-20) above. In this device the thickness of the channel is much smaller than the thickness of the oxide between the gate and the channel, and therefore the gate and channel act as a parallel-plate capacitor with its plates separated by the oxide layer. Measured values of this capacitance show that it is nearly independent of gate voltage except near drain-current cutoff, and hence the linear charge-voltage approximation is a very good approximation for the enhancement MOST. This fact is reflected in the excellent square-law behavior of this device. When C is evaluated as a parallel-plate capacitor with its plates separated by the oxide layer, the coefficient multiplying the bracket in Eq. (A1-20) takes the form given by Eq. (6-47).

REFERENCES

1. Sevin, L. J., Jr.: "Field-effect Transistors," McGraw-Hill Book Company, New York, 1965.
2. Middlebrook, R. D.: A Simple Derivation of Field-effect Transistor Characteristics, *Proc. IEEE*, vol. 51, August, 1963 (Letters).

Analysis of the Diode-biased Amplifier

Figure A2-1 shows the circuit diagram of the diode-biased amplifier used in conjunction with a difference amplifier in Fig. 12-26; the currents i_{C1} and i'_{C1} are the collector currents in the difference amplifier. In the interest of algebraic simplicity it is assumed that $v_{BE} = V_o = \text{const}$ and $i_C = \beta i_B$. Then the node equation at the collector of Q_2 is

$$\left(\frac{1}{R_1} + \frac{2}{R_2}\right) v_{CE2} - \frac{1}{R_1} V_{CC} - \frac{2}{R_2} V_o = -i_{C2} \tag{A2-1}$$

$$= -\beta \left(\frac{v_{CE2} - V_o}{R_2} - i_{C1}\right) \tag{A2-2}$$

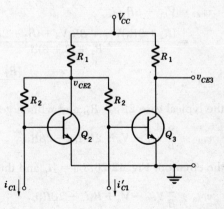

Fig. A2-1 The diode-biased amplifier.

Collecting terms yields

$$\left(\frac{1}{R_1} + \frac{2+\beta}{R_2}\right) v_{CE2} = \frac{1}{R_1} V_{CC} + \frac{2+\beta}{R_2} V_o + \beta i_{C1} \tag{A2-3}$$

Solving this expression for v_{CE2} gives

$$v_{CE2} = \frac{(2+\beta)R_1 V_o + R_2 V_{CC} + \beta R_1 R_2 i_{C1}}{R_2 + (2+\beta)R_1} \tag{A2-4}$$

The base current for Q_3 is now

$$i_{B3} = \frac{v_{CE2} - V_o}{R_2} - i'_{C1} \tag{A2-5}$$

$$= \frac{(2+\beta)R_1 V_o}{R_2[R_2 + (2+\beta)R_1]} + \frac{V_{CC} + \beta R_1 i_{C1}}{R_2 + (2+\beta)R_1} - \frac{V_o}{R_2} - i'_{C1} \tag{A2-6}$$

The collector currents for the difference amplifier can now be expressed as

$$i_{C1} = I_{C1} + i_{c1} \qquad \text{and} \qquad i'_{C1} = I_{C1} - i_{c1} \tag{A2-7}$$

where I_{C1} is the constant quiescent component of current and i_{c1} is the signal component. Substituting these expressions into (A2-6) and collecting terms yields

$$i_{B3} = \frac{V_{CC} - V_o + \beta R_1 I_{C1} + \beta R_1 i_{c1}}{R_2 + (2+\beta)R_1} - I_{C1} + i_{c1} \tag{A2-8}$$

$$= \frac{V_{CC} - V_o - (R_2 + 2R_1)I_{C1} + [R_2 + 2(1+\beta)R_1]i_{c1}}{R_2 + (2+\beta)R_1} \tag{A2-9}$$

The voltage at the collector of Q_3 can now be evaluated as

$$v_{CE3} = V_{CC} - \beta R_1 i_{B3} \tag{A2-10}$$

$$= \frac{(R_2 + 2R_1)V_{CC} + \beta R_1 V_o + (R_2 + 2R_1)\beta R_1 I_{C1}}{R_2 + (2+\beta)R_1}$$

$$- \frac{[R_2 + 2(1+\beta)R_1]\beta R_1 i_{c1}}{R_2 + (2+\beta)R_1} \tag{A2-11}$$

In the typical case $2R_1 \gg R_2$, and with large β, (A2-11) reduces to

$$v_{CE3} \approx \frac{2}{\beta} V_{CC} + V_o + 2R_1 I_{C1} - 2\beta R_1 i_{c1} \tag{A2-12}$$

In the circuit of Fig. 12-26, $I_{C1} = \frac{1}{2}I_o$ and thus

$$v_{CE3} \approx \frac{2}{\beta} V_{CC} + V_o + R_1 I_o - 2\beta R_1 i_{c1} \tag{A2-13}$$

$$= V_{CE3} + v_o \tag{A2-14}$$

where the signal component of the output voltage is

$$v_o = -2\beta R_1 i_{c1} \tag{A2-15}$$

With large β the quiescent collector voltages are

$$V_{CE2} \approx V_{CE3} = \frac{2}{\beta} V_{CC} + V_o + R_1 I_o \tag{A2-16}$$

The dominant term on the right-hand side of Eq. (A2-16) is usually the last one. Thus if the current source I_o in the amplifier of Fig. 12-26 is approximated by a transistor circuit like the one in Fig. 10-14b, I_o varies inversely with the resistance level in the amplifier, and the quiescent collector voltages are essentially independent of variations in transistor parameters and resistance level from one microamplifier to another.

Analysis
of the BJT Feedback Pair

The circuit diagram for the BJT feedback pair of Sec. 16-5, exclusive of
the biasing network, is shown in Fig. A3-1a. The capacitor C is a large
dc blocking capacitor that acts as a short circuit at all signal frequencies.
Thus the impedance of the feedback path from the collector of Q_2 to the
emitter of Q_1 is

$$Z_f = \frac{R_f}{1 + sC_fR_f} \qquad\qquad\qquad (A3\text{-}1)$$

It is convenient to define

$$\omega_f = \frac{1}{C_fR_f} \qquad\qquad\qquad (A3\text{-}2)$$

and to express (A3-1) as

$$Z_f = R_f \cdot \frac{\omega_f}{s + \omega_f} \qquad\qquad\qquad (A3\text{-}3)$$

The principal approximation used in this analysis is based on the
assumption that the feedback impedance Z_f is large compared to R_2
and R_e. Under this assumption Z_f does not load the collector of Q_2, and
hence the feedback path can be disconnected from the collector and connected to a controlled voltage source having the voltage V_o as shown in
Fig. A3-1b. Further simplification is achieved by replacing the circuit
connected between the emitter of Q_1 and ground with its Thévenin
equivalent. The impedance of the Thévenin equivalent circuit is

$$Z_{TH} = \frac{R_eZ_f}{R_e + Z_f} \qquad\qquad\qquad (A3\text{-}4)$$

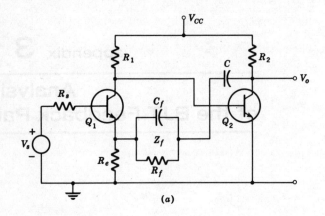

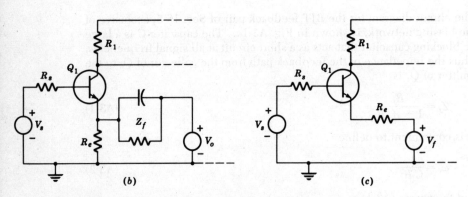

Fig. A3-1 A BJT feedback pair. (*a*) Circuit; (*b*) an approximate equivalent circuit; (*c*) the circuit simplified by Thévenin's theorem.

However, R_e is normally of the order of 100 ohms, and hence it is much smaller than Z_f. Thus the Thévenin equivalent impedance is approximately

$$Z_{TH} = R_e \tag{A3-5}$$

The internal voltage for the Thévenin equivalent circuit, designated as V_f for feedback voltage, is

$$V_f = \frac{R_e}{R_e + Z_f} V_o \approx \frac{R_e}{Z_f} V_o \tag{A3-6}$$

The Thévenin equivalent circuit is shown in Fig. A3-1c.

A small-signal, low-frequency model for the first stage of the am-

plifier, using the Thévenin equivalent circuit developed above, is
shown in Fig. A3-2a. The effects of the bias network, not shown in Fig.
A3-1, are lumped in with V_s' and R_s' with the aid of Thévenin's theorem.
The circuit in Fig. A3-2a can be simplified by replacing the current
source βI_b with an equivalent pair of sources connected to ground (see
Fig. 12-3 and the related discussion) and then applying the reduction
theorem. The simplified circuit is shown in Fig. A3-2b, and when the
components are rearranged, the circuit takes the form shown in Fig.
A3-2c. Since V_f is proportional to the output voltage V_o in accordance
with Eq. (A3-6), the source V_f accounts for the feedback in the circuit.

If the feedback is removed by replacing Z_f with an open circuit,
then V_f becomes zero. Under this condition the complex open-loop
voltage gain, including the effects of R_e, is characterized by two poles
[see Eq. (14-99)], and it can be expressed as

$$\frac{V_o}{V_s'} = A_v' = \frac{K_v'\omega_1'\omega_2'}{(s + \omega_1')(s + \omega_2')} \tag{A3-7}$$

In the usual case the break frequencies ω_1' and ω_2' are widely separated
as in Fig. 14-25 and Example 14-3; thus K_v', ω_1', and ω_2' are easily determined from the measured open-loop gain characteristic.

It follows from Fig. A3-2c that, when the feedback is restored to the

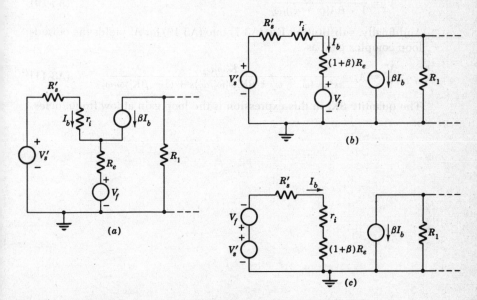

Fig. A3-2 Small-signal models for the first stage of the amplifier in Fig. A3-1.

circuit, the only change is that the feedback voltage V_f subtracts directly from V_s', and thus

$$V_o = A_r'(V_s' - V_f) \tag{A3-8}$$

This expression is based on the low-frequency models in Fig. A3-2, and it is valid at low frequencies. However, when the transistor capacitances are added to the model to account for high-frequency effects, it is seen that the expression involves some approximations. A more detailed analysis supported by experimental studies shows that these approximations are fully justified. The feedback voltage V_f in (A3-8) is given by Eqs. (A3-3) and (A3-6) as

$$V_f = \frac{R_e}{Z_f} V_o = \frac{R_e}{R_f} \frac{s + \omega_f}{\omega_f} V_o$$

The ratio R_e/R_f is the feedback ratio at low frequencies; hence it is convenient to designate it as β and write the feedback voltage as

$$V_f = \beta \frac{s + \omega_f}{\omega_f} V_o \tag{A3-9}$$

Now substituting this expression into Eq. (A3-8) and solving for V_o gives

$$V_o = \frac{A_r' V_s'}{1 + \beta A_r'(s + \omega_f)/\omega_f} \tag{A3-10}$$

And finally, substituting Eq. (A3-7) into (A3-10) for A_r' yields the closed-loop complex gain as

$$\frac{V_o}{V_s'} = A_r = \frac{K_r'\omega_1'\omega_2'}{s^2 + (\omega_1' + \omega_2' + \beta K_r'\omega_1'\omega_2'/\omega_f)s + (1 + \beta K_r')\omega_1'\omega_2'} \tag{A3-11}$$

The quantity $\beta K_r'$ in this expression is the loop gain at low frequencies.

Answers to
Selected Problems

2-2.   980, 9,800, 9.6(10^6)
2-5.   560
2-7.   (b) 25.24 volts; (c) 25.37 volts; (d) 1/41
2-8.   (b) 25.4 volts; (c) 0.255; (d) 5.2 volts

3-2.   4 : 35
3-4.   5 volts, 5 volts, 6.4
3-7.   10 volts, 1 mV, approximately 1 volt
3-10.   855 μF, approximately 9 volts, approximately 9.4 mV, rms
3-12.   (b) 1.5 μF; (c) 30 amp; (d) 75 μF
3-14.   (b) $\frac{1}{11}$; (c) 0.122 volt; (d) 2,500
3-15.   (a) 1,450 pF

4-2.   12.5 ohm-cm, 0.8 cm
4-4.   0.805 volt, 0.326 μm, 0.00323 μm, 0.323 μm

5-2.   322 ohms, 465 mW
5-3.   8.15 volts
5-5.   4.16(10^{17}) cm^{-3}, 25, 0.625 mm

6-1.   (b) -4.5 volts
6-3.   3.4 kilohms, 8.5
6-5.   (b) 25 volts, 7.6 mA; (c) 8.5 volts, 5 mA
6-7.   1.8 kilohms, 9, 2 volts, 7.2 kilohms, 18
6-9.   (b) 20; (c) 0.4 mm; (d) 6.5 volts
6-11.   9.25 volts, 460 ohms, 5.75, 2.9 volts, 470 mW
6-13.   0.14 volt, 14.9, 47.8 mW, 24.8 mW, 22.9 mW, 47.8 mW, 25.1 mW, 22.7 mW
6-16.   (a) 30.5 volts; (c) 32.3 volts

7-2.   (a) 6.8 kilohms; (c) 46.3
7-4.   6 kilohms, 1 kilohm, 3 mA, 18 volts, 2.86 mA, 20 volts, 12
7-6.   2.95 volts, 6 kilohms, 9

7-8.   8.66, 4.45 volts, 15.55 volts, 0.42 mA, 0.42 mA
7-10. 1.5 kilohms, 0.66 volt, 0.66 μA
7-12. 1.26 mmhos, 65
7-14. 129 mV, 88 mV, 12.9, 8.8

8-2.   0.250 μA to 2.5 mA, 0.48 to 0.71 volt
8-4.   100 pcoul, 10^{-8} sec
8-7.   (b) 0.3 volt; (c) 500 μA
8-9.   (c) 123 μA

9-2.   10 volts, 0.75 mA, 0.015 mA, 10, $10 - 0.1 \sin \omega t$ volts
9-4.   (b) 1.9 volts, 18 kilohms, 1.2β
9-6.   15 kilohms, 130 kilohms, 5 mW, 11.5, 5 volts
9-9.   10 pcoul, 0.1 μsec
9-11. 0 mA, 0 volts, −25 volts; −2.5 mA, −0.2 volt, 0 volts; −2 mA, −0.2 volt, − 5 volts

10-1. 8 kilohms, 483 kilohms, 2.5 mA, 0 volts
10-4. 300 mW, 240 mW, 1 watt
10-6. 1.85 mA, 5.2 volts, 1.9 mA, 4.8 volts, 1.96 mA, 4.3 volts, 1.99 mA
10-8. 22.5 kilohms, 69 kilohms, 4 kilohms, 12 kilohms
10-10. (a) 1.35 mA, 1.4 mA, 6.4 volts, 6.6 volts
10-12. 1.9 mA, 4.8 volts, 1.37 kilohms, 64
10-14. 18.9
10-15. 0.062 mA, 2.52 volts, 40.4 kilohms, 40.5

11-1. 2 kilohms, 73 kilohms, 17.6
11-3. 0.715 kilohm, 247 kilohms, 99 kilohms, 198

12-2. 26
12-5. (d) $g_m/2$, 0.2 volt
12-7. $[1 + R_s/(1 + \beta)R_e]V_{CC} + V_o$
12-9. (c) 430 kilohms
12-11. (c) 14.7
12-13. 20 μA, 10.2 volts, 37.5 kilohms, 20
12-15. (c) 27 kilohms

13-2. 0.33 kilohm, 97, 2,940
13-4. 2.22(10^6)
13-6. 12.7 kilohms, 1.95 kilohms; 44,000; 12,300; 1,680
13-8. $v_o = (R_f/R_s)(v_1 + v_2 + v_3)$
13-10. 19.8, 47.5
13-12. 26,000; 1,040
13-14. 19, 314 kilohms, 149

14-2. 13.5 kilohms, 16
14-4. 10 megohms
14-8. 143, 1.64 MHz, 234 MHz
14-10. 76.5 pF, −0.08 dB, 400 MHz, 196 MHz
14-12. (c) 48.8 dB, 45.9 dB, 15.9 Hz, 22.3 Hz
14-14. (b) 540 kHz; (c) 48 MHz
14-16. (b) 59.5 dB, 0.89 MHz, 13.1 MHz, 104 MHz
14-18. (a) 1,000 rad/sec, −1,000 rad/sec

14-20. $v_o = -K_v A[1 - \exp(-\omega_H t)]$
14-22. (c) $v_2 = -[1 + \exp(-200t)]$

15-3. 15.9 kilohms, 20.3 μH, 79.5, 25, 200 kHz, 2.5 MHz
15-6. $c = (1 + R/R_i)/RC_i$, $r = c[R_i/(R + R_i)]^{1/2}$
15-8. 1.8 mH, 376 kHz, 1.2 mH
15-10. 7.5, 1.13, 27 kHz
15-12. 60°, 18.8(10^6) rad/sec, 3 MHz, 3 MHz, 6 MHz
15-14. 208 ohms, 10
15-16. 80 pF, 20 pF

16-1. 0.005, 6.05, 780 kHz, 165
16-3. (b) 4
16-10. 10 kHz
16-12. 5.7 kilohms, 2.5 pF, 51
16-14. 6, 17.5 kilohms, 3.6 pF, 1.18 MHz
16-17. 67.3

17-2. 200 mA, 400 mA, 40 volts, 4 watts, 2 watts
17-4. 27 volts, 20 ohms, 1,870 ohms, 226 ohms
17-6. 18.75 watts, 25 volts, 8.33 ohms
17-8. 50 volts, 62.5 ohms, 20 watts

Index